Contemporary
Abstract Algebra

Contemporary Abstract Algebra

THIRD EDITION

Joseph A. Gallian
University of Minnesota, Duluth

D. C. HEATH AND COMPANY
Lexington, Massachusetts   Toronto

Address editorial correspondence to:

D. C. Heath and Company
125 Spring Street
Lexington, MA 02173

Acquisitions Editor: Charles Hartford
Developmental Editor: Philip Charles Lanza
Production Editor: Jennifer Brett
Designer: Jan Shapiro
Photo Researcher: Billie Ingram
Production Coordinator: Chuck Dutton
Permissions Editor: Margaret Roll

Cover: The picture *Lace by Nine* is reproduced from the book *Symmetry in Chaos* by Michael Field and Martin Golubitsky, published by Oxford University Press, 1992.

International Standard Book Number: 0-669-33907-5

Library of Congress Catalog Number: 93-70728

10   9   8   7   6   5   4

Dedicated to my mother and stepfather, Alvira and Arthur Strauss

Dear Sir or Madam, will you read my book, it
took me years to write, will you take a look?

John Lennon and Paul McCartney, *Paperback Writer*, single

What do I want for students who use this book? They should obtain a solid introduction to the traditional topics. In fact, dozens of books serve this function well. But I want more. I want readers to come away with the view that abstract algebra is a contemporary subject; that its concepts and methodology are being used by working physicists, chemists, and computer scientists as well as mathematicians; that important contributions to the subject are being made *now*.

I want students to enjoy reading this book. We customarily assume that mathematics books, by the nature of the subject, must be humorless, lifeless, and sterile—and often they are. But I have broken this convention by including lines from popular songs, poems, quotations, biographies, historical notes, hundreds of figures, dozens of photographs, and numerous tables and charts. I have also included reproductions of stamps and currency that honor mathematicians.

I want students to be able to do computations and to write proofs. Accordingly, I have included an abundance of exercises to develop both skills. Some computer exercises are included as well.

In general, the scope, style, spirit, and organization of the previous editions remains unchanged. The changes in the third edition include the following: all functions are written on the left; a new chapter on symmetry and counting has been added; the material on internal direct products has been included in the chapter on normal subgroups rather than treated as a separate chapter; two introductory chapters on rings from previous editions have been merged; a section on public key cryptography has been added; and many new exercises are included.

I am indebted to the following people for serving as reviewers for this edition: Elwyn H. Davis, Pittsburg State University; Robert Johnson, Washington and Lee University; Paul Klingsberg, St. Joseph's University; David Kenoyer, State University of New York, Plattsburgh; John Little, College of the Holy Cross; Warren Nichols, Florida State University; Frank Okoh, Wayne State University; Kenneth Smith, Central Michigan University; F. C. Y. Tang, University of Waterloo; and Mary Wright, Southern Illinois University.

I am grateful to the following people who have kindly sent me comments that have been incorporated into this edition: Marston Conder, University of Auckland; Anthony Gaglione, United States Naval Academy; Jonathan Golan, University of Haifa; Robert Lax, Louisiana State University; Robert McFarland, University of Minnesota, Duluth; David Moulton, University of California, Berkeley; Frank Okoh, Wayne State University; R. Deradoss Pandian, North Central College; Doris Schattschneider, Moravian College; Sándor Szabó, University of Bahrain; Shirley Wilson, North Central College; and David Witte, Williams College.

I wish to express my appreciation to Jennifer Brett, Charles Hartford, Philip Lanza, and Jan Shapiro of D. C. Heath for their cooperation and assistance in preparing this edition. Finally, I owe much to Craig Kirkpatrick for a superb job of copyediting.

Joseph A. Gallian

The picture *Lace by Nine* is an example of a symmetric chaotic attractor that is formed on a computer by iterating a symmetric mapping—in this case a mapping with ninefold symmetry. The color picture is obtained by iterating the mapping

$$f(z) = (-2.5 + 8.0|z|^2 - 0.7 \, \text{Re}(z^9) - 0.9|z|)z + 1.0\bar{z}^8$$

of the complex plane thirty million times, counting the number of times each pixel on the computer screen is visited during the iteration process, and then coloring by number. The actual choice of colors is made both to bring out the fine structure of the attractor and to be aesthetically pleasing. The computer software that produced *Lace by Nine* is called *prism* and was developed by Field and Golubitsky.

CONTENTS

Contemporary
Abstract Algebra

PART 1

INTEGERS
AND
EQUIVALENCE
RELATIONS

0

Preliminaries

The whole of science is nothing more than a refinement of everyday thinking.

Albert Einstein, *Physics and Reality*

PROPERTIES OF INTEGERS

Much of abstract algebra involves properties of integers and sets. In this chapter we collect the ones we need for future reference.

An important property of the integers, which we will often use, is the so-called Well Ordering Principle. Since this property cannot be proved from the usual properties of arithmetic, we will take it as an axiom.

Well Ordering Principle

Every nonempty set of positive integers contains a smallest member.

The concept of divisibility plays a fundamental role in the theory of numbers. We say a nonzero integer t is a *divisor* of an integer s if there is an integer u such that $s = tu$. In this case, we write $t \mid s$ (read "t divides s"). When t is not a divisor of s, we write $t \nmid s$. A *prime* is a positive integer greater than 1 whose only positive divisors are 1 and itself. We say an integer s is a *multiple* of an integer t if there is an integer u such that $s = tu$.

As our first application of the Well Ordering Principle, we establish a fundamental property of integers that we will use often.

Division Algorithm

Let a and b be integers with b > 0. Then there exist unique integers q and r with the property that a = bq + r where $0 \le r < b$.

Proof. We begin with the existence portion of the theorem. Consider the set $S = \{a - bk \mid k$ is an integer and $a - bk \ge 0\}$. If $0 \in S$, then b divides a and we may obtain the desired result with $q = a/b$ and $r = 0$. Now assume $0 \notin S$. Since S is nonempty (if $a > 0$, $a - b \cdot 0 \in S$; if $a < 0$, $a - b(2a) = a(1 - 2b) \in S$; $a \ne 0$ since $0 \notin S$), we may apply the Well Ordering Principle to conclude that S has a smallest member, say $r = a - bq$. Then $a = bq + r$ and $r \ge 0$, so all that remains to be proved is that $r < b$.

If $r > b$, then $a - b(q + 1) = a - bq - b = r - b > 0$ so that $a - b(q + 1) \in S$. But $a - b(q + 1) < a - bq$, and $a - bq$ is the *smallest* member of S. Thus $r \le b$. If $r = b$, then $0 \in S$ and this case is already taken care of. So, $r < b$.

To establish the uniqueness of q and r, let us suppose that there are integers q, q', r, and r' such that

$$a = bq + r, \quad 0 \le r < b \quad \text{and} \quad a = bq' + r', \quad 0 \le r' < b.$$

For convenience, we may also suppose that $r' \ge r$. Then $bq + r = bq' + r'$ and $b(q - q') = r' - r$. So, b divides $r' - r$ and $0 \le r' - r \le r' < b$. It follows that $r' - r = 0$ and therefore $r' = r$ and $q = q'$. ∎

The integer q in the division algorithm is called the *quotient* upon dividing a by b; the integer r is called the *remainder* upon dividing a by b.

Example 1   For $a = 17$ and $b = 5$, the division algorithm gives $17 = 5 \cdot 3 + 2$; for $a = -23$ and $b = 6$, the division algorithm gives $-23 = 6(-4) + 1$.   ❑

Several states use linear functions to encode the month and date of birth into a three-digit number that is incorporated into driver's license numbers. If the encoding function is known, the division algorithm can be used to recapture the month and date of birth from the three-digit number. For instance, the last three digits of a Florida male driver's license number are those given by the formula $40(m - 1) + b$, where m is the number of the month of birth and b is the day of birth. Thus, since $177 = 40 \cdot 4 + 17$, a person with these last three digits was born on May 17. For New York licenses issued prior to September of 1992, the last two digits indicate the year of birth and the three preceding digits code the month and date of birth. For a male driver, these three digits are $63m + 2b$, where m denotes the number of the month of birth and b is the date of birth. So, since $701 = 63 \cdot 11 + 2 \cdot 4$, a license that ends with 70174 indicates that the holder is a male born on November 4, 1974. (In cases where the formula for the driver's license number yields the same result for two or more people, a "tie-breaking" digit is inserted before the two digits for the year of birth.) Incidentally, Wisconsin uses the same method as Florida to encode birth information, but the numbers immediately precede the last pair of digits.

DEFINITIONS   Greatest Common Divisor, Relatively Prime Integers
The *greatest common divisor* of two nonzero integers a and b is the largest of all common divisors of a and b. We denote this integer by $\gcd(a, b)$. When $\gcd(a, b) = 1$, we say a and b are *relatively prime.*

We leave it as an exercise (exercise 12) to prove that every common divisor of a and b divides $\gcd(a, b)$.

The following property of the greatest common divisor of two integers plays a critical role in abstract algebra. The proof provides an application of the division algorithm and our second application of the Well Ordering Principle. This result is Proposition 1 in Book Seven of Euclid's *Elements,* written about 300 B.C.

GCD Is a Linear Combination

For any nonzero integers a and b, there exist integers s and t such that $\gcd(a, b) = as + bt$. Moreover, $\gcd(a, b)$ is the smallest positive integer of the form $as + bt$.

Proof.   Consider the set $S = \{am + bn \mid m, n$ are integers and $am + bn > 0\}$. Since S is obviously nonempty (if some choice of m and n makes $am + bn < 0$, then replace m and n by $-m$ and $-n$), the Well Ordering Principle asserts that S has a smallest member, say, $d = as + bt$. We claim that $d = \gcd(a, b)$. To verify this claim, use the division algorithm to write $a = dq + r$, where $0 \le r < d$. If $r > 0$, then $r = a - dq = a - (as + bt)q = a - asq - btq = a(1 - sq) + b(-tq) \in S$, contradicting the fact that d is the smallest member of S. So, $r = 0$ and d divides a. Analogously (or better yet, by symmetry), d divides b as well. This proves that d is a common divisor of a and b. Now suppose d' is another common divisor of a and b and write $a = d'h$ and $b = d'k$. Then $d = as + bt = (d'h)s + (d'k)t = d'(hs + kt)$ so that d' is a divisor of d. Thus, among all common divisors of a and b, d is the greatest.   ∎

Example 2 $\gcd(4, 15) = 1$; $\gcd(4, 10) = 2$; $\gcd(2^2 \cdot 3^2 \cdot 5, 2 \cdot 3^3 \cdot 7^2) = 2 \cdot 3^2$. Note that 4 and 15 are relatively prime whereas 4 and 10 are not. Also, $4 \cdot 4 + 15(-1) = 1$ and $(-2)4 + 1 \cdot 10 = 2$.   □

Although the above result is a powerful theoretical tool, it does not provide a practical method for calculating $\gcd(a, b)$ or specific values for s and t. The next example shows how this can be done.

Example 3   Euclidean Algorithm (Proposition 2 of Book Seven of the *Elements*). For any pair of positive integers a and b, we may find gcd(a, b) by repeated use of division to produce a decreasing sequence of integers: $r_1 > r_2 > \cdots$ as follows.

$$a = bq_1 + r_1 \qquad\qquad 0 < r_1 < b,$$
$$b = r_1q_2 + r_2 \qquad\qquad 0 < r_2 < r_1,$$
$$r_1 = r_2q_3 + r_3 \qquad\qquad 0 < r_3 < r_2,$$
$$\vdots$$
$$r_{k-3} = r_{k-2}q_{k-1} + r_{k-1} \qquad 0 < r_{k-1} < r_{k-2},$$
$$r_{k-2} = r_{k-1}q_k + r_k \qquad\qquad 0 < r_k < r_{k-1},$$
$$r_{k-1} = r_kq_{k+1} + 0.$$

(It is a consequence of the Well Ordering Principle that this process must eventually result in a remainder of 0.) Then r_k, the last nonzero remainder, is gcd(a, b).

To see that $r_k = $ gcd(a, b), observe that since $r_k \mid r_{k-1}$, the next to the last equation implies that $r_k \mid r_{k-2}$. This, in turn, implies $r_k \mid r_{k-3}$. Continuing to work backwards in this fashion, we see that $r_k \mid b$ and $r_k \mid a$. This proves that r_k is a common divisor of a and b. Now, if r is another common divisor of a and b, the first equation above shows that $r \mid r_1$. The second equation then shows that $r \mid r_2$. Continuing, we see that $r \mid r_k$ and indeed r_k is the greatest of all common divisors of a and b.

Let's apply this process to compute gcd(2520, 154).

$$2520 = 154 \cdot 16 + 56,$$
$$154 = 56 \cdot 2 + 42,$$
$$56 = 42 \cdot 1 + 14,$$
$$42 = 14 \cdot 3.$$

Since 14 is the last nonzero remainder, gcd(2520, 154) = 14.

The preceding equations also provide a method of expressing gcd(a, b) as a linear combination of a and b. For instance, we may use the equations to write the remainders in terms of 2520 and 154. Namely,

$$56 = 2520 + 154(-16),$$
$$42 = 154 + 56(-2),$$
$$14 = 56 + 42(-1)$$
$$= 56 + [154 + 56(-2)](-1)$$
$$= 56 \cdot 3 + 154(-1)$$
$$= [2520 + 154(-16)]3 + 154(-1)$$
$$= 2520 \cdot 3 + 154(-49). \qquad\qquad \square$$

The next lemma is frequently used.

Euclid's Lemma* $p \mid ab$ implies $p \mid a$ or $p \mid b$
   If p is a prime that divides ab, then p divides a or p divides b.

---

*This result is Proposition 14 in Book Nine of Euclid's *Elements*.

Proof. Suppose p is a prime that divides ab but does not divide a. We must show that p divides b. Since p does not divide a, there are integers s and t such that $1 = as + pt$. Then $b = abs + ptb$, and since p divides the right-hand side of this equation, p also divides b. ∎

Note that Euclid's Lemma may fail when p is not a prime, since $6 \mid 4 \cdot 3$ but $6 \nmid 4$ and $6 \nmid 3$.

Our next property shows that the primes are the building blocks for all integers. We will often use this property without explicitly saying so.

Theorem 0.1 *Fundamental Theorem of Arithmetic*
Every integer greater than 1 is a prime or a product of primes. This product is unique, except for the order in which the factors appear. Thus, if $n = p_1 p_2 \ldots p_r$ and $n = q_1 q_2 \ldots q_s$ where the p's and q's are primes, then $r = s$ and, after renumbering the q's, we have $p_i = q_i$ for all i.

We will prove the existence portion of Theorem 0.1 later in this chapter. The uniqueness portion is a consequence of Euclid's Lemma.

Another concept that frequently arises is that of the least common multiple of two integers.

DEFINITION   Least Common Multiple
The least common multiple of two nonzero integers a and b is the smallest positive integer that is a multiple of both a and b. We will denote this integer by $\mathrm{lcm}(a, b)$.

We leave it as an exercise (exercise 12) to prove that every common multiple of a and b is a multiple of $\mathrm{lcm}(a, b)$.

Example 4 $\mathrm{lcm}(4, 6) = 12$; $\mathrm{lcm}(4, 8) = 8$; $\mathrm{lcm}(10, 12) = 60$; $\mathrm{lcm}(6, 5) = 30$; $\mathrm{lcm}(2^2 \cdot 3^2 \cdot 5, 2 \cdot 3^3 \cdot 7^2) = 2^2 \cdot 3^3 \cdot 5 \cdot 7^2$. ❑

MODULAR ARITHMETIC

Another application of the division algorithm that will be important to us is modular arithmetic.

DEFINITION $a \bmod n$
Let n be a fixed positive integer. For any integer a, $a \bmod n$ (sometimes read "a modulo n") is the remainder upon dividing a by n.

We will often use the mod concept in combination with addition and multiplication.

Example 5

$$8 \bmod 3 = 2; \qquad -8 \bmod 3 = 1;$$
$$23 \bmod 6 = 5; \qquad -23 \bmod 6 = 1;$$

$$6 \bmod 7 = 6; \qquad -6 \bmod 7 = 1;$$
$$(7 + 4) \bmod 3 = 2; \qquad (7 \cdot 4) \bmod 3 = 1;$$
$$(10 + 5) \bmod 6 = 3; \qquad (10 \cdot 5) \bmod 6 = 2;$$
$$(6 + 5) \bmod 7 = 4; \qquad (6 \cdot 5) \bmod 7 = 2. \qquad \square$$

Throughout this text, we will make frequent use of modular equations.

DEFINITION   Modular Equations

If a and b are integers and n is a positive integer, we write $a = b \bmod n$ when n divides $a - b$.

Example 6

$$13 = 3 \bmod 5 \qquad 22 = 10 \bmod 6 \qquad -10 = 4 \bmod 7. \quad \square$$

Example 7   Consider the equations

1. $x = 1 \bmod 5$.
2. $x = 3 \bmod 5$.
3. $2x = 5 \bmod 6$.
4. $ax = b \bmod n$.

Note that x is a solution of equation 1 if and only if $x - 1 = 5k$ for some integer k. That is, $x - 1 = 0, \pm 5, \pm 10, \ldots$ or x belongs to $\{\ldots, -9, -4, 1, 6, 11, \ldots\}$. Similarly, x is a solution of equation 2 if and only if x belongs to $\{\ldots, -7, -2, 3, 8, 13, \ldots\}$. Equation 3 has no solution, since the left side of the equation $2x - 5 = 6k$ is always odd whereas the right side is always even. In view of equation 3, we see that equation 4 need not have a solution, but one important case in which it does is whenever a and n are relatively prime. For then there are integers s and t such that $as + nt = 1$. So, $asb + ntb = b$. Thus, $asb - b = -ntb$ is divisible by n and sb is a solution to the equation $ax = b \bmod n$. $\square$

Modular arithmetic is an indispensable tool in computer science. One application concerns the generation of "random" numbers: a sequence of integers that is easily produced and passes a number of statistical tests for randomness. A common method for generating such numbers is to use a recursion formula of the form

$$x_{n+1} = (ax_n + c) \bmod m.$$

To use such a formula, one chooses the modulus m, the multiplier a, and the increment c, as well as an initial value for x_0 (often called the *seed*). Over the years, guidelines have evolved for selecting a, c, and m. Typically, m is a large power of 2 (e.g., 2^{32}) and $a = 5$. Of course, numbers generated in this way are completely determined by the choice of these parameters and the initial value, and so are not random at all. Nevertheless, these numbers do possess many properties that we expect of random numbers and so are useful for simulation.

Figure 0.1

Modular arithmetic is often used in assigning an extra digit to identification numbers for the purpose of detecting forgery or errors. We present two such applications.

Example 8   The United States Postal Service money order shown in Figure 0.1 has an identification number consisting of 10 digits together with an extra digit called a *check*. The check digit is the 10-digit number modulo 9. Thus, the number 3953988164 has the check digit 2 since 3953988164 = 2 mod 9. If the number 39539881642 were incorrectly entered into a computer (programmed to calculate the check digit) as, say, 39559881642 (an error in the fourth position), the machine would calculate the check as 4 whereas the entered check digit would be 2. Thus the error would be detected.   □

Example 9   Airline companies, United Parcel Service, and Avis and National rental car companies use the modulo 7 values of identification numbers to assign check digits. Thus, the identification number 00121373147367 (see Figure 0.2) has the check digit 3 appended to it because 121373147367 = 3 mod 7. Similarly, the UPS pickup record number 768113999, shown in Figure 0.3, has the check digit 2 appended to it.

In Chapters 2 and 5, we will examine more sophisticated means of assigning check digits to numbers.   □

MATHEMATICAL INDUCTION

There are two forms of proof by mathematical induction that we will use. Both are equivalent to the Well Ordering Principle. The explicit formulation of the method of mathematical induction came in the 16th century. Francisco Maurolycus (1494–1575), a teacher of Galileo, used

Figure 0.2

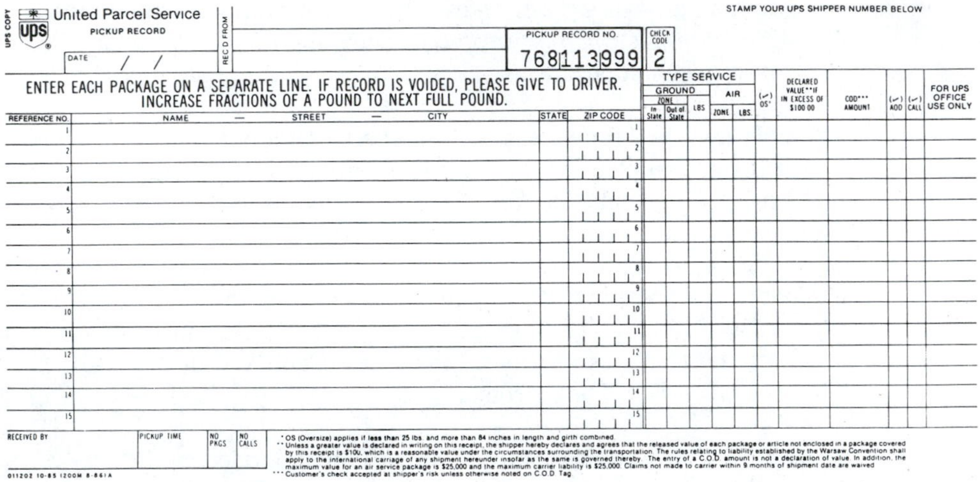

Figure 0.3

it in 1575 to prove that $1 + 3 + 5 + \cdots + (2n - 1) = n^2$, and Blaise Pascal (1623–1662) used it when he presented what we now call Pascal's triangle for the coefficients of the binomial expansion. The term was coined by Augustus De Morgan.

First Principle of Mathematical Induction

Let S be a set of integers containing a. Suppose S has the property that whenever some integer $n \geq a$ belongs to S, then the integer $n + 1$ also belongs to S. Then, S contains every integer greater than or equal to a.

So, to use induction to prove that a statement involving positive integers is true for every positive integer, we must first verify that the statement is true for the integer 1. We then *assume* the statement is true for the integer n and use this assumption to prove that the statement is true for the integer $n + 1$.

Example 10 We will use the First Principle of Mathematical Induction to prove that for every positive integer n, the expression $2^{2n} - 1$ is divisible by 3. Let S be the set of all positive integers k for which $2^{2k} - 1$ is divisible by 3. Clearly, $1 \in S$. Now assume that some integer $n \in S$. We must show that $2^{2(n+1)} - 1$ is divisible by 3, so that $n + 1 \in S$. Since $2^{2(n+1)} - 1 = 2^{2n+2} - 1 = 4 \cdot 2^{2n} - 1 = 4 \cdot 2^{2n} - 4 + 3 = 4(2^{2n} - 1) + 3$ and $n \in S$, it follows that 3 divides this last expression. Thus, by induction, $2^{2n} - 1$ is divisible by 3 for all positive integers n. ◻

Example 11 DeMoivre's Theorem
We use induction to prove that for every positive integer n and every real number θ, $(\cos \theta + i \sin \theta)^n = \cos n\theta + i \sin n\theta$, where i is the complex number $\sqrt{-1}$. Obviously, the statement is true for $n = 1$. Now assume it is true for n. We must prove that $(\cos \theta + i \sin \theta)^{n+1} = \cos(n + 1)\theta + i \sin(n + 1)\theta$. Observe that

$$
\begin{aligned}
(\cos \theta + i \sin \theta)^{n+1} &= (\cos \theta + i \sin \theta)^n(\cos \theta + i \sin \theta) \\
&= (\cos n\theta + i \sin n\theta)(\cos \theta + i \sin \theta) \\
&= \cos n\theta \cos \theta + i(\sin n\theta \cos \theta + \sin \theta \cos n\theta) \\
&\quad - \sin n\theta \sin \theta.
\end{aligned}
$$

Now, using trigonometric identities for $\cos(\alpha + \beta)$ and $\sin(\alpha + \beta)$, we see that this last term is $\cos(n + 1)\theta + i \sin(n + 1)\theta$. So, by induction, the statement is true for all positive integers. ◻

In many instances, the assumption that a statement is true for an integer n does not readily lend itself to a proof that the statement is true for the integer $n + 1$. In such cases, the following equivalent form of induction may be more convenient.

Second Principle of Mathematical Induction

Let S be a set of integers containing a. Suppose S has the property that n belongs to S whenever every integer less than n and greater than or equal to a belongs to S. Then, S contains every integer greater than or equal to a.

To use this form of induction, we first show that the statement is true for the integer a. We then *assume* that the statement is true for *all* integers that are greater than or equal to a and less than n, and use this assumption to prove that the statement is true for all integers greater than or equal to a.

Example 12 We will use the Second Principle of Mathematical Induction with $a = 2$ to prove the existence portion of the Fundamental

Theorem of Arithmetic. Let S be the set of integers greater than 1 that are primes or products of positive primes. Clearly, $2 \in S$. Now we assume that for some integer n, S contains all integers k with $2 \leq k < n$. We must show that $n \in S$. If n is a prime, then $n \in S$ by definition. If n is not a prime, then n can be written in the form ab where $1 < a < n$ and $1 < b < n$. Since we are assuming that both a and b belong to S, we know that each of them is a prime or a product of positive primes. Thus, n is also a product of positive primes. This completes the proof.

$\square$

Notice that it is more natural to prove the Fundamental Theorem of Arithmetic with the Second Principle of Mathematical Induction than with the First Principle. Knowing that a particular integer factors as a product of primes does not tell you something about factoring the next larger integer. (Does knowing that 5280 is a product of primes help you to factor 5281 as a product of primes?)

The following problem appeared in the "Brain Boggler" section of the January 1988 issue of the science magazine *Discover*.

Example 13   The Quakertown Poker Club plays with blue chips worth $5.00 and red chips worth $8.00. What is the largest bet that cannot be made?

To gain insight into this problem, we try various combinations of blue and red chips to obtain: 5, 8, 10, 13, 15, 16, 18, 20, 21, 23, 24, 26, 28, 29, 30, 31, 32, 33, 34, 35, 36, 37, 38, 39, 40. It appears that the answer is 27. But how can we be sure? Well, we need only prove that every integer greater than 27 can be written in the form $a \cdot 5 + b \cdot 8$, where a and b are nonnegative integers. This will solve the problem, since a represents the number of blue chips and b the number of red chips needed to make a bet of $a \cdot 5 + b \cdot 8$. For the purpose of contrast, we will give two proofs—one using the First Principle of Mathematical Induction and one using the Second Principle.

Let S be the set of all integers of the form $a \cdot 5 + b \cdot 8$, where a and b are nonnegative. Obviously, $28 \in S$. Now assume that some integer $n \in S$, say, $n = a \cdot 5 + b \cdot 8$. We must show that $n + 1 \in S$. First, note that since $n \geq 28$, we cannot have both a and b less than 3. If $a \geq 3$, then

$$n + 1 = (a \cdot 5 + b \cdot 8) + (-3 \cdot 5 + 2 \cdot 8)$$
$$= (a - 3) \cdot 5 + (b + 2) \cdot 8.$$

(Regarding chips, this last equation says we may increase a bet from n to $n + 1$ by removing 3 blue chips from the pot and adding 2 red chips.) If $b \geq 3$, then

$$n + 1 = (a \cdot 5 + b \cdot 8) + (5 \cdot 5 - 3 \cdot 8)$$
$$= (a + 5) \cdot 5 + (b - 3) \cdot 8.$$

(The bet can be increased by 1 by removing 3 red chips and adding 5 blue chips.) This completes the proof.

To prove the same statement by the Second Principle, we note that each of the integers 28, 29, 30, 31, and 32 is in S. Now assume that for some integer $n > 32$, S contains all integers k with $28 \leq k < n$. We must show that $n \in S$. Since $n - 5 \in S$, there are nonnegative integers a and b such that $n - 5 = a \cdot 5 + b \cdot 8$. But then $n = (a + 1) \cdot 5 + b \cdot 8$. Thus n is in S. ◻

In the remainder of the text, we shall be less formal about induction and dispense with the phrase "Let S be the set of integers. . . ."

EQUIVALENCE RELATIONS

In mathematics, things that are considered different in one context may be viewed as equivalent in another context. We have already seen one such example. Indeed, the sums $2 + 1$ and $4 + 4$ are certainly different in ordinary arithmetic, but are the same under modulo 5 arithmetic. Congruent triangles that are situated differently in the plane are not the same, but they are considered to be the same in plane geometry. In physics, vectors of the same magnitude and direction can produce different effects—a 10-pound weight placed 2 feet from a fulcrum produces a different effect than a 10-pound weight placed 1 foot from a fulcrum. But in linear algebra, vectors of the same magnitude and direction are considered to be the same. What is needed to make these distinctions precise is an appropriate generalization of the notion of equality; that is, we need a formal mechanism for specifying whether or not two quantities are the same in a given setting. This mechanism is an equivalence relation.

DEFINITION   Equivalence Relation
An *equivalence relation* on a set S is a set R of ordered pairs of elements of S such that

1. $(a, a) \in R$ for all $a \in S$ (reflexive property).
2. $(a, b) \in R$ implies $(b, a) \in R$ (symmetric property).
3. $(a, b) \in R$ and $(b, c) \in R$ imply $(a, c) \in R$ (transitive property).

When R is an equivalence relation on a set S, it is customary to write aRb instead of $(a, b) \in R$. Also, since an equivalence relation is just a generalization of equality, a suggestive symbol such as $\approx$, $\equiv$, or $\sim$ is usually used to denote the relation. If $\sim$ is an equivalence relation on a set S and $a \in S$, then the set $[a] = \{x \in S \mid x \sim a\}$ is called the *equivalence class of S containing a*.

Example 14   Let *S* be the set of all triangles in a plane. If $a, b \in S$, define $a \sim b$ if a and b are similar—that is, if a and b have corresponding angles that are the same. Then, $\sim$ is an equivalence relation on *S*.   ❑

Example 15   Let *S* be the set of all polynomials with real coefficients. If $f, g \in S$, define $f \sim g$ if $f' = g'$, where f' is the derivative of f. Then, $\sim$ is an equivalence relation on *S*. Since two functions with equal derivatives differ by a constant, we see that for any f in S, $[f] = \{f + c \mid c$ is real$\}$.   ❑

Example 16   Let *S* be the set of integers and let n be a positive integer. If $a, b \in S$, define $a \equiv b$, if $a = b$ modulo n (that is, if $a - b$ is divisible by n). Then, $\equiv$ is an equivalence relation on S and $[a] = \{a + kn \mid k \in S\}$. Since this particular relation is important in abstract algebra, we will take the trouble to verify that it is indeed an equivalence relation. Certainly, $a - a$ is divisible by n so that $a \equiv a$ for a in *S*. Next, assume that $a \equiv b$, say, $a - b = rn$. Then, $b - a = (-r)n$ and therefore $b \equiv a$. Finally, assume $a \equiv b$ and $b \equiv c$, say, $a - b = rn$ and $b - c = sn$. Then, we have $a - c = (a - b) + (b - c) = rn + sn = (r + s)n$, so that $a \equiv c$.   ❑

Example 17   Let $\equiv$ be as in Example 16 and let $n = 7$. Then we have $16 \equiv 2; 9 \equiv -5; 24 \equiv 3$. Also, $[1] = \{\ldots, -20, -13, -6, 1, 8, 15, \ldots\}$ and $[4] = \{\ldots -17, -10, -3, 4, 11, 18, \ldots\}$.   ❑

Example 18   Let $S = \{(a, b) \mid a, b$ are integers, $b \neq 0\}$. If (a, b), $(c, d) \in S$, define $(a, b) \approx (c, d)$ if $ad = cb$. Then $\approx$ is an equivalence relation on *S*. [The motivation for this example comes from fractions. In fact, the pairs (a, b) and (c, b) are equivalent if the fractions a/b and c/d are equal.]   ❑

DEFINITION   Partition
A *partition* of a set *S* is a collection of nonempty disjoint subsets of *S* whose union is *S*. Figure 0.4 illustrates a partition of a set into four subsets.

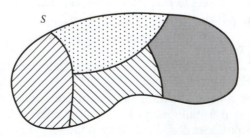

Figure 0.4   Partition of *S* into four subsets.

Example 19 The sets $\{0\}$, $\{1, 2, 3, \ldots\}$, and $\{\ldots, -3, -2, -1\}$ constitute a partition of the set of integers. ◻

Example 20 The set of nonnegative integers and the set of nonpositive integers do not partition the integers since both contain 0. ◻

The next theorem reveals that equivalence relations and partitions are intimately intertwined.

Theorem 0.2 *Equivalence Classes Partition*

The equivalence classes of an equivalence relation on a set S constitute a partition of S. Conversely, for any partition P of S, there is an equivalence relation on S whose equivalence classes are the elements of P.

Proof. Let $\sim$ be an equivalence relation on a set S. For any $a \in S$, the reflexive property shows $a \in [a]$. So, $[a]$ is nonempty. Also, the union of all equivalence classes is S. Now, suppose that $[a]$ and $[b]$ are distinct equivalence classes. We must show that $[a] \cap [b] = \emptyset$. On the contrary, assume $c \in [a] \cap [b]$. We will show that $[a] \subseteq [b]$. To this end, let $x \in [a]$. We then have $c \sim a$, $c \sim b$, and $x \sim a$. By the symmetric property, we also have $a \sim c$. Thus, by transitivity, $x \sim c$, and by transitivity again, $x \sim b$. This proves $[a] \subseteq [b]$. Analogously, $[b] \subseteq [a]$. Thus, $[a] = [b]$, in contradiction to our assumption that $[a]$ and $[b]$ are distinct equivalence classes.

We leave the proof of the converse portion of the theorem as an exercise. ∎

FUNCTIONS (MAPPINGS)

Although the concept of a function plays a central role in nearly every branch of mathematics, the terminology and notation associated with functions vary quite a bit. In this section, we fix ours.

DEFINITION Function (Mapping)

A *function* ϕ (or *mapping*) *from a set A to a set B* is a rule that assigns to each element a of A exactly one element b of B. The set A is called the *domain of* ϕ and B is called the *codomain of* ϕ. If ϕ assigns b to a, then b is called the *image of a under* ϕ. The subset of B comprised of all the images of elements of A is called the *image of A under* ϕ.

We use the shorthand $\phi\colon A \to B$ to mean that ϕ is a mapping from A to B. We will write $\phi(a) = b$ or $\phi\colon a \to b$ to indicate that ϕ carries a to b.

DEFINITION Composition of Functions

Let $\phi\colon A \to B$ and $\psi\colon B \to C$. The *composition* $\psi\phi$ is the mapping from A to C defined by $(\psi\phi)(a) = \psi(\phi(a))$ for all a in A. The composition function $\psi\phi$ can be visualized as in Figure 0.5.

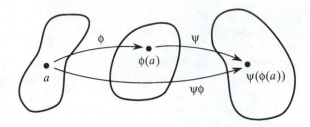

Figure 0.5   Composition of functions ϕ and ψ.

In calculus courses, the composition of f with g is written $(f \circ g)(x)$ and is defined as $f(g(x))$. When we compose functions, we omit the "circle."

There are several kinds of functions that occur often enough to be given names.

DEFINITION   One-to-One Function
A function ϕ from a set A to a set B is called *one-to-one* if $\phi(a_1) = \phi(a_2)$ implies $a_1 = a_2$.

The term *one-to-one* is suggestive since the definition ensures that one element of B can be the image of only one element of A. Alternatively, ϕ is one-to-one if $a_1 \neq a_2$ implies $\phi(a_1) \neq \phi(a_2)$. That is, different elements of A map to different elements of B. See Figure 0.6.

DEFINITION   Function from A onto B
A function ϕ from a set A to a set B is said to be *onto B* if each element of B is the image of at least one element of A. In symbols, $\phi: A \to B$ is onto if for each b in B there is at least one a in A such that $\phi(a) = b$. See Figure 0.7.

The next theorem summarizes the facts about functions we will need.

Theorem 0.3   *Properties of Functions*
Given functions $\alpha: A \to B$, $\beta: B \to C$, and $\gamma: C \to D$. Then

1. *$\gamma(\beta\alpha) = (\gamma\beta)\alpha$ (associativity)*
2. *If α and β are one-to-one, then $\beta\alpha$ is one-to-one.*

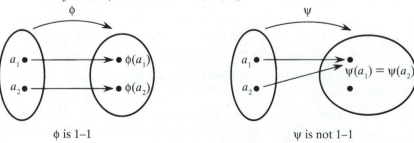

ϕ is 1–1 ψ is not 1–1

Figure 0.6

φ

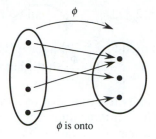

ψ

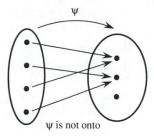

φ is onto

ψ is not onto

Figure 0.7

3. *If α and β are onto, then βα is onto.*
4. *If α is one-to-one and onto, then there is a function $α^{-1}$ from B onto A such that $(α^{-1}α)(a) = a$ for all a in A and $(αα^{-1})(b) = b$ for all b in B.*

Proof. We prove only part 1. The remaining parts are left as exercises. Let $a ∈ A$. Then $(γ(βα))(a) = γ((βα)(a)) = γ(β(α(a)))$. On the other hand, $((γβ)(α))(a) = (γβ)(α(a)) = γ(β(α(a)))$. So, $γ(βα) = (γβ)α$. ∎

Example 21 Let **Z** denote the set of integers, **R** the set of real numbers, and **N** the set of nonnegative integers. The following table illustrates the properties of one-to-one and onto.

Domain	Codomain	Rule	1–1	Onto
Z	**Z**	$x → x^3$	yes	no
R	**R**	$x → x^3$	yes	yes
Z	**N**	$x → |x|$	no	yes
Z	**Z**	$x → x^2$	no	no

To verify that $x → x^3$ is one-to-one in the first two cases, notice that if $x^3 = y^3$, we may take the cube root of both sides of the equation to obtain $x = y$. Clearly, the mapping from **Z** to **Z** given by $x → x^3$ is not onto since 2 is the cube of no integer. However, $x → x^3$ defines an onto function from **R** to **R** since every real number is the cube of its cube root (that is, $\sqrt[3]{b} → b$). The remaining verifications are left as exercises. ❑

EXERCISES

Failure is instructive. The person who really thinks learns quite as much from his failures as from his successes.

John Dewey

1. For each value of *n* listed, find all integers less than *n* and relatively prime to *n*. *n* = 8, 12, 20, 25.

2. Determine $\gcd(2^4 \cdot 3^2 \cdot 5 \cdot 7^2, 2 \cdot 3^3 \cdot 7 \cdot 11)$ and $\text{lcm}(2^3 \cdot 3^2 \cdot 5, 2 \cdot 3^3 \cdot 7 \cdot 11)$.

3. Calculate $(7 \cdot 3) \bmod 5, (7 + 3) \bmod 5, (15 \cdot 4) \bmod 7, (15 + 4) \bmod 7$.

4. Find integers s and t so that $1 = 7 \cdot s + 11 \cdot t$. Show that s and t are not unique.

5. In Florida, the fourth and fifth digits from the end of a driver's license number give the year of birth. The last three digits for a male with birth month m and birth date b are represented by $40(m - 1) + b$. For females the digits are, $40(m - 1) + b + 500$. Determine the dates of birth of people that have last five digits 42218 and 53953.

6. For driver's license numbers issued in New York prior to September of 1992, the three digits preceding the last two of the number of a male with birth month m and birth date b are represented by $63m + 2b$. For females the digits are $63m + 2b + 1$. Determine the dates of birth and sex(es) corresponding to the numbers 248 and 601.

7. Show that if a and b are positive integers, then $ab = \text{lcm}(a, b) \cdot \gcd(a, b)$. (This exercise is referred to in Chapter 7.)

8. Suppose a and b are integers that divide the integer c. If a and b are relatively prime, show that ab divides c. Show, by example, that if a and b are not relatively prime, then ab need not divide c.

9. If there are integers a, b, s, and t with the property that $at + bs = 1$, show that $\gcd(a, b) = 1$.

10. Let $d = \gcd(a, b)$. If $a = da'$ and $b = db'$, show that $\gcd(a', b') = 1$.

11. Show that $5n + 3$ and $7n + 4$ are relatively prime for all n.

12. Let a and b be positive integers and let $d = \gcd(a, b)$ and $m = \text{lcm}(a, b)$. If t divides both a and b, prove that t divides d. If s is a multiple of both a and b, prove that s is a multiple of m.

13. Let n and a be positive integers and let $d = \gcd(a, n)$. Show that the equation $ax = 1 \bmod n$ has a solution if and only if $d = 1$. (This exercise is referred to in Chapter 2.)

14. Let n be a fixed positive integer greater than 1. For any integers i and j, prove that $(i \bmod n + j \bmod n) \bmod n = (i + j) \bmod n$. (This exercise is referred to in Chapters 6 and 7.)

15. Let a, b, s, and t be integers. If $a = b \bmod st$, show that $a = b \bmod s$ and $a = b \bmod t$. What condition on s and t is needed to make the converse true? (This exercise is referred to in Chapter 7.)

16. Show that $\gcd(a, bc) = 1$ if and only if $\gcd(a, b) = 1$ and $\gcd(a, c) = 1$. (This exercise is referred to in Chapter 7.)

17. Use the Euclidean algorithm to find $\gcd(34, 126)$ and write it as a linear combination of 34 and 126.

18. Let $p_1, p_2, \ldots, p_n$ be distinct positive primes. Show that $p_1 p_2 \cdots p_n + 1$ is divisible by none of these primes.

19. Prove that there are infinitely many primes. (*Hint:* Use the previous exercise.)

20. For every positive integer n, prove that $1 + 2 + \cdots + n = n(n + 1)/2$.

21. For every positive integer n, prove that a set with exactly n elements has exactly 2^n subsets (counting the empty set and the entire set).

22. (Generalized Euclid's Lemma) If p is a prime and p divides $a_1 a_2 \cdots a_n$, prove that p divides a_i for some i.

23. Use the Generalized Euclid's Lemma (see exercise 22) to establish the uniqueness portion of the Fundamental Theorem of Arithmetic.

24. What is the largest bet that cannot be made with chips worth $7.00 and $9.00? Verify that your answer is correct with both forms of induction.

25. The Fibonacci numbers are: 1, 1, 2, 3, 5, 8, 13, 21, 34, ... In general, the Fibonacci numbers are defined by $f_1 = 1, f_2 = 1$, and $f_{n+2} = f_{n+1} + f_n$ for $n = 1, 2, 3, \ldots$. Prove that the nth Fibonacci number f_n satisfies $f_n < 2^n$.

26. Prove that the First Principle of Mathematical Induction is a consequence of the Well Ordering Principle.

27. In the cut "As" from *Songs in the Key of Life,* Stevie Wonder mentions the equation $8 \times 8 \times 8 \times 8 = 4$. Find all integers n for which this statement is true, modulo n.

28. Determine the check digit for a money order with identification number 7234541780.

29. Suppose that in one of the noncheck positions of a money order number, the digit 0 is substituted for the digit 9 or vice versa. Prove that this error will not be detected by the check digit. Prove that all other errors involving a single position are detected.

30. Suppose a money order with the identification number and check digit 21720421168 is erroneously copied as 27750421168. Will the check digit detect the error?

31. A transposition error involving distinct adjacent digits is one of the form $\ldots ab \ldots \rightarrow \ldots ba \ldots$ with $a \neq b$. Prove that the money order check digit scheme will not detect such errors unless the check digit itself is transposed.

32. Determine the check digit for the Avis rental car with identification number 540047. (See Example 9.)

33. Show that a substitution of a digit a_i' for the digit a_i ($a_i' \neq a_i$) in a noncheck position of a UPS number is detected if and only if $|a_i - a_i'| \neq 7$.

34. Determine which transposition errors involving adjacent digits are detected by the UPS check digit. $9, 2 \qquad 8 \, 1$

35. Prove that for every integer n, $n^3 = n \bmod 6$.

36. If it were 2:00 A.M. now, what time would it be 3736 hours from now?

37. If n is an odd integer, prove that $n^2 = 1 \bmod 8$.

38. If the odometer of an automobile read 97,000 now, what would it read 12,000 miles later?

39. Let $S = \{(x, y) \mid x, y \text{ are real}\}$. If (a, b) and (c, d) belong to S, define $(a, b)R(c, d)$ if $a^2 + b^2 = c^2 + d^2$. Prove that R is an equivalence relation on S. Give a geometrical description of the equivalence classes of S.

40. Let S be the set of real numbers. If $a, b \in S$, define $a \sim b$ if $a - b$ is an integer. Show that $\sim$ is an equivalence relation on S. Describe the equivalence classes of S.

41. Let S be the set of integers. If $a, b \in S$, define aRb if $ab \geq 0$. Is R an equivalence relation on S?

42. Let S be the set of integers. If $a, b \in S$, define aRb if $a + b$ is even. Prove that R is an equivalence relation and determine the equivalence classes of S.

43. A *relation* on a set S is a set of ordered pairs of elements of S. Find an example of a relation that is reflexive and symmetric, but not transitive.

44. Find an example of a relation that is reflexive and transitive, but not symmetric.

45. Find an example of a relation that is symmetric and transitive, but not reflexive.

SUGGESTED READINGS

Mary Joan Collison, "The Unique Factorization Theorem: From Euclid to Gauss," *Mathematics Magazine* 53 (1980): 96–100.

This article examines the history of the Fundamental Theorem of Arithmetic.

Linda Deneen, "Secret Encryption with Public Keys," *The UMAP Journal* 8 (1987): 9–29.

This well-written article describes several ways in which modular arithmetic can be used to code secret messages. They range from a simple scheme used by Julius Caesar to a highly sophisticated scheme invented in 1978 and based on modular n arithmetic, where n has more than 200 digits.

J. A. Gallian, "Breaking the Missouri License Code," *The UMAP Journal* 13 (1992): 37–42.

This article gives the formula used by New York and Missouri (up until 1992) to assign driver's license numbers.

Ian Richards, "The Invisible Prime Factor," *American Scientist* 70 (1982): 176–179.

The author explains how elementary number theory such as Euclid's Lemma and modular arithmetic can be used to test whether an integer is prime. He then discusses how prime numbers can be used to create secret codes that are extremely difficult to break.

PART 2

GROUPS

1

Introduction to Groups

Symmetry is a vast subject, significant in art and nature.
Mathematics lies at its root, and it would be hard to find a
better one on which to demonstrate the working of the
mathematical intellect.

Hermann Weyl, *Symmetry*

SYMMETRIES OF A SQUARE

Suppose we remove a square from a plane, move it in some way, then
put the square back into the space it originally occupied. Our goal in this
chapter is to describe in some reasonable fashion all possible ways this
can be done. More specifically, we want to describe the possible relationships between the starting position of the square and its final position in terms of motions. However, we are interested in the net effect of
a motion, rather than the motion itself. Thus, for example, we consider
a 90° rotation and a 450° rotation as equal, since the relationship
between the starting position and the final position of the square is the
same for each of these motions. With this simplifying convention, it is
an easy matter to achieve our goal.

To begin, we can think of the square as being transparent (glass,
say), with the corners marked on one side with the colors blue, white,
pink, and green. This makes it easy to distinguish between motions that
have different effects. With this marking scheme, we are now in a position to describe, in simple fashion, all possible ways a square object can

R_0 = Rotation of 0° (no change in position)

$$\begin{array}{cc} P & W \\ G & B \end{array} \quad \xrightarrow{\ R_0\ } \quad \begin{array}{cc} P & W \\ G & B \end{array}$$

R_{90} = Rotation of 90° (counterclockwise)

$$\begin{array}{cc} P & W \\ G & B \end{array} \quad \xrightarrow{\ R_{90}\ } \quad \begin{array}{cc} W & B \\ P & G \end{array}$$

R_{180} = Rotation of 180°

$$\begin{array}{cc} P & W \\ G & B \end{array} \quad \xrightarrow{\ R_{180}\ } \quad \begin{array}{cc} B & G \\ W & P \end{array}$$

R_{270} = Rotation of 270°

$$\begin{array}{cc} P & W \\ G & B \end{array} \quad \xrightarrow{\ R_{270}\ } \quad \begin{array}{cc} G & P \\ B & W \end{array}$$

H = Rotation of 180° about a horizontal axis

$$\begin{array}{cc} P & W \\ G & B \end{array} \quad \xrightarrow{\ H\ } \quad \begin{array}{cc} G & B \\ P & W \end{array}$$

V = Rotation of 180° about a vertical axis

$$\begin{array}{cc} P & W \\ G & B \end{array} \quad \xrightarrow{\ V\ } \quad \begin{array}{cc} W & P \\ B & G \end{array}$$

D = Rotation of 180° about the main diagonal

$$\begin{array}{cc} P & W \\ G & B \end{array} \quad \xrightarrow{\ D\ } \quad \begin{array}{cc} P & G \\ W & B \end{array}$$

D' = Rotation of 180° about the other diagonal

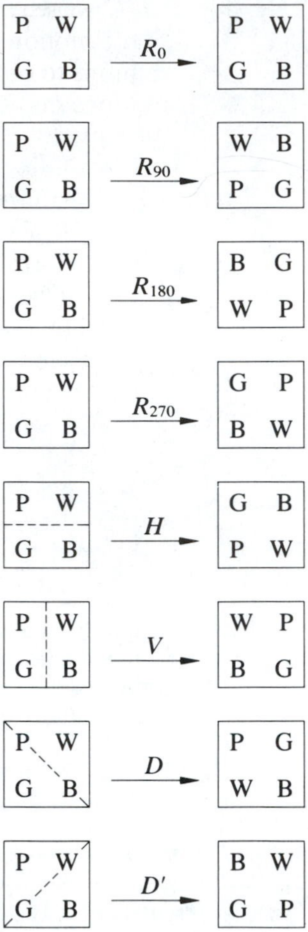

$$\begin{array}{cc} P & W \\ G & B \end{array} \quad \xrightarrow{\ D'\ } \quad \begin{array}{cc} B & W \\ G & P \end{array}$$

Figure 1.1

be repositioned. See Figure 1.1. We now claim that any motion—no matter how complicated—is equivalent to one of these eight. To verify this claim, observe that the final position of the square is completely determined by the location and orientation (that is, face up or face down) of any particular corner. But, clearly, there are only four locations and two orientations for a given corner, so there are exactly eight distinct final positions for the corner.

Let's investigate some consequences of the fact that every motion is equal to one of the eight listed in Figure 1.1. Suppose a square is repositioned by a rotation of 90° followed by a rotation of 180° about the horizontal axis of symmetry. In pictures,

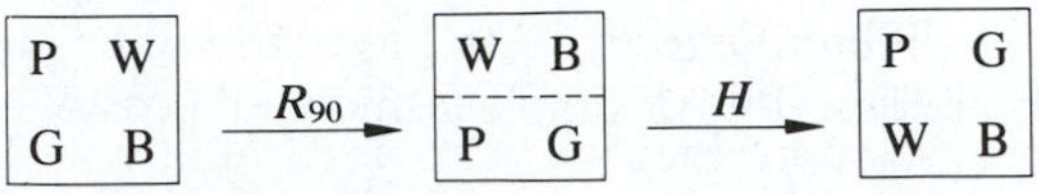

$$\begin{array}{cc} P & W \\ G & B \end{array} \quad \xrightarrow{\ R_{90}\ } \quad \begin{array}{cc} W & B \\ P & G \end{array} \quad \xrightarrow{\ H\ } \quad \begin{array}{cc} P & G \\ W & B \end{array}$$

Thus, we see that this pair of motions—taken together—is equal to the single motion D. This observation suggests that we can compose two motions to obtain a single motion. And indeed we can, since the eight motions may be viewed as functions from the set of points making up the square to itself and as such we can combine them using function composition.

With this in mind, we may now write $HR_{90} = D$. The eight motions R_0, R_{90}, R_{180}, R_{270}, H, V, D, and D', together with the operation composition, form a mathematical system called the *dihedral group of order 8* (the order of a group is the number of elements it contains). It is denoted by D_4. Rather than introduce the formal definition of a group here, let's look at some properties of groups by way of the example D_4.

To facilitate future computations, we construct an *operation table* or *Cayley table* (so named in honor of the prolific English mathematician Arthur Cayley, who first introduced them in 1854) for D_4 below. The circled entry represents the fact that $D = HR_{90}$. (In general, *ab* denotes the entry at the intersection of the row with *a* at the left and the column with *b* at the top.)

	R_0	R_{90}	R_{180}	R_{270}	H	V	D	D'
R_0	R_0	R_{90}	R_{180}	R_{270}	H	V	D	D'
R_{90}	R_{90}	R_{180}	R_{270}	R_0	D'	D	H	V
R_{180}	R_{180}	R_{270}	R_0	R_{90}	V	H	D'	D
R_{270}	R_{270}	R_0	R_{90}	R_{180}	D	D'	V	H
H	H	(D)	V	D'	R_0	R_{180}	R_{90}	R_{270}
V	V	D'	H	D	R_{180}	R_0	R_{270}	R_{90}
D	D	V	D'	H	R_{270}	R_{90}	R_0	R_{180}
D'	D'	H	D	V	R_{90}	R_{270}	R_{180}	R_0

Notice how beautiful this table looks! This is no accident! Perhaps the most important feature of this table is that it is completely filled in without introducing any new motions. Of course, this is because, as we have already pointed out, any sequence of motions turns out to be the same as one of these eight. Algebraically, this says that if A and B are in D_4, then so is AB. This property is called *closure,* and it is one of the requirements for a mathematical system to be a group. Next, notice that if A is any element of D_4, then $AR_0 = R_0A = A$. Thus, combining any element A on either side with R_0 yields A back again. An element R_0 with this property is called an *identity,* and every group must have one. Moreover, we see that for each element A in D_4, there is exactly one element B in D_4 so that $AB = BA = R_0$. In this case, B is said to be the *inverse* of A and vice versa. For example, R_{90} and R_{270} are inverses of each other, and H is its own inverse. The term *inverse* is a descriptive one, for if A and B are inverses of each other, then B "undoes" whatever A "does," in the sense that A and B taken together in either order produce R_0 representing no change. Another striking feature of the table is that every element of D_4 appears exactly once in each row and column. This feature is something that all groups must have, and, indeed, it is quite useful to keep this fact in mind when constructing the table in the first place.

Another property of D_4 deserves special comment. Observe that $HD \neq DH$ but $R_{90}R_{180} = R_{180}R_{90}$. Thus, in a group, AB may or may not be the same as BA. If it happens that $AB = BA$ for *all* choices of group elements A and B, we say the group is *commutative* or—better yet— *Abelian* (in honor of the great Norwegian mathematician Niels Abel). Otherwise, we say the group is *non-Abelian*.

Thus far, we have illustrated, by way of D_4, three of the four conditions that define a group—namely, closure, existence of an identity, and existence of inverses. The remaining condition required for a group is *associativity;* that is, $(ab)c = a(bc)$ for all a, b, c in the set. To be sure that D_4 is indeed a group, we should check this equation for each of the $8^3 = 512$ possible choices of a, b, and c in D_4. In practice, however, this is rarely done! Here, for example, we simply observe that the eight motions are functions and the operation is function composition. Then, since function composition is associative, we do not have to check the equations.

THE DIHEDRAL GROUPS

The analysis carried out above for a square can similarly be done for an equilateral triangle or regular pentagon or, indeed, any regular n-gon ($n \geq 3$). The corresponding group is denoted by D_n and is called the *dihedral group of order 2n.*

The dihedral groups arise frequently in art and nature. Many of the decorative designs used on floor coverings, pottery, and buildings have one of the dihedral groups as a group of symmetry. Corporation logos are rich sources of dihedral symmetry [1]. Chrysler's logo has D_5 as a symmetry group, and Mercedes-Benz has D_3. The ubiquitous five-pointed star has symmetry group D_5. Sea animals of the family that includes the starfish, sea cucumbers, feather stars, and sand dollars exhibit patterns with D_5 symmetry group.

Mineralogists determine the internal structures of crystals (that is, rigid bodies in which the particles are arranged in three-dimensional repeating patterns—table salt and table sugar are two examples) by studying two-dimensional x-ray projections of the atomic makeup of the crystals. The symmetry present in the projections reveals the internal symmetry of the crystals themselves. Commonly occurring symmetry patterns are D_4 and D_6 (see Figure 1.2). Interestingly, it is mathematically impossible for a crystal to possess a D_n symmetry pattern for $n = 5$ or $n > 6$.

Perhaps the most spectacular manifestations of the dihedral groups

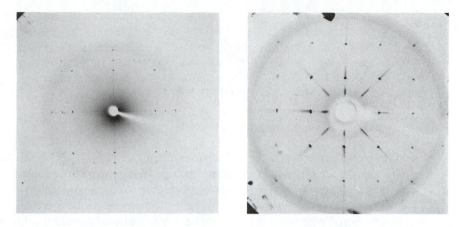

Figure 1.2 X-ray diffraction photos revealing D_4 symmetry patterns in crystals.

are the huge circular stained glass rose windows in the Gothic cathedrals of Europe. In the beautiful book *Rose Windows* [2], one can find magnificent representations of D_{12}, D_{16}, D_{24}, D_{30}, and many others. These windows exhibit an intricate blend of religion, art, geometry, architecture, and science. The groups D_{12} and D_{24} are the most common ones associated with rose windows.*

The dihedral group of order $2n$ is often called the *group of symmetries of a regular n-gon*. A *plane symmetry* of a figure F in a plane is a function from the plane to itself that carries F onto F and preserves distances; that is, for any points p and q in the plane, the distance from the image of p to the image of q is the same as the distance from p to q. (The term "symmetry" is from the Greek word *symmetron* meaning "well-ordered.") The *symmetry group* of a plane figure is the set of all symmetries of the figure. Symmetries in three dimensions are defined analogously. Obviously, a rotation of a plane about a point in the plane is a symmetry of the plane and a rotation about a line in three dimensions is a symmetry in three-dimensional space. Similarly, any translation of a plane or of three-dimensional space is a symmetry. A *reflection across a line L* is that function that leaves every point of L fixed and takes any point q, not on L, to the point q' so that L is the perpendicular bisector of the line segment joining q and q' (see Figure 1.3). A reflection across a plane in three dimensions is defined analogously. Notice that the restriction of a 180° rotation about a line L in three dimensions to a plane containing L is a reflection across L in the plane. Thus, in the dihedral groups, the motions that we described as 180° rotations about

---

*The number 12 had special significance to medieval Christians. It was considered the number of perfection, of the universe, of the Logos of Time, the Apostles, the zodiac, the tribes of Israel, and the precious stones in the foundation of New Jerusalem. There are 24 elders of the Apocalypse.

Figure 1.3

axes of symmetry in three dimensions (for example, H, V, D, D') are reflections across lines in two dimensions. Just as a reflection across a line is a plane symmetry that cannot be achieved by a physical motion of the plane in two dimensions, a reflection across a plane is a three-dimensional symmetry that cannot be achieved by a physical motion of three-dimensional space. A cup, for instance, has reflective symmetry across the plane bisecting the cup, but this symmetry cannot be duplicated with a physical motion in three dimensions.

Many objects and figures have rotational symmetry but not reflective symmetry. A symmetry group consisting of the rotational symmetries of $0°$, $360°/n$, $2(360°)/n$, ..., $(n-1)360°/n$ and no other symmetries is called a *cyclic rotation group of order n* and is denoted by $\langle R_{360/n} \rangle$. Cyclic rotation groups along with dihedral groups are favorites of artists, designers, and nature. Figure 1.4 illustrates with corporate logos the cyclic rotation groups of orders 2, 3, 4, 5, 6, 8, 16, and 20.

Further examples of the occurrence of dihedral groups and cyclic groups in art and nature can be found in the references. A study of symmetry in greater depth is done in Chapters 27 and 28.

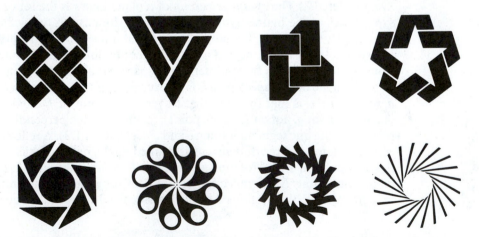

Figure 1.4   Logos with cyclic rotation symmetry groups.

EXERCISES

If you think you can or can't, you are right.

Henry Ford

1. With pictures and words, describe each symmetry in D_3 (the set of symmetries of an equilateral triangle).
2. Write out a complete multiplication table for D_3.
3. Is D_3 Abelian?
4. Describe in pictures or words the elements of D_5 (symmetries of a regular pentagon).
5. For $n \geq 3$, describe the elements of D_n. (*Hint:* You will need to consider two cases—n even and n odd.) How many elements does D_n have?
6. In D_n, explain geometrically why a reflection followed by a reflection must be a rotation.
7. In D_n, explain geometrically why a rotation followed by a rotation must be a rotation.
8. In D_n, explain geometrically why a rotation and a reflection taken together in either order must be a reflection.
9. Associate the number $+1$ with a rotation and the number -1 with a reflection. Describe an analogy between multiplying these two numbers and multiplying elements of D_n.
10. If r_1, r_2, and r_3 represent rotations from D_n and f_1, f_2, and f_3 represent reflections from D_n, determine whether $r_1 r_2 f_1 r_3 f_2 f_3 r_3$ is a rotation or reflection.
11. Find elements A, B, and C in D_4 such that $AB = BC$ but $A \neq C$. (Thus, "cross" cancellation is not valid.)
12. Explain what the following diagram proves about the group D_n.

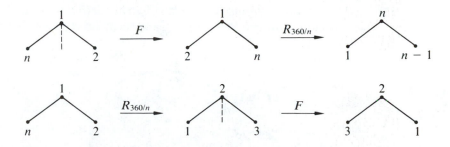

13. Describe the symmetries of a nonsquare rectangle. Construct the corresponding Cayley table.

14. Describe the symmetries of a parallelogram that is neither a rectangle nor a rhombus. Describe the symmetries of a rhombus that is not a rectangle.

15. Describe the symmetries of a noncircular ellipse. Do the same for a hyperbola.

16. Consider an infinitely long strip of equally spaced H's:

$$\cdots \text{H H H H} \cdots$$

Describe the symmetries of this strip. Is the group of symmetries of the strip Abelian?

17. For each of the snowflakes in the figure, find the symmetry group and locate the axes of reflective symmetry (disregard imperfections).

Photographs of snowflakes from the Bentley and Humphrey atlas.

18. Does the agitator of a washing machine have a cyclic symmetry group or a dihedral symmetry group?

19. Does an airplane propeller have a cyclic symmetry group or a dihedral symmetry group?

20. Bottle caps that are pried off typically have 22 ridges around the rim. Find the symmetry group of such a cap.

21. Find the symmetry group for the Canadian hendecagonal one dollar coin shown below. Disregard the printing and scene on the coin.

22. For each design below, determine the symmetry group.

23. Determine the symmetry group of the outer shell of the cross section of the AIDS virus shown below.

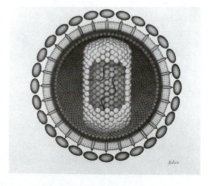

24. Determine the symmetry group of the design on the cover of this book.

REFERENCES

1. B. B. Capitman, *American Trademark Designs,* New York: Dover, 1976.

2. P. Cowen, *Rose Windows,* San Francisco: Chronicle Books, 1979.
3. I. Hargittai and M. Hargittai, *Symmetry Through the Eyes of a Chemist,* New York: VCH, 1986.
4. Caroline MacGillavry, *Fantasy and Symmetry: The Periodic Drawings of M. C. Escher,* New York: Harry N. Abrams, 1976.
5. A. V. Shubnikov and V. A. Koptsik, *Symmetry in Science and Art,* New York: Plenum Press, 1974.
6. P. S. Stevens, *Handbook of Regular Patterns,* Cambridge, Mass: MIT Press, 1980.
7. H. Weyl, *Symmetry,* Princeton: Princeton University Press, 1952.

Niels Abel

He [Abel] has left mathematicians something to keep them busy for five hundred years.

Charles Hermite

Norway has issued five stamps to honor Abel. This one was issued in 1929 to commemorate the 100th anniversary of his death.

A 500-kroner bank note first issued by Norway in 1948.

NIELS HENRIK ABEL, one of the foremost mathematicians of the 19th century, was born in Norway on August 5, 1802. At the age of 16, he began reading the classic mathematical works of Newton, Euler, Lagrange, and Gauss. When Abel was 18 years old, his father died and the burden of supporting the family fell upon him. He took in private pupils and did odd jobs, while continuing to do mathematical research. At the age of 19, Abel solved a problem that had vexed leading mathematicians for hundreds of years. He proved that, unlike the situation for equations of degree 4 or less, there is no finite (closed) formula for the solution of the general fifth-degree equation.

Although Abel died long before the advent of the subjects that now comprise abstract algebra, his solution to the quintic problem laid groundwork for many of these subjects. In addition to his work in the theory of equations, Abel made outstanding contributions to the theory of elliptic functions, elliptic integrals, Abelian integrals, and infinite series. Just when his work was beginning to receive the attention it deserved, Abel contracted tuberculosis. He died on April 6, 1829, at the age of 26. In 1870, Camille Jordan introduced the term *Abelian group* to honor Abel. Norway has issued five stamps and a 500-kroner bank note to honor Abel.

2

Groups

A good stock of examples, as large as possible, is
indispensable for a thorough understanding of any concept,
and when I want to learn something new, I make it my first job
to build one.

Paul R. Halmos

Facts about groups of symmetries have been known for hundreds of years. In the 13th century, the Moors demonstrated a sophisticated understanding of symmetry with their decorative wall patterns at the Alhambra, a fortress-palace in Southern Spain. The Alhambra is the crowning glory of Moorish architecture (see [2, pp. 169–176]). In his study of buildings, Leonardo da Vinci (1452–1519) determined all possible symmetry groups of planar objects.

DEFINITION AND EXAMPLES OF GROUPS

The term *group* was coined by Galois about 160 years ago to describe sets of one-to-one functions on finite sets that could be grouped together to form a closed set. As is the case with most fundamental concepts in mathematics, the modern definition of a group that follows is the result of a long evolutionary process. Although this definition was given by both Heinrich Weber and Walther von Dyck in 1882, it did not gain universal acceptance until this century.

DEFINITION   Binary Operation

Let *G* be a set. A *binary operation* on *G* is a function that assigns each ordered pair of elements of *G* an element of *G*.

A binary operation on a set *G*, then, is simply a method (or formula) by which the members of an ordered pair from *G* combine to yield a new member of *G*. The most familiar binary operations are ordinary addition, subtraction, and multiplication of integers. Division of integers is not a binary operation on the integers (because an integer divided by an integer need not be an integer).

The binary operations addition modulo *n* and multiplication modulo *n* on the set $\{0, 1, 2, \ldots, n - 1\}$, which we denote by Z_n, play an extremely important role in abstract algebra. In certain situations we will want to combine the elements of Z_n by addition modulo *n* only; in other situations we will want to use both addition modulo *n* and multiplication modulo *n* to combine the elements. It will be clear from the context whether we are using addition only or addition and multiplication. For example, when multiplying matrices with entries from Z_n, we will need both addition modulo *n* and multiplication modulo *n*.

DEFINITION   Group

Let *G* be a nonempty set together with a binary operation (usually called multiplication) that assigns to each ordered pair (a, b) of elements of *G* an element in *G* denoted by *ab*. We say *G* is a *group* under this operation if the following three properties are satisfied.

1. *Associativity.* The operation is associative; that is, $(ab)c = a(bc)$ for all *a, b, c* in *G*.
2. *Identity.* There is an element *e* (called the *identity*) in *G*, such that $ae = ea = a$ for all *a* in *G*.
3. *Inverses.* For each element *a* in *G*, there is an element *b* in *G* (called the *inverse* of *a*) such that $ab = ba = e$.

In words, then, a group is a set together with an associative operation such that every element has an inverse and any pair of elements can be combined without going outside the set. This latter condition is called *closure*. Be sure to verify closure when testing for a group. (See Example 5.)

If a group has the property that $ab = ba$ for every pair of elements *a* and *b*, we say the group is *Abelian*. A group is *non-Abelian* if there is some pair of elements *a* and *b* for which $ab \neq ba$. When encountering a particular group for the first time, one should determine whether or not it is Abelian.

Now that we have the formal definition of a group, our first job is to build a good stock of examples. These examples will be used throughout the text to illustrate the theorems. (The best way to grasp the meat of a theorem is to see what it says in specific cases.) As we progress, the reader is bound to have hunches and conjectures that can be tested against the stock of examples. To develop a complete understanding of these examples, the reader should supply the missing details.

Example 1   The set of integers Z (so denoted because the German word for integers is *Zahlen*), the set of rational numbers Q (for quotient), and the set of real numbers **R** are all groups under ordinary addition. In each case the identity is 0 and the inverse of a is $-a$.   ❑

Example 2   The set of integers under ordinary multiplication is not a group. Property (3) fails. For example, there is no *integer b* such that $5b = 1$.   ❑

Example 3   The subset $\{1, -1, i, -i\}$ of the complex numbers is a group under complex multiplication. Note that -1 is its own inverse, whereas the inverse of i is $-i$.   ❑

Example 4   The set Q^+ of positive rationals is a group under ordinary multiplication. The inverse of any a is $1/a = a^{-1}$.   ❑

Example 5   The set S of positive irrational numbers together with 1 under multiplication satisfies the three properties given in the definition of a group but is not a group. Indeed, $\sqrt{2} \cdot \sqrt{2} = 2$, so S is not closed under multiplication.   ❑

Example 6   A rectangular array of the form $\begin{bmatrix} a & b \\ c & d \end{bmatrix}$ is called a 2 × 2 *matrix*. The set of all 2 × 2 matrices with real entries is a group under componentwise addition. That is,

$$\begin{bmatrix} a_1 & b_1 \\ c_1 & d_1 \end{bmatrix} + \begin{bmatrix} a_2 & b_2 \\ c_2 & d_2 \end{bmatrix} = \begin{bmatrix} a_1 + a_2 & b_1 + b_2 \\ c_1 + c_2 & d_1 + d_2 \end{bmatrix}.$$

The identity is $\begin{bmatrix} 0 & 0 \\ 0 & 0 \end{bmatrix}$ and the inverse of $\begin{bmatrix} a & b \\ c & d \end{bmatrix}$ is $\begin{bmatrix} -a & -b \\ -c & -d \end{bmatrix}$.   ❑

Example 7   The set $Z_n = \{0, 1, \ldots, n - 1\}$ for $n \geq 1$ is a group under addition modulo n. For any j in Z_n, the inverse of j is $n - j$. This group is usually referred to as the *group of integers modulo n*.   ❑

As we have seen, the real numbers, the 2 × 2 matrices with real entries, and the integers modulo n are all groups under the appropriate addition. But what about multiplication? In each case the existence of some elements that do not have inverses prevents the set from being a group under the usual multiplication. However, we can form a group in each case by simply throwing out the rascals. Examples 8, 9, and 11 illustrate this.

Example 8   The set $\mathbf{R}^{\#}$ of nonzero real numbers is a group under ordinary multiplication. The identity is 1. The inverse of a is $1/a$.   ◻

Example 9*   The *determinant* of 2×2 matrix $\begin{bmatrix} a & b \\ c & d \end{bmatrix}$ is the number $ad - bc$. If A is a 2×2 matrix, det A denotes the determinant of A.

The set

$$GL(2, \mathbf{R}) = \left\{ \begin{bmatrix} a & b \\ c & d \end{bmatrix} \middle| a, b, c, d \in \mathbf{R}, ad - bc \neq 0 \right\}$$

of 2×2 matrices with real entries and nonzero determinant is a non-Abelian group under the operation

$$\begin{bmatrix} a_1 & b_1 \\ c_1 & d_1 \end{bmatrix} \begin{bmatrix} a_2 & b_2 \\ c_2 & d_2 \end{bmatrix} = \begin{bmatrix} a_1 a_2 + b_1 c_2 & a_1 b_2 + b_1 d_2 \\ c_1 a_2 + d_1 c_2 & c_1 b_2 + d_1 d_2 \end{bmatrix}.$$

The first step in verifying that this set is a group is to show that the product of two matrices with nonzero determinant also has nonzero determinant. This follows from the fact that for any pair of 2×2 matrices A and B, det $(AB) = $ (det A)(det B). (See exercise 5.)

Associativity can be verified by direct (but cumbersome) calculations. The identity is $\begin{bmatrix} 1 & 0 \\ 0 & 1 \end{bmatrix}$; the inverse of $\begin{bmatrix} a & b \\ c & d \end{bmatrix}$ is

$$\begin{bmatrix} \dfrac{d}{ad - bc} & \dfrac{-b}{ad - bc} \\ \dfrac{-c}{ad - bc} & \dfrac{a}{ad - bc} \end{bmatrix}$$

(explaining the requirement that $ad - bc \neq 0$). This very important non-Abelian group is called the *general linear group* of 2×2 matrices over $\mathbf{R}$.   ◻

Example 10   The set of 2×2 matrices with real number entries is not a group under the operation defined in Example 9. Inverses do not exist when the determinant is 0.   ◻

Now that we have shown how to make subsets of the real numbers and subsets of the set of 2×2 matrices into multiplicative groups, we next consider the integers under multiplication modulo n.

Example 11   (L. Euler, 1761)
By exercise 13 in Chapter 0, an integer a has a multiplicative inverse modulo n if and only if a and n are relatively prime. So, for each $n > 1$,

---

*For simplicity, we have restricted our matrix examples to the 2×2 case. However, readers who have had linear algebra can readily generalize to $n \times n$ matrices.

we define $U(n)$ to be the set of all positive integers less than n and relatively prime to n. Then $U(n)$ is a group under multiplication modulo n. (We leave as an exercise the proof that this set is closed under this operation.)

For $n = 10$, we have $U(10) = \{1, 3, 7, 9\}$. The Cayley table for $U(10)$ is

mod 10	1	3	7	9
1	1	3	7	9
3	3	9	1	7
7	7	1	9	3
9	9	7	3	1

(Recall that $ab \bmod n$ is defined as the unique integer r with the property $a \cdot b = nq + r$, where $0 \le r < n$ and $a \cdot b$ is ordinary multiplication.) In the case that n is a prime, $U(n) = \{1, 2, \ldots, n - 1\}$. ❑

In his classic book *Lehrbuch der Algebra,* published in 1899, Heinrich Weber gave an extensive treatment of the groups $U(n)$ and described them as the most important examples of finite Abelian groups.

Example 12   The set $\{0, 1, 2, 3\}$ is not a group under multiplication modulo 4. Although 1 and 3 have inverses, the elements 0 and 2 do not. ❑

Example 13   The set of integers under subtraction is not a group, since the operation is not associative. ❑

With the examples given this far as a guide, it is wise for the reader to pause here and think of his or her own examples. Study actively! Don't just read along and be spoon-fed by the book.

Example 14   For all integers $n \ge 1$, the set of complex roots of unity

$$\left\{ \cos \frac{k \cdot 360}{n} + i \sin \frac{k \cdot 360}{n} \,\middle|\, k = 0, 1, 2, \ldots, n - 1 \right\}$$

(i.e., complex zeros of $x^n - 1$) is a group under multiplication. (See De Moivre's Theorem—Example 11 in Chapter 0.) Compare this group with the one in Example 3. ❑

The complex number $a + bi$ can be represented geometrically as the point (a, b) in a plane coordinatized by a horizontal real axis and a vertical i (or imaginary) axis. The distance from the point $a + bi$ to the

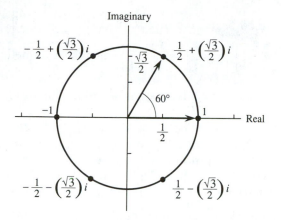

Figure 2.1

origin is $\sqrt{a^2 + b^2}$ and is often denoted by $|a + bi|$. For any angle θ, the line segment joining the complex number $\cos \theta + i \sin \theta$ and the origin forms an angle of θ with the positive real axis. Thus, the six complex zeros of $x^6 = 1$ are located at points around the circle of radius 1, 60° apart, as shown in Figure 2.1.

Example 15 The set $\mathbf{R}^n = \{(a_1, a_2, \ldots, a_n) \mid a_1, a_2, \ldots, a_n \in \mathbf{R}\}$ is a group under componentwise addition [i.e., $(a_1, a_2, \ldots, a_n) + (b_1, b_2, \ldots, b_n) = (a_1 + b_1, a_2 + b_2, \ldots, a_n + b_n)$]. ◻

Example 16 For a fixed point (a, b) in $\mathbf{R}^2$, define $T_{a,b}: \mathbf{R}^2 \rightarrow \mathbf{R}^2$ by $(x, y) \rightarrow (x + a, y + b)$. Then $T(\mathbf{R}^2) = \{T_{a,b} \mid a, b \in \mathbf{R}\}$ is a group under function composition. Straightforward calculations show that $T_{a,b}T_{c,d} = T_{a+c,b+d}$. From this formula we may observe that $T(\mathbf{R}^2)$ is closed, $T_{0,0}$ is the identity, the inverse of $T_{a,b}$ is $T_{-a,-b}$, and $T(\mathbf{R}^2)$ is Abelian. Function composition is always associative. The elements of $T(\mathbf{R}^2)$ are called *translations*. ◻

Example 17 The set of all 2×2 matrices with determinant 1 with entries from Q (rationals), $\mathbf{R}$ (reals), $\mathbf{C}$ (complex numbers), or Z_p (p a prime) is a non-Abelian group under matrix multiplication. This group is called the *special linear group* of 2×2 matrices over Q, $\mathbf{R}$, $\mathbf{C}$, or Z_p, respectively. If the entries are from F, where F is any of the above, we denote this group by $SL(2, F)$. The formula given in Example 9 for the inverse of $\begin{bmatrix} a & b \\ c & d \end{bmatrix}$ simplifies to $\begin{bmatrix} a & -b \\ -c & a \end{bmatrix}$. When the matrix entries are from Z_p, we use modulo p arithmetic to compute determinants, matrix products, and inverses. To illustrate the case $SL(2, Z_5)$, consider the element $A = \begin{bmatrix} 3 & 4 \\ 4 & 4 \end{bmatrix}$. Then $\det A = 3 \cdot 4 - 4 \cdot 4 = -4 =$

1 mod 5 and the inverse of A is $\begin{bmatrix} 4 & -4 \\ -4 & 3 \end{bmatrix} = \begin{bmatrix} 4 & 1 \\ 1 & 3 \end{bmatrix}$. Note that $\begin{bmatrix} 3 & 4 \\ 4 & 4 \end{bmatrix}\begin{bmatrix} 4 & 1 \\ 1 & 3 \end{bmatrix} = \begin{bmatrix} 1 & 0 \\ 0 & 1 \end{bmatrix}$ when the arithmetic is done modulo 5. ❑

Example 9 is a special case of the following general construction.

Example 18. Let F be any of Q, $\mathbf{R}$, $\mathbf{C}$, or Z_p (p a prime). The set $GL(2, F)$ of all 2×2 matrices with nonzero determinant and entries from F is a non-Abelian group under matrix multiplication. As in Example 17, when F is Z_p, modulo p arithmetic is used to calculate determinants, the matrix products, and inverses. The formula given in Example 9 for the inverse of $\begin{bmatrix} a & b \\ c & d \end{bmatrix}$ remains valid for elements from $GL(2, Z_p)$ provided we interpret division by $ad - bc$ as multiplication by the inverse of $ad - bc$ modulo p. For example, in $GL(2, Z_7)$, consider $\begin{bmatrix} 4 & 5 \\ 6 & 3 \end{bmatrix}$. Then the determinant $ad - bc$ is $12 - 30 = -18 = 3$ mod 7 and the inverse of 3 is 5 (since $3 \cdot 5 = 1$ mod 7). So, the inverse of $\begin{bmatrix} 4 & 5 \\ 6 & 3 \end{bmatrix}$ is $\begin{bmatrix} 3 \cdot 5 & 2 \cdot 5 \\ 1 \cdot 5 & 4 \cdot 5 \end{bmatrix} = \begin{bmatrix} 1 & 3 \\ 5 & 6 \end{bmatrix}$. [The reader should check that $\begin{bmatrix} 4 & 5 \\ 6 & 3 \end{bmatrix}\begin{bmatrix} 1 & 3 \\ 5 & 6 \end{bmatrix} = \begin{bmatrix} 1 & 0 \\ 0 & 1 \end{bmatrix}$ in $GL(2, Z_7)$]. ❑

Example 19 The set $\{1, \ldots, n - 1\}$ is a group under multiplication modulo n if and only if n is prime. ❑

Example 20 The set of all symmetries of the infinite ornamental pattern in which the arrowheads are spaced uniformly a unit apart along a

line is an Abelian group under composition. Let T denote a translation to the right by one unit, T^{-1} a translation to the left by one unit, and H a reflection across the horizontal line of the figure. Then, every member of the group is of the form $x_1 x_2 \cdots x_n$, where each $x_i \in \{T, T^{-1}, H\}$. In this case, we say that T, T^{-1}, and H *generate* the group. ❑

Table 2.1 summarizes many of the specific groups that we have presented thus far.

As the above examples demonstrate, the notion of a group is a very broad one indeed. The goal of the axiomatic approach is to find a set of properties that are general enough to permit many diverse examples that possess these properties and that are specific enough to allow one to deduce many interesting consequences of these properties.

The goal of abstract algebra is to discover truths about algebraic sys-

Table 2.1 Summary of Group Examples (F can be any of Q, $\mathbf{R}$, $\mathbf{C}$, or Z_p; L is a reflection)

Group	Operation	Identity	Form of Element	Inverse	Abelian
Z	addition	0	k	$-k$	yes
Q^+	multiplication	1	m/n $m, n > 0$	n/m	yes
Z_n	addition mod n	0	k	$n - k$	yes
$\mathbf{R}^\#$	multiplication	1	x	$1/x$	yes
$GL(2, F)$	matrix multiplication	$\begin{bmatrix} 1 & 0 \\ 0 & 1 \end{bmatrix}$	$\begin{bmatrix} a & b \\ c & d \end{bmatrix}$ $ad - bc \neq 0$	$\begin{bmatrix} \dfrac{d}{ad-bc} & \dfrac{-b}{ad-bc} \\ \dfrac{-c}{ad-bc} & \dfrac{a}{ad-bc} \end{bmatrix}$	no
$U(n)$	multiplication mod n	1	k $\gcd(k, n) = 1$	solution to $kx = 1 \bmod n$	yes
$\mathbf{R}^n$	componentwise addition	$(0, 0, \ldots, 0)$	$(a_1, a_2, \ldots, a_n)$	$(-a_1, -a_2, \ldots, -a_n)$	yes
$SL(2, F)$	matrix multiplication	$\begin{bmatrix} 1 & 0 \\ 0 & 1 \end{bmatrix}$	$\begin{bmatrix} a & b \\ c & d \end{bmatrix}$ $ad - bc = 1$	$\begin{bmatrix} d & -b \\ -c & a \end{bmatrix}$	no
D_n	composition	R_0	R_α, L	$R_{360-\alpha}, L$	no

tems (that is, sets with one or more binary operations) that are independent of the specific nature of the operations. All one knows or needs to know is that these operations, whatever they may be, have certain properties. We then seek to deduce consequences of these properties. This is why this branch of mathematics is called *abstract* algebra. It must be remembered, however, that when a specific group is being discussed, a specific operation must be given (at least implicitly).

ELEMENTARY PROPERTIES OF GROUPS

Now that we have seen many diverse examples of groups, we wish to deduce some properties they share. The definition itself raises some fundamental questions. Every group has *an* identity. Could a group have more than one? Every group element has *an* inverse. Could an element have more than one? The examples suggest not. But examples can only suggest. One cannot prove that every group has a unique identity by looking at examples, because each example inherently has properties that may not be shared by all groups. We are forced to restrict ourselves to the properties that all groups have. That is, we must view groups as abstract entities rather than argue by example. The next three theorems illustrate the abstract approach.

Theorem 2.1 *Uniqueness of the Identity*
        In a group G, there is only one identity element.

Proof. Suppose both e and e' are identities of G. Then,

1. $ae = a$ for all a in G, and
2. $e'a = a$ for all a in G.

The choice of $a = e'$ in (1) and $a = e$ in (2) yields $e'e = e'$ and $e'e = e$. Thus, e and e' are both equal to $e'e$ and so are equal to each other. ∎

Because of this theorem, we may unambiguously speak of "the identity" of a group and denote it by "e" (because the German word for identity is *Einheit*).

Theorem 2.2   *Cancellation*
In a group G, the right and left cancellation laws hold; that is, ba = ca implies b = c, and ab = ac implies b = c.

Proof. Suppose $ba = ca$. Let a' be an inverse of a. Then, multiplying on the right by a' gives $(ba)a' = (ca)a'$. Associativity yields $b(aa') = c(aa')$. Then, $be = ce$ and, therefore, $b = c$ as desired. Similarly, one can prove that $ab = ac$ implies $b = c$ by multiplying by a' on the left. ∎

A consequence of the cancellation property is the fact that in a Cayley table for a group, each group element occurs exactly once in each row and column. (See exercise 28.) Another consequence of the cancellation property is the uniqueness of inverses.

Theorem 2.3   *Uniqueness of Inverses*
For each element a in a group G, there is a unique element b in G such that ab = ba = e.

Proof. Suppose b and c are both inverses of a. Then $ab = e$ and $ac = e$, so that $ab = ac$. Now cancel a. ∎

As was the case with the identity element, it is reasonable, in view of Theorem 2.3, to speak of "the inverse" of an element g of a group; and, in fact, we may unambiguously denote it by g^{-1}. This notation is suggested by that used for ordinary real numbers under multiplication. Similarly, when n is a positive integer, g^n is used to denote the product

$$\underbrace{gg \cdots g}_{n \text{ factors}};$$

we define $g^0 = e$; when n is negative, we define $g^n = (g^{-1})^{-n}$ [for example, $g^{-3} = (g^{-1})^3$]. With this notation, the familiar laws of exponents hold for groups; that is, for all integers m and n and any group element g, we have $g^m g^n = g^{m+n}$ and $(g^m)^n = g^{mn}$. Although the way one manipulates the group expressions $g^m g^n$ and $(g^m)^n$ coincides with the

Table 2.2

Multiplicative Group		Additive Group	
$a \cdot b$ or ab	multiplication	$a + b$	addition
e or 1	identity or one	0	zero
a^{-1}	multiplicative inverse of a	$-a$	additive inverse of a
a^n	power of a	na	multiple of a
ab^{-1}	quotient	$a - b$	difference

laws of exponents for real numbers, the laws of exponents fail to hold for expressions involving two group elements. Thus, for groups in general, $(ab)^n \neq a^n b^n$ (see exercise 19).

Also, one must be careful with this notation when dealing with a specific group whose binary operation is addition and is denoted by "$+$." In this case, the definitions and group properties expressed in multiplicative notation must be translated to additive notation. For example, the inverse of g is written as $-g$. Likewise, for example, g^3 means $g + g + g$ and is usually written as $3g$, whereas g^{-3} means $(-g) + (-g) + (-g)$ and is written as $-3g$. When additive notation is used, do not interpret "ng" as combining n and g under the group operation; n may not even be an element of the group! Unlike the case for real numbers, in an abstract group, we do not permit noninteger exponents such as $g^{1/2}$. Table 2.2 shows the common notation and corresponding terminology for groups under multiplication and groups under addition. As is the case for real numbers, we use $a - b$ as an abbreviation for $a + (-b)$.

Because of the associative property, we may unambiguously write the expression abc, for this can be reasonably interpreted as only $(ab)c$ or $a(bc)$, which are equal. In fact, by using induction and repeated application of the associative property, one can prove a general associative property that essentially means parentheses can be inserted or deleted at will without affecting the value of a product involving any number of group elements. Thus,

$$a^2(bcdb^2) = a^2 b(cd)b^2 = (a^2 b)(cd)b^2 = a(abcdb)b,$$

and so on.

APPLICATIONS OF MODULAR ARITHMETIC

With the terminology and notation introduced thus far, we can describe some interesting applications of modular arithmetic. In Chapter 0 we discussed the methods used by the U.S. Postal Service and United Parcel Service to append check digits to identification numbers. We observed that neither method was able to detect all possible single-digit errors. Detection of all single-digit errors, as well as nearly all errors involving the transposition of two adjacent digits, is easily achieved, however. In this section we present several such methods that are in use.

Most products sold in supermarkets have an identification number coded with bars that are read by optical scanners. See Figure 2.2. This code is called the Universal Product Code (UPC). Each coded item is assigned a 12-digit number. The first six digits identify the manufacturer, the next five identify the product, and the last is a check. (For many items, the 12th digit is not printed but it is always bar coded.) In Figure 2.2, the check digit is 8.

To explain how the check digit is calculated, it is convenient to introduce the dot product notation for two k-tuples:

$$(a_1, a_2, \ldots, a_k) \cdot (w_1, w_2, \ldots, w_k) = a_1 w_1 + a_2 w_2 + \cdots + a_k w_k.$$

An item with the UPC identification number $a_1 a_2 \cdots a_{12}$ satisfies the condition

$$(a_1, a_2, \ldots, a_{12}) \cdot (3, 1, 3, 1, \ldots, 3, 1) = 0 \bmod 10.$$

In particular, the check digit is

$$-(a_1, a_2, \ldots, a_{11}) \cdot (3, 1, 3, 1, \ldots, 3) \bmod 10.$$

To verify that the number in Figure 2.2 satisfies the above condition, we calculate

$$0 \cdot 3 + 2 \cdot 1 + 1 \cdot 3 + 0 \cdot 1 + 0 \cdot 3 + 0 \cdot 1 + 6 \cdot 3 + 5 \cdot 1$$
$$+ 8 \cdot 3 + 9 \cdot 1 + 7 \cdot 3 + 8 \cdot 1 = 90 = 0 \bmod 10.$$

The fixed k-tuple used in the calculation of check digits is called the *weighting vector*.

Now, suppose a single error is made in entering the number in Figure 2.2 into a computer. Say, for instance, that 021000958978 is entered (notice that the seventh digit is incorrect). Then the computer calculates

$$0 \cdot 3 + 2 \cdot 1 + 1 \cdot 3 + 0 \cdot 1 + 0 \cdot 3 + 0 \cdot 1 + 9 \cdot 3$$
$$+ 5 \cdot 1 + 8 \cdot 3 + 9 \cdot 1 + 7 \cdot 3 + 8 \cdot 1 = 99.$$

Figure 2.2

Since $99 \neq 0 \bmod 10$, the entered number cannot be correct. In general, as we will see in Theorem 2.4, when the numbers used in the weighting vector are relatively prime to n, any error involving a single digit will be detected by a check digit computed modulo n.

But why should a weighting vector be used at all? One could detect all single-digit errors in a string $a_1 a_2 \cdots a_k$ by simply choosing a check digit a_{k+1} so that $a_1 + a_2 + \cdots + a_k + a_{k+1} = 0 \bmod 10$. For surely any error in exactly one position will result in a sum that is not 0 modulo 10. (In fact, this method is used on U.S. mail with bar-coded zip codes—see [3]). The advantage of the UPC scheme is that it will detect nearly all errors involving the transposition of two adjacent digits as well as errors involving one digit.* For doubters, let us say that the identification number given in Figure 2.2 is entered as 021000658798. Notice that the last two digits preceding the check digit have been transposed. But by calculating the dot product, we obtain $94 \neq 0 \bmod 10$, so we have detected an error. In fact, the only undetected transposition errors of adjacent digits a and b are those where $|a - b| = 5$. To verify this, we observe that a transposition error of the form

$$a_1 a_2 \cdots a_i a_{i+1} \cdots a_{12} \rightarrow a_1 a_2 \cdots a_{i+1} a_i \cdots a_{12}$$

is undetected if and only if

$$(a_1, a_2, \ldots, a_{i+1}, a_i, \ldots, a_{12}) \cdot (3, 1, 3, 1, \ldots, 3, 1) = 0 \bmod 10.$$

That is, the error is undetected if and only if

$$(a_1, a_2, \ldots, a_{i+1}, a_i, \ldots, a_{12}) \cdot (3, 1, 3, 1, \ldots, 3, 1)$$
$$= (a_1, a_2, \ldots, a_i, a_{i+1}, \ldots, a_{12}) \cdot (3, 1, 3, 1, \ldots, 3, 1) \bmod 10.$$

This equality simplifies to either

$$3a_{i+1} + a_i = 3a_i + a_{i+1} \bmod 10$$

or

$$a_{i+1} + 3a_i = a_i + 3a_{i+1} \bmod 10$$

depending on whether i is even or odd. Both cases reduce to $2(a_{i+1} - a_i) = 0 \bmod 10$.

Identification numbers printed on bank checks (on the bottom left between the two colons) consist of an eight-digit number $a_1 a_2 \cdots a_8$ and a check digit a_9 so that

$$(a_1, a_2, \ldots, a_9) \cdot (7, 3, 9, 7, 3, 9, 7, 3, 9) = 0 \bmod 10.$$

---

*A highly publicized error of this type occurred when Lt. Col. Oliver North gave United States Assistant Secretary of State Elliott Abrams an incorrect Swiss bank account number for depositing $10 million for the Contras. The correct account number began with "386"; the number North gave Abrams began with "368."

As was the case for the UPC scheme, this method detects all single-digit errors and all errors involving the transposition of adjacent digits a and b except when $|a - b| = 5$. But it also detects most errors of the form $\cdots abc \cdots \rightarrow \cdots cba \cdots$, whereas the UPC method detects no errors of this form.

The next theorem reveals the relationship between the weighting vector and its ability to detect errors. In particular, it shows why it is not possible to detect all single-digit errors and all transposition errors with a dot product modulo 10 scheme.

Theorem 2.4   *Error-Detecting Capability*
Suppose an identification number $a_1 a_2 \cdots a_k$ satisfies

$$(a_1, a_2, \ldots, a_k) \cdot (w_1, w_2, \ldots, w_k) = 0 \bmod n$$

where $0 \le a_i < n$ for each i. Then all single-digit errors in the ith position are detected if and only if w_i is relatively prime to n and all errors of the form

$$\cdots a_i a_{i+1} \cdots a_j a_{j+1} \cdots \rightarrow \cdots a_j a_{i+1} \cdots a_i a_{j+1} \cdots$$

(that is, the digits in the ith and jth positions are interchanged) are detected if and only if $w_i - w_j$ is relatively prime to n.

Proof.   Consider a single error in the ith position, say, a_i' is substituted for a_i. Then the dot product of the correct number and the incorrect number differ by $(a_i - a_i')w_i$. Thus, the error is undetected if and only if $(a_i - a_i')w_i = 0 \bmod n$. If w_i is relatively prime to n, then w_i belongs to $U(n)$ and therefore, $w_i^{-1} \bmod n$ exists. So $(a_i - a_i')w_i w_i^{-1} = 0 \bmod n$ and $a_i' = a_i$. If w_i is not relatively prime to n, then an error a_i' with $|a_i - a_i'|$ divisible by $n/\gcd(w_i, n)$ will not be detected.

Now consider an error of the form

$$\cdots a_i a_{i+1} \cdots a_j a_{j+1} \cdots \rightarrow \cdots a_j a_{i+1} \cdots a_i a_{j+1} \cdots .$$

In this case, the dot products of the correct number and the incorrect number differ by

$$(a_i w_i + a_j w_j) - (a_j w_i + a_i w_j) = (a_i - a_j)(w_i - w_j).$$

Thus, the error is undetected if and only if

$$(a_i - a_j)(w_i - w_j) = 0 \bmod n.$$

The conclusion now follows as before.   ∎

In light of Theorem 2.4, we see that to detect all single-digit errors and all transposition errors, the modulus n must be a prime. However, since the standard number system is base 10, a penalty must be paid for using a prime modulus. For instance, if we select $n = 7$, we must restrict our identification numbers to at most five digits, excluding the check digit itself, and each digit must be between 0 and 6. This gives at most $7^5 = 16,807$ identification numbers. Another possibility is to choose $n = 11$. In this case, all the digits from 0 to 9 are available and we can have up to nine digits in length, excluding the check digit. Thus there are $10^9 = 1,000,000,000$ possible identification numbers. The drawback here is that the check digit may turn out to be 10, which is actually *two* digits. In practice, this is handled by ad hoc methods. For example, the 10-digit International Standard Book Number (ISBN) used throughout the world has the property $(a_1, a_2, \ldots, a_{10}) \cdot (10, 9, 8, 7, 6, 5, 4, 3, 2, 1) = 0 \bmod 11$. The first nine digits identify the language of the publishing company's country, the publisher, and the title, and the tenth digit is the check. When a_{10} is required to be 10 to make the dot product 0, the character X is used as the check digit.

In Chapter 5, we will present a check digit scheme, based on the dihedral group of order 10, that detects all single-digit errors and all transposition errors without introducing any new characters or restricting the number of digits involved.

HISTORICAL NOTE

We conclude this chapter with a bit of history concerning the noncommutativity of matrix multiplication. In 1925, quantum theory was replete with annoying and puzzling ambiguities. It was Werner Heisenberg who recognized the cause. He observed that the product of the quantum-theoretical analogs of the classical Fourier series did not necessarily commute. For all his boldness, this shook Heisenberg. As he later recalled [4, p. 94]:

> In my paper the fact that XY was not equal to YX was very disagreeable to me. I felt this was the only point of difficulty in the whole scheme, otherwise I would be perfectly happy. But this difficulty had worried me and I was not able to solve it.

Heisenberg asked his teacher, Max Born, if his ideas were worth publishing. Born was fascinated and deeply impressed by Heisenberg's new approach. Born wrote [1, p. 217]:

> After having sent off Heisenberg's paper to the *Zeitschrift für Physik* for publication, I began to ponder over his symbolic multiplication, and was soon so involved in it that I thought about it for the whole day and could hardly sleep at night. For I felt there was something fundamental behind it, the consummation of our endeavors of many years. And one morning, about the 10 July 1925, I suddenly saw light: Heisenberg's symbolic multiplication was nothing but the matrix calculus, well-known to me since my student days from Rosanes' lectures in Breslau.

Born and his student, Pascual Jordan, reformulated Heisenberg's ideas in terms of matrices, but it was Heisenberg who was credited with the formulation. In his autobiography, Born laments [1, p. 219]:

> Nowadays the textbooks speak without exception of Heisenberg's matrices, Heisenberg's commutation law, and Dirac's field quantization.
>
> In fact, Heisenberg knew at that time very little of matrices and had to study them.

Upon learning in 1933 that he was to receive the Nobel Prize with Dirac and Schrödinger for this work, Heisenberg wrote to Born [1, p. 220]:

> If I have not written to you for such a long time, and have not thanked you for your congratulations, it was partly because of my rather bad conscience with respect to you. The fact that I am to receive the Nobel Prize alone, for work done in Göttingen in collaboration—you, Jordan, and I—this fact depresses me and I hardly know what to write to you. I am, of course, glad that our common efforts are now appreciated, and I enjoy the recollection of the beautiful time of collaboration. I also believe that all good physicists know how great was your and Jordan's contribution to the structure of quantum mechanics—and this remains unchanged by a wrong decision from outside. Yet I myself can do nothing but thank you again for all the fine collaboration, and feel a little ashamed.

The story has a happy ending, however, because Born received the Nobel Prize in 1954 for his fundamental work in quantum mechanics.

EXERCISES

The only way to learn mathematics is to do mathematics.

Paul Halmos, *Hilbert Space Problem Book*

1. Give two reasons why the set of odd integers under addition is not a group.
2. Referring to Example 13, verify the assertion that subtraction is not associative.
3. Show that $\begin{bmatrix} 2 & 2 \\ 1 & 1 \end{bmatrix}$ does not have a multiplicative inverse in $GL(2, \mathbf{R})$.
4. Show that the group $GL(2, \mathbf{R})$ of Example 9 is non-Abelian, by exhibiting a pair of matrices A and B in $GL(2, \mathbf{R})$ such that $AB \neq BA$.
5. If A and B are 2×2 matrices with real entries, prove that $\det (AB) = (\det A)(\det B)$.
6. Give an example of group elements a and b with the property that $a^{-1}ba \neq b$.

7. Translate each of the following multiplicative expressions into its additive counterpart.
   a. a^2b^3,
   b. $a^{-2}(b^{-1}c)^2$,
   c. $(ab^2)^{-3}c^2 = e$.

8. For any elements a and b from a group and any integer n, prove that $(a^{-1}ba)^n = a^{-1}b^na$.

9. Write out a complete operation table for Z_4 (see Example 7).

10. Is the binary operation defined by the following table associative? Is it commutative?

	a	b	c	d
a	a	b	c	d
b	b	a	d	c
c	c	d	a	b
d	a	d	b	c

11. Show that $\{1, 2, 3\}$ under multiplication modulo 4 is not a group but that $\{1, 2, 3, 4\}$ under multiplication modulo 5 is a group.

12. Referring to Example 20, describe in words the symmetry corresponding to $T^{-4}H$.

13. Find the inverse of the element $\begin{bmatrix} 2 & 6 \\ 3 & 5 \end{bmatrix}$ in $GL(2, Z_{11})$.

14. In the notation of Example 16, verify that $T_{a,b}T_{c,d} = T_{a+c,b+d}$.

15. Prove that the set of all 2×2 matrices with entries from **R** and determinant $+1$ is a group under matrix multiplication.

16. For any integer $n > 2$, show that there are at least two elements in $U(n)$ that satisfy $x^2 = 1$.

17. An abstract algebra teacher intended to give a typist a list of nine integers that form a group under multiplication modulo 91. Instead, one of the nine integers was inadvertently left out so that the list appeared as 1, 9, 16, 22, 53, 74, 79, 81. Which integer was left out? (This really happened!)

18. Let G be a group with the following property: If a, b, and c belong to G and $ab = ca$, then $b = c$. Prove that G is Abelian.

19. (Law of Exponents for Abelian groups) Let a and b be elements of an Abelian group and let n be any integer. Show that $(ab)^n = a^nb^n$. Is this also true for non-Abelian groups?

20. (Socks-Shoes Property) In a group, prove that $(ab)^{-1} = b^{-1}a^{-1}$. Find an example that shows it is possible to have $(ab)^{-2} \neq b^{-2}a^{-2}$. Find a non-Abelian example that shows it is possible to have $(ab)^{-1} = a^{-1}b^{-1}$ for some distinct nonidentity elements a and b. Draw an analogy between the statement $(ab)^{-1} = b^{-1}a^{-1}$ and the act of putting on and taking off your socks and shoes.

21. Prove that a group G is Abelian if and only if $(ab)^{-1} = a^{-1}b^{-1}$ for all a and b in G.

22. In a group, prove that $(a^{-1})^{-1} = a$ for all a.

23. Show that the set $\{5, 15, 25, 35\}$ is a group under multiplication modulo 40. What is the identity element of this group? Can you see any relationship between this group and $U(8)$?

24. If $a_1, a_2, \ldots, a_n$ belong to a group, what is the inverse of $a_1 a_2 \cdots a_n$?

25. The integers 5 and 15 are among a collection of 12 integers that form a group under multiplication modulo 56. List all 12.

26. Give an example of a group with 105 elements. Give two examples of groups with 42 elements.

27. Construct a Cayley table for $U(12)$.

28. Prove that every group table is a *Latin square**; that is, each element of the group appears exactly once in each row and each column. (This exercise is referred to in this chapter.)

29. Suppose the table below is a group table. Fill in the blank entries.

	e	a	b	c	d
e	e	—	—	—	—
a	—	b	—	—	e
b	—	c	d	e	—
c	—	d	—	a	b
d	—	—	—	—	—

30. Prove that if $(ab)^2 = a^2 b^2$ in a group G, then $ab = ba$.

31. Let a, b, and c be elements of a group. Solve the equation $axb = c$ for x. Solve $a^{-1}xa = c$ for x.

32. Prove that the set of all rational numbers of the form $3^m 6^n$, where m and n are integers, is a group under multiplication.

33. Let G be a finite group. Show that the number of elements x of G such that $x^3 = e$ is odd. Show that the number of elements x of G such that $x^2 \neq e$ is even.

34. Let G be a group and let $g \in G$. Define a function ϕ_g from G to G by $\phi_g(x) = gxg^{-1}$ for all x in G. Show that ϕ_g is one-to-one and onto. [That is, $\phi_g(x) = \phi_g(y)$ implies $x = y$ and, for each y in G, there is an x in G such that $\phi_g(x) = y$.]

35. Let G be a group and g, $h \in G$. Define ϕ_g, ϕ_h, ϕ_{gh} as in the previous problem (that is, $\phi_h(x) = hxh^{-1}$ and $\phi_{gh}(x) = (gh)x(gh))^{-1}$. Show that $\phi_g \circ \phi_h = \phi_{gh}$.

---

*Latin squares are useful in designing statistical experiments. There is also a close connection between Latin squares and finite geometries.

36. Explain why division is not a binary operation on the set of real numbers.

37. Consider the set $G = \{0, 1, 2, 3, 4, 5, 6, 7\}$. Suppose there is a group operation $*$ on G that satisfies the following two conditions:

a. $a * b \leq a + b$ for all a, b in G,

b. $a * a = 0$ for all a in G.

Construct the multiplication table for G. (This group is sometimes called the *Nim group*.)

38. Prove that if G is a group with the property that the square of every element is the identity, then G is Abelian. (This exercise is referred to in Chapters 8, 22, and 26.)

39. In D_n, let $r = R_{360/n}$ and let f be any reflection. Use a diagram to verify that $frf = r^{-1}$. Use this relation to write the following elements in the form r^i or $r^i f$ where $0 \leq i < n$.

a. In D_4, $fr^{-2}fr^5$,

b. In D_5, $r^{-3}fr^4fr^{-2}$,

c. In D_6, $fr^5fr^{-2}f$.

40. Prove that the set of all 3×3 matrices with real entries of the form

$$
\begin{bmatrix}
1 & a & b \\
0 & 1 & c \\
0 & 0 & 1
\end{bmatrix}
$$

is a group. (Multiplication is defined by

$$
\begin{bmatrix}
1 & a & b \\
0 & 1 & c \\
0 & 0 & 1
\end{bmatrix}
\begin{bmatrix}
1 & a' & b' \\
0 & 1 & c' \\
0 & 0 & 1
\end{bmatrix}
=
\begin{bmatrix}
1 & a + a' & b' + ac' + b \\
0 & 1 & c' + c \\
0 & 0 & 1
\end{bmatrix}.
$$

This group, sometimes called the *Heisenberg group* after the Nobel Prize-winning physicist Werner Heisenberg, is intimately related to the Heisenberg Uncertainty Principle of Quantum Physics.)

41. Prove the assertion made in Example 19 that the set $\{1, 2, \ldots, n - 1\}$ is a group under multiplication modulo n if and only if n is prime.

42. In a finite group, show that the number of nonidentity elements that satisfy the equation $x^5 = e$ is a multiple of 4.

43. Use the UPC scheme to determine the check digit for the number 07312400508.

44. Use the bank scheme to determine the check digit for the number 09190204.

45. Verify the check digit for the ISBN assigned to this book.

46. The ISBN 0-669-03925-4 is the result of a transposition of two adjacent digits not involving the first or last digit. Determine the correct ISBN.

47. The State of Utah appends a ninth digit a_9 to their eight-digit driver's license number $a_1 a_2 \cdots a_8$ so that $(a_1, a_2, \ldots, a_8, a_9) \cdot (9, 8, 7, 6, 5, 4,$ $3, 2, 1) = 0 \mod 10$.

   a. If the first eight digits of a Utah license number are 14910573, what is the ninth digit?

   b. Suppose a legitimate Utah license number 149105767 is miscopied as 149105267. How would you know a mistake was made? Is there any way you could determine the correct number? If you knew that the error was in the seventh position, could you correct the mistake?

   c. If a legitimate Utah number 149105767 were miscopied as 199105767, would you be able to tell a mistake was made? Explain.

   d. Explain why any transposition error involving adjacent digits of a Utah number would be detected.

48. The Canadian province of Quebec uses the weighting vector $(12, 11, 10, \ldots, 2, 1)$ and modulo 10 arithmetic to append a check digit to driver's license numbers. Criticize this method. Describe all single-digit errors that are undetected by this scheme. How does the transposition of two adjacent digits of a number affect the check digit of the number?

49. (IBM Check Digit Method) Major credit cards, banks in West Germany, many libraries in the United States, and South Dakota driver's license numbers employ the following check digit method. To a number $a_1 a_2 \cdots a_k$, where k is odd, append

$$-((a_1, a_2, \ldots, a_k) \cdot (2, 1, 2, 1, \ldots) + r) \mod 10$$

where r is the number of a_i, i odd, that exceed 4. [If k is even, the weighting vector $(1, 2, 1, 2, \ldots)$ is used instead and r is the number of a_i, i even, that exceed 4.] For example, the number 3215698917546 yields $r = 4$ since the digits in the odd positions 5, 7, 11, and 13 exceed 4. Calculate the check digit for the number 3125600196431. Prove that this method detects all single-digit errors. Why does this not contradict Theorem 2.4? Determine which transposition errors involving adjacent digits go undetected by this method.

50. Many identification number schemes utilize both numbers and letters. In these cases the "check" can be a numeral or a letter. One of the most common of these systems is the so-called Code 39. For the purpose of computing the check character, the letter A is assigned the value 10, B the value 11, and so on. The check character is the number or letter α such that the sum of α and the dot product of the identification number (not including the check character) and the weighting vector $(\ldots, 3, 2)$ is 0 mod 36. For example, the check character of the number 2705A0086164ZE is N since $(2, 7, 0, 5, 10, 0, 0, 8, 6, 1, 6, 4, 35, 14) \cdot (15, 14, 13, 12, 11, 10, 9, 8, 7, 6, 5, 4, 3, 2) = 589 = 13 \mod 36$ and $13 + 23 = 0 \mod 36$. Use the Code 39 scheme to determine the check character for the number 2105A0055186ZA.

51. Suppose the weighting vector for ISBN numbers were changed to $(1, 2, 3, 4, 5, 6, 7, 8, 9, 10)$. Explain how this would affect the check digit.

PROGRAMMING EXERCISES

Almost immediately after the war, Johnny [Von Neumann] and I also began to discuss the possibilities of using computers heuristically to try to obtain insights into questions of pure mathematics. By producing examples and by observing the properties of special mathematical objects, one could hope to obtain clues as to the behavior of general statements which have been tested on examples.

S. M. Ulam, *Adventures of a Mathematician*

1. Write a program for printing out the following information about $U(n)$ (see Example 11). Assume $n < 100$.
   a. The elements of $U(n)$.
   b. The inverse of each member of $U(n)$.
   Run your program for $n = 12, 15, 30, 36, 63$.
2. Determine the size of $U(k)$ for $k = 9, 27, 81, 243, 25, 125, 49, 121$. On the basis of this information, try to guess a formula for the size of $U(p^n)$ as a function of the prime p and the integer n.
3. Determine the size of $U(k)$ for $k = 18, 54, 162, 486, 50, 250, 98, 242$. Make a conjecture about the relationship between the size of $U(2p^n)$ and the size of $U(p^n)$ where p is a prime other than 2.
4. Write a program that calculates the UPC check digit. Check your program with the numbers 011300101373 and 038000001277.
5. Write a program that calculates the bank number check digit.
6. Write a program that calculates the ISBN check digit.
7. Write a program that calculates the IBM check digit.

REFERENCES

1. Max Born, *My Life: Recollections of a Nobel Laureate,* New York: Charles Scribner's Sons, 1978.
2. J. Bronowski, *The Ascent of Man,* Boston: Little, Brown and Company, 1973.
3. J. A. Gallian, "The Zip Code Bar Code," *UMAP Journal* 7 (1986): 191–195.
4. J. Mehra and H. Rechenberg, *The Historical Development of Quantum Theory,* Vol. 3, New York: Springer-Verlag, 1982.

SUGGESTED READINGS

P. J. Denning, "Computer Viruses," *American Scientist* 76 (1988): 236–238.
    The author discusses computer viruses (programs that attack other programs) and possible immunization schemes. One of these employs check digits.

Joseph A. Gallian, "Assigning Driver's License Numbers," *Mathematics Magazine* 64 (1991): 13–22.

> This article describes various methods used by the states to assign driver's license numbers. Several include check digits for error detection.

J. A. Gallian, "The Mathematics of Identification Numbers," *The College Mathematics Journal* 22 (1991): 194–202.

> This article is a comprehensive survey of check digit schemes that are associated with identification numbers.

J. A. Gallian and S. Winters, "Modular Arithmetic in the Marketplace," *The American Mathematical Monthly* 95 (1988): 548–551.

> This article provides a more detailed analysis of the check digit schemes presented in this chapter and in Chapter 0. In particular, the error detection rates for the various schemes are given.

J. E. White, "Introduction to Group Theory for Chemists," *Journal of Chemical Education* 44 (1967): 128–135.

> Students interested in the physical sciences may find this article worthwhile. It begins with easy examples of groups and builds up to applications of group theory concepts and terminology to chemistry.

SUGGESTED FILM

Dihedral Kaleidoscopes with H. S. M. Coxeter, International Film Bureau, $13\frac{1}{2}$ minutes, in color.

> A dihedral kaleidoscope is formed by two plane mirrors intersecting at an angle π/n. When an object is placed before the mirrors of such a kaleidoscope, $2n$ objects are seen: the original and $2n - 1$ reflections. This illustrates the fact that a dihedral kaleidoscope generates, by two reflections, the dihedral group of order $2n$. The film discusses the special cases for $n = 1, 2, 3, 4$, makes some observations in the general case, and proceeds to the limiting case in which the two mirrors are parallel. Dihedral kaleidoscopes are then used to show some interesting regular figures, their stellations, and tessellations of the plane. This film was awarded first prize for mathematics at the International Festival of Scientific and Technical Films.

3

Finite Groups; Subgroups

In our own time, in the period 1960–1980, we have seen particle physics emerge as the playground of group theory.

Freeman Dyson

TERMINOLOGY AND NOTATION

As we will soon discover, finite groups—that is, groups with finitely many elements—have interesting arithmetic properties. To facilitate the study of finite groups, it is convenient to introduce some terminology and notation.

DEFINITION   Order of a Group
The number of elements of a group (finite or infinite) is called its *order*. We will use $|G|$ to denote the order of G.

Thus, the group Z of integers under addition has infinite order, whereas the group $U(10) = \{1, 3, 7, 9\}$ under multiplication modulo 10 has order 4.

DEFINITION   Order of an Element
The *order* of an element g in a group G is the smallest positive integer n such that $g^n = e$. (In additive notation, this would be $ng = 0$.) If no such integer exists, we say g has *infinite order*. The order of an element g is denoted by $|g|$.

So, to find the order of a group element g, you need only compute the sequence of products $g, g^2, g^3, \ldots$, until you reach the identity for the first time. The exponent of this product (or coefficient if the operation is addition) is the order of g. If the identity never appears in the sequence, then g has infinite order.

Example 1   Consider $U(15) = \{1, 2, 4, 7, 8, 11, 13, 14\}$ under multiplication modulo 15. To find the order of 7, say, we compute the sequence $7^1 = 7, 7^2 = 4, 7^3 = 13, 7^4 = 1$, so $|7| = 4$. To find the order of 11, we compute $11^1 = 11, 11^2 = 1$, so $|11| = 2$. Similar computations show that $|1| = 1, |2| = 4, |4| = 2, |8| = 4, |13| = 4, |14| = 2$. [Here is a trick that makes these calculations easier. Rather than compute the sequence $13^1, 13^2, 13^3, 13^4$, we may observe that $13 = -2$ modulo 15 (since $13 + 2 = 0$ mod 15) so that $13^2 = (-2)^2 = 4, 13^3 = -2 \cdot 4 = -8, 13^4 = (-2)(-8) = 1$.]   ❑

Example 2   Consider Z_{10} under addition modulo 10. Since $1 \cdot 2 = 2$, $2 \cdot 2 = 4, 3 \cdot 2 = 6, 4 \cdot 2 = 8, 5 \cdot 2 = 0$, we know that $|2| = 5$. Similar computations show that $|0| = 1, |7| = 10, |5| = 2, |6| = 5$.   ❑

Example 3   Consider Z under ordinary addition. Here every nonzero element has infinite order, since the sequence $a, 2a, 3a, \ldots$ never includes 0 when $a \neq 0$.   ❑

The perceptive reader may have noticed among our examples of groups in Chapter 2 that some are subsets of others with the same binary operation. The group in Example 17 with real entries, for instance, is a subset of the group in Example 9. Similarly, the group of complex numbers $\{1, -1, i, -i\}$ is a subset of the group described in Example 14 for n equal to any multiple of 4. This situation arises so often that we introduce a special term to describe it.

DEFINITION   Subgroup

If a subset H of a group G is itself a group under the operation of G, we say H is a *subgroup* of G.

We use the notation $H \leq G$ to mean H is a subgroup of G. If we want to indicate that H is a subgroup of G, but not equal to G itself, we write $H < G$. Such a subgroup is called a *proper subgroup*. The subgroup $\{e\}$ is called the *trivial subgroup* of G; a subgroup that is not $\{e\}$ is called a *nontrivial subgroup* of G.

Notice that Z_n under addition modulo n is *not* a subgroup of Z under addition, since addition modulo n is not the operation of Z.

SUBGROUP TESTS

When determining whether or not a subset H of a group G is a subgroup of G, one need not directly verify the group axioms. The next three results provide simple tests that suffice to show that a subset of a group is a subgroup.

Theorem 3.1 *One-Step Subgroup Test*

> *Let G be a group and H a nonempty subset of G. Then, H is a subgroup of G if ab^{-1} is in H whenever a and b are in H.*

Proof. Since the operation of H is the same as that of G, it is clear that this operation is associative. Next, we show that e is in H. Since H is nonempty, we may pick some x in H. Then, letting $a = x$ and $b = x$ in the hypothesis, we have $e = xx^{-1} = ab^{-1}$ is in H. To verify that x^{-1} is in H whenever x is in H, all we need to do is to choose $a = e$ and $b = x$ in the statement of the theorem. Finally, the proof will be complete when we show that H is closed; that is, if x, y belong to H, we must show that xy is in H also. Well, we have already shown that y^{-1} is in H whenever y is; so letting $a = x$ and $b = y^{-1}$, we have $xy = x(y^{-1})^{-1} = ab^{-1}$ is in H. ∎

Although we have dubbed Theorem 3.1 the "One-Step Subgroup Test," there are actually four steps involved in applying the theorem. (After you gain some experience, the first three steps are routine.) Notice the similarity between the last three steps listed below and the three steps involved in the Principle of Mathematical Induction.

1. Identify the property P that distinguishes the elements of H; that is, identify a defining condition.
2. Prove that the identity has property P. (This verifies that H is nonempty.)
3. *Assume* that two elements a and b have property P.
4. Use the assumption about a and b to show that ab^{-1} has property P.

Of course, when the operation of G is addition, step 4 becomes $a - b$. The procedure is illustrated in Examples 4 and 5.

Example 4 Let G be an Abelian group with identity e. Then $H = \{x \in G \mid x^2 = e\}$ is a subgroup of G. Here, the defining property of H is the condition $x^2 = e$. So, we first note that $e^2 = e$ so that H is nonempty. Now we assume that a and b belong to H. This means $a^2 = e$ and $b^2 = e$. Finally, we must show that $(ab^{-1})^2 = e$. Since G is Abelian, $(ab^{-1})^2 = ab^{-1}ab^{-1} = a^2(b^{-1})^2 = a^2(b^2)^{-1} = ee^{-1} = e$. Therefore, ab^{-1} belongs to H and, by the One-Step Subgroup Test, H is a subgroup of G. □

In many instances, a subgroup will consist of all elements that have a particular form. Here, the property P is the particular form.

Example 5 Let G be an Abelian group under multiplication with identity e. Then $H = \{x^2 \mid x \in G\}$ is a subgroup of G. (In words, H is the set of all "squares.") Since $e^2 = e$, the identity has the correct form. Next we write two elements of H in the correct form, say, a^2 and b^2. We must show that $a^2(b^2)^{-1}$ also has the correct form; that is, $a^2(b^2)^{-1}$ is the square of some element. Since G is Abelian, we may write $a^2(b^2)^{-1}$ as $(ab^{-1})^2$, which is the correct form. Thus, H is a subgroup of G. ❏

How do you prove that a subset of a group is *not* a subgroup? Here are three possible ways, any one of which guarantees that the subset is not a subgroup:

1. Show that the identity is not in the set.
2. Exhibit an element of the set whose inverse is not in the set.
3. Exhibit two elements of the set whose product is not in the set.

Example 6   Let G be the group of nonzero real numbers under multiplication, $H = \{x \in G \mid x = 1 \text{ or } x \text{ is irrational}\}$ and $K = \{x \in G \mid x \geq 1\}$. Then H is not a subgroup of G since $\sqrt{2} \in H$ but $\sqrt{2} \cdot \sqrt{2} = 2 \notin H$. Also, K is not a subgroup since $2 \in K$ but $2^{-1} \notin K$. ❏

Beginning students often prefer to use the next theorem instead of Theorem 3.1.

Theorem 3.2   *Two-Step Subgroup Test*
Let G be a group and H a nonempty subset of G. Then, H is a subgroup of G if $ab \in H$ whenever $a, b \in H$ (closed under multiplication), and $a^{-1} \in H$ whenever $a \in H$ (closed under inverse).

Proof.   By Theorem 3.1, it suffices to show that $a, b \in H$ implies $ab^{-1} \in H$. So, we suppose that $a, b \in H$. Since H is closed under inverse, we also have $b^{-1} \in H$. Thus, $ab^{-1} \in H$ by closure under multiplication. ∎

When dealing with finite groups, it is easier to use the following subgroup test.

Theorem 3.3   *Finite Subgroup Test*
Let H be a nonempty finite subset of a group G. Then, H is a subgroup of G if H is closed under the operation of G.

Proof.   In view of Theorem 3.2, we need only prove that $a^{-1} \in H$ whenever $a \in H$. If $a = e$, then $a^{-1} = a$ and we are done. If $a \neq e$, consider the sequence $a, a^2, a^3, \ldots$. Since H is finite and closure implies that all positive powers of a are in H, not all of these elements are distinct. Say, $a^i = a^j$ and $i > j$. Then $a^{i-j} = e$; and since $a \neq e$, $i - j > 1$. Thus, $a^{i-j} = a \cdot a^{i-j-1} = e$ and, therefore, $a^{i-j-1} = a^{-1}$. But, $i - j - 1 \geq 1$ implies $a^{i-j-1} \in H$ and we are done. ∎

EXAMPLES OF SUBGROUPS

The proofs of the next few theorems show how our subgroup tests work. We first introduce an important notation. For any element a from a group we let $\langle a \rangle$ denote the set $\{a^n \mid n \in Z\}$. In particular, observe that the exponents of a include all negative integers as well as 0 and the positive integers (a^0 is defined to be the identity).

Theorem 3.4 $\langle a \rangle$ *Is a Subgroup*
Let G be a group, and let a be any element of G. Then, $\langle a \rangle$ is a subgroup of G.

Proof.   Let a^n, $a^m \in \langle a \rangle$. Then, $a^n \cdot (a^m)^{-1} = a^{n-m} \in \langle a \rangle$; so, by Theorem 3.1, $\langle a \rangle$ is a subgroup of G.   ∎

The subgroup $\langle a \rangle$ is called the *cyclic subgroup of G generated by a*. In the case that $G = \langle a \rangle$ we say G is *cyclic* and a is a *generator of G*. (A cyclic group may have many generators.) Notice that although the list $\ldots,\ a^{-2},\ a^{-1},\ a^0,\ a^1,\ a^2,\ \ldots$ has infinitely many entries, the set $\{a^n \mid n \in Z\}$ might have only finitely many elements. Also note that, since $a^i \cdot a^j = a^{i+j} = a^{j+i} = a^j$, every cyclic group is Abelian.

Example 7   In $U(10)$, $\langle 3 \rangle = \{3, 9, 7, 1\} = U(10)$, for $3^1 = 3$, $3^2 = 9$, $3^3 = 7$, $3^4 = 1$, $3^5 = 3^4 \cdot 3 = 1 \cdot 3$, $3^6 = 3^4 \cdot 3^2 = 9, \ldots$; $3^{-1} = 7$ (since $3 \cdot 7 = 1$), $3^{-2} = 9$, $3^{-3} = 3$, $3^{-4} = 1$, $3^{-5} = 3^{-4} \cdot 3^{-1} = 7$, $3^{-6} = 3^{-4} \cdot 3^{-2} = 1 \cdot 9 = 9, \ldots$.   ☐

Example 8   In Z_{10}, $\langle 2 \rangle = \{2, 4, 6, 8, 0\}$. Remember, a^n means na when the operation is addition.   ☐

Example 9   In Z, $\langle -1 \rangle = Z$. Here each entry in the list $\ldots, -2(-1)$, $-1(-1)$, $0(-1)$, $1(-1)$, $2(-1), \ldots$ represents a distinct group element.   ☐

Example 10   In D_n, the dihedral group of order $2n$, let R denote a rotation of $360/n$ degrees. Then,

$$R^n = R_{360°} = e, \qquad R^{n+1} = R, \qquad R^{n+2} = R^2, \ldots.$$

Similarly, $R^{-1} = R^{n-1}$, $R^{-2} = R^{n-2}, \ldots$ so that $\langle R \rangle = \{e, R, \ldots, R^{n-1}\}$. We see, then, that the powers of R "cycle back" periodically with period n. Diagrammatically, raising R to successive positive powers is the same as moving counterclockwise around the following circle one node at a time, whereas raising R to successive negative powers is the same as moving around the circle clockwise one node at a time.

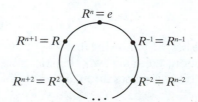

☐

In Chapter 4 we will show that $|\langle a \rangle| = |a|$; that is, the order of the subgroup generated by a is the order of a itself. (Actually, the definition of $|a|$ was chosen to ensure the validity of this equation.)

We next consider one of the most important subgroups.

DEFINITION   Center of a Group

The *center, $Z(G)$,* of a group G is the subset of elements in G that commute with every element of G. In symbols,

$$Z(G) = \{a \in G \mid ax = xa \text{ for all } x \text{ in } G\}.$$

[The notation $Z(G)$ comes from the fact that the German word for center is *Zentrum*. The term was coined by J. A. de Seguier in 1904.]

Theorem 3.5   *Center Is a Subgroup*

The center of a group G is a subgroup of G.

Proof.   For variety, we shall use Theorem 3.2 to prove this result. Clearly, $e \in Z(G)$, so $Z(G)$ is nonempty. Now, suppose $a, b \in Z(G)$. Then $(ab)x = a(bx) = a(xb) = (ax)b = (xa)b = x(ab)$ for all x in G; and, therefore, $ab \in Z(G)$.

Next, assume that $a \in Z(G)$. Then, we have $ax = xa$ for all x in G. What we want is $a^{-1}x = xa^{-1}$ for all x in G. Informally, all we need do to obtain the second equation from the first one is simultaneously to bring the a's across the equal sign:

$$ax = xa$$

becomes $xa^{-1} = a^{-1}x$. (Be careful here; groups need not be commutative. The a on the left comes across as a^{-1} on the left and the a on the right comes across as a^{-1} on the right.) Formally, the desired equation can be obtained from the original one by multiplying it on the left and right by a^{-1}, like so:

$$a^{-1}(ax)a^{-1} = a^{-1}(xa)a^{-1},$$
$$(a^{-1}a)xa^{-1} = a^{-1}x(aa^{-1}),$$
$$exa^{-1} = a^{-1}xe,$$
$$xa^{-1} = a^{-1}x.$$

This shows that $a^{-1} \in Z(G)$ whenever a is.   ∎

For practice, let's determine the centers of the dihedral groups.

Example 11   For $n \geq 3$,

$$Z(D_n) = \begin{cases} \{R_0, R_{180}\} & \text{when } n \text{ is even,} \\ \{R_0\} & \text{when } n \text{ is odd.} \end{cases}$$

We begin by showing that $Z(D_n)$ cannot contain a reflection. If F is a reflection, there are two possible cases for the reflection axis for F. Either this axis passes through a vertex of the n-gon, or it joins the midpoints of two opposite sides of the n-gon. Let's assume first that the axis passes through a vertex. Label the n-gon as shown below.

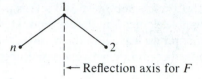

Reflection axis for F

Now, $R_{360/n}F$ gives

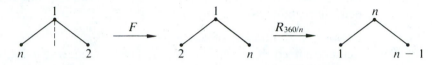

whereas $FR_{360/n}$ gives

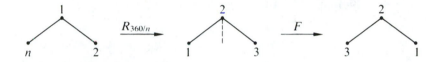

Thus, $R_{360/n}F$ sends vertex 1 to vertex n, whereas $FR_{360/n}$ sends vertex 1 to vertex 2. Since $n \geq 3$, we have $R_{360/n}F \neq FR_{360/n}$, so that F is not in the center of D_n. A similar argument on the following diagram rules out reflections that join midpoints of opposite sides (this case arises when n is even).

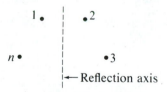

Reflection axis

We have proved, then, that no reflection is in the center of D_n.

Next, consider a rotation $R = R_{k \cdot 360/n}$ ($1 \leq k < n$) in D_n. Let's assume that $0° < k \cdot 360°/n < 180°$. Label the n-gon as shown in the following figure, and let F denote a reflection across the axis passing through vertex 1.

Now, FR sends vertex 1 to a vertex on the right side of the reflection axis, whereas RF sends vertex 1 to a vertex on the left side of the reflection axis. Thus, $FR \neq RF$. A similar argument shows that $FR \neq RF$ when $180° < k \cdot 360°/n < 360°$. This proves that R_0 and R_{180} are the only possible elements in the center of D_n. When n is odd, D_n has no $180°$ rotation; and we leave it to the reader to show that when n is even, R_{180} does indeed commute with every member of D_n. ☐

Although an element from a non-Abelian group need not necessarily commute with every element of the group, there are always some elements with which it will commute. This observation prompts the next definition and theorem.

DEFINITION · Centralizer of a in G
Let a be a fixed element of a group G. The *centralizer of a in G, $C(a)$*, is the set of all elements in G that commute with a. In symbols, $C(a) = \{g \in G \mid ga = ag\}$.

Example 12   In D_4, we have the following centralizers.

$$C(R_0) = D_4 = C(R_{180}),$$
$$C(R_{90}) = \{R_0, R_{90}, R_{180}, R_{270}\} = C(R_{270}),$$
$$C(H) = \{R_0, H, R_{180}, V\} = C(V),$$
$$C(D) = \{R_0, D, R_{180}, D'\} = C(D').$$ ☐

Notice that each of the centralizers in Example 12 is actually a subgroup of D_4. The next theorem shows that this was not a coincidence.

Theorem 3.6   *C(a) Is a Subgroup*
For each a in a group G, the centralizer of a is a subgroup of G.

Proof.   A proof similar to that of Theorem 3.5 is left to the reader to supply.   ∎

EXERCISES

If I were to prescribe one process in the training of men which is fundamental to success in any direction, it would be thoroughgoing training in the habit of accurate observation. It is a habit which every one of us should be seeking ever more to perfect.

Eugene G. Grace

1. For each group in the following list, find the order of the group and the order of each element in the group. In each case, how are the orders of the elements of the group related to the order of the group?

$$Z_{12}, \quad U(10), \quad U(12), \quad U(20), \quad D_4$$

2. Let Q be the group of rational numbers under addition and let Q^* be the group of nonzero rational numbers under multiplication. In Q, list the elements in $\langle \frac{1}{2} \rangle$. In Q^*, list the elements in $\langle \frac{1}{2} \rangle$.

3. Let Q and Q^* be as in exercise 2. Find the order of each element in Q and in Q^*.

4. Prove that in any group, an element and its inverse have the same order.

5. Without actually computing the orders, explain why the two elements in each of the following pairs of elements from Z_{30} must have the same order: $\{2, 28\}, \{8, 22\}$. Do the same for the following pairs of elements from $U(15)$: $\{2, 8\}, \{7, 13\}$.

6. Let x belong to a group. If $x^2 \neq e$ and $x^6 = e$, prove that $x^4 \neq e$ and $x^5 \neq e$. What can we say about the order of x?

7. Show that $U(14) = \langle 3 \rangle = \langle 5 \rangle$. [Hence, $U(14)$ is cyclic.] Is $U(14) = \langle 11 \rangle$?

8. Show that $Z_{10} = \langle 3 \rangle = \langle 7 \rangle = \langle 9 \rangle$. Is $Z_{10} = \langle 2 \rangle$?

9. Show that $U(20) \neq \langle k \rangle$ for any k in $U(20)$. [Hence, $U(20)$ is not cyclic.]

10. Prove that an Abelian group with two elements of order 2 must have a subgroup of order 4.

11. Find a group that contains elements a and b such that $|a| = |b| = 2$ and

    a. $|ab| = 3$,   **b.** $|ab| = 4$,   **c.** $|ab| = 5$.

    Can you see any relationship between $|a|$, $|b|$ and $|ab|$?

12. For each divisor k of n, let $U_k(n) = \{x \in U(n) \mid x = 1 \bmod k\}$. (For example, $U_3(21) = \{1, 4, 10, 13, 16, 19\}$ and $U_7(21) = \{1, 8\}$.) List the elements of $U_4(20)$, $U_5(20)$, $U_5(30)$, and $U_{10}(30)$. Prove that $U_k(n)$ is a subgroup of $U(n)$. (This exercise is referred to in Chapter 7.)

13. Suppose m is a divisor of k and k is a divisor of n. Show that $U_k(n)$ is a subgroup of $U_m(n)$. (This notation is explained in exercise 12.)

14. If H and K are subgroups of G, show that $H \cap K$ is a subgroup of G. (Can you see that the same proof shows that the intersection of any number of subgroups of G, finite or infinite, is again a subgroup of G?)

15. Let G be a group. Show that $Z(G) = \cap_{a \in G} C(a)$. [This means the intersection of all subgroups of the form $C(a)$.]

16. Let G be a group, and let $a \in G$. Prove that $C(a) = C(a^{-1})$.

17. Suppose G is the group defined by the following Cayley table.

	1	2	3	4	5	6	7	8
1	1	2	3	4	5	6	7	8
2	2	1	8	7	6	5	4	3
3	3	4	5	6	7	8	1	2
4	4	3	2	1	8	7	6	5
5	5	6	7	8	1	2	3	4
6	6	5	4	3	2	1	8	7
7	7	8	1	2	3	4	5	6
8	8	7	6	5	4	3	2	1

a. Find the centralizer of each member of G.

b. Find $Z(G)$.

c. Find the order of each element of G. How are these orders arithmetically related to the order of the group?

18. Let G be a group, and let $a \in Z(G)$. In a Cayley table for G, how does the row headed by a compare with the column headed by a?

19. Prove Theorem 3.6.

20. If H is a subgroup of G, then by the *centralizer $C(H)$ of H* we mean the set $\{x \in G \mid xh = hx$ for all $h \in H\}$. Prove that $C(H)$ is a subgroup of G.

21. Must the centralizer of an element of a group be Abelian?

22. Must the center of a group be Abelian?

23. Let G be an Abelian group with identity e and let n be some integer. Prove that the set of all elements of G that satisfy the equation $x^n = e$ is a subgroup of G. Give an example of a group G in which the set of all elements of G that satisfy the equation $x^2 = e$ does not form a subgroup of G. (This exercise is referred to in Chapter 11.)

24. Suppose a belongs to a group and $|a| = 5$. Prove that $C(a) = C(a^3)$. Find an element a from some group such that $|a| = 6$ and $C(a) \neq C(a^3)$.

25. Show that a group of order 6 cannot have a subgroup of order 4. (*Hint:* Suppose there is a group G of order 6 that has a subgroup H of order 4. Let x belong to G but not to H. Now show that $xH = \{xh \mid h \in H\}$ and H have no elements in common.) Is the proof you used also valid for a subgroup H of order 5? Generalize to the case where G has order n.

26. Suppose n is an even positive integer and H is a subgroup of Z_n. Prove that either every member of H is even or exactly half of the members of H are even.

27. Suppose a group contains elements a and b such that $|a| = 4$, $|b| = 2$, and $a^3 b = ba$. Find $|ab|$.

28. Consider the elements $A = \begin{bmatrix} 0 & -1 \\ 1 & 0 \end{bmatrix}$ and $B = \begin{bmatrix} 0 & 1 \\ -1 & -1 \end{bmatrix}$ from $SL(2, \mathbf{R})$. Find $|A|$, $|B|$, and $|AB|$. Does your answer surprise you?

29. Consider the element $A = \begin{bmatrix} 1 & 1 \\ 0 & 1 \end{bmatrix}$ in $SL(2, \mathbf{R})$. What is the order of A? If we view $A = \begin{bmatrix} 1 & 1 \\ 0 & 1 \end{bmatrix}$ as a member of $SL(2, Z_p)$ (p is a prime), what is the order of A?

30. For any positive integer n and any angle θ, show that in the group $SL(2, \mathbf{R})$

$$\begin{bmatrix} \cos \theta & -\sin \theta \\ \sin \theta & \cos \theta \end{bmatrix}^n = \begin{bmatrix} \cos n\theta & -\sin n\theta \\ \sin n\theta & \cos n\theta \end{bmatrix}.$$

Use this formula to find the order of

$$\begin{bmatrix} \cos 60° & -\sin 60° \\ \sin 60° & \cos 60° \end{bmatrix} \text{ and } \begin{bmatrix} \cos \sqrt{2}° & -\sin \sqrt{2}° \\ \sin \sqrt{2}° & \cos \sqrt{2}° \end{bmatrix}.$$

(Geometrically, $\begin{bmatrix} \cos\theta & -\sin\theta \\ \sin\theta & \cos\theta \end{bmatrix}$ represents a rotation of the plane θ degrees.)

31. Let G be the symmetry group of a circle. Show that G has elements of every finite order as well as elements of infinite order.

32. Let $|x| = 6$. Find $|x^2|$, $|x^3|$, $|x^4|$, $|x^5|$. Let $|y| = 9$. Find $|y^i|$ for $i = 2, 3, \ldots, 8$. Do these examples suggest any relationship between the order of the power of an element and the order of the element?

33. D_4 has seven cyclic subgroups. List them. Find a subgroup of D_4 of order 4 that is not cyclic.

34. $U(15)$ has six cyclic subgroups. List them.

35. If $|a| = n$ and k divides n, prove that $|a^{n/k}| = k$.

36. Suppose G is a group that has exactly eight elements of order 3. How many subgroups of order 3 does G have?

37. Let $H = \{x \in U(20) \mid x = 1 \bmod 3\}$. Is H a subgroup of $U(20)$?

38. Compute the orders of the following groups:
    a. $U(3)$, $U(4)$, $U(12)$;
    b. $U(5)$, $U(7)$, $U(35)$;
    c. $U(4)$, $U(5)$, $U(20)$;
    d. $U(3)$, $U(5)$, $U(15)$.

39. On the basis of your answers to exercise 38, make a conjecture about the relationship among $|U(r)|$, $|U(s)|$, and $|U(rs)|$.

40. Compute $|U(4)|$, $|U(10)|$, and $|U(40)|$. Do these groups provide a counterexample to your answer to exercise 39? If so, revise your conjecture.

41. Find a cyclic subgroup of order 4 in $U(40)$.

42. Find a noncyclic subgroup of order 4 in $U(40)$.

43. Let $G = \left\{ \begin{bmatrix} a & b \\ c & d \end{bmatrix} \mid a, b, c, d \in Z \right\}$ under addition. Let $H = \left\{ \begin{bmatrix} a & b \\ c & d \end{bmatrix} \mid a + b + c + d = 0 \right\}$. Prove that H is a subgroup of G. What if 0 is replaced by 1?

44. Let $G = GL(2, \mathbf{R})$. Let $H = \{A \in G \mid \det A$ is a power of $2\}$. Show that H is a subgroup of G.

45. Let H be a subgroup of $\mathbf{R}$ under addition. Let $K = \{2^a \mid a \in H\}$. Prove that K is a subgroup of $\mathbf{R}^{\#}$ under multiplication.

46. Let G be a group of functions from $\mathbf{R}$ to $\mathbf{R}^{\#}$ under multiplication. Let $H = \{f \in G \mid f(1) = 1\}$. Prove that H is a subgroup of G.

47. Let $G = GL(2, \mathbf{R})$ and $H = \left\{ \begin{bmatrix} a & 0 \\ 0 & b \end{bmatrix} \mid a$ and b are nonzero integers $\right\}$. Prove or disprove that H is a subgroup of G.

48. Let $H = \{a + bi \mid a, b \in \mathbf{R}, ab \geq 0\}$. Prove or disprove that H is a subgroup of $\mathbf{C}$ under addition.

49. Let $H = \{a + bi \mid a, b \in \mathbf{R}, a^2 + b^2 = 1\}$. Prove or disprove that H is a subgroup of $\mathbf{C}^*$ under multiplication. Describe the elements of H geometrically.

50. Find the smallest subgroup of Z containing
    a. 8 and 14 (the notation for this is $\langle 8, 14 \rangle$),
    b. 8 and 13,
    c. 6 and 15,
    d. m and n,
    e. 12, 18, and 45.
    In each part, find an integer k such that the subgroup is $\langle k \rangle$.

51. Let $G = GL(2, \mathbf{R})$.

    a. Find $C\left(\begin{bmatrix} 1 & 1 \\ 0 & 0 \end{bmatrix}\right)$.

    b. Find $C\left(\begin{bmatrix} 0 & 1 \\ 1 & 0 \end{bmatrix}\right)$.

    c. Find $Z(G)$.

52. Let G be a finite group with more than one element. Show that G has an element of prime order.

PROGRAMMING EXERCISES

A Programmer's Lament

I really hate this damned machine;
I wish that they would sell it
It never does quite what I want
But only what I tell it.

Dennie L. Van Tassel,
The Compleat Computer

1. Write a program to print out the cyclic subgroups of $U(n)$ generated by each k in $U(n)$. Assume $n < 100$. Run the program for $n = 12, 15$, and 30. Compare the order of the subgroups with the order of the group itself. What arithmetic relationship do these integers have?

2. Repeat exercise 1 for Z_n. Have the program list the elements of Z_n that generate all of Z_n—that is, those elements $k, 0 \le k \le n - 1$, for which $Z_n = \langle k \rangle$. How does this set compare with $U(n)$? (See programming exercise 1a in Chapter 2.) Make a conjecture.

3. Write a program that does the following. For each pair of elements a and b from $U(n)$, print out $|a|$, $|b|$, and $|ab|$ on the same line. Assume $n < 100$. Run your program for $n = 15, 30$, and 42. What is the arithmetic relationship between $|ab|$ and $|a|$ and $|b|$? Make a conjecture.

4. Repeat exercise 3 for Z_n using $a + b$ in place of ab.

Write a program that prints out the elements of $U_k(n)$. Assume $n < 100$. (See exercise 12 for notation.) Run your program for the following choices of n and k: (75, 3), (50, 5), (40, 5), (40, 8), (44, 4), (60, 4), and (40, 4).

SUGGESTED READINGS

J. Gallian and M. Reid, "Abelian Forcing Sets," *American Mathematical Monthly* 100 (1993): 580–582.

A set S is called *Abelian forcing* if the only groups that satisfy $(ab)^n = a^n b^n$ for all a and b in the group and all n in S are the Abelian ones. This paper characterizes the Abelian forcing sets.

Gina Kolata, "Perfect Shuffles and Their Relation to Math," *Science* 216 (1982): 505–506.

This is a delightful nontechnical article that discusses how group theory and computers were used to solve a difficult problem about shuffling a deck of cards. Serious work on the problem was begun by an undergraduate student as part of a programming course.

SUGGESTED SOFTWARE

L. D. Geissinger, *Exploring Small Groups in Abstract Algebra,* Chicago: Harcourt Brace Jovanovich, 1988.

This award-winning program helps students visualize and understand many of the basic properties of groups that are covered in this text. Among these are subgroups, order of an element, centralizer, conjugacy classes, cosets, and factor groups. The system requirements are IBM PC, XT, or AT with 256K RAM, one double-sided disk drive, color graphics adaptor, and color monitor. The software requires DOS version 2.0 or higher.

4

Cyclic Groups

The notion of a "group," viewed only 30 years ago as the epitome of sophistication, is today one of the mathematical concepts most widely used in physics, chemistry, biochemistry, and mathematics itself.

Alexey Sosinsky, 1991

PROPERTIES OF CYCLIC GROUPS

Recall from Chapter 3 that a group G is called *cyclic* if there is an element a in G such that $G = \{a^n \mid n \in Z\}$. Such an element a is called a *generator* of G.

In view of the notation introduced in the previous chapter, we may indicate that G is a cyclic group generated by a by writing $G = \langle a \rangle$.

In this chapter, we examine cyclic groups in detail and determine their important characteristics. We begin with a few examples.

Example 1 The set of integers Z under ordinary addition is cyclic. Both 1 and -1 are generators. (Recall that, when the operation is addition, 1^n is interpreted as

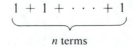

$$\underbrace{1 + 1 + \cdots + 1}_{n \text{ terms}}$$

when n is positive and as

$$(-1) + (-1) + \cdots + (-1)$$

$$|n| \text{ terms}$$

when n is negative.) ☐

Example 2   The set $Z_n = \{0, 1, \ldots, n - 1\}$ for $n \geq 1$ is a cyclic group under addition modulo n. Again, 1 and $-1 = n - 1$ are generators. ☐

Unlike Z, which has only two generators, Z_n may have many generators (depending on which n we are given).

Example 3 $Z_8 = \langle 1 \rangle = \langle 3 \rangle = \langle 5 \rangle = \langle 7 \rangle$. To verify, for instance, that $Z_8 = \langle 3 \rangle$, we note that $\langle 3 \rangle = \{3, (3 + 3) \bmod 8, (3 + 3 + 3) \bmod 8, \ldots\}$ is the set $\{3, 6, 1, 4, 7, 2, 5, 0\} = Z_8$. Thus, 3 is a generator of Z_8. On the other hand, 2 is not a generator since $\langle 2 \rangle = \{0, 2, 4, 6\} \neq Z_8$. ☐

Example 4   (See Example 11 in Chapter 2.)
$U(10) = \{1, 3, 7, 9\} = \{3^0, 3^1, 3^3, 3^2\} = \langle 3 \rangle$. Also, $\{1, 3, 7, 9\} = \{7^0, 7^3, 7^1, 7^2\} = \langle 7 \rangle$. So both 3 and 7 are generators for $U(10)$. ☐

Quite often in mathematics, a "nonexample" is as helpful in understanding a concept as an example. With regard to cyclic groups, $U(8)$ serves this purpose; that is, $U(8)$ is not a cyclic group. How can we verify this? Well, note that $U(8) = \{1, 3, 5, 7\}$. But,

$$\langle 1 \rangle = \{1\}$$
$$\langle 3 \rangle = \{3, 1\}$$
$$\langle 5 \rangle = \{5, 1\}$$
$$\langle 7 \rangle = \{7, 1\}$$

so that $U(8) \neq \langle a \rangle$ for any a in $U(8)$.

With these examples under our belts, we should now be ready to tackle cyclic groups in an abstract way and state their key properties.

Theorem 4.1   *Criterion for $a^i = a^j$*

Let G be a group, and let a belong to G. If a has infinite order, then all distinct powers of a are distinct group elements. If a has finite order, say, n, then $\langle a \rangle = \{e, a, a^2, \ldots, a^{n-1}\}$ and $a^i = a^j$ if and only if n divides $i - j$.

Proof.   If a has infinite order, there is no nonzero n such that a^n is the identity. Since $a^i = a^j$ implies $a^{i-j} = e$, we must have $i - j = 0$, and the first statement of the theorem is proved.

Now assume that $|a| = n$. We will prove that $\langle a \rangle = \{e, a, \ldots, a^{n-1}\}$. Certainly, the elements $e, a, \ldots, a^{n-1}$ are distinct. For if $a^i = a^j$

with $0 \le j < i \le n - 1$, then $a^{i-j} = e$. But this contradicts the fact that n is the least positive integer such that a^n is the identity.

Now, suppose that a^k is an arbitrary member of $\langle a \rangle$. By the division algorithm, there exist integers q and r such that

$$k = qn + r \quad \text{with} \quad 0 \le r < n.$$

Then $a^k = a^{qn+r} = a^{qn} \cdot a^r = (a^n)^q \cdot a^r = e \cdot a^r = a^r$, so that $a^k \in \{e, a, a^2, \ldots, a^{n-1}\}$. This proves that $\langle a \rangle = \{e, a, a^2, \ldots, a^{n-1}\}$.

Next, we assume that $a^i = a^j$ and prove that n divides $i - j$. We begin by observing that $a^i = a^j$ implies $a^{i-j} = e$. Again, by the division algorithm, there are integers q and r such that

$$i - j = qn + r \quad \text{with} \quad 0 \le r < n.$$

Then $a^{i-j} = a^{qn+r}$ and, therefore, $e = ea^r = a^r$. Since n is the least positive integer such that a^n is the identity, we must have $r = 0$ so that n divides $i - j$.

Conversely, if n divides $i - j$, then $a^{i-j} = a^{nq} = e^q = e$ so that $a^i = a^j$. ∎

One special case of Theorem 4.1 occurs so often it deserves singling out.

Corollary $a^k = e$ *Implies that* $|a|$ *Divides* k
Let G be a group and let a be an element of G of order n. If $a^k = e$, then n divides k.

Proof.   Since $a^k = e = a^0$, we know by Theorem 4.1, that n divides $k - 0$. ∎

Theorem 4.1 and its corollary for the case $|a| = 6$ are illustrated in Figure 4.1.

What is important about Theorem 4.1 in the finite case is that it says that multiplication in $\langle a \rangle$ is essentially done by *addition* modulo n. That is, if $(i + j) \bmod n = k$, then $a^i \cdot a^j = a^k$. Thus, no matter what

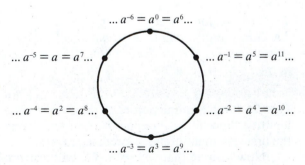

Figure 4.1

4 / Cyclic Groups

group G is, or how the element a is chosen, multiplication in $\langle a \rangle$ works the same as addition in Z_n whenever $|a| = n$. Similarly, if a has infinite order, then multiplication in $\langle a \rangle$ works the same as addition in Z, since $a^i \cdot a^j = a^{i+j}$ and no modular arithmetic is done.

For these reasons, the cyclic groups Z_n and Z serve as prototypes for all cyclic groups, and algebraists say that there is essentially only one cyclic group of each order. What is meant by this is that, although there may be many different sets of the form $\{a^n \mid n \in Z\}$, there is essentially only one way to operate on these sets, depending on the order of a. Algebraists do not really care what the elements of a set are; they care only about the algebraic properties of the set—that is, the ways the elements of a set can be combined. We will return to this theme in the chapter on isomorphisms.

In Example 3, we saw that 3 was a generator for Z_8 whereas 2 was not. Similarly, 3 and 7 are generators for $U(10)$ whereas 9 is not. It would be nice to be able to "eyeball" the generators for Z_n and for cyclic groups in general. Theorem 4.2 and its corollary give us a simple arithmetic method for identifying generators.

Theorem 4.2 *Generators of Cyclic Groups*
Let $G = \langle a \rangle$ be a cyclic group of order n. Then $G = \langle a^k \rangle$ if and only if $gcd(k, n) = 1$.

Proof. If $gcd(k, n) = 1$, we may write $1 = ku + nv$ for some integers u and v. Then $a = a^{ku+nv} = a^{ku} \cdot a^{nv} = a^{ku}$. Thus, a belongs to $\langle a^k \rangle$ and therefore all powers of a belong to $\langle a^k \rangle$. So, $G = \langle a^k \rangle$ and a^k is a generator of G.

Now suppose that $gcd(k, n) = d > 1$. Write $k = td$ and $n = sd$. Then $(a^k)^s = (a^{td})^s = (a^{sd})^t = (a^n)^t = e$, so that $|a^k| \leq s < n$. This shows that a^k is not a generator of G. ∎

Taking $G = Z_n$ and $a = 1$ in Theorem 4.2, we have the following useful result. [In particular, note that the generators of Z_n are precisely the elements of $U(n)$.]

Corollary *Generators of Z_n*
An integer k is a generator of Z_n if and only if $gcd(k, n) = 1$.

The value of Theorem 4.2 is that once one generator of a cyclic group has been found, all generators of the cyclic group can easily be determined. For example, consider the subgroup of all rotations in D_6. Clearly one generator is R_{60}. And, since $|R_{60}| = 6$, we see by Theorem 4.2 that the only other generator is $(R_{60})^5 = R_{300}$. Of course, we could have readily deduced this information without the aid of Theorem 4.2 by direct calculations. So, to illustrate the real power of Theorem 4.2, let us use it to find all generators of the cyclic group $U(50)$. First, note that direct computations show that $|U(50)| = 20$ and that 3 is one of its generators. Thus, in view of Theorem 4.2, the complete list of generators for $U(50)$ is

$$3 \bmod 50 = 3, \qquad 3^{11} \bmod 50 = 47,$$
$$3^3 \bmod 50 = 27, \qquad 3^{13} \bmod 50 = 23,$$
$$3^7 \bmod 50 = 37, \qquad 3^{17} \bmod 50 = 13,$$
$$3^9 \bmod 50 = 33, \qquad 3^{19} \bmod 50 = 17.$$

Admittedly, we had to do some arithmetic here, but it certainly entailed much less work than finding all the generators by simply determining the order of each element of $U(50)$ one by one.

Classification of Subgroups of Cyclic Groups

The next theorem tells us how many subgroups a finite cyclic group has and how to find them.

Theorem 4.3   *Fundamental Theorem of Cyclic Groups*
Every subgroup of a cyclic group is cyclic. Moreover, if $|\langle a \rangle| = n$, then the order of any subgroup of $\langle a \rangle$ is a divisor of n; and, for each divisor k of n, the group $\langle a \rangle$ has exactly one subgroup of order k—namely, $\langle a^{n/k} \rangle$.

Before we prove this theorem, let's see what it means. Understanding what a theorem means is a prerequisite to understanding its proof. Suppose $G = \langle a \rangle$ and G has order 30. The first part of the theorem says that if H is any subgroup of G, then H has the form $\langle a^k \rangle$ for some k. The second part of the theorem says that G has one subgroup of each of the orders 1, 2, 3, 5, 6, 10, 15, and 30—and no others. The proof will also show how to find these subgroups.

Proof.   Let $G = \langle a \rangle$ and suppose that H is a subgroup of G. We must show that H is cyclic. If it consists of the identity alone, then clearly H is cyclic. So we may assume that $H \neq \{e\}$. We now claim that H contains an element of the form a^t, where t is positive. Since $G = \langle a \rangle$, every element of H has the form a^t; and when a^t belongs to H with $t < 0$, then a^{-t} belongs to H also and $-t$ is positive. Thus, our claim is verified. Now let m be the least positive integer such that $a^m \in H$. By closure, $\langle a^m \rangle \leq H$. We next claim that $H = \langle a^m \rangle$. To prove this claim, it suffices to let b be an arbitrary member of H and show that b is in $\langle a^m \rangle$. Since $b \in G = \langle a \rangle$, we have $b = a^k$ for some k. Now, apply the division algorithm to k and m to obtain integers q and r such that $k = mq + r$ where $0 \leq r < m$. Then $a^k = a^{mq+r} = a^{mq} \cdot a^r$ so that $a^r = a^{-mq}a^k$. Since $a^k = b \in H$ and $a^{-mq} = (a^m)^{-q}$ is in H also, $a^r \in H$. But, m is the *least* positive integer such that $a^m \in H$, and $0 \leq r < m$, so r must be 0. Thus, $a^{-mq}a^k = e$, and therefore $b = a^k = a^{mq} = (a^m)^q \in \langle a^m \rangle$. This proves the assertion of the theorem that every subgroup of a cyclic group is cyclic.

To prove the next portion of the theorem, suppose that $|\langle a \rangle| = n$ and H is any subgroup of $\langle a \rangle$. We have already shown that $H = \langle a^m \rangle$ for some m. And, since $(a^m)^n = (a^n)^m = e^m = e$, we know from the corollary to Theorem 4.1 that $|a^m|$ is a divisor of n. Thus, $|H| = |a^m|$ is a divisor of n.

Finally, let k be any divisor of n. Clearly, $(a^{n/k})^k = a^n = e$ and $(a^{n/k})^t \neq e$ for any positive $t < k$, so $\langle a^{n/k} \rangle$ has order k. We next show that $\langle a^{n/k} \rangle$ is the only subgroup of order k. To this end, let H be any subgroup of order k. We have previously shown that $H = \langle a^m \rangle$, where m is the least positive integer such that a^m is in H. Now, writing $n = mq + r$, where $0 \leq r < m$, we have $e = a^n = a^{mq+r} = a^{mq} \cdot a^r$ so that $a^r = a^{-mq} = (a^m)^{-q} \in H$. Thus, $r = 0$ and $n = mq$. So, $k = |H| = |\langle a^m \rangle| = n/m$. It follows that $m = n/k$ and $H = \langle a^m \rangle = \langle a^{n/k} \rangle$. ∎

Returning for a moment to our discussion of the cyclic group $\langle a \rangle$ where a has order 30, we may conclude from Theorem 4.3 that the subgroups of $\langle a \rangle$ are precisely those of the form $\langle a^m \rangle$ where m is a divisor of 30. Moreover, if k is a divisor of 30, the subgroup of order k is $\langle a^{30/k} \rangle$. So the list of subgroups of $\langle a \rangle$ is:

$$\langle a \rangle = \{e, a, a^2, \ldots, a^{29}\} \qquad \text{order } 30,$$
$$\langle a^2 \rangle = \{e, a^2, a^4, \ldots, a^{28}\} \qquad \text{order } 15,$$
$$\langle a^3 \rangle = \{e, a^3, a^6, \ldots, a^{27}\} \qquad \text{order } 10,$$
$$\langle a^5 \rangle = \{e, a^5, a^{10}, a^{15}, a^{20}, a^{25}\} \qquad \text{order } 6,$$
$$\langle a^6 \rangle = \{e, a^6, a^{12}, a^{18}, a^{24}\} \qquad \text{order } 5,$$
$$\langle a^{10} \rangle = \{e, a^{10}, a^{20}\} \qquad \text{order } 3,$$
$$\langle a^{15} \rangle = \{e, a^{15}\} \qquad \text{order } 2,$$
$$\langle a^{30} \rangle = \{e\} \qquad \text{order } 1.$$

In general, if $\langle a \rangle$ has order n and k divides n, then $\langle a^{n/k} \rangle$ is the unique subgroup of order k.

Taking the group in Theorem 4.3 to be Z_n and a to be 1, we obtain the following important special case.

Corollary *Subgroups of Z_n*

For each divisor k of n, the set $\langle n/k \rangle$ is the unique subgroup of Z_n of order k; moreover, these are the only subgroups of Z_n.

Example 5 The list of subgroups of Z_{30} is

$$\langle 1 \rangle = \{0, 1, 2, \ldots, 29\} \qquad \text{order } 30,$$
$$\langle 2 \rangle = \{0, 2, 4, \ldots, 28\} \qquad \text{order } 15,$$
$$\langle 3 \rangle = \{0, 3, 6, \ldots, 27\} \qquad \text{order } 10,$$
$$\langle 5 \rangle = \{0, 5, 10, 15, 20, 25\} \qquad \text{order } 6,$$
$$\langle 6 \rangle = \{0, 6, 12, 18, 24\} \qquad \text{order } 5,$$
$$\langle 10 \rangle = \{0, 10, 20\} \qquad \text{order } 3,$$
$$\langle 15 \rangle = \{0, 15\} \qquad \text{order } 2,$$
$$\langle 30 \rangle = \{0\} \qquad \text{order } 1.$$ ❑

By combining Theorems 4.2 and 4.3, we can easily count the number of elements of each order in a finite cyclic group. For convenience, we introduce an important number-theoretic function called the *Euler phi function*. Let $\phi(1) = 1$, and for any integer $n > 1$, let $\phi(n)$ denote the number of positive integers less than n and relatively prime to n. Notice that $|U(n)| = \phi(n)$.

Theorem 4.4   *Number of Elements of Each Order in a Cyclic Group*
If d is a divisor of n, the number of elements of order d in a cyclic group of order n is $\phi(d)$.

Proof.   By Theorem 4.3, there is exactly one subgroup of order d—call it $\langle a \rangle$. Then every element of order d also generates the subgroup $\langle a \rangle$ and, by Theorem 4.2, an element a^k generates $\langle a \rangle$ if and only if $\gcd(k,d) = 1$. The number of such elements is precisely $\phi(d)$. ∎

The relationship between the various subgroups of a group can be illustrated with a *subgroup lattice* of the group. This is a diagram that includes all the subgroups of the group and connects a subgroup H at one level to a subgroup K at a higher level with a sequence of line segments if and only if H is a proper subgroup of K. Although there are many ways to draw such a diagram, the connections between the subgroups must be the same. Typically one attempts to present the diagram in an eye-pleasing fashion. The lattice diagram for Z_{30} is shown in Figure 4.2.

The precision of Theorem 4.3 can be appreciated by comparing the ease with which we are able to identify the subgroups of Z_{30} with that of, say, doing the same for $U(30)$ or D_{30}. And these groups have relatively simple structures among noncyclic groups.

We will prove in Chapter 8 that a certain portion of Theorem 4.3

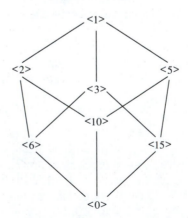

Figure 4.2

extends to arbitrary finite groups; namely, the order of a subgroup divides the order of the group itself. We will also see, however, that a finite group need not have exactly one subgroup corresponding to each divisor of the order of the group. For some divisors, there may be none at all, whereas for other divisors, there may be many.

One final remark about the importance of cyclic groups is appropriate. Although cyclic groups constitute a very narrow class of finite groups, we will see in Chapter 11 that they play the role of building blocks for all finite Abelian groups in much the same way that primes are the building blocks for the integers and that chemical elements are the building blocks for the chemical compounds.

EXERCISES

A mathematician, like a painter or a poet, is a maker of patterns. If his patterns are more permanent than theirs, it is because they are made with ideas. . . . The mathematician's patterns, like the painter's or poet's, must be beautiful; the ideas, like the colors or the words, must fit together in a harmonious way. Beauty is the first test; there is no permanent place in the world for ugly mathematics.

G. H. Hardy

1. Find all generators of Z_6, Z_8, and Z_{20}.
2. Suppose that $\langle a \rangle$, $\langle b \rangle$, and $\langle c \rangle$ are cyclic groups of orders 6, 8, and 20, respectively. Find all generators of $\langle a \rangle$, $\langle b \rangle$, and $\langle c \rangle$.
3. List the elements of the subgroups $\langle 20 \rangle$ and $\langle 10 \rangle$ in Z_{30}.
4. List the elements of the subgroups $\langle 3 \rangle$ and $\langle 15 \rangle$ in Z_{18}.
5. List the elements of the subgroups $\langle 3 \rangle$ and $\langle 7 \rangle$ in $U(20)$.
6. What do exercises 3, 4, and 5 have in common? Try to make a generalization that includes these three cases.
7. Find an example of a noncyclic group, all of whose proper subgroups are cyclic.
8. Let a be an element of a group and let $|a| = 15$. Compute the orders of the following elements of G.
   a. a^3, a^6, a^9, a^{12};
   b. a^5, a^{10};
   c. a^2, a^4, a^8, a^{14}.
9. How many subgroups does Z_{20} have? List a generator for each of these subgroups. Suppose that $G = \langle a \rangle$ and $|a| = 20$. How many subgroups does G have? List a generator for each of these subgroups.
10. Let $G = \langle a \rangle$ and let $|a| = 24$. List all generators for the subgroup of order 8.

11. Let G be a group and let $a \in G$. Prove that $\langle a^{-1} \rangle = \langle a \rangle$.

12. Suppose that a has infinite order. Find all generators of the subgroup $\langle a^3 \rangle$.

13. Suppose that $|a| = 24$. Find a generator for $\langle a^{21} \rangle \cap \langle a^{10} \rangle$. In general, what is a generator for the subgroup $\langle a^m \rangle \cap \langle a^n \rangle$?

14. Suppose that a cyclic group G has exactly three subgroups: G itself, $\{e\}$, and a subgroup of order 7. What is $|G|$?

15. Let G be an Abelian group and let $H = \{g \in G \mid |g| \text{ divides } 12\}$. Prove that H is a subgroup of G. Is there anything special about 12 here? Would your proof be valid if 12 were replaced by some other positive integer? State the general result.

16. If a cyclic group has an element of infinite order, how many elements of finite order does it have?

17. List the cyclic subgroups of $U(30)$.

18. Let G be a group and let a be an element of G of order n. For each integer k between 1 and n, show that $|a^k| = |a^{n-k}|$.

19. Let G be a group and let a be an element of G.
    a. If $a^{12} = e$, what can we say about the order of a?
    b. If $a^m = e$, what can we say about the order of a?
    c. Suppose that $|G| = 24$ and that G is cyclic. If $a^8 \neq e$ and $a^{12} \neq e$, show that $\langle a \rangle = G$.

20. Prove that a group of order 3 must be cyclic.

21. Let Z denote the group of integers under addition. Is every subgroup of Z cyclic? Why? Describe all the subgroups of Z.

22. For any element a in any group G, prove that $\langle a \rangle$ is a subgroup of $C(a)$ (the centralizer of a).

23. If $|a| = n$, show that $\langle a^k \rangle = \langle a^{\gcd(n,k)} \rangle$ and that $|a^k| = n/\gcd(n, k)$.

24. Find all generators of Z.

25. Let a be an element of a group and suppose that a has infinite order. How many generators does $\langle a \rangle$ have?

26. For each value of n listed below, determine whether or not $U(n)$ is a cyclic group. When it is cyclic, list all of the generators of $U(n)$, $n = 5, 9, 10, 14, 15, 18, 20, 22, 25$. Make a conjecture about the prime power decomposition of integers n for which $U(n)$ is cyclic. Are $n = 8$ and $n = 16$ counterexamples of your conjecture? (Try them.) If so, modify your conjecture.

27. List all the elements of order 8 in $Z_{8000000}$. How do you know your list is complete?

28. Let i denote the complex number $\sqrt{-1}$. Why might it make sense in certain physical situations to interpret i geometrically as a $90°$ rotation?

29. Let G be a finite group. Show that there exists a fixed positive integer n such that $a^n = e$ for all a in G. (Note that n is independent of a.)

30. Determine the subgroup lattice for Z_{12}.
31. Determine the subgroup lattice for Z_{p^2q}, where p and q are distinct primes.
32. Determine the subgroup lattice for Z_8.
33. Determine the subgroup lattice for Z_{p^n}, where p is a prime and n is some positive integer.
34. Determine the subgroup lattice for $U(12)$.
35. Show that the group of positive rational numbers under multiplication is not cyclic.
36. Consider the set {4, 8, 12, 16}. Show that this set is a group under multiplication modulo 20 by constructing its Cayley table. What is the identity element? Is the group cyclic? If so, find all of its generators.
37. Consider the set {7, 35, 49, 77}. Show that this set is a group under multiplication modulo 84 by constructing its Cayley table. What is the identity element? Is the group cyclic?
38. Let m and n be elements of the group Z. Find a generator for the group $\langle m \rangle \cap \langle n \rangle$.
39. Suppose that a and b are group elements that commute and have orders m and n. If $\langle a \rangle \cap \langle b \rangle = \{e\}$, prove that the group contains an element whose order is the least common multiple of m and n. Show that this need not be true if a and b do not commute.
40. Prove that an infinite group must have an infinite number of subgroups.
41. Let p be a prime. If a group has more than $p - 1$ elements of order p, why can't the group be cyclic?
42. Suppose that G is a cyclic group and that 6 divides $|G|$. How many elements of order 6 does G have? If 8 divides $|G|$, how many elements of order 8 does G have? If a is one element of order 8, list the other elements of order 8.
43. List all the elements of Z_{40} that have order 10.
44. Let $|x| = 40$. List all the elements of $\langle x \rangle$ that have order 10.
45. Determine the orders of the elements of D_{33} and how many there are of each.
46. How many elements of order 4 does D_{12} have? How many elements of order 4 does D_{4n} have?
47. If G is an Abelian group and contains cyclic subgroups of orders 4 and 5, what other sizes of cyclic subgroups must G contain?
48. If G is an Abelian group and contains cyclic subgroups of orders 4 and 6, what other sizes of cyclic subgroups must G contain?
49. If G is an Abelian group and contains a pair of subgroups of order 2, show that G must contain a subgroup of order 4. Must this subgroup be cyclic?
50. Given the fact that $U(49)$ is cyclic and has 42 elements, deduce the number of generators that $U(49)$ has without actually finding any of the generators.

51. Let a and b be elements of a group. If $|a| = 10$ and $|b| = 21$, show $\langle a \rangle \cap \langle b \rangle = \{e\}$.

52. Let a and b be elements of a group. If $|a| = m$, $|b| = n$, and m and n are relatively prime, show that $\langle a \rangle \cap \langle b \rangle = \{e\}$.

53. Let a and b belong to a group. If $|a| = 24$ and $|b| = 10$, what are the possibilities for $\langle a \rangle \cap \langle b \rangle$?

54. Prove that $U(2^n)$ $(n \geq 3)$ is not cyclic.

55. Suppose that G is a group of order 16 and that, by direct computation, you know that G has at least nine elements x such that $x^8 = e$. Can you conclude that G is not cyclic? What if G has at least five elements x such that $x^4 = e$? Generalize.

56. Prove that Z_n has an even number of generators if $n > 2$.

57. Bertrand's Postulate from number theory says that for any integer $N > 1$, there is always a prime between N and $2N$. Use this fact to prove that Z_n has more than two generators whenever $n > 6$.

58. Suppose that $|x| = n$. Find a necessary and sufficient condition on r and s such that $\langle x^r \rangle \subseteq \langle x^s \rangle$.

59. Let $|x| = n$. Show that $\langle x^r \rangle = \langle x^s \rangle$ if and only if $\gcd(n, r) = \gcd(n, s)$.

60. Prove that $H = \left\{ \begin{bmatrix} 1 & n \\ 0 & 1 \end{bmatrix} \mid n \in Z \right\}$ is a cyclic subgroup of $GL(2, \mathbf{R})$.

61. Let G be a group, let H be a subgroup of G, and let a be an element of G. Suppose that $|a| = n$ and that, for each proper divisor k of n, $a^k \notin H$. Show that $\langle a \rangle \cap H = \{e\}$.

62. Suppose that G is a finite group with the property that every nonidentity element has prime order (for example, D_3 and D_5). If $Z(G)$ is not trivial, prove that every nonidentity element of G has the same order.

PROGRAMMING EXERCISES

It seems to me the impact and role of the electronic computer will significantly affect pure mathematics also, just as it has already done so in the mathematical sciences, principally physics, astronomy and chemistry.

S. M. Ulam, *Adventures of a Mathematician*

1. For all $1 < n < 100$, have the computer determine whether $U(n)$ is cyclic. When $U(n)$ is cyclic, list all of the generators and all of the subgroups. Run your program for $n = 8, 9, 18, 20, 25, 30, 49, 50,$ and 60.

2. For any pair of positive integers m and n, let $Z_m \oplus Z_n = \{(a, b) \mid a \in Z_m, b \in Z_n\}$. For any pair of elements (a, b) and (c, d) in $Z_m \oplus Z_n$, define $(a, b) + (c, d) = ((a + c) \bmod m, (b + d) \bmod n)$. [For example, in $Z_3 \oplus Z_4$, we have $(1, 2) + (2, 3) = (0, 1)$.] Write a program to check

whether or not $Z_m \oplus Z_n$ is cyclic. Run your program for the following choices for m and n: (2, 2), (2, 3), (2, 4), (2, 5), (3, 4), (3, 5), (3, 6), (3, 7), (3, 8), (3, 9), and (4, 6). On the basis of this output, guess how m and n must be related for $Z_m \oplus Z_n$ to be cyclic.

3. In this exercise we assume $a, b \in U(n)$. Define $\langle a, b \rangle = \{a^i b^j \mid 0 \le i < |a|, 0 \le j < |b|\}$. Write a program to compute the orders of $\langle a, b \rangle$, $\langle a \rangle$, $\langle b \rangle$, and $\langle a \rangle \cap \langle b \rangle$. Run your program for the following choices for a, b, and n: (21, 101, 550), (21, 49, 550), (7, 11, 100), (21, 31, 100), and (63, 77, 100). On the basis of your output, make a conjecture about the arithmetic relationship among $|\langle a, b \rangle|$, $|\langle a \rangle|$, $|\langle b \rangle|$, and $|\langle a \rangle \cap \langle b \rangle|$.

SUGGESTED READINGS

S. R. Cavior, "The Subgroups of the Dihedral Group," *Mathematics Magazine* 48 (1975): 107.

> For each positive integer n, let $d(n)$ denote the number of positive divisors of n and let $\sigma(n)$ denote the sum of the positive divisors of n. This paper gives a short proof that the number of subgroups of the dihedral group D_n ($n \ge 3$) is $d(n) + \sigma(n)$.

Deborah L. Massari, "The Probability of Generating a Cyclic Group," *Pi Mu Epsilon Journal* 7 (1979): 3–6.

> In this easy-to-read paper, it is shown that the probability of a randomly chosen element from a cyclic group being a generator of the group depends only on the set of prime divisors of the order of the group, and not on the order itself. The article, written by an undergraduate student, received first prize in a Pi Mu Epsilon Paper Contest.

J. J. Sylvester

I really love my subject.

J. J. Sylvester

JAMES JOSEPH SYLVESTER was the most colorful and influential mathematician in America in the 19th century. Sylvester was born on September 3, 1814, in London and showed his mathematical genius early. At the age of 14, he studied under De Morgan and won several prizes for his mathematics, and at the unusually young age of 25, he was elected a Fellow of the Royal Society.

After receiving B.A. and M.A. degrees from Trinity College in Dublin in 1841, Sylvester began a professional life that would include academics, law, and actuarial careers. In 1876, at the age of 62, he was appointed to a prestigious position at the newly founded Johns Hopkins University. During his seven years at Johns Hopkins, Sylvester pursued research in pure mathematics, the first ever done in America, with tremendous vigor and enthusiasm. He also founded the *American Journal of Mathematics,* the first journal in America devoted to mathematical research. Sylvester returned to England in 1884 to a professorship at Oxford, a position he held until his death on March 15, 1897.

Sylvester's major contributions to mathematics were in the theory of equations, matrix theory, determinant theory, and invariant theory (which he founded with Cayley). His writings and lectures—flowery and eloquent, pervaded with poetic flights, emotional expressions, bizarre utterances, and paradoxes—reflected the personality of this sensitive, excitable, and enthusiastic man. We quote three of his students.* E. W. Davis commented on Sylvester's teaching methods.

> Sylvester's methods! He had none. "Three lectures will be delivered on a New Universal Algebra," he would say; then, "The course must be extended to twelve." It did last all the rest of that year. The following year the course was to be *Substitutions-Theorie,* by Netto. We all got the text. He lectured about three times, following the text closely and stopping sharp at the end of the hour. Then he began to think about matrices again. "I must give one lecture a week on those," he said. He could not confine himself to the hour, nor to the one lecture a week. Two weeks were passed, and Netto was forgotten entirely and never mentioned again. Statements like the following were not infrequent in his lectures: "I haven't proved this, but I am as sure as I can be of anything that it must be so. From this it will follow, etc." At the next lecture it turned out that what he was so sure of was false. Never mind, he kept on forever guessing and trying, and presently a wonderful discovery followed, then another and another. Afterward he would go back and work it all over again, and surprise us with all sorts of side lights. He then made another leap in the dark, more treasures were discovered, and so on forever.

Sylvester's enthusiasm for teaching and his influence on his students are captured in the following passage written by Sylvester's first student at Johns Hopkins, G. B. Halsted.

> A short, broad man of tremendous vitality, . . . Sylvester's capacious head was ever lost in the highest cloud-lands of pure mathematics. Often in the dead of night he would get his favorite pupil, that he might communicate the very last product of his creative thought. Everything he saw suggested to him something new in the higher algebra. This transmutation of everything into new mathematics was a revelation to those who knew him intimately. They began to do it themselves.

Another characteristic of Sylvester, which is very unusual among mathematicians, was his apparent inability to remember mathematics! W. P. Durfee had the following to say:

> Sylvester had one remarkable peculiarity. He seldom remembered theorems, propositions, etc., but had always to deduce them when he wished to use them. In this he was the very antithesis of Cayley, who was thoroughly conversant with everything that had been done in every branch of mathematics.
>
> I remember once submitting to Sylvester some investigations that I had been engaged on, and he immediately denied my first statement, saying that such a proposition had never been heard of, let alone proved. To his astonishment, I showed him a paper of his own in which he had proved the proposition; in fact, I believe the object of his paper had been the very proof which was so strange to him.

---

*F. Cajori, *Teaching and History of Mathematics in the U.S.,* Washington, 1890, 265–266.

SUPPLEMENTARY EXERCISES FOR CHAPTERS 1–4

It is better to wear out than to rust out.

Bishop Richard Cumberland

1. Let G be a group and let H be a subgroup of G. For any x in G, define $xHx^{-1} = \{xhx^{-1} \mid h \in H\}$. Prove that:
   a. xHx^{-1} is a subgroup of G.
   b. If H is cyclic, then xHx^{-1} is cyclic.
   c. If H is Abelian, then xHx^{-1} is Abelian.

   The group xHx^{-1} is called a *conjugate* of H. (Note that conjugation preserves structure.)

2. Let G be a group and let H be a subgroup of G. Define $N(H) = \{x \in G \mid xHx^{-1} = H\}$. Prove that $N(H)$ (called the *normalizer* of H) is a subgroup of G.*

3. Let G be a group. For each $a \in G$, define $\text{cl}(a) = \{xax^{-1} \mid x \in G\}$. Prove that these subsets of G partition G. [$\text{cl}(a)$ is called the *conjugacy class* of a.]

4. The group defined by the following table is called the *group of quaternions*. Use the table to determine each of the following:
   a. the center
   b. $\text{cl}(a)$
   c. $\text{cl}(b)$
   d. all cyclic subgroups

	e	a	a^2	a^3	b	ba	ba^2	ba^3
e	e	a	a^2	a^3	b	ba	ba^2	ba^3
a	a	a^2	a^3	e	ba^3	b	ba	ba^2
a^2	a^2	a^3	e	a	ba^2	ba^3	b	ba
a^3	a^3	e	a	a^2	ba	ba^2	ba^3	b
b	b	ba	ba^2	ba^3	a^2	a^3	e	a
ba	ba	ba^2	ba^3	b	a	a^2	a^3	e
ba^2	ba^2	ba^3	b	ba	e	a	a^2	a^3
ba^3	ba^3	b	ba	ba^2	a^3	e	a	a^2

5. Prove that, in any group, $|ab| = |ba|$.

6. (Conjugation preserves order.) Prove that, in any group, $|xax^{-1}| = |a|$. (This exercise is referred to in Chapter 24.)

7. Prove that if a is the only element of order 2 in a group, then a lies in the center of the group.

---

*This very important subgroup was first used by L. Sylow in 1872 to prove the existence of certain kinds of subgroups in a group. His work is discussed in Chapter 24.

8. Let G be the plane symmetry group of the infinite strip of equally spaced H's shown below.

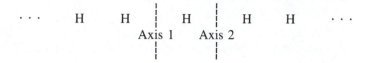

Let x be the reflection about Axis 1 and let y be the reflection about Axis 2. Calculate $|x|$, $|y|$, and $|xy|$. Must the product of elements of finite order have finite order?

9. What are the orders of the elements of D_{15}? How many elements have each of these orders?

10. Prove that a group of order 4 is Abelian.

11. Prove that a group of order 5 must be cyclic.

12. Prove that an Abelian group of order 6 must be cyclic.

13. Let G be an Abelian group and let n be a fixed positive integer. Let $G^n = \{g^n \mid g \in G\}$. Prove that G^n is a subgroup of G. Give an example showing that G^n need not be a subgroup of G when G is non-Abelian. (This exercise is referred to in Chapter 11.)

14. Let $G = \{a + b\sqrt{2}\}$, where a and b are rational numbers not both 0. Prove that G is a group under ordinary multiplication.

15. (1969 Putnam Competition) Prove that no group is the union of two proper subgroups. Does the statement remain true if "two" is replaced by "three"?

16. Prove that the subset of elements of finite order in an Abelian group forms a subgroup. (This subgroup is called the *torsion subgroup*.) Is the same thing true for non-Abelian groups?

17. Let p be a prime and let G be an Abelian group. Show that the set of all elements whose orders are powers of p is a subgroup of G.

18. Let x and y belong to a group G. Assume $x \neq e$, $|y| = 2$, and $yxy^{-1} = x^2$. Find $|x|$.

19. Suppose that a group is generated by two elements a and b (that is, every element of the group can be expressed as some product of a's and b's). Given that $a^3 = b^2 = e$ and $ba^2 = ab$, construct the Cayley table for the group. We have already seen an example of a group that satisfies these conditions. Name it.

20. Suppose that a group is generated by two elements a and b. Given that $a^4 = b^2 = e$ and $ba = a^3b$, construct the Cayley table for the group. Give an example of a group that satisfies these conditions.

21. Let x, y belong to a group G. If $xy \in Z(G)$, prove that $xy = yx$.

22. Suppose that H and K are nontrivial subgroups of Q under addition. Show that $H \cap K$ is a nontrivial subgroup of Q. Is this true if Q is replaced by $\mathbf{R}$?

23. Let H be a subgroup of G and let g be an element of G. Prove that $N(gHg^{-1}) = gN(H)g^{-1}$.

24. Let H be a subgroup of a group G and let $|g| = n$. If g^m belongs to H and m and n are relatively prime, prove that g belongs to H.

25. Find a group that contains elements a and b such that $|a| = 2$, $|b| = 11$, and $|ab| = 2$.

26. Suppose that G is a group with exactly eight elements of order 10. How many cyclic subgroups of order 10 does G have?

27. (From the 1989 Putnam Competition) Let S be a nonempty set with an associative operation that is left and right cancellative ($xy = xz$ implies $y = z$, and $yx = zx$ implies $y = z$). Assume that for every a in S the set $\{a^n \mid n = 1, 2, 3, \ldots\}$ is finite. Must S be a group?

28. Let $H_1, H_2, H_3, \ldots$ be a sequence of subgroups of a group with the property that $H_1 \subseteq H_2 \subseteq H_3 \ldots$. Prove that the union of the sequence is a subgroup.

29. Let $\mathbf{R}^*$ be the group of nonzero real numbers under multiplication and let $H = \{x \in \mathbf{R}^* \mid x^2 \text{ is rational}\}$. Prove that H is a subgroup of $\mathbf{R}^*$.

30. Let G be a group and let n be a fixed positive integer. Show that $H = \{a \in G \mid (ag)^n = g^n \text{ for all } g \text{ in } G\}$ is a subgroup of G.

31. Let $G = GL(2, \mathbf{R})$ and let $H = \{A \in G \mid \det A \text{ is rational}\}$. Prove or disprove that H is a subgroup of G. What if we replace "rational" by "integer"?

32. Suppose that G is a group that has exactly one nontrivial proper subgroup. Prove that G is cyclic and $|G| = p^2$ where p is prime.

33. Suppose that G is a group and G has exactly two nontrivial proper subgroups. Prove that G is cyclic and $|G| = pq$ where p and q are distinct primes or that G is cyclic and $|G| = p^3$ where p is prime.

34. If $|a^2| = |b^2|$, prove or disprove that $|a| = |b|$.

35. Let x and y be integers such that $9x + 5y$ is divisible by 11. For which of the following values of k must $10x + ky$ be divisible by 11?
    a. 0
    b. 1
    c. 3
    d. 7
    e. 8

36. Suppose that a and b belong to a group, a and b commute, and $|a|$ and $|b|$ are relatively prime. Prove that $|ab| = |a| \, |b|$. Give an example showing that $|ab|$ need not be $|a| \, |b|$ when a and b commute but $|a|$ and $|b|$ are not relatively prime. (Don't use $b = a^{-1}$.)

5

Permutation Groups

Wigner's discovery about the electron permutation group was just the beginning. He and others found many similar applications and nowadays group theoretical methods—especially those involving characters and representations—pervade all branches of quantum mechanics.

George Mackey, Proceedings of the American Philosophical Society

DEFINITION AND NOTATION

In this chapter, we study certain groups of functions, called permutation groups, from a set A to itself. In the early and mid-19th century, groups of permutations were the only groups investigated by mathematicians. It was not until around 1850 that the notion of an abstract group was introduced by Cayley, and it took another quarter century before the idea firmly took hold.

DEFINITIONS Permutation of A, Permutation Group of A
A *permutation* of a set A is a function from A to A that is both one-to-one and onto. A *permutation group of a set A* is a set of permutations of A that forms a group under function composition.

Although groups of permutations of any nonempty set A of objects exist, we will focus on the case where A is finite. Furthermore, it is customary, as well as convenient, to take A to be a set of the form

$\{1, 2, 3, \ldots, n\}$ for some positive integer n. Unlike in calculus, where most functions are defined on infinite sets and are given by formulas, in algebra, permutations of finite sets are usually given by an explicit listing of each element of the domain and its corresponding functional value. For example, we define a permutation α of the set $\{1, 2, 3, 4\}$ by specifying

$$\alpha(1) = 2, \qquad \alpha(2) = 3, \qquad \alpha(3) = 1, \qquad \alpha(4) = 4.$$

A more convenient way to express this correspondence is to write α in array form as

$$\alpha = \begin{bmatrix} 1 & 2 & 3 & 4 \\ 2 & 3 & 1 & 4 \end{bmatrix}.$$

Here $\alpha(j)$ is placed directly below j for each j. Similarly, the permutation β of the set $\{1, 2, 3, 4, 5, 6\}$ given by

$$\beta(1) = 5, \quad \beta(2) = 3, \quad \beta(3) = 1, \quad \beta(4) = 6, \quad \beta(5) = 2, \quad \beta(6) = 4$$

is expressed in array form as

$$\beta = \begin{bmatrix} 1 & 2 & 3 & 4 & 5 & 6 \\ 5 & 3 & 1 & 6 & 2 & 4 \end{bmatrix}.$$

Composition of permutations expressed in array notation is carried out from right to left by going from top to bottom, then again from top to bottom. For example, let

$$\sigma = \begin{bmatrix} 1 & 2 & 3 & 4 & 5 \\ 2 & 4 & 3 & 5 & 1 \end{bmatrix}$$

and

$$\gamma = \begin{bmatrix} 1 & 2 & 3 & 4 & 5 \\ 5 & 4 & 1 & 2 & 3 \end{bmatrix};$$

then

$$\gamma\sigma = \begin{bmatrix} 1 & 2 & 3 & 4 & 5 \\ 5 & 4 & 1 & 2 & 3 \end{bmatrix}\begin{bmatrix} 1 & 2 & 3 & 4 & 5 \\ 2 & 4 & 3 & 5 & 1 \end{bmatrix} = \begin{bmatrix} 1 & 2 & 3 & 4 & 5 \\ 4 & 2 & 1 & 3 & 5 \end{bmatrix}.$$

On the right we have 4 under 1, since $(\gamma\sigma)(1) = \gamma(\sigma(1)) = \gamma(2) = 4$, so $\gamma\sigma$ sends 1 to 4. The remainder of the bottom row $\gamma\sigma$ is obtained in a similar fashion.

We are now ready to give some examples of permutation groups.

Example 1   Symmetric Group S_3
    Let S_3 denote the set of all one-to-one functions from $\{1, 2, 3\}$ to itself. Then S_3, under function composition, is a group with six ele-

ments. The six elements are

$$\varepsilon = \begin{bmatrix} 1 & 2 & 3 \\ 1 & 2 & 3 \end{bmatrix}, \qquad \alpha = \begin{bmatrix} 1 & 2 & 3 \\ 2 & 3 & 1 \end{bmatrix}, \qquad \alpha^2 = \begin{bmatrix} 1 & 2 & 3 \\ 3 & 1 & 2 \end{bmatrix},$$

$$\beta = \begin{bmatrix} 1 & 2 & 3 \\ 1 & 3 & 2 \end{bmatrix}, \qquad \alpha\beta = \begin{bmatrix} 1 & 2 & 3 \\ 2 & 1 & 3 \end{bmatrix}, \qquad \alpha^2\beta = \begin{bmatrix} 1 & 2 & 3 \\ 3 & 2 & 1 \end{bmatrix}.$$

Note that $\beta\alpha = \begin{bmatrix} 1 & 2 & 3 \\ 3 & 2 & 1 \end{bmatrix} \neq \alpha\beta$, so that S_3 is non-Abelian.   ❏

Example 1 can be generalized as follows.

Example 2 Symmetric Group S_n
   Let $A = \{1, 2, \ldots, n\}$. The set of all permutations of A is called the *symmetric group of degree n* and is denoted by S_n. Elements of S_n have the form

$$\alpha = \begin{bmatrix} 1 & 2 & \cdots & n \\ \alpha(1) & \alpha(2) & \cdots & \alpha(n) \end{bmatrix}.$$

It is easy to compute the order of S_n. There are n choices of $\alpha(1)$. Once $\alpha(1)$ has been determined, there are $n - 1$ possibilities for $\alpha(2)$ [since α is one-to-one, we must have $\alpha(1) \neq \alpha(2)$]. After choosing $\alpha(2)$, there are exactly $n - 2$ possibilities for $\alpha(3)$. Continuing along in this fashion, we see that S_n must have $n(n - 1) \cdots 3 \cdot 2 \cdot 1 = n!$ elements. We leave it to the reader to prove that S_n is non-Abelian when $n \geq 3$.   ❏

Example 3 Symmetries of a Square
   As a third example, we associate each motion in D_4 with the permutation of the locations of each of the four corners of a square. For example, if we label the four corner positions as in the figure below and keep these labels fixed for reference, we may describe a 90° rotation by the permutation

$$\rho = \begin{bmatrix} 1 & 2 & 3 & 4 \\ 2 & 3 & 4 & 1 \end{bmatrix},$$

whereas a reflection across a horizontal axis yields

$$\phi = \begin{bmatrix} 1 & 2 & 3 & 4 \\ 2 & 1 & 4 & 3 \end{bmatrix}.$$

These two elements generate the entire group (that is, every element is some combination of the ρ's and ϕ's.)   ❏

CYCLE NOTATION

There is another notation commonly used to specify permutations. It is called *cycle notation* and was first introduced by the great French mathematician Cauchy in 1815. Cycle notation has theoretical advantages in that certain important properties of the permutation can be readily determined when cyclic notation is used.

As an illustration of cycle notation, let us consider the permutation

$$\alpha = \begin{bmatrix} 1 & 2 & 3 & 4 & 5 & 6 \\ 2 & 1 & 4 & 6 & 5 & 3 \end{bmatrix}.$$

This assignment of values could be presented schematically as follows:

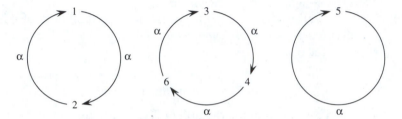

Although mathematically satisfactory, such diagrams are cumbersome. Instead, we leave out the arrows and simply write $\alpha = (1, 2)(3, 4, 6)(5)$. As a second example, consider

$$\beta = \begin{bmatrix} 1 & 2 & 3 & 4 & 5 & 6 \\ 5 & 3 & 1 & 6 & 2 & 4 \end{bmatrix}.$$

In cycle notation, β can be written $(2, 3, 1, 5)(6, 4)$ or $(4, 6)(3, 1, 5, 2)$, since both of these unambiguously specify the function β. An expression of the form $(a_1, a_2, \ldots, a_m)$ is called a *cycle of length m* or an *m-cycle*.

A multiplication of cycles can be introduced by thinking of a cycle as a permutation that fixes any symbol not appearing in the cycle. Thus, the cycle $(4, 6)$ can be thought of as representing the permutation $\begin{bmatrix} 1 & 2 & 3 & 4 & 5 & 6 \\ 1 & 2 & 3 & 6 & 5 & 4 \end{bmatrix}$. In this way, we can multiply cycles by thinking of them as permutations given in array form. Consider the following example from S_8. Let $\alpha = (13)(27)(456)(8)$ and $\beta = (1237)(648)(5)$. (When the domain consists of single-digit integers, it is common practice to omit the commas between the digits.) What is the cycle form of $\alpha\beta$? Of course, one could say that $\alpha\beta = (13)(27)(456)(8)(1237)(648)(5)$, but it is usually more desirable to express a permutation in a *disjoint cycle form* (that is, the various cycles have no number in common). Well, keeping in mind that function composition is done from right to left and that each cycle that does not contain a symbol fixes the symbol, we observe that: (5) fixes 1; (648) fixes 1; (1237) sends 1 to 2; (8) fixes 2; (456) fixes 2; (27) sends 2 to 7; and (13) fixes 7. So the net effect of $\alpha\beta$ is to send 1 to 7. Thus we begin $\alpha\beta = (17 \cdot \cdot \cdot) \cdot \cdot \cdot$. Now, repeating the entire process beginning with 7, we have, cycle by cycle, right to left, $7 \to 7 \to 7 \to 1 \to 1 \to 1 \to 1 \to 3$, so that $\alpha\beta = (173 \cdot \cdot \cdot) \cdot \cdot \cdot$. Ultimately, we have $\alpha\beta = (1732)(48)(56)$. The important thing to bear in mind when multiplying cycles is to "keep moving" from one cycle to the next from right to left. (*Warning:* Some authors compose cycles from left to right. When reading another text, be sure to determine which convention is being used.)

To be sure you understand how to switch from one notation to the other and how to multiply permutations, we will do one more example of each.

If array notations for α and β, respectively, are

$$\begin{bmatrix} 1 & 2 & 3 & 4 & 5 \\ 2 & 1 & 3 & 5 & 4 \end{bmatrix} \quad \text{and} \quad \begin{bmatrix} 1 & 2 & 3 & 4 & 5 \\ 5 & 4 & 1 & 2 & 3 \end{bmatrix}$$

then, in cycle notation, $\alpha = (12)(3)(45)$, $\beta = (153)(24)$, and $\alpha\beta = (14)(253)$.

One can convert $\alpha\beta$ back to array form without converting each cycle of $\alpha\beta$ into array form by simply observing that (14) means 1 goes to 4 and 4 goes to 1; (253) means $2 \to 5$, $5 \to 3$, $3 \to 2$.

One final remark about the cycle notation: most mathematicians prefer not to write cycles that have only one entry. In this case, it is understood that any missing element is mapped to itself. With this convention, the permutation α above can be written as (12)(45). Similarly,

$$\alpha = \begin{bmatrix} 1 & 2 & 3 & 4 & 5 \\ 3 & 2 & 4 & 1 & 5 \end{bmatrix}$$

can be written $\alpha = (134)$. Of course, the identity permutation consists only of cycles with one entry, so we cannot omit all of these! In this case, one usually writes just one cycle. For example,

$$\varepsilon = \begin{bmatrix} 1 & 2 & 3 & 4 & 5 \\ 1 & 2 & 3 & 4 & 5 \end{bmatrix}$$

could be written as $\varepsilon = (5)$ or $\varepsilon = (1)$. Just remember that missing elements are mapped to themselves.

PROPERTIES OF PERMUTATIONS

We are now ready to state several theorems about permutations and cycles. The proof of the first theorem is implicit in our discussion of writing permutations in cycle form.

Theorem 5.1   *Products of Disjoint Cycles*

Every permutation of a finite set can be written as a cycle or as a product of disjoint cycles.

Proof. Let α be a permutation on $A = \{1, 2, \ldots, n\}$. To write α in disjoint cycle form, we start by choosing any member of A, say a_1, and let

$$a_2 = \alpha(a_1), \qquad a_3 = \alpha(\alpha(a_1)) = \alpha^2(a_1),$$

and so on, until we arrive at $a_1 = \alpha^m(a_1)$ for some m. We know such an m exists because the sequence $a_1, \alpha(a_1), \alpha^2(a_1), \cdots$ must be finite; so there must eventually be a repetition, say $\alpha^i(a_1) = \alpha^j(a_1)$ for some i and j with $i < j$. Then $a_1 = \alpha^m(a_1)$, where $m = j - i$. We express this relationship among $a_1, a_2, \ldots, a_m$ as

$$\alpha = (a_1, a_2, \ldots, a_m) \cdots.$$

The three dots at the end indicate the possibility that we may not have exhausted the set A in this process. In such a case, we merely choose any element b_1 of A not appearing in the first cycle and proceed to create a new cycle as before. That is, we let $b_2 = \alpha(b_1), b_3 = \alpha^2(b_1)$, and so on, until we reach $b_1 = \alpha^k(b_1)$ for some k. This new cycle will have no elements in common with the previously constructed cycle. For, if so, then $\alpha^i(a_1) = \alpha^j(b_1)$ for some i and j. But then $\alpha^{i-j}(a_1) = b_1$ and therefore $b_1 = a_t$ for some t. This contradicts the way b_1 was chosen. Continuing this process until we run out of elements of A, our permutation will appear as

$$\alpha = (a_1, a_2, \ldots, a_m)(b_1, b_2, \ldots, b_k) \cdots (c_1, c_2, \ldots, c_t).$$

In this way, we see that every permutation can be written as a product of disjoint cycles. ∎

Theorem 5.2 *Disjoint Cycles Commute*

If the pair of cycles $\alpha = (a_1, a_2, \ldots, a_m)$ and $\beta = (b_1, b_2 \ldots, b_n)$ have no entries in common, then $\alpha\beta = \beta\alpha$.

Proof. For definiteness, let us say that α and β are permutations of the set

$$S = \{a_1, a_2, \ldots, a_m, b_1, b_2, \ldots, b_n, c_1, c_2, \ldots, c_k\}$$

where the c's are the members of S left fixed by both α and β. To prove that $\alpha\beta = \beta\alpha$, we must show that $(\alpha\beta)(x) = (\beta\alpha)(x)$ for all x in S. If x is one of the a elements, say a_i, then

$$(\alpha\beta)(a_i) = \alpha(\beta(a_i)) = \alpha(a_i) = a_{i+1}$$

since β fixes all a elements. (We interpret a_{i+1} as a_1 if $i = m$.) For the same reason,

$$(\beta\alpha)(a_i) = \beta(\alpha(a_i)) = \beta(a_{i+1}) = a_{i+1}.$$

Hence, the functions $\alpha\beta$ and $\beta\alpha$ agree on a elements. A similar argument shows that $\alpha\beta$ and $\beta\alpha$ agree on the b elements as well. Finally, suppose that x is a c element, say c_i. Then, since both α and β fix c elements, we have

$$(\alpha\beta)(c_i) = \alpha(\beta(c_i)) = \alpha(c_i) = c_i$$

and

$$(\beta\alpha)(c_i) = \beta(\alpha(c_i)) = \beta(c_i) = c_i.$$

This completes the proof.   ∎

In demonstrating how to multiply cycles, we showed that the product $(13)(27)(456)(8)(1237)(648)(5)$ can be written in disjoint cycle form as $(1732)(48)(56)$. Is economy in expression the only advantage to writing a permutation in disjoint cycle form? No. The next theorem shows that the disjoint cycle form has the enormous advantage of allowing us to "eyeball" the order of the permuation.

Theorem 5.3   *Order of a Permutation (Ruffini—1799)*
The order of a permutation of a finite set written in disjoint cycle form is the least common multiple of the lengths of the cycles.

Proof.   First, observe that a cycle of length n has order n. (Verify this yourself.) Next, suppose that α and β are disjoint cycles of lengths m and n, and let k be the least common multiple of m and n. It follows from Theorem 4.1 that both α^k and β^k are the identity permutation ε and, since α and β commute, $(\alpha\beta)^k = \alpha^k\beta^k$ is also the identity. Thus, we know by the corollary to Theorem 4.1 ($a^k = e$ implies that $|a|$ divides k) that the order of $\alpha\beta$—let us call it t—must divide k. But then $(\alpha\beta)^t = \alpha^t\beta^t = \varepsilon$, so that $\alpha^t = \beta^{-t}$. However, it is clear that if α and β have no common symbol, the same is true for α^t and β^{-t}, since raising a cycle to a power does not introduce new symbols. But, if α^t and β^{-t} are equal and have no common symbols, they must both be the identity, because every symbol in α^t is fixed by β^{-t} and vice versa (remember that a symbol not appearing in a permutation is fixed by the permutation). It follows, then, that both m and n must divide t. This means that k, the least common multiple of m and n, divides t also. This shows that $k = t$.

Thus far, we have proved that the corollary is true in the cases where the permutation is a single cycle or a product of two disjoint cycles. The general case involving more than two cycles can be handled in an analogous way.   ∎

As we will soon see, a particularly important kind of permutation is a cycle of length 2—that is, a permutation of the form (ab). Many authors call these permutations *transpositions,* since the effect of (ab) is to interchange or transpose a and b.

Theorem 5.4   *Product of 2-Cycles*
Every permutation in S_n, $n > 1$, is a product of 2-cycles.

Proof.   First, note that the identity can be expressed as $(12)(12)$, and so it is a product of 2-cycles. By Theorem 5.1, we know that every permutation can be written in the form

$$(a_1a_2 \cdots a_k)(b_1b_2 \cdots b_t) \cdots (c_1c_2 \cdots c_s).$$

A direct computation shows that this is the same as

$$(a_1a_k)(a_1a_{k-1}) \cdots (a_1a_2)(b_1b_t)(b_1b_{t-1}) \cdots (b_1b_2)(c_1c_s)(c_1c_{s-1}) \cdots (c_1c_2)$$

This completes the proof.   ∎

The example below demonstrates this technique. It also shows that the decomposition of a permutation into a product of 2-cycles is not unique.

Example 4

$$\begin{aligned}
(12345) &= (15)(14)(13)(12) \\
&= (45)(53)(25)(15) \\
&= (21)(25)(24)(23) \\
&= (54)(52)(21)(25)(23)(13)
\end{aligned}$$ ❑

Example 4 even shows that the *number* of 2-cycles may vary from one decomposition to the next. Theorem 5.5 (due to Cauchy) says, however, that there is one aspect of a decomposition that never varies.
We isolate a special case of Theorem 5.5 as a lemma.

Lemma   *If $\varepsilon = \beta_1\beta_2 \cdots \beta_r$, where the β's are 2-cycles, then r is even.*

Proof.   Clearly, $r \neq 1$, since a 2-cycle is not the identity. If $r = 2$, we are done. So, we suppose that $r > 2$ and we proceed by induction. Since $(ij) = (ji)$, the product $\beta_1\beta_2$ can be expressed in one of the following forms shown on the left:

$$\begin{aligned}
(ab)(ab) &= \varepsilon \\
(ab)(ac) &= (bc)(ab) \\
(ab)(cd) &= (cd)(ab) \\
(ab)(bc) &= (bc)(ac).
\end{aligned}$$

If the first case occurs, we may delete $\beta_1\beta_2$ from the original product to obtain $\varepsilon = \beta_3 \cdots \beta_r$ and therefore, by the Second Principle of Mathematical Induction, $r - 2$ is even. In the other three cases, we replace the form of $\beta_1\beta_2$ on the left by its counterpart on the right to obtain a new product of r 2-cycles that is still the identity, but where the first occurrence of the integer a is in the second 2-cycle of the product instead of the first. We now repeat the procedure just described with $\beta_2\beta_3$, and, as before, we obtain a product of $(r - 2)$ 2-cycles equal to the identity or a new product of r 2-cycles, where the first occurrence of a is in the third 2-cycle. Continuing this process, we must obtain a product of $(r - 2)$

2-cycles equal to the identity, because otherwise we have a product equal to the identity in which the first occurrence of the integer a is in the last 2-cycle, and such a product does not fix a whereas the identity does. Hence, by induction, $r - 2$ is even and r is even as well. ∎

Theorem 5.5 *Always Even or Always Odd*
If a permutation α can be expressed as a product of an even number of 2-cycles, then every decomposition of α into a product of 2-cycles must have an even number of 2-cycles. In symbols, if

$$\alpha = \beta_1\beta_2 \cdots \beta_r \quad and \quad \alpha = \gamma_1\gamma_2 \cdots \gamma_s,$$

where the β's and the γ's are 2-cycles, then r and s are both even or both odd.

Proof. Observe that $\beta_1\beta_2 \cdots \beta_r = \gamma_1\gamma_2 \cdots \gamma_s$ implies

$$\varepsilon = \gamma_1\gamma_2 \cdots \gamma_s\beta_r^{-1} \cdots \beta_2^{-1}\beta_1^{-1}$$
$$= \gamma_1\gamma_2 \cdots \gamma_s\beta_r \cdots \beta_2\beta_1,$$

since a 2-cycle is its own inverse. Thus, the lemma above guarantees that $s + r$ is even. It follows that r and s are both even or both odd. ∎

DEFINITION Even and Odd Permutations
A permutation that can be expressed as a product of an even number of 2-cycles is called an *even* permutation. A permutation that can be expressed as a product of an odd number of 2-cycles is called an *odd* permutation.

Theorems 5.4 and 5.5 together show that every permutation can be unambiguously classified as either even or odd, but not both. At this point, it is natural to ask what significance this observation has. The answer is given in Theorem 5.6.

Theorem 5.6 *Even Permutations Form a Group*
The set of even permutations in S_n forms a subgroup of S_n.

Proof. This proof is left to the reader. ∎

The subgroup of even permutations in S_n arises so often that we give it a special name and notation.

DEFINITION Alternating Group of Degree n
The group of even permutations of n symbols is denoted by A_n and is called the *alternating group of degree n*.

The alternating groups are among the most important examples of groups. The groups A_4 and A_5 will arise on several occasions in later chapters. In particular, A_5 has great historical significance. It follows from exercise 19 that A_n has order $n!/2$ (when $n > 1$). Thus, we see that exactly half of the members of S_n are even permutations (when $n > 1$).

Table 5.1   The Alternating Group A_4 of Even Permutations of {1, 2, 3, 4}

(In this table, the permutations of A_4 are designated as $\alpha_1, \alpha_2, \ldots, \alpha_{12}$ and an entry k inside the table represents α_k. For example, $\alpha_3 \, \alpha_8 = \alpha_6$.)

	α_1	α_2	α_3	α_4	α_5	α_6	α_7	α_8	α_9	α_{10}	α_{11}	α_{12}
$(1) = \alpha_1$	1	2	3	4	5	6	7	8	9	10	11	12
$(12)(34) = \alpha_2$	2	1	4	3	6	5	8	7	10	9	12	11
$(13)(24) = \alpha_3$	3	4	1	2	7	8	5	6	11	12	9	10
$(14)(23) = \alpha_4$	4	3	2	1	8	7	6	5	12	11	10	9
$(123) = \alpha_5$	5	8	6	7	9	12	10	11	1	4	2	3
$(243) = \alpha_6$	6	7	5	8	10	11	9	12	2	3	1	4
$(142) = \alpha_7$	7	6	8	5	11	10	12	9	3	2	4	1
$(134) = \alpha_8$	8	5	7	6	12	9	11	10	4	1	3	2
$(132) = \alpha_9$	9	11	12	10	1	3	4	2	5	7	8	6
$(143) = \alpha_{10}$	10	12	11	9	2	4	3	1	6	8	7	5
$(234) = \alpha_{11}$	11	9	10	12	3	1	2	4	7	5	6	8
$(124) = \alpha_{12}$	12	10	9	11	4	2	1	3	8	6	5	7

A geometric interpretation of A_4 is given in Example 5, and a multiplication table for A_4 is given as Table 5.1.

Example 5   Rotations of a Tetrahedron

The 12 rotations of a regular tetrahedron can be conveniently described with the elements of A_4. The top row of Figure 5.1 illustrates the identity and three 180° "edge" rotations about axes joining midpoints of two edges. The second row consists of 120° "face" rotations about axes joining a vertex to the center of the opposite face. The third row consists of $-120°$ (or 240°) "face" rotations. Notice that the four rotations in the second row can be obtained from those in the first row by left-multiplying the four in the first row by the rotation (123), whereas those in the third row can be obtained from those in the first row by left-multiplying the ones in the first row by (132).   ❑

A CHECK-DIGIT SCHEME BASED ON D_5

In Chapters 0 and 2 we presented several schemes for appending a check digit to an identification number. Among these schemes, only the International Standard Book Number method was capable of detecting all single-digit errors and all transposition errors involving adjacent digits. However, recall that this success was achieved by introducing the alphabetical character X to handle the case where 10 was required to make the dot product 0 modulo 11. In the late 1960s, a scheme was devised that detects all single-digit errors and all transposition errors involving adjacent digits without the introduction of a new character (see [2] and [3]).

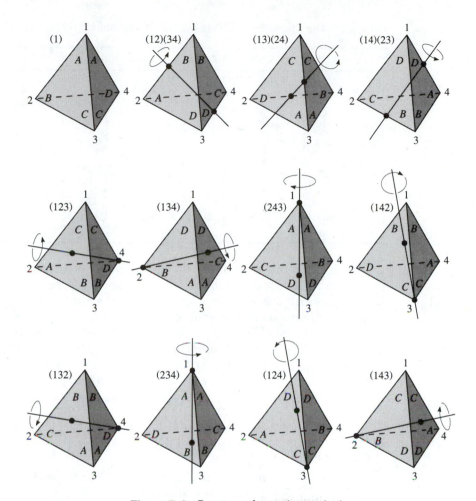

Figure 5.1 Rotations of a regular tetrahedron.

To describe this method, we need the permutation $\sigma = (0)(14)(23)(58697)$ and the dihedral group of order 10 as represented in Table 5.2. (Here we use 0 through 4 for the rotations and 5 through 9 for the reflections.) To append a check digit to any string of digits, we weight the digits of the string with successive powers of σ, starting with the right digit and, using the operation $*$ defined in Table 5.2, multiply the result and take the inverse of the product. To illustrate, consider 793. Beginning with the digit 3, we apply successive powers of σ to transform each digit of 793. That is, $\sigma(3) = 2$; $\sigma^2(9) = \sigma(\sigma(9)) = \sigma(7) = 5$; $\sigma^3(7) = \sigma^2(\sigma(7)) = \sigma^2(5) = \sigma(\sigma(5)) = \sigma(8) = 6$. Then we compute $(6 * 5 * 2)^{-1}$ using Table 5.2 to obtain $3^{-1} = 2$, the check digit. For the number 17326, we determine $(\sigma^5(1) * \sigma^4(7) * \sigma^3(3) * \sigma^2(2) * \sigma(6))^{-1} = (4 * 9 * 2 * 2 * 9)^{-1} = 0^{-1} = 0$.

Table 5.2   The Multiplication Table of D_5

*	0	1	2	3	4	5	6	7	8	9
0	0	1	2	3	4	5	6	7	8	9
1	1	2	3	4	0	6	7	8	9	5
2	2	3	4	0	1	7	8	9	5	6
3	3	4	0	1	2	8	9	5	6	7
4	4	0	1	2	3	9	5	6	7	8
5	5	9	8	7	6	0	4	3	2	1
6	6	5	9	8	7	1	0	4	3	2
7	7	6	5	9	8	2	1	0	4	3
8	8	7	6	5	9	3	2	1	0	4
9	9	8	7	6	5	4	3	2	1	0

To see that this scheme detects all single-digit errors, we observe that an error-free number $a_n a_{n-1} \cdots a_1 a_0$ (where a_0 is the check digit) has the property that $\sigma^n(a_n) * \sigma^{n-1}(a_{n-1}) * \cdots * \sigma(a_1) * a_0 = 0$ and, therefore, any particular factor in this product is uniquely determined by all of the others. Thus, a single-digit error does not result in a product of 0. The fact that all transposition errors of adjacent digits are detected can be verified by showing that for all distinct a and b, $\sigma(a) * b \neq \sigma(b) * a$. Then, for all i, $\sigma^{i+1}(a) * \sigma^i(b) \neq \sigma^{i+1}(b) * \sigma^i(a)$, and consequently a transposition of distinct adjacent digits, will not result in a product of 0 (see exercise 44).

In addition to being foolproof in detecting single-digit errors and transpositions of adjacent digits, this method will detect approximately 90% of all other types of errors.

The permutation σ we have used is not the only one possible. Indeed, (36)(07249851) and (0)(14)(23)(56789) will also achieve the desired effect. Our check digit scheme based on D_5 can be adapted to provide a check digit scheme based on D_n for any $n \geq 3$. Thus, for example, D_{13} could be used to assign a check "letter" to any string of alphabetical characters. Likewise, D_{18} could be used to assign a check "character" to any string of alphanumeric characters. See [1] for more information.

EXERCISES

When you feel how depressingly
slowly you climb,
it's well to remember that
Things Take Time.

     Piet Hein, "T. T. T.," *Grooks*
                   (1966)*

---

*Hein is a Danish engineer and poet, and is the inventor of the game *Hex*.

1. Find the order of each of the following permutations.
   a. (14)   **b.** (147)   **c.** (14762)

2. What is the order of a k-cycle $(a_1a_2 \cdots a_k)$?

3. What is the order of each of the following permutations?
   a. (124)(357)   **b.** (124)(356)
   c. (124)(35)   **d.** (124)(3578)

4. What is the order of each of the following permutations?

   a. $\begin{bmatrix} 1 & 2 & 3 & 4 & 5 & 6 \\ 2 & 1 & 5 & 4 & 6 & 3 \end{bmatrix}$ **b.** $\begin{bmatrix} 1 & 2 & 3 & 4 & 5 & 6 & 7 \\ 7 & 6 & 1 & 2 & 3 & 4 & 5 \end{bmatrix}$

5. What is the order of the product of a pair of disjoint cycles of lengths 4 and 6?

6. What are the possible orders for the elements of S_6 and A_6? What about S_7 and A_7? (This exercise is referred to in Chapter 25.)

7. Show that A_8 contains an element of order 15.

8. What is the maximum order of any element in A_{10}?

9. Determine whether the following permuations are even or odd.
   a. (135)   **b.** (1356)   **c.** (13567)
   d. (12)(134)(152)   **e.** (1243)(3521)

10. Show that a function from a finite set S to itself is one-to-one if and only if it is onto. Is this true when S is infinite?

11. Let n be a positive integer. If n is odd, is an n-cycle an odd or an even permutation? If n is even, is an n-cycle an odd or an even permutation?

12. If α is even, prove that α^{-1} is even. If α is odd, prove that α^{-1} is odd.

13. Prove that A_n is a subgroup of S_n.

14. Prove that the product of an even permutation and an odd permutation is odd.

15. Is the product of two odd permutations an even or an odd permutation?

16. Associate an even permutation with the number $+1$ and an odd permutation with the number -1. Draw an analogy between the result of multiplying two permutations and the result of multiplying their corresponding numbers $+1$ or -1.

17. Let

    $$\alpha = \begin{bmatrix} 1 & 2 & 3 & 4 & 5 & 6 \\ 2 & 1 & 3 & 5 & 4 & 6 \end{bmatrix} \quad \text{and} \quad \beta = \begin{bmatrix} 1 & 2 & 3 & 4 & 5 & 6 \\ 6 & 1 & 2 & 4 & 3 & 5 \end{bmatrix}.$$

    Compute each of the following.
    a. α^{-1}
    b. $\beta\alpha$
    c. $\alpha\beta$

18. Let

    $$\alpha = \begin{bmatrix} 1 & 2 & 3 & 4 & 5 & 6 & 7 & 8 \\ 2 & 1 & 3 & 5 & 4 & 7 & 6 & 8 \end{bmatrix} \quad \text{and}$$

    $$\beta = \begin{bmatrix} 1 & 2 & 3 & 4 & 5 & 6 & 7 & 8 \\ 1 & 3 & 8 & 7 & 6 & 5 & 2 & 4 \end{bmatrix}.$$

Write α and β as
a. products of disjoint cycles,
b. products of 2-cycles.

19. Show that if H is a subgroup of S_n, then either every member of H is an even permutation or exactly half of them are even. (This exercise is referred to in Chapter 25.)

20. Compute the order of each member of A_4. What arithmetic relation do these orders have with the order of A_4?

21. Do the odd permutations in S_n form a group? Why?

22. Let α and β belong to S_n. Prove that $\alpha^{-1}\beta^{-1}\alpha\beta$ is an even permutation.

23. Use Table 5.1 to compute the following.
    a. The centralizer of $\alpha_3 = (13)(24)$.
    b. The centralizer of $\alpha_{12} = (124)$.

24. Determine the subgroup lattice for S_3. (*Hint:* Exercise 25 in Chapter 3 is relevant here.)

25. Show that $(123)^{-1} = (321)$ and that $(1478)^{-1} = (8741)$.

26. What cycle is $(a_1a_2 \cdot \cdot \cdot a_n)^{-1}$?

27. Let G be a group of permutations on a set X. Let $a \in X$ and define stab$(a) = \{\alpha \in G \,|\, \alpha(a) = a\}$. We call stab$(a)$ the *stabilizer of a in G* (since it consists of all members of G that leave a fixed). Prove that stab(a) is a subgroup of G. (This subgroup was introduced by Galois in 1832.) Exercise 27 is referred to in Chapter 8.

28. Let $\beta = (1, 3, 5, 7, 9, 8, 6)(2, 4, 10)$. What is the smallest positive integer n for which $\beta^n = \beta^{-5}$?

29. Let $\alpha = (1, 3, 5, 7, 9)(2, 4, 6)(8, 10)$. If α^m is a 5-cycle, what can you say about m?

30. Let $H = \{\beta \in S_5 \,|\, \beta(1) = 1 \text{ and } \beta(3) = 3\}$. Prove that H is a subgroup of S_5.

31. How many elements of order 5 are there in A_6?

32. In S_4, find a cyclic subgroup of order 4 and a noncyclic subgroup of order 4.

33. Suppose that β is a 10-cycle. For which integers i between 2 and 10 is β^i also a 10-cycle?

34. In S_3, find elements α and β so that $|\alpha| = 2$, $|\beta| = 2$, and $|\alpha\beta| = 3$.

35. Find group elements α and β so that $|\alpha| = 3$, $|\beta| = 3$, and $|\alpha\beta| = 5$.

36. Prove that S_n is non-Abelian for all $n \geq 3$.

37. Represent the symmetry group of an equilateral triangle as a group of permutations of its vertices (see Example 3).

38. Let α and β belong to S_n. Prove that $\beta\alpha\beta^{-1}$ and α are both even or both odd.

39. Let G be the set of all permutations of the positive integers. Let H be the subset of elements of G that can be expressed as a product of a finite number of cycles. Prove that H is a subgroup of G.

40. Show that A_5 has 24 elements of order 5, 20 elements of order 3, and 15 elements of order 2. (This exercise is referred to in Chapter 25.)

41. Show that every element in A_n for $n \geq 3$ can be expressed as a 3-cycle or a product of three cycles.

42. Show that for $n \geq 3$, $Z(S_n) = \{\varepsilon\}$.

43. Use the check digit scheme based on D_5 to append a check digit to 45723. ✓

44. Verify the statement made in the discussion of the check digit scheme based on D_5 that $\sigma(a) * b \neq \sigma(b) * a$ for distinct a and b. Use this to prove that $\sigma^{i+1}(a) * \sigma^i(b) \neq \sigma^{i+1}(b) * \sigma^i(a)$ for all i. Prove that this implies that all transposition errors involving adjacent digits are detected.

45. Let $\sigma = (124875)(36)$. For a string $a_1 a_2 \cdots a_k$ (k odd), assign the check digit $-(\sigma(a_1) + a_2 + \sigma(a_3) + a_4 + \sigma(a_5) + \cdots + \sigma(a_k))$ mod 10. Calculate the check digit for the number 3125600196431. Prove that this method detects all single-digit errors. Determine which transposition errors involving adjacent digits go undetected by this method. How does this method compare with the method given in exercise 49 in Chapter 2?

46. (From the Indiana College Mathematics Competition) A card-shuffling machine always rearranges cards in the same way relative to the order in which they were given to it. All of the hearts arranged in order from ace to king were put into the machine, and then the shuffled cards were put into the machine again to be shuffled. If the cards emerged in the order 10, 9, Q, 8, K, 3, 4, A, 5, J, 6, 2, 7, in what order were the cards after the first shuffle?

47. Show that a permutation with odd order must be an even permutation.

48. Let G be a group. Prove or disprove that $H = \{g^2 | g \in G\}$ is a subgroup of G.

49. Why does the fact that the orders of the elements of A_4 are 1, 2, and 3 imply that $|Z(A_4)| = 1$?

PROGRAMMING EXERCISE

ASCII and ye shall receive.
> **Jeffrey Armstrong**

1. Write a program to implement the check digit scheme discussed in this chapter.

REFERENCES

1. J. A. Gallian, "The Mathematics of Identification Numbers," *The College Mathematics Journal* 22 (1991): 194–202.

2. H. P. Gumm, "A New Class of Check-Digit Methods for Arbitrary Number Systems," *IEEE Transactions on Information Theory* 31 (1985): 102–105.

3. J. Verhoeff, *Error Detecting Decimal Codes,* Amsterdam: Mathematisch Centrum, 1969.

SUGGESTED READINGS

J. Alperin, "Groups and Symmetry," *Mathematics Today,* New York: Springer-Verlag, 1978.

> This beautifully written article is intended to convey to the intelligent non-mathematician something of the nature and development of group theory. It succeeds admirably. In a manner that is accessible to all, Alperin discusses symmetry groups, Galois theory, Lie groups, and simple groups.

Dmitry Fomin, "Getting It Together with 'Polynominoes,'" *Quantum,* Nov./Dec. 1991: 20–23.

> In this article, permutation groups are used to analyze various sorts of checkerboard tiling problems.

H. P. Gumm, "Encoding of Numbers to Detect Typing Errors." *International Journal of Applied Engineering Education* 2 (1986): 61–65.

> Gumm discusses some of the standard methods for assigning check digits. He includes a program for implementing the scheme presented in this chapter.

I. N. Herstein and I. Kaplansky, *Matters Mathematical,* New York: Chelsea, 1978.

> Chapter 3 of this book discusses several interesting applications of permutations to games.

Douglas Hofstadter, "The Magic Cube's Cubies Are Twiddled by Cubists and Solved by Cubemeisters," *Scientific American* 244 (1981): 20–39.

> This article, written by a Pulitzer Prize recipient, discusses the group theory involved in the solution of the Magic (Rubik's) Cube. In particular, permutation groups, subgroups, conjugates (elements of the form xyx^{-1}), commutators (elements of the form $xyx^{-1}y^{-1}$), and the "always-even-always-odd" theorem (Theorem 5.5) are prominently mentioned. At one point, Hofstadter says, "It is this kind of marvelously concrete illustration of an abstract notion of group theory that makes the Magic Cube one of the most amazing things ever invented for teaching mathematical ideas."

N. Wagner and P. Putter, "Error Detecting Decimal Digits," *Communications of the Association for Computing Machinery* 32 (1989): 106–110.

> This article describes the experience had by two mathematicians who were hired by a large mail-order company to make recommendations for an error-correction scheme for the company's account numbers. They recommended a four-digit method.

S. Winters, "Error-Detecting Schemes Using Dihedral Groups," *UMAP Journal* 11, no. 4 (1990): 299–308.

> This article discusses error-detection schemes based on D_n for n odd. Schemes for both one and two check digits are analyzed.

Augustin Cauchy

You see that little young man? Well! He will supplant all of us in so far as we are mathematicians.

Spoken by Lagrange to Laplace about the 11-year-old Cauchy

This stamp was issued by France in Cauchy's honor.

AUGUSTIN LOUIS CAUCHY was born on August 21, 1789, in Paris, the eldest of six children. By the time he was 11, both Laplace and Lagrange had recognized Cauchy's extraordinary talent for mathematics. In school he won prizes for Greek, Latin, and the humanities. At the age of 21, he was given a commission in Napoleon's army as a civil engineer. For the next few years, Cauchy attended to his engineering duties while carrying out brilliant mathematical research on the side.

In 1815, at the age of 26, Cauchy was made Professor of Mathematics at the École Polytechnique and was recognized as the leading mathematician in France. Cauchy and his contemporary Gauss were the last men to know the whole of mathematics as known at their time, and both made important contributions to nearly every branch, both pure and applied, as well as to physics and astronomy.

Cauchy introduced a new level of rigor into mathematical analysis. We owe our contemporary notions of limit and continuity to him. He gave the first proof of the Fundamental Theorem of Calculus. Cauchy was the founder of complex function theory and a pioneer in the theory of permutation groups and determinants. His total output of mathematics fills 24 large quarto volumes and is second only to that of Euler. He wrote over 500 research papers after the age of 50. Cauchy died at the age of 67 on May 23, 1857.

6

Isomorphisms

Mathematicians do not study objects, but relations among
objects; they are indifferent to the replacement of objects by
others as long as relations do not change. Matter is not
important, only form interests them.

Henri Poincaré

MOTIVATION

Suppose two students, one American and another German, are asked to
count a handful of objects. The American student says, "one, two, three,
four, five . . ." whereas the German student says: "ein, zwei, drei, vier,
fünf. . . ." Are the two students doing different things? No. They are
both counting the objects, but they are using different terminology to do
so. Similarly, when one person says: "two plus three is five" and another
says: "zwei und drei ist fünf," the two are in agreement on the *concept*
they are describing, but they are using different terminology to describe
the concept. An analogous situation often occurs with groups; the same
group is described with different terminology. We have seen two examples of this so far. In Chapter 1, we described the symmetries of a square
in geometric terms (e.g., H, V, R_{90}), whereas in Chapter 5 we described
the *same* group by way of permutations of the corners. In both cases,
the underlying group was the symmetries of a square. In Chapter 4, we

observed that when we have a cyclic group of order n generated by a, the operation turns out to be essentially that of addition modulo n, since $a^r \cdot a^s = a^k$, where $k = (r + s) \bmod n$.

DEFINITION AND EXAMPLES

In this chapter, we give a formal method for determining whether two groups defined in different terms are really the *same*. When this is the case, we say that there is an isomorphism between the two groups. This notion was first introduced by Galois about a century and a half ago. The term *isomorphism* is derived from the Greek words *isos,* "same" or "equal," and *morphe,* "form." R. Allenby has colorfully defined an algebraist as "a person who can't tell the difference between isomorphic systems."

DEFINITION Group Isomorphism

An *isomorphism* ϕ from a group of G to a group $\overline{G}$ is a one-to-one mapping (or function) from G onto $\overline{G}$ that preserves the group operation. That is,

$$\phi(ab) = \phi(a)\phi(b) \qquad \text{for all } a, b \text{ in } G.$$

If there is an isomorphism from G onto $\overline{G}$, we say that G and $\overline{G}$ are *isomorphic* and write $G \approx \overline{G}$.

This definition can be visualized as shown in Figure 6.1. The pairs of dashed arrows represent the group operations.

It is implicit in the definition of isomorphism that the operation on the left side of the equal sign is that of G whereas the operation on the right side is that of $\overline{G}$. The four cases involving $\cdot$ and $+$ are shown in Table 6.1. It is also implicit in the definition of isomorphism that isomorphic groups must have the same order.

There are four separate steps in proving that a group G is isomorphic to a group $\overline{G}$.

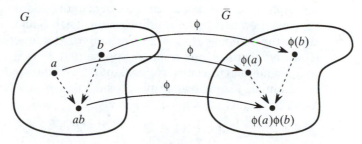

Figure 6.1

Table 6.1

G Operation	$\overline{G}$ Operation	Operation Preservation
$\cdot$	$\cdot$	$\phi(a \cdot b) = \phi(a) \cdot \phi(b)$
$\cdot$	$+$	$\phi(a \cdot b) = \phi(a) + \phi(b)$
$+$	$\cdot$	$\phi(a + b) = \phi(a) \cdot \phi(b)$
$+$	$+$	$\phi(a + b) = \phi(a) + \phi(b)$

STEP 1   "Mapping." Define a candidate for the isomorphism; that is, define a function ϕ from G to $\overline{G}$.

STEP 2   "1–1." Prove that ϕ is one-to-one; that is, assume that $\phi(a) = \phi(b)$ and prove that $a = b$.

STEP 3   "Onto." Prove that ϕ is onto; that is, for any element $\overline{g}$ in $\overline{G}$, find an element g in G such that $\phi(g) = \overline{g}$.

STEP 4   "O.P." Prove that ϕ is operation-preserving; that is, show that $\phi(ab) = \phi(a)\phi(b)$ for all a and b in G.

None of these steps is unfamiliar to you. The only one that may appear novel is the fourth one. It requires that one be able to obtain the same result by multiplying two elements and then mapping, or by mapping two elements and then multiplying. Roughly speaking, this says that the two processes—operating and mapping—can be done in either order without affecting the result. This same concept arises in calculus when we say

$$\lim_{x \to a} (f(x) \cdot g(x)) = \lim_{x \to a} f(x) \lim_{x \to a} g(x)$$

or

$$\int_a^b (f + g)\, dx = \int_a^b f\, dx + \int_a^b g\, dx.$$

Before going any further, let's consider some examples.

Example 1   Let G be the real numbers under addition and $\overline{G}$ the positive real numbers under multiplication. Then G and $\overline{G}$ are isomorphic under the mapping $\phi(x) = 2^x$. Certainly, ϕ is a function from G to $\overline{G}$. To prove that it is one-to-one, suppose that $2^x = 2^y$. Then $\log_2 2^x = \log_2 2^y$ and therefore $x = y$. For "onto," we must find for any positive real number y some real number x such that $\phi(x) = y$; that is, $2^x = y$. Well, solving for x gives $\log_2 y$. Finally,

$$\phi(x + y) = 2^{x+y} = 2^x \cdot 2^y = \phi(x)\phi(y)$$

for all x and y in G so that ϕ is operation-preserving as well.   □

Example 2   Any finite cyclic group of order n is isomorphic to Z_n. Any infinite cyclic group is isomorphic to Z. Indeed, in either case, if a is a generator of the cyclic group, the mapping $a^k \rightarrow k$ is an isomorphism. That this correspondence is a function and is one-to-one is the essence of Theorem 4.1. Obviously, the mapping is onto. That the mapping is operation-preserving follows from exercise 14 in Chapter 0 in the finite case and from the definitions in the infinite case.   ❏

Example 3   The mapping from **R** under addition to itself given by $\phi(x) = x^3$ is *not* an isomorphism. Although ϕ is one-to-one and onto, it is not operation-preserving since it is not true that $(x + y)^3 = x^3 + y^3$ for all x and y.   ❏

Example 4 $U(10) \approx Z_4 \approx U(5)$. To verify this, one need only observe that both $U(10)$ and $U(5)$ are cyclic of order 4. Then appeal to example 2.   ❏

Example 5 $U(10) \not\approx U(12)$. This is a bit more tricky to prove. First, note that $x^2 = 1$ for all x in $U(12)$. Now, suppose that ϕ is an isomorphism from $U(10)$ onto $U(12)$. Then,

$$\phi(9) = \phi(3 \cdot 3) = \phi(3)\phi(3) = 1$$

and

$$\phi(1) = \phi(1 \cdot 1) = \phi(1)\phi(1) = 1.$$

Thus, $\phi(9) = \phi(1)$, but $9 \neq 1$, which is a contradiction to the supposed one-to-one character of ϕ.   ❏

Example 6   There is no isomorphism from Q, the group of rational numbers under addition, to Q^*, the group of nonzero rational numbers under multiplication. If ϕ were such a mapping, there would be a rational number a such that $\phi(a) = -1$. But then

$$-1 = \phi(a) = \phi(\tfrac{1}{2}a + \tfrac{1}{2}a) = \phi(\tfrac{1}{2}a)\phi(\tfrac{1}{2}a) = [\phi(\tfrac{1}{2}a)]^2.$$

However, no rational number squared is -1.   ❏

Example 7   Let $G = SL(2, \mathbf{R})$, the group of 2×2 real matrices with determinant 1. Let M be any 2×2 real matrix with nonzero determinant. Then we can define a mapping from G to G itself by $\phi_M(A) = MAM^{-1}$ for all A in G. To verify that ϕ_M is an isomorphism, we carry out the four steps.

STEP 1 ϕ_M is a function from G to G. Here, we must show that $\phi_M(A)$ is indeed an element of G whenever A is. This follows from properties of determinants:

$$\det(MAM^{-1}) = (\det M)(\det A)(\det M)^{-1} = \det A = 1.$$

Thus, MAM^{-1} is in G.

STEP 2 ϕ_M is one-to-one. Suppose that $\phi_M(A) = \phi_M(B)$. Then MAM^{-1} $= MBM^{-1}$ and, by left and right cancellation, $A = B$.

STEP 3 ϕ_M is onto. Let B belong to G. We must find a matrix A in G such that $\phi_M(A) = B$. How shall we do this? If such a matrix A is to exist, it must have the property that $MAM^{-1} = B$. But this tells us exactly what A must be! For we can solve for A to obtain $A = M^{-1}BM$.

STEP 4 ϕ_M is operation-preserving. Let A and B belong to G. Then,

$$\phi_M(AB) = M(AB)M^{-1} = MA(M^{-1}M)BM^{-1}$$
$$= (MAM^{-1})(MBM^{-1}) = \phi_M(A)\phi_M(B).$$

The mapping ϕ_M is called *conjugation* by M.   ☐

CAYLEY'S THEOREM

Our next example is a classic theorem of Cayley. An important generalization of it will be given in Chapter 25.

Example 8   Cayley's Theorem (1854)
Every group is isomorphic to a group of permutations.
    To prove this, let G be any group. We must find a group $\overline{G}$ of permutations that we believe is isomorphic to G. Since G is all we have to work with, we will have to use it to construct $\overline{G}$. For any g in G, define a function T_g from G to G by

$$T_g(x) = gx \qquad \text{for all } x \text{ in } G.$$

(In words, T_g is just multiplication by g on the left.) We leave it as an exercise to prove that T_g is a permutation on the set of elements of G. Now, let $\overline{G} = \{T_g \mid g \in G\}$. Then, $\overline{G}$ is a group under the operation of function composition. To verify this, we first observe that for any g and h in G we have $T_g T_h(x) = T_g(T_h(x)) = T_g(hx) = g(hx) = (gh)x = T_{gh}(x)$, so that $T_g T_h = T_{gh}$. From this it follows that T_e is the identity and $(T_g)^{-1} = T_{g^{-1}}$ (see exercise 8). Since function composition is associative, we have verified all the conditions for $\overline{G}$ to be a group.
    The isomorphism ϕ between G and $\overline{G}$ is now ready-made. For every g in G, define $\phi(g) = T_g$. Clearly, $g = h$ implies $T_g = T_h$, so that ϕ is a function from G to $\overline{G}$. On the other hand, if $T_g = T_h$, then $T_g(e) = T_h(e)$ or $ge = he$. Thus, ϕ is one-to-one. By the way $\overline{G}$ was constructed, we see that ϕ is onto. The only condition that remains to be checked is that ϕ is operation-preserving. To this end, let x and y belong to G. Then

$$\phi(xy) = T_{xy} = T_x T_y = \phi(x)\phi(y).$$ ☐

    The group $\overline{G}$ constructed above is called the *left regular representation of G*. For concreteness, let us calculate the left regular representation $\overline{U(12)}$ for $U(12) = \{1, 5, 7, 11\}$. Writing the permutations of $U(12)$ in array form, we have (remember, T_x is just multiplication by x)

$$T_1 = \begin{bmatrix} 1 & 5 & 7 & 11 \\ 1 & 5 & 7 & 11 \end{bmatrix}, \qquad T_5 = \begin{bmatrix} 1 & 5 & 7 & 11 \\ 5 & 1 & 11 & 7 \end{bmatrix},$$

$$T_7 = \begin{bmatrix} 1 & 5 & 7 & 11 \\ 7 & 11 & 1 & 5 \end{bmatrix}, \qquad T_{11} = \begin{bmatrix} 1 & 5 & 7 & 11 \\ 11 & 7 & 5 & 1 \end{bmatrix}.$$

It is instructive to compare the Cayley table for $U(12)$ and its left regular representation $\overline{U(12)}$.

$U(12)$	1	5	7	11
1	1	5	7	11
5	5	1	11	7
7	7	11	1	5
11	11	7	5	1

$\overline{U(12)}$	T_1	T_5	T_7	T_{11}
T_1	T_1	T_5	T_7	T_{11}
T_5	T_5	T_1	T_{11}	T_7
T_7	T_7	T_{11}	T_1	T_5
T_{11}	T_{11}	T_7	T_5	T_1

It should be abundantly clear from these tables that $U(12)$ and $\overline{U(12)}$ are only notationally different.

Perhaps the most important aspect of Cayley's Theorem is that it shows that the present-day set of axioms we have adopted for a group is the correct abstraction of its much earlier predecessor—a group of permutations. Indeed, Cayley's Theorem tells us that abstract groups are not different from permutation groups. Rather, it is the viewpoint that is different. It is this difference of viewpoint that has stimulated the tremendous progress in group theory and many other branches of mathematics in the 20th century.

It is sometimes very difficult to prove or disprove, whichever the case may be, that two particular groups are isomorphic. For example, it requires somewhat sophisticated techniques to prove the surprising fact that the group of real numbers under addition is isomorphic to the group of complex numbers under addition. Likewise, it is not easy to prove that the group of nonzero complex numbers under multiplication is isomorphic to the group of complex numbers with absolute value of 1 under multiplication. In geometric terms, this says that, as groups, the punctured plane and the unit circle are isomorphic.

PROPERTIES OF ISOMORPHISMS

Our first theorem in this chapter gives a catalog of properties of isomorphisms and isomorphic groups.

Theorem 6.1 *Properties of Isomorphisms*
    Suppose that ϕ is an isomorphism from a group G onto a group $\overline{G}$. Then

  1. *ϕ carries the identity of G to the identity of $\overline{G}$.*
  2. *For every integer n and for every group element a in G, $\phi(a^n) = [\phi(a)]^n$.*

3. *For any elements a and b in G, a and b commute if and only if $\phi(a)$ and $\phi(b)$ commute.*
4. *G is Abelian if and only if $\overline{G}$ is Abelian.*
5. *$|a| = |\phi(a)|$ for all a in G (isomorphisms preserve orders).*
6. *G is cyclic if and only if $\overline{G}$ is cyclic.*
7. *For a fixed integer k and a fixed group element b in G, the equation $x^k = b$ has the same number of solutions in G as does the equation $x^k = \phi(b)$ in $\overline{G}$.*
8. *ϕ^{-1} is an isomorphism from $\overline{G}$ onto G.*
9. *If K is a subgroup of G, then $\phi(K) = \{\phi(k) \mid k \in K\}$ is a subgroup of $\overline{G}$.*

Proof.   We will restrict ourselves to proving only properties 1 and 5. Note, however, that property 4 follows directly from property 3, and property 6 directly from property 5. For convenience, let us denote the identity in G by e_G, and the identity in $\overline{G}$ by $e_{\overline{G}}$. Then $e_G = e_G e_G$ so that

$$\phi(e_G) = \phi(e_G e_G) = \phi(e_G)\phi(e_G).$$

But $\phi(e_G) \in \overline{G}$, so that $\phi(e_G) = e_{\overline{G}}\phi(e_G)$, as well. Thus, by cancellation, we have $e_{\overline{G}} = \phi(e_G)$. This proves property 1.

To prove property 5, we note that $a^n = e$ if and only if $\phi(a^n) = \phi(e)$. So, by properties 1 and 2, $a^n = e$ if and only if $(\phi(a))^n = e$. Thus, a has infinite order if and only if $\phi(a)$ has infinite order, and a has finite order n if and only if $\phi(a)$ has order n.   ∎

Property 7 is quite useful for showing that two groups are *not* isomorphic. Often b is picked to be the identity. For example, consider $\mathbf{C}^*$ and $\mathbf{R}^*$. Because the equation $x^4 = 1$ has four solutions in $\mathbf{C}^*$ but only two in $\mathbf{R}^*$, no matter how one attempts to define an isomorphism from $\mathbf{C}^*$ to $\mathbf{R}^*$, property 7 cannot hold.

Theorem 6.1 shows that isomorphic groups have many properties in common. Actually, the definition is precisely formulated so that isomorphic groups have *all* group-theoretic properties in common. By this we mean that if two groups are isomorphic, then any property that can be expressed in the language of group theory is true for one if and only if it is true for the other. This is why algebraists speak of isomorphic groups as "equal" or "the same." Admittedly, calling such groups equivalent, rather than the same, might be more appropriate, but we bow to long-standing tradition.

AUTOMORPHISMS

Certain kinds of isomorphisms are referred to so often that they have been given special names.

DEFINITION Automorphism

An isomorphism from a group G onto itself is called an *automorphism* of G.

The isomorphism of Example 7 is an automorphism of $SL(2, \mathbf{R})$. Two more examples follow.

Example 9 The function ϕ from $\mathbf{C}$ to $\mathbf{C}$ given by $\phi(a + bi) = a - bi$ is an automorphism of the group of complex numbers under addition. The restriction of ϕ to $\mathbf{C}^\#$ is also an automorphism of the group of the nonzero complex numbers under multiplication. ☐

Example 10 Let $\mathbf{R}^2 = \{(a, b) \mid a, b \in \mathbf{R}\}$. Then $\phi(a, b) = (b, a)$ is an automorphism of the group $\mathbf{R}^2$ under componentwise addition. Geometrically, ϕ reflects each point in the plane across the line $y = x$. More generally, any reflection across a line passing through the origin or any rotation of the plane about the origin is an automorphism of $\mathbf{R}^2$. ☐

The situation where the element M in Example 7 is in $SL(2, \mathbf{R})$ is a particular instance of an automorphism that arises often enough to warrant a name and notation of its own.

DEFINITION Inner Automorphism Induced by a

Let G be a group, and let $a \in G$. The function ϕ_a defined by $\phi_a(x) = axa^{-1}$ for all x in G is called the *inner automorphism of G induced by a.*

We leave it as an exercise to show that ϕ_a is actually an automorphism of G. (Use Example 7 as a model.)

Example 11 The action of the inner automorphism of D_4 induced by R_{90} is given below.

$$
\begin{array}{ll}
 & \phi_{R_{90}} \\
x & \to R_{90}\, x\, R_{90}^{-1} \\
\hline
R_0 & \to R_{90}R_0R_{90}^{-1} = R_0 \\
R_{90} & \to R_{90}R_{90}R_{90}^{-1} = R_{90} \\
R_{180} & \to R_{90}R_{180}R_{90}^{-1} = R_{180} \\
R_{270} & \to R_{90}R_{270}R_{90}^{-1} = R_{270} \\
H & \to R_{90}HR_{90}^{-1} = V \\
V & \to R_{90}VR_{90}^{-1} = H \\
D & \to R_{90}DR_{90}^{-1} = D' \\
D' & \to R_{90}D'R_{90}^{-1} = D \\
\hline
\end{array}
$$

☐

When G is a group, we use $\mathrm{Aut}(G)$ to denote the set of all automorphisms of G and $\mathrm{Inn}(G)$ to denote the set of all inner automorphisms of G. The reason these sets are noteworthy is demonstrated by the next theorem.

Theorem 6.2 Aut(*G*) *and* Inn(*G*) *Are Groups**
The set of automorphisms of a group and the set of inner automorphisms of a group are both groups.

Proof.   The proof of Theorem 6.2 is left as an exercise.   ∎

The determination of Inn(*G*) is routine. If $G = \{e, a, b, c, \ldots\}$, then $\text{Inn}(G) = \{\phi_e, \phi_a, \phi_b, \phi_c, \ldots\}$. This latter list may have duplications, however, since ϕ_a may be equal to ϕ_b even though $a \neq b$ (see exercise 32). Thus, the only work involved in determining Inn(*G*) is in deciding which distinct elements give the distinct automorphisms. On the other hand, the determination of Aut(*G*) is, in general, quite involved.

Example 12   Inn(D_4)
To determine Inn(D_4), we first observe that the complete list of inner automorphisms is $\phi_{R_0}, \phi_{R_{90}}, \phi_{R_{180}}, \phi_{R_{270}}, \phi_H, \phi_V, \phi_D$, and $\phi_{D'}$. Our job is to determine the repetitions in this list. Since $R_{180} \in Z(D_4)$, we have $\phi_{R_{180}}(x) = R_{180}xR_{180}^{-1} = x$, so that $\phi_{R_{180}} = \phi_{R_0}$. Also, $\phi_{R_{270}}(x) = R_{270}xR_{270}^{-1} = R_{90}R_{180}xR_{180}^{-1}R_{90}^{-1} = R_{90}xR_{90}^{-1} = \phi_{R_{90}}(x)$. Similarly, since $H = R_{180}V$ and $D' = R_{180}D$, we have $\phi_H = \phi_V$ and $\phi_D = \phi_{D'}$. This proves that the previous list can be pared down to $\phi_{R_0}, \phi_{R_{90}}, \phi_H$, and ϕ_D. We leave it to the reader to show that these are distinct.   □

Example 13   Aut(Z_{10})
To compute Aut(Z_{10}), we assume that α is an element of Aut(Z_{10}) and try to discover enough information about α to determine how α must be defined. Because Z_{10} is so simple, this is not difficult to do. To begin with, observe that once we know $\alpha(1)$ we know $\alpha(k)$ for any k, because

$$\alpha(k) = \alpha(\underbrace{1 + 1 + \cdots + 1}_{k \text{ terms}})$$

$$= \underbrace{\alpha(1) + \alpha(1) + \cdots + \alpha(1)}_{k \text{ terms}} = k\alpha(1).$$

So, we need only determine the choices for $\alpha(1)$ that make α an automorphism of Z_{10}. Since property 5 of Theorem 6.1 tells us that $|\alpha(1)| = 10$, there are four candidates for $\alpha(1)$:

$$\alpha(1) = 1; \quad \alpha(1) = 3; \quad \alpha(1) = 7; \quad \alpha(1) = 9.$$

To distinguish among the four possibilities, we refine our notation by denoting the mapping that sends 1 to 1 by α_1, 1 to 3 by α_3, 1 to 7 by α_7, and 1 to 9 by α_9. So the only possibilities for Aut(Z_{10}) are $\alpha_1, \alpha_3, \alpha_7$, and α_9. But are all these automorphisms? Clearly, α_1 is the identity. Let us check α_3. Since $\alpha_3(1) = 3$ is a generator of Z_{10}, it follows that α_3 is

*The group Aut(*G*) was first studied by O. Hölder in 1893 and, independently, by E. H. Moore in 1894.

onto (and, by exercise 10 in Chapter 5, it is also one-to-one). Finally, since $\alpha_3(a + b) = 3(a + b) = 3a + 3b = \alpha_3(a) + \alpha_3(b)$, we see that α_3 is operation-preserving as well. Thus, $\alpha_3 \in \text{Aut}(Z_{10})$. The same argument shows that α_7 and α_9 are also automorphisms.

This gives us the elements of $\text{Aut}(Z_{10})$ but not the structure. For instance, what is $\alpha_3\alpha_3$? Well, $(\alpha_3\alpha_3)(1) = \alpha_3(3) = 3 \cdot 3 = 9 = \alpha_9(1)$, so $\alpha_3\alpha_3 = \alpha_9$. A similar calculation shows that $\alpha_3^4 = \alpha_1$, so that $|\alpha_3| = 4$. Thus $\text{Aut}(Z_{10})$ is cyclic. Actually, the following Cayley tables reveal that $\text{Aut}(Z_{10})$ is isomorphic to $U(10)$.

$U(10)$	1	3	7	9
1	1	3	7	9
3	3	9	1	7
7	7	1	9	3
9	9	7	3	1

$\text{Aut}(Z_{10})$	α_1	α_3	α_7	α_9
α_1	α_1	α_3	α_7	α_9
α_3	α_3	α_9	α_1	α_7
α_7	α_7	α_1	α_9	α_3
α_9	α_9	α_7	α_3	α_1

□

With Example 13 as a guide, we are now ready to tackle the group $\text{Aut}(Z_n)$. The result is particularly nice since it relates the two kinds of groups we have most frequently encountered thus far—the cyclic groups Z_n and the U-groups $U(n)$.

Theorem 6.3 $\text{Aut}(Z_n) \approx U(n)$
For every positive integer n, $\text{Aut}(Z_n)$ is isomorphic to $U(n)$.

Proof.   As in Example 13, any automorphism α is determined by the value of $\alpha(1)$ and $\alpha(1) \in U(n)$. Now consider the correspondence from $\text{Aut}(Z_n)$ to $U(n)$ given by $T: \alpha \rightarrow \alpha(1)$. The fact that $\alpha(k) = k\alpha(1)$ (see Example 13) implies that T is a one-to-one mapping. For if α and β belong to $\text{Aut}(Z_n)$ and $\alpha(1) = \beta(1)$, then $\alpha(k) = \beta(k)$ for all k in Z_n and therefore $\alpha = \beta$.

To prove that T is onto, let $r \in U(n)$ and consider the mapping α from Z_n to Z_n defined by $\alpha(s) = sr(\text{mod } n)$ for all s in Z_n. We leave it as an exercise to verify that α is an automorphism of Z_n (see exercise 16). Then, since $T(\alpha) = \alpha(1) = r$, T is onto $U(n)$.

Finally, we establish the fact that T is operation-preserving. Let α, $\beta \in \text{Aut}(Z_n)$. We then have

$$T(\alpha\beta) = (\alpha\beta)(1) = \alpha(\beta(1)) = \alpha(\underbrace{1 + 1 + \cdots + 1}_{\beta(1)})$$

$$= \underbrace{\alpha(1) + \alpha(1) + \cdots + \alpha(1)}_{\beta(1)} = \alpha(1)\beta(1)$$

$$= T(\alpha)T(\beta).$$

This completes the proof. ∎

EXERCISES

No pain, no gain.
      Anonymous

1. Find an isomorphism from the group of integers under addition to the group of even integers under addition.
2. Find Aut(Z).
3. Let $\mathbf{R}^+$ be the group of positive real numbers under multiplication. Show that the mapping $\phi(x) = \sqrt{x}$ is an automorphism of $\mathbf{R}^+$.
4. Show that $U(8)$ is not isomorphic to $U(10)$.
5. Show that $U(8)$ is isomorphic to $U(12)$.
6. Prove that the relation isomorphism is an equivalence relation. (See Chapter 0 for definition of equivalence relation.)
7. Prove that S_4 is not isomorphic to D_{12}.
8. In the notation of Example 8, prove that $(T_g)^{-1} = T_{g^{-1}}$.
9. Explain why the three parts of exercise 1 of the Supplementary Exercises, Chapter 1–4, follow immediately from Theorem 6.1.
10. Let G be a group. Prove that the mapping $\alpha(g) = g^{-1}$ for all g in G is an automorphism if and only if G is Abelian.
11. Find two groups G and H such that $G \not\approx H$, but Aut(G) $\approx$ Aut(H).
12. Let G be the group given in exercise 4 in the Supplementary Exercises for Chapters 1–4. Find Inn(G).
13. Let G be a group and let a belong to G. Prove that the mapping of ϕ_a defined by $\phi_a(x) = axa^{-1}$ is an automorphism of G.
14. Find Aut(Z_6).
15. If G is a group, prove that Aut(G) and Inn(G) are groups.
16. Let $r \in U(n)$. Prove that the mapping $\alpha: Z_n \to Z_n$ defined by $\alpha(s) = sr \bmod n$ for all s in Z_n is an automorphism of Z_n (this exercise is referred to in this chapter).
17. Prove that the mapping from $U(16)$ to itself given by $x \to x^3$ is an automorphism. What about $x \to x^5$ and $x \to x^7$? Generalize.
18. Suppose that ϕ is an automorphism of D_4 that takes R_{90} to itself and H to V. Prove that ϕ is the inner automorphism induced by R_{90}.
19. Show that S_7 (the symmetric group on the integers $\{1, 2, 3, 4, 5, 6, 7\}$) is isomorphic to the subgroup of all those elements of S_8 that send 8 to 8.
20. Prove that the quaternion group (see exercise 4, Supplementary Exercises for Chapters 1–4) is not isomorphic to the dihedral group D_4.
21. Prove or disprove that $U(20)$ and $U(24)$ are isomorphic.
22. Referring to Example 8, prove that T_g is indeed a permutation on the set G.

23. Show that the mapping $\phi(a + bi) = a - bi$ is an automorphism of the group of complex numbers under addition. Show that ϕ preserves complex multiplication as well—that is, $\phi(xy) = \phi(x)\phi(y)$ for all x and y in **C**.

24. Let

$$G = \{a + b\sqrt{2} \mid a, b \text{ rational}\}$$

and

$$H = \left\{ \begin{bmatrix} a & 2b \\ b & a \end{bmatrix} \, \middle| \, a, b \text{ rational} \right\}.$$

Show that G and H are isomorphic under addition. Prove that G and H are closed under multiplication. Does your isomorphism preserve multiplication as well as addition? (G and H are examples of rings—a topic we will take up in Part 3.)

25. Prove that Z under addition is not isomorphic to Q under addition.

26. Look up the words *isobar*, *isomer*, and *isotope* in a dictionary. Relate their meanings to the meaning of isomorphism.

27. Let **C** be the complex numbers and

$$M = \left\{ \begin{bmatrix} a & -b \\ b & a \end{bmatrix} \, \middle| \, a, b \in \mathbf{R} \right\}.$$

Prove that **C** and M are isomorphic under addition and that $\mathbf{C}^{\#}$ and $M^{\#}$ are isomorphic under multiplication.

28. Let $\mathbf{R}^n = \{(a_1, a_2, \ldots, a_n) \mid a_i \in \mathbf{R}\}$. Show that the mapping $\phi: (a_1, a_2, \ldots, a_n) \rightarrow (-a_1, -a_2, \ldots, -a_n)$ is an automorphism of the group $\mathbf{R}^n$ under componentwise addition. This automorphism is called *inversion*. Describe the action of ϕ geometrically.

29. Consider the following statement: The order of a subgroup divides the order of the group. Suppose you could prove this for finite permutation groups. Would the statement then be true for all finite groups? Explain.

30. Prove that for any integer n there are only a finite number of nonisomorphic groups of order n.

31. Show that the mapping $a \rightarrow \log_{10} a$ is an isomorphism from $\mathbf{R}^+$ under multiplication to $\mathbf{R}$ under addition. Explain how this isomorphism is implicitly employed when one uses a slide rule to multiply or divide real numbers.

32. Let G be a group and let $g \in G$. If $z \in Z(G)$, show that the inner automorphism induced by g is the same as the inner automorphism induced by zg (that is, that the mappings ϕ_g and ϕ_{zg} are equal).

33. Suppose that g and h induce the same inner automorphism of a group G. Prove that $h^{-1}g \in Z(G)$.

34. Combine the results of exercises 32 and 33 into a single "if and only if" theorem.

35. Explain why $S_n(n \geq 3)$ contains a subgroup isomorphic to D_n.

36. Let $G = \{0, \pm 2, \pm 4, \pm 6, \ldots\}$ and $H = \{0, \pm 3, \pm 6, \pm 9, \ldots\}$. Show that G and H are isomorphic groups under addition.

Suggested Film

Nim and Other Oriented Graph Games with Andrew Gleason, International Film Bureau, Inc., 63 minutes.

This filmed lecture describes Nim and related games and constructs an algebraic theory that gives information about games more complicated than Nim. The film gives a good example of a nontrivial, nonobvious application of group theory.

Arthur Cayley

Cayley is forging the weapons for future generations of physicists.

Peter Tait

ARTHUR CAYLEY was born on August 16, 1821, in England. His genius showed itself at an early age. He published his first research paper while an undergraduate of 20, and the next year he published eight papers. While still in his early twenties, he originated the concept of n-dimensional geometry.

After graduating from Trinity College, Cambridge, Cayley stayed on for three years as a tutor. At the age of 25, he began a 14-year career as a lawyer. During this period, he published approximately 200 mathematical papers, many of which are now classics.

In 1863, Cayley accepted the newly established Sadlerian professorship of mathematics at Cambridge University. He spent the rest of his life in that position. One of his notable nonmathematical accomplishments was his role in the successful effort to have women admitted to Cambridge.

Among Cayley's many innovations in mathematics were the notions of an abstract group and a group algebra, and the matrix concept. He made major contributions to geometry and linear algebra. Cayley and his life-long friend and collaborator J. J. Sylvester were the founders of the theory of invariants, which was later to play an important role in the theory of relativity.

115

Cayley's collected works comprise 13 volumes, each about 600 pages in length. He died on January 26, 1895. One of his students wrote of him:

> But he was more than a mathematician with a singleness of aim, which Words-worth would have chosen for his "Happy Warrior," he persevered to the last in his nobly lived ideal. His life had a significant influence on those who knew him; they admired his character as much as they respected his genius; and they felt that, at his death, a great man had passed from the world.

7

External Direct Products

The universe is an enormous direct product of representations of symmetry groups.

Steven Weinberg*

DEFINITION AND EXAMPLES

In this chapter, we show how one may piece together groups to make larger groups. In Chapter 9, we will show that one can often start with one large group and decompose it into a product of smaller groups in much the same way as a composite positive integer can be broken down into a product of primes. These methods will later be used to give us a simple way to construct all finite Abelian groups.

DEFINITION   External Direct Product

Let $G_1, G_2, \ldots, G_n$ be a finite collection of groups. The *external direct product* of $G_1, G_2, \ldots, G_n$, written as $G_1 \oplus G_2 \oplus \cdots \oplus G_n$, is the set of all n-tuples for which the ith component is an element of G_i, and the operation is componentwise. In symbols,

$$G_1 \oplus G_2 \oplus \cdots \oplus G_n = \{(g_1, g_2, \ldots, g_n) \mid g_i \in G_i\},$$

---

*Weinberg received the 1979 Nobel Prize in physics with Sheldon Glashow and Abdus Salam for their construction of a single theory incorporating weak and electromagnetic interactions.

where $(g_1, g_2, \ldots, g_n)(g_1', g_2', \ldots, g_n')$ is defined to be $(g_1 g_1', g_2 g_2', \ldots, g_n g_n')$. It is understood that each product $g_i g_i'$ is performed with the operation of G_i. We leave it to the reader to show that the external direct product of groups is itself a group.

This construction is not new to students who have had linear algebra or physics. Indeed, $\mathbf{R}^2 = \mathbf{R} \oplus \mathbf{R}$ and $\mathbf{R}^3 = \mathbf{R} \oplus \mathbf{R} \oplus \mathbf{R}$—the operation being componentwise addition. Of course, there is also scalar multiplication, but we ignore this for the time being, since we are interested only in the group structure at this point.

Example 1

$$U(8) \oplus U(10) = \{(1, 1), (1, 3), (1, 7), (1, 9), (3, 1), (3, 3), (3, 7),$$
$$(3, 9), (5, 1), (5, 3), (5, 7), (5, 9), (7, 1), (7, 3),$$
$$(7, 7), (7, 9)\}.$$

The product $(3, 7)(7, 9) = (5, 3)$, since the first two components are combined by multiplication modulo 8, whereas the second two components are combined by multiplication modulo 10. $\square$

Example 2

$$Z_2 \oplus Z_3 = \{(0, 0), (0, 1), (0, 2), (1, 0), (1, 1), (1, 2)\}.$$

Clearly, this is an Abelian group of order 6. Is this group related to another Abelian group of order 6 that we know, namely, Z_6? Consider the subgroup of $Z_2 \oplus Z_3$ generated by $(1, 1)$. Since the operation in each component is addition, we have $(1, 1) = (1, 1)$, $2(1, 1) = (0, 2)$, $3(1, 1) = (1, 0)$, $4(1, 1) = (0, 1)$, $5(1, 1) = (1, 2)$, and $6(1, 1) = (0, 0)$. Hence $Z_2 \oplus Z_3$ is cyclic. It follows then that $Z_2 \oplus Z_3$ is isomorphic to Z_6. $\square$

On the basis of Example 2, one might be tempted to conjecture that $Z_m \oplus Z_n \approx Z_{mn}$ for all positive integers m and n. This is easily seen to be false by looking at $Z_2 \oplus Z_2$. Theorem 7.2 shows, however, that these groups are isomorphic in certain cases.

PROPERTIES OF EXTERNAL DIRECT PRODUCTS

Our first theorem gives a simple method for computing the order of an element in a direct product in terms of the orders of the component pieces.

Theorem 7.1 *Order of an Element in a Direct Product*
The order of an element of a direct product of a finite number of finite groups is the least common multiple of the orders of the components of

the element. In symbols,

$$|(g_1, g_2, \ldots, g_n)| = \mathrm{lcm}(|g_1|, |g_2|, \ldots, |g_n|).$$

Proof. To simplify matters, we first consider the special case for which the direct product has only two factors. Let (g_1, g_2) be an arbitrary element of $G_1 \oplus G_2$. Let $s = \mathrm{lcm}(|g_1|, |g_2|)$ and let $t = |(g_1, g_2)|$. Clearly then,

$$(g_1, g_2)^s = (g_1^s, g_2^s) = (e, e),$$

so, by the corollary to Theorem 4.1, t must divide s. In particular, $t \le s$. But, $(g_1^t, g_2^t) = (g_1, g_2)^t = (e, e)$, and it follows, for the same reason, that both $|g_1|$ and $|g_2|$ must divide t. Thus, t is a common multiple of $|g_1|$ and $|g_2|$; and, therefore, $s \le t$, since s is the *least* common multiple of $|g_1|$ and $|g_2|$. Putting this information about s and t together, we find that $s = t$.

We leave it to the reader to verify that the argument extends to the general case. ∎

The next two examples are applications of Theorem 7.1.

Example 3 We determine the number of elements of order 5 in $Z_{25} \oplus Z_5$. By Theorem 7.1, we may count the number of elements (a, b) in $Z_{25} \oplus Z_5$ with the property that $5 = |(a, b)| = \mathrm{lcm}(|a|, |b|)$. Clearly this requires that either $|a| = 5$ and $|b| = 1$ or 5, or $|b| = 5$ and $|a| = 1$ or 5. We consider three mutually exclusive cases.

CASE 1 $|a| = 5$ and $|b| = 5$.
Here there are four choices for a and four choices for b. This gives 16 elements of order 5.

CASE 2 $|a| = 5$ and $|b| = 1$.
In this case there are four choices for a and only one for b. This gives four more elements of order 5.

CASE 3 $|a| = 1$ and $|b| = 5$.
This time there is one choice for a and there are four choices for b, so we obtain four more elements of order 5.

Thus, $Z_{25} \oplus Z_5$ has 24 elements of order 5. ❑

Example 4 We determine the number of cyclic subgroups of order 10 in $Z_{100} \oplus Z_{25}$. We begin by counting the number of elements (a, b) of order 10.

CASE 1 $|a| = 10$ and $|b| = 1$ or 5.
Since Z_{100} has a unique cyclic subgroup of order 10 and any cyclic group of order 10 has four generators (Theorem 4.4), there are four choices for a. Similarly, there are five choices for b. This gives 20 possibilities for (a, b).

CASE 2 $|a| = 2$ and $|b| = 5$.

Since any finite cyclic group of even order has a unique subgroup of order 2, there is only one choice for a. Obviously, there are four choices for b. So, this case yields four more possibilities for (a, b).

Thus, $Z_{100} \oplus Z_{25}$ has 24 elements of order 10. Because each cyclic subgroup of order 10 has four elements of order 10 and no two of them can have an element of order 10 in common, there must be $24/4 = 6$ cyclic subgroups of order 10. (This method is analogous to determining the number of sheep in a flock by counting legs and dividing by 4.)   ◻

The direct product notation is convenient for specifying certain subgroups of a direct product.

Example 5   Since 5 has order 6 in Z_{30} and 3 has order 4 in Z_{12}, $\langle 5 \rangle \oplus \langle 3 \rangle$ is a subgroup of order 24 in $Z_{30} \oplus Z_{12}$.   ◻

The next theorem and its corollary characterize those direct products of cyclic groups that are themselves cyclic.

Theorem 7.2   *Criterion for $G \oplus H$ to Be Cyclic*
Let G and H be finite cyclic groups. Then $G \oplus H$ is cyclic if and only if $|G|$ and $|H|$ are relatively prime.

Proof.   Let $|G| = m$ and $|H| = n$ so that $|G \oplus H| = mn$. To prove the first half of the theorem, we assume $G \oplus H$ is cyclic and show that m and n are relatively prime. Suppose $\gcd(m, n) = t$ and $t \neq 1$. Let $g \in G$ with $|g| = m$ and $h \in H$ with $|h| = n$. Then both $g^{m/t}$ and $h^{n/t}$ have order t and $\langle (g^{m/t}, e) \rangle$ and $\langle (e, h^{n/t}) \rangle$ are distinct subgroups of order t in the cyclic group $G \oplus H$. Since this contradicts Theorem 4.3, we have $\gcd(m, n) = 1$.

To prove the other half of the theorem, let $G = \langle g \rangle$ and $H = \langle h \rangle$ and suppose $\gcd(m, n) = 1$. Then, $|(g, h)| = \text{lcm}(m, n) = mn = |G \oplus H|$ so that (g, h) is a generator of $G \oplus H$.   ■

As a consequence of Theorem 7.2 and an induction argument, we obtain the following extension of Theorem 7.2.

Corollary 1   *Criterion for $G_1 \oplus G_2 \oplus \cdots \oplus G_n$ to Be Cyclic*
An external direct product $G_1 \oplus G_2 \oplus \cdots \oplus G_n$ of a finite number of finite cyclic groups is cyclic if and only if $|G_i|$ and $|G_j|$ are relatively prime when $i \neq j$.

Corollary 2   *Criterion for $Z_{n_1 n_2 \cdots n_k} \approx Z_{n_1} \oplus Z_{n_2} \oplus \cdots \oplus Z_{n_k}$*
Let $m = n_1 n_2 \cdots n_k$. Then Z_m is isomorphic to $Z_{n_1} \oplus Z_{n_2} \oplus \cdots \oplus Z_{n_k}$ if and only if n_i and n_j are relatively prime when $i \neq j$.

By using the above results in an iterative fashion, one can express the same group (up to isomorphism) in many different forms. For example, we have

$$Z_2 \oplus Z_2 \oplus Z_3 \oplus Z_5 \approx Z_2 \oplus Z_6 \oplus Z_5 \approx Z_2 \oplus Z_{30}.$$

Similarly,

$$Z_2 \oplus Z_2 \oplus Z_3 \oplus Z_5 \approx Z_2 \oplus Z_6 \oplus Z_5$$
$$\approx Z_2 \oplus Z_3 \oplus Z_2 \oplus Z_5 \approx Z_6 \oplus Z_{10}.$$

Thus, $Z_2 \oplus Z_{30} \approx Z_6 \oplus Z_{10}$. Note, however, that $Z_2 \oplus Z_{30} \not\approx Z_{60}$.

THE GROUP OF UNITS MODULO n AS AN EXTERNAL DIRECT PRODUCT

The U-groups provide a convenient way to illustrate the preceding ideas. We first introduce some notation. If k is a divisor of n, let

$$U_k(n) = \{x \in U(n) \mid x = 1 \bmod k\}.$$

For example, $U_7(105) = \{1, 8, 22, 29, 43, 64, 71, 92\}$. It can be readily shown that $U_k(n)$ is indeed a subgroup of $U(n)$. (See exercise 12 in Chapter 3.)

Theorem 7.3 $U(n)$ *as an External Direct Product*
Suppose s and t are relatively prime. Then $U(st)$ is isomorphic to the external direct product of $U(s)$ and $U(t)$. In short,

$$U(st) \approx U(s) \oplus U(t).$$

Moreover, $U_s(st)$ is isomorphic to $U(t)$ and $U_t(st)$ is isomorphic to $U(s)$.

Proof.   An isomorphism from $U(st)$ to $U(s) \oplus U(t)$ is $x \rightarrow (x \bmod s, x \bmod t)$; an isomorphism from $U_s(st)$ to $U(t)$ is $x \rightarrow x \bmod t$; an isomorphism from $U_t(st)$ to $U(s)$ is $x \rightarrow x \bmod s$. We leave the verification that these mappings are well-defined, operation-preserving, and one-to-one to the reader. (See exercises 14, 15, and 16 in Chapter 0; see also [1].)   ∎

As a consequence of Theorem 7.3, we have the following result.

Corollary   Let $m = n_1 n_2 \cdots n_k$, where $\gcd(n_i, n_j) = 1$ for $i \neq j$. Then,

$$U(m) \approx U(n_1) \oplus U(n_2) \oplus \cdots \oplus U(n_k).$$

To see how these results work, let's apply them to $U(105)$. We obtain

$$U(105) \approx U(7) \oplus U(15)$$
$$U(105) \approx U(21) \oplus U(5)$$
$$U(105) \approx U(3) \oplus U(5) \oplus U(7)$$

Moreover,

$U(7) \approx U_{15}(105) = \{1, 16, 31, 46, 61, 76\}$

$U(15) \approx U_7(105) = \{1, 8, 22, 29, 43, 64, 71, 92\}$

$U(21) \approx U_5(105) = \{1, 11, 16, 26, 31, 41, 46, 61, 71, 76, 86, 101\}$

$U(5) \approx U_{21}(105) = \{1, 22, 43, 64\}$

$U(3) \approx U_{35}(105) = \{1, 71\}$

Among all groups, surely the cyclic groups Z_n have the simplest structures and, at the same time, are the easiest groups with which to compute. Direct products of groups of the form Z_n are only slightly more complicated in structure and computability. Because of this, algebraists endeavor to describe a finite Abelian group as such a direct product. Indeed, we shall soon see that every finite Abelian group can be so represented. With this goal in mind, let us reexamine the U-groups. Using the corollary to Theorem 7.3 and the fact (see [2, p. 93]), first proved by Carl Gauss in 1801, that

$$U(2) \approx \{1\}, \qquad U(4) \approx Z_2, \qquad U(2^n) \approx Z_2 \oplus Z_{2^{n-2}} \qquad \text{for } n \geq 3,$$

and

$$U(p^n) \approx Z_{p^n - p^{n-1}} \qquad \text{for } p \text{ an odd prime,}$$

we now can write U-groups as an external direct product of cyclic groups. For example,

$$U(105) = U(3 \cdot 5 \cdot 7) \approx U(3) \oplus U(5) \oplus U(7)$$
$$\approx Z_2 \oplus Z_4 \oplus Z_6$$

and

$$U(720) = U(16 \cdot 9 \cdot 5) \approx U(16) \oplus U(9) \oplus U(5)$$
$$\approx Z_2 \oplus Z_4 \oplus Z_6 \oplus Z_4.$$

What is the advantage of expressing a group in this form? Well, for one thing, we immediately see that the orders of the elements $U(720)$ can only be 1, 2, 3, 4, 6, or 12. For another thing, we can readily determine the number of elements of order 12, say, that $U(720)$ has. Because $U(720)$ is isomorphic to $Z_2 \oplus Z_4 \oplus Z_6 \oplus Z_4$, it suffices to calculate the number of elements of order 12 in $Z_2 \oplus Z_4 \oplus Z_6 \oplus Z_4$, and this is easy. By Theorem 7.1, an element (a, b, c, d) has order 12 if and only if $|c| = 3$ or 6, and $|b| = 4$ or $|c| = 3$ or 6, and $|d| = 4$. So, in the first case, we may take c to be 1, 2, 4, or 5 and b to be 1 or 3. Thus, there are eight choices in all for b and c. Once these are made, we may choose a and d arbitrarily. This is eight more choices. Thus, there are 64 elements of the form (a, b, c, d) where $|c| = 3$ or 6 and $|b| = 4$.

In addition to these 64, we must include elements of the form (a, b, c, d) where $|c| = 3$ or 6, $|d| = 4$, $|b| = 1$ or 2, and a is arbitrary. Here there are $4 \cdot 2 \cdot 2 \cdot 2 = 32$ choices. So, altogether, $U(720)$ has

96 elements of order 12. These calculations tell us more. Since Aut(Z_{720}) is isomorphic to $U(720)$, we also know that there are 96 automorphisms of Z_{720} of order 12. Imagine trying to deduce this information directly from $U(720)$ or, worse yet, from Aut(Z_{720})! These results beautifully illustrate the advantage of being able to represent a finite Abelian group as a direct product of cyclic groups. They also illustrate the value of our theorems about Aut(Z_n) and $U(n)$. After all, theorems are labor-saving devices. If you want to convince yourself of this, try to prove directly from the definitions that Aut(Z_{720}) has exactly 96 elements of order 12.

APPLICATIONS

Here is an application of U-groups to number theory.

Example 6   We determine the last two digits of 23^{123}. First observe that the value we seek is simply 23^{123} mod 100 and that 23 belongs to $U(100)$. Since $U(100) \approx U(4) \oplus U(25) \approx Z_2 \oplus Z_{20}$, we see that $x^{20} = 1$ for all x in $U(100)$. Thus, modulo 100, $23^{123} = (23)^{120}23^3 = 23^3 = 23^2 \cdot 23 = 29 \cdot 23 = 67$. ❑

In many situations there is a desire for security against unauthorized interpretation of coded data, the most obvious being military and diplomatic transmissions. Premium television services such as Home Box Office (HBO), Showtime, and the Disney channel also have a need to protect their television signals to local cable operators and satellite dish subscribers from being received free by dish owners.

We conclude this chapter with two applications of the material presented here to cryptography—the science of sending and deciphering secret messages.

DATA SECURITY

Because computers are built from two-state electronic components, it is natural to represent information as strings of 0's and 1's called *binary strings*. A binary string of length n can naturally be thought of as an element of $Z_2 \oplus Z_2 \oplus \cdots \oplus Z_2$ (n copies) where the parentheses and the commas have been deleted. Thus the binary string 11000110 corresponds to the element $(1, 1, 0, 0, 0, 1, 1, 0)$ in $Z_2 \oplus Z_2 \oplus Z_2 \oplus Z_2 \oplus Z_2 \oplus Z_2 \oplus Z_2 \oplus Z_2$. Similarly, two binary strings $a_1a_2 \cdots a_n$ and $b_1b_2 \cdots b_n$ are added componentwise modulo 2 just as their corresponding elements in $Z_2 \oplus Z_2 \oplus \cdots \oplus Z_2$ are. For example,

$$11000111 + 01110110 = 10110001$$

and

$$10011100 + 10011100 = 00000000.$$

The fact that the sum of two binary sequences $a_1 a_2 \cdots a_n + b_1 b_2 \cdots b_n = 00 \cdots 0$ if and only if the sequences are identical is the basis for a data security system used by HBO to protect its television signals.

Beginning in 1984, HBO scrambled its signal. To unscramble the signal, a cable system operator or dish owner who pays a monthly fee has to have a password that is changed monthly. The password is transmitted along with the scrambled signal and scrambled versions of the password. The technical term for the scrambling process is *encryption*. Although HBO uses binary sequences of length 56, we will illustrate the method with sequences of length 8. Let us say that the password for this month is p. Each authorized user of the service is assigned a sequence uniquely associated with him or her. Let us call these the *keys* and label them $k_1, k_2, \ldots$. HBO transmits the password p, the scrambled signal, and the encrypted sequences $k_1 + p, k_2 + p, \cdots$ (that is, one sequence for each authorized user). A microprocessor in each subscriber's decoding box simply adds its key, say k_i, to each of the encrypted sequences. That is, it calculates $k_i + (k_1 + p), k_i + (k_2 + p), \ldots$. The microprocessor then compares each of these calculated sequences with the correct password p. When one of the sequences matches p, the microprocessor can unscramble the signal. Notice that the correct password p will be produced precisely when k_i is added to $k_i + p$ since $k_i + (k_i + p) = (k_i + k_i) + p = 00 \cdots 0 + p = p$ and $k_i + (k_j + p) \neq p$ when $k_j \neq k_i$. If a subscriber with key k_i fails to pay the monthly bill, HBO can terminate the subscriber's service by not transmitting the sequence $k_i + p$ the next month.

To illustrate, let us say that the password for this month is $p = 10101100$ and your key is $k = 00111101$. One of the sequences transmitted by HBO is $k + p$: $00111101 + 10101100 = 10010001$. Your decoder box adds your key $k = 00111101$ to all the sequences received. Eventually, it finds the sequence obtained by adding the password to your key (namely, $p + k = 10010001$) and calculates $00111101 + 10010001 = 10101100$ to obtain the password p. This password then permits the decoder to unscramble the TV signal.

One might suspect that a computer hacker could find the password by simply trying a large number of possible keys until one "unlocks" the password. However, with sequences of length 56, there are 2^{56} possible keys, though perhaps only a few million are used by HBO to unlock its monthly password, and the number 2^{56} is so large (it exceeds 72 quadrillion) that, even if one tries a billion possible keys, the chance of finding one that works is essentially 0.

APPLICATION TO PUBLIC KEY CRYPTOGRAPHY

In the mid-1970s, Ron Rivest, Adi Shamir, and Len Adelman devised an ingenious method that permits each person that is to receive a secret message to publicly tell how to scramble messages sent to him or her. And even though the method used to scramble the message is known publicly, only the person for whom it is intended will be able to unscramble the message. The idea is based on the facts that there exist efficient methods for finding very large prime numbers (say about 100 digits long) and for multiplying large numbers, but no one knows an efficient algorithm for factoring large integers (say about 200 digits long). So, the person who is to receive the message finds a pair of large primes p and q and chooses an integer r with $1 < r < m$, where $m = \text{lcm}(p - 1, q - 1)$ so that r is relatively prime to m (any such r will do). This person calculates $n = pq$ and announces that a message M is to be sent to him or her publicly as M^r mod n. Although r, n, and M^r are available to everyone, only the person who knows how to factor n as pq will be able to decipher the message.

To present a simple example that nevertheless illustrates the principal features of the method, say we wish to send the message "YES." We convert the message into a string of digits by replacing A by 01, B by 02, . . . , Z by 26, and a blank by 00. So, the message YES becomes 250519. To keep the numbers involved from becoming too unwieldy, we send the message in blocks of four digits and fill in with blanks when needed. Thus, the message YES is represented by the two blocks 2505 and 1900. The person to whom the message is to be sent has picked two primes p and q, say $p = 37$ and $q = 73$ (in actual practice, p and q would have 100 or so digits), and a number r that has no prime divisors in common with $\text{lcm}(p - 1, q - 1) = 72$, say $r = 5$, and has published this information in a public directory. We will send the "scrambled" numbers $(2505)^5$ mod 2701 and $(1900)^5$ mod 2701 rather than 2505 and 1900 and the receiver will unscramble them. We show the work involved for us and the receiver only for the block 2505. The arithmetic involved in computing these numbers is simplified as follows:

$$2505 \bmod 2701 = 2505$$
$$(2505)^2 \bmod 2701 = 602$$
$$(2505)^4 \bmod 2701 = (602)(602) \bmod 2701 = 470.$$

So, $(2505)^5 \bmod 2701 = (2505)(470) \bmod 2701 = 2415$.

Thus the number 2415 is sent to the receiver. Now the receiver must take the number he or she receives, 2415, and convert it back to 2505. To do so, the receiver takes the two factors of 2701, $p = 37$ and $q = 73$, and calculates the least common multiple of $p - 1 = 36$ and $q - 1 = 72$, which is 72. (This is where the knowledge of p and q is necessary.) Next, the receiver must find $s = r^{-1}$ in $U(72)$—that is, solve the equation $5 \cdot s = 1 \mod 72$. This number is 29. (There is a simple algorithm for finding this number.) Then the receiver takes the number received, 2415, and calculates $(2415)^{29} \mod 2701$. This calculation can be simplified as follows:

$$2415 \mod 2701 = 2415$$
$$(2415)^2 \mod 2701 = 766 \mod 2701$$
$$(2415)^4 \mod 2701 = (766)^2 \mod 2701 = 639 \mod 2701$$
$$(2415)^8 \mod 2701 = (639)^2 \mod 2701 = 470 \mod 2701$$
$$(2415)^{16} \mod 2701 = (470)^2 \mod 2701 = 2119 \mod 2701.$$

So, $(2415)^{29} \mod 2701 = (2415)^{16}(2415)^8(2415)^4(2415) \mod 2701 = (2119)(470)(639)(2415) \mod 2701 = (2119)(470) \mod 2701 \times (639)(2415) \mod 2701 = (1962)(914) \mod 2701 = 2505$. [We compute the product $(2119)(470)(639)(2415)$ in two stages so that we may use a hand calculator.]

Thus the receiver correctly determines the code for "YE." On the other hand, without knowing how pq factors, one cannot find the modulus (in our case, 72) that is needed to determine the intended message.

The procedure described above is called the *RSA public key encryption scheme* in honor of the three people (Rivest, Shamir, and Adelman) who discovered the method. The algorithm is summarized below.

Receiver

1. Pick very large primes p and q and compute $n = pq$.
2. Compute the least common multiple of $p - 1$ and $q - 1$, let us call it m.
3. Pick r relatively prime to m.
4. Find s so that $rs = 1$ modulo m.
5. Disregard p and q.
6. Publicly announce n and r.

Sender

1. Convert the message to a string of digits. (In practice, the ASCII code is used.)
2. Break up the message into uniform blocks of digits; call them M_1, $M_2, \ldots, M_k$.
3. Check to see that the greatest common divisor of each M_i and n is 1. If not, n can be factored and our code is broken.
4. Calculate and send $R_i = M_i^r \mod n$.

Receiver

1. For each received message R_i, calculate $R_i^s \mod n$.
2. Convert the string of digits back to a string of characters.

Why does this method work? Well, we know that $U(n) \approx U(p) \oplus U(q) \approx Z_{p-1} \oplus Z_{q-1}$. Thus an element of the form x^m in $U(n)$ corresponds under an isomorphism to one of the form (mx_1, mx_2) in $Z_{p-1} \oplus Z_{q-1}$. Since m is the least common multiple of $p - 1$ and

$q - 1$, we may write $m = u(p - 1)$ and $m = v(q - 1)$ for so
v. Then $(mx_1, mx_2) = (u(p - 1)x_1, v(q - 1)x_2) =$
$Z_{p-1} \oplus Z_{q-1}$ and it follows that $x^m = 1$ for all x in $U(n)$. So, becau~~
each message M_i is an element of $U(n)$ and r was chosen so that $rs = 1$
$+ tm$ for some t, we have, modulo n,

$$R_i^s = (M_i^r)^s = M_i^{rs} = M_i^{1+tm} = (M_i^m)^t M_i = 1^t M_i = M_i.$$

EXERCISES

It is true that Fourier has the opinion that the principal object of mathematics is the public utility and the explanation of natural phenomena; but a scientist like him ought to know that the unique object of science is the honor of the human spirit and on this basis a question of the theory of numbers is worth as much as a question about the planetary system.

C. J. Jacobi

1. Find the order of each element in $Z_2 \oplus Z_4$.
2. Show that $G \oplus H$ is Abelian if and only if G and H are Abelian.
3. Show that $Z_2 \oplus Z_2 \oplus Z_2$ has seven subgroups of order 2.
4. Determine the subgroup lattice of $Z_2 \oplus Z_2$.
5. Prove or disprove that $Z \oplus Z$ is a cyclic group.
6. Prove, by comparing orders of elements, that $Z_8 \oplus Z_2$ is not isomorphic to $Z_4 \oplus Z_4$.
7. Prove that $G_1 \oplus G_2$ is isomorphic to $G_2 \oplus G_1$.
8. Is $Z_3 \oplus Z_9$ isomorphic to Z_{27}? Why?
9. Is $Z_3 \oplus Z_5$ isomorphic to Z_{15}? Why?
10. How many elements of order 9 does $Z_3 \oplus Z_9$ have? (Do not do this exercise by brute force.)
11. Explain why $Z_8 \oplus Z_4$ and $Z_{8000000} \oplus Z_{4000000}$ must have the same number of elements of order 4.
12. The dihedral group D_n of order $2n$ has a subgroup of n rotations and a subgroup of order 2. Explain why D_n cannot be isomorphic to the external direct product of two such groups.
13. Prove that the group of complex numbers under addition is isomorphic to $\mathbf{R} \oplus \mathbf{R}$.
14. Suppose that $G_1 \approx G_2$ and $H_1 \approx H_2$. Prove that $G_1 \oplus H_1 \approx G_2 \oplus H_2$.
15. Construct a Cayley table for $Z_2 \oplus Z_3$.
16. How many elements of order 4 does $Z_4 \oplus Z_4$ have? (Do not do this exercise by checking each member of $Z_4 \oplus Z_4$.)
17. What is the order of any nonidentity element of $Z_3 \oplus Z_3 \oplus Z_3$?
18. How many subgroups of order 4 does $Z_4 \oplus Z_2$ have?

19. Let M be the group of all real 2×2 matrices under addition. Let $N = \mathbf{R} \oplus \mathbf{R} \oplus \mathbf{R} \oplus \mathbf{R}$ under componentwise addition. Prove that M and N are isomorphic. What is the corresponding theorem for the group of $n \times n$ matrices under addition?

20. The group $S_3 \oplus Z_2$ is isomorphic to which of the following groups: Z_{12}, $Z_6 \oplus Z_2$, A_4, D_6?

21. Let G be a group, and let $H = \{(g, g) \mid g \in G\}$. Show that H is a subgroup of $G \oplus G$. (This subgroup is called the *diagonal* of $G \oplus G$.) When G is the set of real numbers under addition, describe $G \oplus G$ and H geometrically.

22. Find a subgroup of $Z_4 \oplus Z_2$ that is not of the form $H \oplus K$ where H is a subgroup of Z_4 and K is a subgroup of Z_2.

23. Find all subgroups of order 3 in $Z_9 \oplus Z_3$.

24. Find all subgroups of order 4 in $Z_4 \oplus Z_4$.

25. What is the largest order of any element in $Z_{30} \oplus Z_{20}$?

26. How many elements of order 2 are in $Z_{2000000} \oplus Z_{4000000}$?

27. Find a subgroup of $Z_{800} \oplus Z_{200}$ that is isomorphic to $Z_2 \oplus Z_4$.

28. Find a subgroup of $Z_{12} \oplus Z_4 \oplus Z_{15}$ that is of order 9.

29. Show that G is isomorphic to a subgroup of $G \oplus H$.

30. Let

$$H = \left\{ \begin{bmatrix} 1 & a & b \\ 0 & 1 & 0 \\ 0 & 0 & 1 \end{bmatrix} \Bigg| \, a, b \in Z_3 \right\}.$$

(See exercise 40 in Chapter 2 for the definition of multiplication.) Show that H is an Abelian group of order 9. Is H isomorphic to Z_9 or $Z_3 \oplus Z_3$?

31. Let $G = \{3^m 6^n \mid m, n \in Z\}$ under multiplication. Prove that G is isomorphic to $Z \oplus Z$.

32. Let $(a_1, a_2, \ldots, a_n) \in G_1 \oplus G_2 \oplus \cdots \oplus G_n$. Give a necessary and sufficient condition for $|(a_1, a_2, \ldots, a_n)| = \infty$.

33. Prove that $D_3 \oplus D_4 \not\approx D_{24}$.

34. Determine the number of cyclic subgroups of order 15 in $Z_{90} \oplus Z_{36}$.

35. For any Abelian group G and any positive integer n, let $G^n = \{g^n \mid g \in G\}$ (see exercise 13, Supplementary Exercises for Chapters 1–4). If H and K are Abelian, show that $(H \oplus K)^n = H^n \oplus K^n$.

36. Suppose G is a group of order 4 and $x^2 = e$ for all x in G. Prove that G is isomorphic to $Z_2 \oplus Z_2$. (This exercise is referred to in Chapter 8.)

37. If a finite Abelian group has exactly 24 elements of order 6, how many cyclic subgroups of order 6 does it have?

38. The group $Z_2 \oplus D_3$ is isomorphic to one of the following: Z_{12}, $Z_2 \oplus Z_2 \oplus Z_3$, A_4, D_6. Determine which one by elimination.

39. Determine the number of elements of order 15 and the number of cyclic subgroups of order 15 in $Z_{30} \oplus Z_{20}$.

40. Express $U(165)$ as an external direct product of cyclic additive groups of the form Z_n.

41. Express $U(165)$ as an external direct product of U-groups in three different ways.

42. Without doing any calculations in $\text{Aut}(Z_{20})$, determine how many elements of $\text{Aut}(Z_{20})$ have order 4. How many have order 2?

43. Without doing any calculations in $\text{Aut}(Z_{720})$, determine how many elements of $\text{Aut}(Z_{720})$ have order 6.

44. Without doing any calculating in $U(27)$, decide how many subgroups $U(27)$ has.

45. What is the largest order of any element in $U(900)$?

46. Let p and q be odd primes and m and n be positive integers. Explain why $U(p^m) \oplus U(q^n)$ is not cyclic.

47. Use the results presented in this chapter to prove that $U(55)$ is isomorphic to $U(75)$.

48. Use the results presented in this chapter to prove that $U(144)$ is isomorphic to $U(140)$.

49. For every $n > 2$, prove that $U(n)^2 = \{x^2 \mid x \in U(n)\}$ is a proper subgroup of $U(n)$.

50. Show that $U(55)^3 = \{x^3 \mid x \in U(55)\}$ is $U(55)$.

51. Find an integer n so that $U(n)$ contains a subgroup isomorphic to $Z_5 \oplus Z_5$.

52. Find a subgroup of order 6 in $U(700)$.

53. Show that there is a U-group containing a subgroup isomorphic to $Z_3 \oplus Z_3$.

54. Show that no U-group has order 14.

55. Show that there is a U-group containing a subgroup isomorphic to Z_{14}.

56. Show that no U-group is isomorphic to $Z_4 \oplus Z_4$.

57. Show that there is a U-group containing a subgroup isomorphic to $Z_4 \oplus Z_4$.

58. Using the RSA scheme with $p = 37$, $q = 73$, and $r = 5$, what number would be sent for the message "RM"?

59. Assuming that a message has been sent via the RSA scheme with $p = 37$, $q = 73$, and $r = 5$, decode the received message "34."

PROGRAMMING EXERCISES

> In a few minutes, a computer can make a mistake so great that it would take many men many months to equal it.
>
> **Merle L. Meacham**

1. Write a program that will print out the Cayley table for $Z_m \oplus Z_n$. Assume $m \leq 8$ and $n \leq 8$. Run your program for $m = 5$, $n = 6$; $m = 3$, $n = 5$; and $m = 4$, $n = 8$.

2. Write a program that will compute the order of any element of $Z_m \oplus Z_n$. Run your program for $(2, 3)$, $(4, 6)$, and $(3, 2)$ from $Z_{12} \oplus Z_{15}$.

3. Let $U(n)^k = \{x^k \mid x \in U(n)\}$. Write a program that computes $U(n)^k$. Make reasonable assumptions about n and k. Run your program for $(n, k) = (45, 3)$, $(65, 3)$, $(144, 3)$, $(200, 2)$, and $(200, 5)$. Do you see a relationship between $|U(n)|$ and k? Use the theory developed in this chapter to analyze these groups as external direct products of cyclic groups of the form Z_n.

4. Implement the algorithm given on page 122 to express $U(n)$ as an external direct product of groups of the form Z_k. Assume that n is given in prime-power factorization form. Run your program for $3 \cdot 5 \cdot 7$, $16 \cdot 9 \cdot 5$, $8 \cdot 3 \cdot 25$, $9 \cdot 5 \cdot 11$, and $2 \cdot 27 \cdot 125$.

REFERENCES

1. J. A. Gallian and D. Rusin, "Factoring Groups of Integers Modulo n," *Mathematics Magazine* 53 (1980): 33–36.
2. D. Shanks, *Solved and Unsolved Problems in Number Theory,* 2d ed., New York: Chelsea, 1978.

SUGGESTED READINGS

Richard W. Ball, "On the Order of an Element in a Group," *American Mathematical Monthly* 71 (1964): 784–785.

Several results regarding the orders of elements of a group are given.

Y. Cheng, "Decompositions of U-groups," *Mathematics Magazine* 62 (1989): 271–273.

This article explores the decomposition of $U(st)$, where s and t are relatively prime, in greater detail than we have provided.

SUPPLEMENTARY EXERCISES FOR CHAPTERS 5–7

My mind rebels at stagnation. Give me problems, give me work, give me the most obstruse cryptogram, or the most intricate analysis, and I am in my own proper atmosphere.

Sherlock Holmes, *The Sign of the Four*

1. A subgroup N of a group G is called a *characteristic subgroup* if $\phi(N) = N$ for all automorphisms ϕ of G. (The term *characteristic* was coined by G. Frobenius in 1895.) Prove that every subgroup of a cyclic group is characteristic.

2. Prove that the center of a group is characteristic.

3. The *commutator subgroup* G' of a group G is the subgroup generated by the set $\{x^{-1}y^{-1}xy \mid x, y \in G\}$. (That is, every element of G' has the form $a_1^{i_1} a_2^{i_2} \cdots a_k^{i_k}$, where each a_j has the form $x^{-1}y^{-1}xy$, each $i_j = \pm 1$, and k is any positive integer.) Prove that G' is a characteristic subgroup of G. This subgroup was first introduced by G. A. Miller in 1898.

4. Prove that the characteristic property is transitive. That is, if N is a characteristic subgroup of K and K is a characteristic subgroup of G, then N is a characteristic subgroup of G.

5. Let $G = Z_3 \oplus Z_3 \oplus Z_3$ and let H be the subgroup of $SL(3, Z_3)$, consisting of

$$
\left\{ \begin{bmatrix} 1 & a & b \\ 0 & 1 & c \\ 0 & 0 & 1 \end{bmatrix} \;\middle|\; a, b, c \in Z_3 \right\}.
$$

(See exercise 40 in Chapter 2 for the definition of multiplication.) Determine the number of elements of each order in G and H. Are G and H isomorphic?

6. Let H and K be subgroups of G. Prove that HK is a group if and only if $HK = KH$.

7. Let H and K be subgroups of a finite group G. Prove that

$$
|HK| = \frac{|H|\,|K|}{|H \cap K|}.
$$

(This exercise is referred to in Chapters 11 and 24.)

8. The *exponent* of a group is the smallest positive integer n such that $x^n = e$ for all x in the group. Prove that every finite Abelian group has an exponent that divides the order of the group.

9. Determine all U-groups of exponent 2.

10. Can a group have more subgroups than it has elements?

11. Let $\mathbf{R}^+$ denote the multiplicative group of positive reals and let $T = \{z \in \mathbf{C} \mid |z| = 1\}$ be the multiplicative group of complex numbers of norm 1. (Recall that $|a + bi| = \sqrt{a^2 + b^2}$.) Show that every element of $\mathbf{C}^\#$ can be uniquely expressed in the form of rz where $r \in \mathbf{R}^+$ and $z \in T$.

12. Prove that $Q^\#$ under multiplication is not isomorphic to $\mathbf{R}^\#$ under multiplication.

13. Prove that Q under addition is not isomorphic to $\mathbf{R}$ under addition.

14. Prove that $\mathbf{R}$ under addition is not isomorphic to $\mathbf{R}^\#$ under multiplication.

15. Show that Q^+ (the set of positive rational numbers) under multiplication is not isomorphic to Q under addition.

16. Suppose that $G = \{e, x, x^2, y, yx, yx^2\}$ is a non-Abelian group with $|x| = 3$ and $|y| = 2$. Show that $xy = yx^2$.

17. Give an example of a noncyclic group G with the property that every non-identity element of G has order 5.

18. Let G be an Abelian group under addition. Let n be a fixed positive integer and let $H = \{(g, ng) \mid g \in G\}$. Show that H is a subgroup of $G \oplus G$. When G is the set of real numbers under addition, describe H geometrically.

19. Find a subgroup of $Z_{12} \oplus Z_{20}$ isomorphic to $Z_4 \oplus Z_5$.

20. Suppose that $G = \oplus_{i=1}^n G_i$. Prove that $Z(G) = \oplus_{i=1}^n Z(G_i)$.

21. Give an example of a group G with a proper subgroup H such that G and H are isomorphic.

22. What is the order of the largest cyclic subgroup in $\mathrm{Aut}(Z_{720})$? (*Hint:* It is not necessary to consider automorphisms of Z_{720}.)

23. Let $\beta = (1, 3, 5, 7, 9)(2, 4, 6)(8, 10)$. If β^m is a 5-cyclic, what can be said about m?

24. Let G be a group and let $g \in G$. Show that $Z(G)\langle g \rangle$ is a subgroup of G.

25. Show that $D_{11} \oplus Z_3 \not\approx D_3 \oplus Z_{11}$. (This exercise is referred to in Chapter 24.)

26. Show that $D_{33} \not\approx D_{11} \oplus Z_3$. (This exercise is referred to in Chapter 24.)

27. Show that $D_{33} \not\approx D_3 \oplus Z_{11}$. (This exercise is referred to in Chapter 24.)

28. Exhibit four nonisomorphic groups of order 66. (This exercise is referred to in Chapter 24.)

29. Prove that $|\mathrm{Inn}(G)| = 1$ if and only if G is Abelian.

30. Prove that $x^{100} = 1$ for all x in $U(1000)$.

31. Find a subgroup of order 6 in $U(450)$.

32. List four elements of $Z_{20} \oplus Z_5 \oplus Z_{60}$ that form a noncyclic subgroup.

33. In S_{10}, let $\beta = (13)(17)(265)(289)$. Find an element in S_{10} that commutes with β but is not a power of β.

34. Prove or disprove that $Z_4 \oplus Z_{15} \approx Z_6 \oplus Z_{10}$.

35. Prove or disprove that $D_{12} \approx Z_3 \oplus D_4$.

36. Describe a three-dimensional solid whose symmetry group is isomorphic to D_5.

37. Let $G = U(15) \oplus Z_{10} \oplus S_5$. Find the order of $(2, 3, (123)(15))$. Find the inverse of $(2, 3, (123)(15))$.

38. Let $G = Z \oplus Z_{10}$ and let $H = \{g \in G \mid |g| = \infty \text{ or } |g| = 1\}$. Prove or disprove that H is a subgroup of G.

8

Cosets and Lagrange's Theorem

The next theorem, attributed to J. L. Lagrange, is of
fundamental importance for it introduces arithmetic
relationships into group theory.

Richard A. Dean, *Elements of Abstract Algebra*

PROPERTIES OF COSETS

In this chapter, we will prove the single most important theorem in finite
group theory—Lagrange's theorem. But first, we introduce a new and
powerful tool for analyzing a group—the notion of a coset. This notion
was invented by Galois in 1830, although the term was coined by G. A.
Miller in 1910.

DEFINITION   Coset of H in G

Let G be a group and let H be a subgroup of G. For any $a \in G$, the set
$aH = \{ah \mid h \in H\}$ is called the *left coset of H in G containing a*. Analogously, $Ha = \{ha \mid h \in H\}$ is called the *right coset of H in G containing
a*. The element a is called the *coset representative of aH* (or Ha).

Example 1   Let $G = S_3$ and $H = \{(1), (13)\}$. Then the left cosets of H are:

$$(1)H = H,$$
$$(12)H = \{(12), (12)(13)\} = \{(12), (132)\} = (132)H,$$
$$(13)H = \{(13), (1)\} = H,$$
$$(23)H = \{(23), (23)(13)\} = \{(23), (123)\} = (123)H. \qquad \square$$

Example 2   Let $\mathcal{H} = \{R_0, R_{180}\}$ in D_4, the dihedral group of order 8. Then,

$$R_0\mathcal{H} = \mathcal{H},$$
$$R_{90}\mathcal{H} = \{R_{90}, R_{270}\} = R_{270}\mathcal{H},$$
$$R_{180}\mathcal{H} = \{R_{180}, R_0\} = \mathcal{H},$$
$$V\mathcal{H} = \{V, VR_{180}\} = \{V, H\} = H\mathcal{H},$$
$$D\mathcal{H} = \{D, D'\} = D'\mathcal{H}. \qquad \square$$

Example 3   Let $H = \{0, 3, 6\}$ in Z_9 under addition. In the case that the group operation is addition, we use the notation $a + H$ instead of aH. Then the cosets of H in Z_9 are

$$0 + H = \{0, 3, 6\} = 3 + H = 6 + H,$$
$$1 + H = \{1, 4, 7\} = 4 + H = 7 + H,$$
$$2 + H = \{2, 5, 8\} = 5 + H = 8 + H. \qquad \square$$

The three examples above illustrate a few facts about cosets that are worthy of our attention. First, cosets are usually not subgroups. Second, aH may be the same as bH, even though a is not the same as b. Third, since $(12)H = \{(12), (132)\}$ whereas $H(12) = \{(12), (123)\}$, aH need not be the same as Ha.

These examples and observations raise many questions. When does $aH = bH$? Do aH and bH have any elements in common? When does $aH = Ha$? Which cosets are subgroups? Why are cosets important? The next lemma and theorem answer these questions. (Analogous results hold for right cosets.)

Lemma   *Properties of Cosets*
Let H be a subgroup of G, and let a and b belong to G. Then,

1. *$a \in aH$,*
2. *$aH = H$ if and only if $a \in H$,*
3. *$aH = bH$ or $aH \cap bH = \emptyset$,*
4. *$aH = bH$ if and only if $a^{-1}b \in H$,*
5. *$|aH| = |bH|$,*
6. *$aH = Ha$ if and only if $H = aHa^{-1}$,*
7. *aH is a subgroup of G if and only if $a \in H$.*

Proof.

1. $a = ae \in aH$.
2. Property 2 is left as an exercise.
3. To prove property 3, we suppose that $aH \cap bH \neq \emptyset$ and prove that $aH = bH$. Let $x \in aH \cap bH$. Then there exist h_1, h_2 in H such that $x = ah_1$ and $x = bh_2$. Thus, $a = bh_2h_1^{-1}$ and $aH = bh_2h_1^{-1}H = bH$, by property 2.
4. Observe that $aH = bH$ if and only if $H = a^{-1}bH$. The result now follows from property 2.
5. We leave it as an exercise for the student to prove that the correspondence $ah \to bh$ for all h in H is a one-to-one, onto function from aH to bH.
6. Note that $aH = Ha$ if and only if $(aH)a^{-1} = (Ha)a^{-1}$—that is, if and only if $aHa^{-1} = H$.
7. If aH is a subgroup, then it contains the identity e. Thus, $aH \cap eH \neq \emptyset$; and, by property 3, we have $aH = eH = H$. Thus, from property 2, we have $a \in H$. Conversely, if $a \in H$, then, again by property 2, $aH = H$. ∎

Although most mathematical theorems are written in symbolic form, one should also know what they say *in words*. In the preceding lemma, property 1 says simply that the left coset of H containing a *does* contain a. Property 2 says that the H "absorbs" an element if and only if the element belongs to H. Property 3 says—and this is very important—that two left cosets of H are either identical or disjoint. Thus, a left coset of H is uniquely determined by any one of its elements. In particular, any element of a left coset can be used to represent the coset. Property 4 shows how we may transfer a question about equality of left cosets of H to a question about H itself and vice versa. Property 5 says that all left cosets of H have the same size. Property 6 is analogous to property 4 in that it shows how a question about the equality of the left and right cosets of H containing a is equivalent to a question about the equality of two subgroups of G. The last property of the lemma says that H itself is the only coset of H that is a subgroup of G.

Note that properties 1, 3, and 5 of the lemma guarantee that the left cosets of a subgroup H of G partition G into blocks of equal size. In practice, the subgroup H is often chosen so that the cosets partition the group in some highly desirable fashion. For example, if G is $\mathbf{R}^3$ (i.e., $\mathbf{R} \oplus \mathbf{R} \oplus \mathbf{R}$) and H is a plane through the origin, then the coset $(a, b, c) + H$ is the plane passing through the point (a, b, c) and parallel to H. Thus, the cosets of H constitute a partition of 3-space into planes parallel to H. If $G = GL(2, \mathbf{R})$ and $H = SL(2, \mathbf{R})$, then for any matrix A in G, the coset AH is the set of *all* 2×2 matrices with the same determinant as A. Thus,

$$\begin{bmatrix} 2 & 0 \\ 0 & 1 \end{bmatrix} H \qquad \text{is the set of all } 2 \times 2 \text{ matrices of determinant 2}$$

and

$$\begin{bmatrix} 1 & 2 \\ 2 & 1 \end{bmatrix} H \qquad \text{is the set of all } 2 \times 2 \text{ matrices of determinant } -3.$$

LAGRANGE'S THEOREM AND CONSEQUENCES

We are now ready to prove a theorem that has been around for over 200 years—longer than group theory itself! (This theorem was not originally stated in group theoretic terms.) At this stage, it should come as no surprise.

Theorem 8.1   *Lagrange's Theorem*: $|H|$ Divides $|G|$*
> *If G is a finite group and H is a subgroup of G, then $|H|$ divides $|G|$. Moreover, the number of distinct left (right) cosets of H in G is $|G|/|H|$.*

> **Proof.**   Let $a_1H, a_2H, \ldots, a_rH$ denote the distinct left cosets of H in G. Then, for each a in G, we have $aH = a_iH$ for some i. Also, by property 1 of the lemma, $a \in aH$. Thus, each member of G belongs to one of the cosets a_iH. In symbols,
>
> $$G = a_1H \cup a_2H \cup \cdots \cup a_rH.$$
>
> Now, property 3 of the lemma shows that this union is disjoint, so that
>
> $$|G| = |a_1H| + |a_2H| + \cdots + |a_rH|.$$
>
> Finally, since $|a_iH| = |H|$ for each i, we have $|G| = r|H|$.   ∎

We pause to emphasize that Lagrange's theorem is a candidate criterion; that is, it provides a list of candidates for the orders of the subgroups of a group. Thus, a group of order 12 may have subgroups of orders 12, 6, 4, 3, 2, 1 but no others. *Warning!* The converse of Lagrange's theorem is false. For example, a group of order 12 need not have a subgroup of order 6. We prove this in Example 5.

A special name and notation have been adopted for the number of left (or right) cosets of a subgroup in a group. The *index* of a subgroup H in G is the number of left cosets of H in G. This number is denoted by $|G:H|$. When G is finite, Lagrange's Theorem tells us that $|G:H| = |G|/|H|$.

Corollary 1   *$|a|$ Divides $|G|$*
> *In a finite group, the order of each element of the group divides the order of the group.*

> **Proof.**   Recall that the order of an element is the order of the subgroup generated by that element.   ∎

---

*Lagrange stated his version of this theorem in 1770, but the first complete proof was given by Pietro Abbati some 30 years later.

Corollary 2   *Groups of Prime Order Are Cyclic*
A group of prime order is cyclic.

Proof.   Suppose that G has prime order. Let $a \in G$ and $a \neq e$. Then, $|\langle a \rangle|$ divides $|G|$ and $|\langle a \rangle| \neq 1$. Thus, $|\langle a \rangle| = |G|$ and the corollary follows.   ∎

Corollary 3 $a^{|G|} = e$
Let G be a finite group, and let $a \in G$. Then, $a^{|G|} = e$.

Proof.   By Corollary 1, $|G| = |a|k$. Thus, $a^{|G|} = a^{|a|k} = e^k = e$.   ∎

Corollary 4   *Fermat's Little Theorem*
For every integer a and every prime p, a^p modulo $p = a$ modulo p.

Proof.   By the division algorithm, $a = pm + r$, where $0 \leq r < p$. Thus, $a = r$ modulo p, and it suffices to prove that $r^p = r$ modulo p. If $r = 0$, the result is trivial, so we may assume that $r \in U(p)$. [Recall that $U(p) = \{1, 2, \ldots, p - 1\}$ under multiplication modulo p.] Then, by the previous corollary, $r^{p-1} = 1$ and, therefore, $r^p = r$.   ∎

Fermat's Little Theorem has been used in conjunction with computers to test for primality of certain numbers. One case concerned the number $p = 2^{257} - 1$. If p is a prime, then we know from Fermat's Little Theorem that $10^p = 10 \bmod p$ and, therefore, $10^{p+1} = 100 \bmod p$. Using multiple precision and a simple loop, a computer was able to calculate $10^{p+1} = 10^{2^{257}} \bmod p$ in a few seconds. The result was not 100, and so p is not prime.

As an immediate consequence of the above corollaries, we may classify all groups of order at most 7.

Example 4   Classification of Groups of Order at Most 7
The only groups (up to isomorphism) of order at most 7 are Z_1, Z_2, Z_3, Z_4, $Z_2 \oplus Z_2$, Z_5, Z_6, D_3, and Z_7. Clearly, any group of order 1 is isomorphic to Z_1. Now, Corollary 2 shows that we need discuss only groups of order 4 and 6. If G is a noncyclic group of order 4, then, by Corollary 1, for any x in G, we have $x^2 = e$. Thus, by exercise 38 in Chapter 2, G is Abelian. Proving that G is isomorphic to $Z_2 \oplus Z_2$ is an easy exercise (exercise 36 in Chapter 7).

Now let G be a group of order 6. We claim that G has an element of order 3. By Corollary 1 of Lagrange's Theorem, any nonidentity element of G has order 2, 3, or 6. If there is an element a of order 6, then $|a^2| = 3$. Thus, to verify our claim, we may assume that every nonidentity element of G has order 2. In this case, exercise 38 in Chapter 2 says that G is Abelian. Let $a, b \in G$ and $a \neq b$. Then $\{e, a, b, ab\}$ is a subgroup of order 4 of a group of order 6. Since this is impossible, G must have an element of order 3; call it x. Now, let $y \in G$, but $y \notin \langle x \rangle$; then, $|y| = 2$, for if $|y| = 3$, then $\langle x \rangle$, $y\langle x \rangle$, and $y^2\langle x \rangle$ are distinct cosets containing nine elements. It follows that $G = \{e, x, x^2, y, yx, yx^2\}$. If G is Abelian, then $|xy| = 6$ and G is isomorphic to Z_6. If G is non-Abelian, we will show that there is only one possible way to define multiplication for G. This will prove that there is only one non-Abelian group of order 6 (up to isomorphism).

To this end, consider the product xy. Which of the six elements of G is it? Well, since $y \notin \langle x \rangle$, we see that xy cannot be equal to e, x, or x^2; cancellation shows that $xy \neq y$; and the fact that G is non-Abelian shows that $xy \neq yx$. Thus, $xy = yx^2$. But, this relation completely determines the multiplication table for G. For example, consider the row headed by x. We know that $x \cdot e = x$, $x \cdot x = x^2$, $x \cdot x^2 = e$, $xy = yx^2$, $x(yx) = (xy)x = (yx^2)x = y$, and $x(yx^2) = (xy)x^2 = (yx^2)x^2 = yx$. The rest of the multiplication table can be completed in a similar fashion (see exercise 46). ❑

Observe that it follows from the classification of groups of order 6 that S_3 is isomorphic to D_3.

Example 5 The Converse of Lagrange's Theorem is False.*
The group A_4 of order 12 has no subgroups of order 6. To verify this, recall that A_4 has eight elements of order 3 (see page 94) and suppose that H is a subgroup of order 6. Let a be any element of order 3 in A_4. Since H has index 2 in A_4, at most two of the cosets H, aH, and a^2H are distinct. But equality of any pair of these three implies that $aH = H$, so that $a \in H$. (For example, if $H = a^2H$, multiply on the left by a.) Thus, a subgroup of A_4 of order 6 would have to contain eight elements of order 3, which is absurd. ❑

AN APPLICATION OF COSETS TO PERMUTATION GROUPS

Lagrange's Theorem and its corollaries dramatically demonstrate the fruitfulness of the coset concept. We next consider an application of cosets to permutation groups.

DEFINITION Stabilizer of a Point
Let G be a group of permutations of a set S. For each i in S, let $\text{stab}_G(i) = \{\phi \in G \mid \phi(i) = i\}$. We call $\text{stab}_G(i)$ the *stabilizer of i in G*. The student should verify that $\text{stab}_G(i)$ is a subgroup of G. (See exercise 27 in Chapter 5.)

DEFINITION Orbit of a Point
Let G be a group of permutations of a set S. For each s in S, let $\text{orb}_G(s) = \{\phi(s) \mid \phi \in G\}$. The set $\text{orb}_G(s)$ is a subset of S called the *orbit of s under G*.

---

*The first counterexample to the converse of Lagrange's Theorem was given by Paolo Ruffini in 1799.

Example 6 should clarify these two definitions.

Example 6   Let

$$G = \{(1), (132)(465)(78), (132)(465), (123)(456), (123)(456)(78), (78)\}$$

Then,

$$\mathrm{orb}_G(1) = \{1, 3, 2\}, \qquad \mathrm{stab}_G(1) = \{(1), (78)\},$$
$$\mathrm{orb}_G(2) = \{2, 1, 3\}, \qquad \mathrm{stab}_G(2) = \{(1), (78)\},$$
$$\mathrm{orb}_G(4) = \{4, 6, 5\}, \qquad \mathrm{stab}_G(4) = \{(1), (78)\},$$
$$\mathrm{orb}_G(7) = \{7, 8\}, \qquad \mathrm{stab}_G(7) = \{(1), (132)(465), (123)(456)\}. \quad \square$$

Example 7   We may view D_4 as a group of permutations of the points enclosed by a square. Figure 8.1(a) illustrates the orbit of the point p, and Figure 8.1(b) illustrates the orbit of the point q under D_4. Observe that $\mathrm{stab}_{D_4}(p) = \{R_0, D\}$, whereas $\mathrm{stab}_{D_4}(q) = \{R_0\}$. $\square$

The preceding two examples also illustrate the following theorem.

Theorem 8.2   *Orbit-Stabilizer Theorem*
Let G be a finite group of permutations of a set S. Then for any i from S,
$$|G| = |\mathrm{orb}_G(i)| \, |\mathrm{stab}_G(i)|.$$

Proof.   By Lagrange's Theorem, $|G|/|\mathrm{stab}_G(i)|$ is the number of distinct left cosets of $\mathrm{stab}_G(i)$ in G. Thus, it suffices to establish a one-to-one correspondence between the left cosets of $\mathrm{stab}_G(i)$ and the integers in the orbit of i. To do this, we define a correspondence by $\phi\mathrm{stab}_G(i) \overset{T}{\to} \phi(i)$. To show that T is a well-defined function, we must show that $\alpha\mathrm{stab}_G(i) = \beta\mathrm{stab}_G(i)$ implies $\alpha(i) = \beta(i)$. But, $\alpha\mathrm{stab}_G(i) = \beta\mathrm{stab}_G(i)$ implies $\alpha^{-1}\beta \in \mathrm{stab}_G(i)$, so that $(\alpha^{-1}\beta)(i) = i$ and, therefore, $\beta(i) = \alpha(i)$. Reading this argument backwards shows that T is also one-to-one. We complete the proof by showing that T is onto $\mathrm{orb}_G(i)$. Let $j \in \mathrm{orb}_G(i)$. Then $\alpha(i) = j$ for some $\alpha \in G$ and clearly $T(\alpha\mathrm{stab}_G(i)) = \alpha(i) = j$, so that T is onto. ∎

We leave as an exercise the proof of the important fact that the orbits of the elements of a set S under a group partition S (exercise 33).

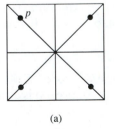   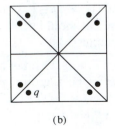

(a)   (b)

Figure 8.1

THE ROTATION GROUP OF A CUBE AND A SOCCER BALL

It cannot be overemphasized that Theorem 8.2 and Lagrange's Theorem (Theorem 8.1) are *counting* theorems.* They enable one to determine the numbers of elements in various sets. To see how Theorem 8.2 works, we will determine the order of the rotation group of a cube and a soccer ball. That is, we wish to find the number of essentially different ways that we can take a cube or a soccer ball in a certain location in space, physically rotate it, and then return it to its original position in space.

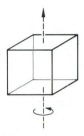

Figure 8.2 Axis of rotation of a cube.

Example 8   Let G be the rotation group of a cube. Label the six faces of the cube 1 through 6. Since any rotation of the cube must carry each face of the cube to exactly one other face of the cube and different rotations induce different permutations of the faces, G can be viewed as a group of permutations on the set $\{1, 2, 3, 4, 5, 6\}$. Clearly, there is some rotation that carries face number 1 to any other face so that $|\text{orb}_G(1)|$ = 6. Next, we consider $\text{stab}_G(1)$. Here, we are asking for all rotations of a cube that leave face number 1 where it is. Surely, there are only four such motions—rotations of 0°, 90°, 180°, and 270°—about the line perpendicular to the face and passing through its center (see Figure 8.2). Thus, by Theorem 8.2, $|G| = |\text{orb}_G(1)| \, |\text{stab}_G(1)| = 6 \cdot 4 = 24$.   ☐

Now that we know how many rotations a cube has, it is simple to determine the actual structure of the rotation group of a cube. Recall that S_4 is the symmetric group of degree 4.

Theorem 8.3   *The Rotation of a Cube*
The group of rotations of a cube is isomorphic to S_4.

Proof.   Since the group of rotations of a cube has the same order as S_4, we need only prove that the group of rotations is isomorphic to a subgroup of S_4. To this end, observe that a cube has four diagonals and any

---

*People who don't count, won't count (Anatole France).

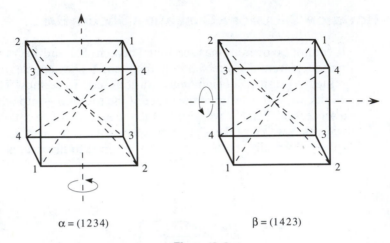

$\alpha = (1234)$ $\beta = (1423)$

Figure 8.3

rotation group induces a group of permutations on the four diagonals. But we must be careful not to assume that different rotations correspond to different permutations. To see that this is so, all we need do is show that all 24 permutations of the diagonals arise from rotations. Labeling the consecutive diagonals 1, 2, 3, and 4, it is obvious that there is a 90° rotation that gives the permutation $\alpha = (1234)$; another 90° rotation perpendicular to our first one gives the permutation $\beta = (1423)$. See Figure 8.3. So, the group of permutations induced by the rotations contains the eight-element subgroup $\{\varepsilon, \alpha, \alpha^2, \alpha^3, \beta^2, \beta^2\alpha, \beta^2\alpha^2, \beta^2\alpha^3\}$ and $\alpha\beta$, which has order 3. Clearly, then, the rotations yield all 24 permutations. ■

Taking a bit of geometry for granted, we can also use Theorem 8.2 to determine the number of rotations of a four-dimensional cube (sometimes called a *hypercube*). A four-dimensional cube has eight three-dimensional cubes as its faces.* Let G denote the rotation group of a four-dimensional cube, and label one of the eight three-dimensional bounding cubes with the number 1. Analogous to the three-dimensional case, there is a rotation that takes the three-dimensional cube number 1 to any of the eight three-dimensional cubes. So we have $|\text{orb}_G(1)| = 8$. But $\text{stab}_G(1)$ is the subgroup of G that does not change the location of the three-dimensional cube number 1, and our previous argument shows that there are 24 such rotations. Thus,

$$|G| = |\text{orb}_G(1)| \ |\text{stab}_G(1)| = 8 \cdot 24 = 192.$$

---

*You might persuade yourself of this by observing that a two-dimensional "cube"—that is, a square—has $2 \cdot 2$ line segments as its faces; a three-dimensional cube has $2 \cdot 3$ squares as its faces; and, in general, an n-dimensional cube has $2 \cdot n$ cubes of dimension $n - 1$ as its faces. Better yet, read the delightful book *Flatland* [1].

For completeness, we will include the following information about the *n*-dimensional cube.

Dimension	Number of vertices	Number of $(n-1)$-dimensional faces	Order of rotation group
n	2^n	$2 \cdot n$	$n!\ 2^{n-1}$

Example 9　A soccer ball has 20 faces that are regular hexagons and 12 faces that are regular pentagons. (The technical term for this solid is *truncated icosahedron.*) To determine the number of rotational symmetries of a soccer ball using Theorem 8.2, we may choose our set *S* to be the 20 hexagons or the 12 pentagons. Let us say that *S* is the set of 12 pentagons. Since any pentagon can be carried to any other pentagon by some rotation, the orbit of the rotation group is *S*. Also, there are five rotations that fix (stabilize) any particular pentagon. Thus, by the orbit-stabilizer theorem, there are $12 \cdot 5 = 60$ rotational symmetries. (In case you are interested, the rotation group of a soccer ball is isomorphic to A_5.)　❑

In nature, the helix is the structure that occurs most often. Running a close second are polyhedrons made from pentagons and hexagons, such as the dodecahedron and the truncated icosahedron. Although it is impossible to enclose space with hexagons alone, adding 12 pentagons will be sufficient to enclose the space. Many viruses have this kind of structure (see [2]).

EXERCISES

If but the will be firmly bent, no stuff resists the mind's intent.
Oliver St. John Gogarty, *The Image-Maker*

1. Let $H = \{(1), (12)(34), (13)(24), (14)(23)\}$. Find the left cosets of *H* in A_4 (see Table 5.1 on page 94).
2. Let *H* be as in exercise 1. How many left cosets of *H* in S_4 are there? (Determine this without listing them.)

3. Let $H = \{0, \pm 3, \pm 6, \pm 9, \ldots\}$. Find all the left cosets of H in Z.

4. Rewrite the condition $a^{-1}b \in H$ given in property 4 of the lemma in additive notation. Assume that the group is Abelian.

5. Let H be as in exercise 3. Use the previous exercise to decide whether or not the following cosets of H are the same.
   a. $11 + H$ and $17 + H$
   b. $-1 + H$ and $5 + H$
   c. $7 + H$ and $23 + H$

6. Let n be an integer greater than 1. Let $H = \{0, \pm n, \pm 2n, \pm 3n, \ldots\}$. Find all left cosets of H in Z. How many are there? $\cap$

7. Find all of the left cosets of $\{1, 11\}$ in $U(30)$.

8. Suppose that a has order 15. Find all of the left cosets of $\langle a^5 \rangle$ in $\langle a \rangle$.

9. Let $|a| = 30$. How many left cosets of $\langle a^4 \rangle$ in $\langle a \rangle$ are there? List them.

10. Let G be a group and let H be a subgroup of G. Let $a \in G$. Prove that $aH = H$ if and only if $a \in H$.

11. Let G be a group and let H be a subgroup of G. Let $a, b \in G$. Prove that the number of elements in aH is the same as the number of elements in bH.

12. In $\mathbf{R} \oplus \mathbf{R}$ under componentwise addition, let $H = \{(x, 3x) \mid x \in \mathbf{R}\}$. (Note that H is the subgroup of all points on the line $y = 3x$.) Show that $(2, 5) + H$ is a straight line passing through the point $(2, 5)$ and parallel to the line $y = 3x$.

13. In $\mathbf{R} \oplus \mathbf{R}$, suppose that H is the subgroup of all points lying on a line through the origin. Show that any left coset of H is just a line parallel to H.

14. In $\mathbf{R} \oplus \mathbf{R} \oplus \mathbf{R}$, let H be a subgroup of points lying in a plane through the origin. Show that every left coset of H is a plane parallel to H.

15. The set of all solutions of the linear system

$$3x + 2y - 3z = 1$$
$$5x + y + 4z = -3$$

is a coset of some subgroup of $\mathbf{R} \oplus \mathbf{R} \oplus \mathbf{R}$. Describe this subgroup.

16. Let G be the group of nonzero complex numbers under multiplication, and let $H = \{x \in G \mid |x| = 1\}$. (Recall that $|a + bi| = \sqrt{a^2 + b^2}$.) Give a geometric description of the cosets of H.

17. Let G be a group of order 60. What are the possible orders for the subgroups of G?

18. Suppose that K is a proper subgroup of H, and H is a proper subgroup of G. If $|K| = 42$ and $|G| = 420$, what are the possible orders of H?

19. Suppose that $|G| = pq$, where p and q are prime. Prove that every proper subgroup of G is cyclic.

20. Recall that, for any integer n greater than 1, $\phi(n)$ denotes the number of integers less than n and relatively prime to n. Prove that if a is any integer relatively prime to n, then $a^{\phi(n)} = 1$ modulo n.

21. Compute 5^{15} modulo 7 and 7^{13} modulo 11.

22. Use Corollary 1 of Lagrange's Theorem to prove that the order of $U(n)$ is even when $n > 2$.

23. Prove that a non-Abelian group of order 10 must have five elements of order 2. Generalize this to the case of a non-Abelian group of order $2p$ where p is prime and $p \neq 2$. How many elements of order 2 does an Abelian group of order $2p$ have when p is prime and $p \neq 2$?

24. Without checking the group axioms, explain why the following table cannot be a group table.

	a	b	c	d	f
a	a	b	c	d	f
b	b	a	d	f	c
c	c	f	a	b	d
d	d	c	f	a	b
f	f	d	b	c	a

25. Find all the left cosets of $\{(0, 1), (1, 2), (2, 4), (3, 3)\}$ in $Z_4 \oplus U(5)$.

26. Suppose that G is a group with more than one element and G has no proper, nontrivial subgroups. Prove that $|G|$ is prime. (Do not assume at the outset that G is finite.)

27. Let $|G| = 15$. If G has only one subgroup of order 3 and only one of order 5, prove that G is cyclic. Generalize to $|G| = pq$, where p and q are prime.

28. Let G be a group of order 25. Prove that G is cyclic or $g^5 = e$ for all g in G.

29. Let $|G| = 33$. What are the possible orders for the elements of G? Show that G must have an element of order 3.

30. Let $|G| = 8$. Show that G must have an element of order 2. Show by example that G need not have an element of order 4.

31. Let $\mathbf{R}^\#$ be the group of nonzero real numbers under multiplication and $\mathbf{R}^+$ the subgroup of positive real numbers. Show that $\mathbf{R}^+$ is the only proper subgroup of $\mathbf{R}^\#$ of finite index.

32. Show that Q, the group of rational numbers under addition, has no proper subgroup of finite index.

33. Let G be a group of permutations of a set S. Prove that the orbits of the members of S constitute a partition of S. (This exercise is referred to in this chapter and in Chapter 24.)

34. Explain why S_3 is not listed in Example 4 as one of the groups of order at most 7.

35. Let $G = \{(1), (12)(34), (1234)(56), (13)(24), (1432)(56), (56)(13), (14)(23), (24)(56)\}$.
   a. Find the stabilizer of 1 and the orbit of 1.
   b. Find the stabilizer of 3 and the orbit of 3.
   c. Find the stabilizer of 5 and the orbit of 5.

36. Prove that 3, 5 and 7 are the only three consecutive odd integers that are prime.

37. Determine the complete subgroup lattice for D_5, the dihedral group of order 10.

38. Prove that a group of order 10 must have an element of order 2.

39. Suppose a group contains elements of orders 1 through 10. What is the minimum possible order of the group?

40. Let G be a finite Abelian group and n a positive integer that is relatively prime to $|G|$. Show that the mapping $a \rightarrow a^n$ is an automorphism of G.

41. Show that in a group of odd order, the equation $x^2 = a$ has a unique solution for all a in G.

42. Let G be the group of plane rotations about a point P in the plane. Thinking of G as a group of permutations of the plane, describe the orbit of a point Q in the plane. (This is the motivation for the name "orbit.")

43. Let G be the rotation group of a cube. Label the faces of the cube 1 through 6, and let H be the subgroup of elements of G that carry face 1 to itself. If σ is a rotation that carries face 2 to face 1, give a physical description of the coset $H\sigma$.

44. The group D_4 acts as a group of permutations of the points enclosed by the squares shown below. (The axes of symmetry are drawn for reference purposes.) For each square, locate the points in the orbit of the indicated point under D_4. In each case, determine the stabilizer of the indicated point.

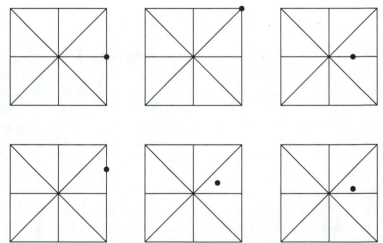

45. Let $G = GL(2, \mathbf{R})$, the group of 2×2 matrices over $\mathbf{R}$ with nonzero determinant. Let H be the subgroup of matrices of determinant ± 1. If $a, b \in G$ and $aH = bH$, what can be said about $\det(a)$ and $\det(b)$? Is the converse true? [Determinants have the property that $\det(xy) = \det(x) \det(y)$.]

46. Complete the multiplication table for the group of order 6 in Example 4.

47. If G is a finite group with fewer than 100 elements and G has subgroups of orders 10 and 25, what is the order of G?

48. Calculate the orders of the following (refer to Figure 27.5 for illustrations):

    a. The group of rotations of a regular tetrahedron (a solid with four congruent triangles as faces)

    b. The group of rotations of a regular octahedron (a solid with eight congruent triangles as faces)

    c. The group of rotations of a regular dodecahedron (a solid with 12 congruent pentagons as faces)

    d. The group of rotations of a regular icosahedron (a solid with 20 congruent triangles as faces)*

49. A soccer ball has 20 faces that are regular hexagons and 12 faces that are regular pentagons. Use Theorem 8.2 to explain why a soccer ball cannot have a 60° rotational symmetry about a line through the centers of two opposite hexagonal faces.

50. Determine how many rotation symmetries each of the following solids has. (The figure below each solid is an "unfolded" version of the solid.)

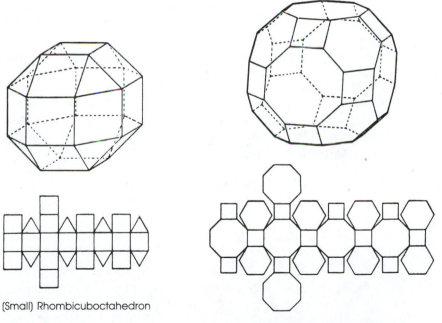

(Small) Rhombicuboctahedron

Great Rhombicuboctahedron or
Truncated Cuboctahedron

---

*The icosahedron is the logo of the Mathematical Association of America.

Optional

51. What is the order of the group of rotations of a volleyball?

52. Let n be an odd integer. Prove that every subgroup of D_n of odd order is cyclic.

References

1. E. A. Abbott, *Flatland,* 6th ed., New York: Dover, 1952.
2. John Galloway, "Nature's Second-Favourite Structure," *New Scientist* 114 (March 1988): 36–39.

Suggested Films

The Hypercube: Projections and Slicing, Banchoff/Strauss Productions, 12 min., color.

> In this award-winning computer-generated film, the four-dimensional cube appears first as a square that rotates about axes to appear as a 3-cube, then about different planes in 4-space to show the full structure of the 4-cube. The same sequence then appears in perspective. Finally, the computer calculates and displays the slices of a square by parallel lines, the slices of a 3-cube by three sets of parallel planes, and the slices of a 4-cube by four different sets of parallel hyperplanes.

Symmetries of the Cube with H. S. M. Coxeter and W. O. J. Moser, International Film Bureau, $13\frac{1}{2}$ min., color.

> The aim of this film is to exhibit the interesting symmetry properties of the square, cube, and octahedron. The viewer is shown how the cube can be generated by means of three reflections, and that the group of symmetries of a cube has order 48. This is done by showing that an orthoscheme (which is 1/48 of the cube) placed inside a kaleidoscope made of three mirrors produces the entire cube. The film also demonstrates that the cube and the octahedron have the same symmetry group.

Joseph Lagrange

Lagrange is the Lofty Pyramid of the Mathematical Sciences.

Napoleon Bonaparte

This stamp was issued by France in Lagrange's honor in 1958.

JOSEPH LOUIS LAGRANGE was born in Italy of French ancestry on January 25, 1736. He became captivated by mathematics at an early age when he read an essay by Halley on Newton's calculus. At the age of 19 he became a professor of mathematics at the Royal Artillery School in Turin. Lagrange made significant contributions to many branches of mathematics and physics, among them the theory of numbers, the theory of equations, ordinary and partial differential equations, the calculus of variations, analytic geometry, fluid dynamics, and celestial mechanics. His methods for solving third- and fourth-degree polynomial equations by radicals laid the groundwork for the group-theoretic approach to solving polynomials taken by Galois. Lagrange was a very careful writer with a clear and elegant style.

At the age of 40, Lagrange was appointed Head of the Berlin Academy, succeeding Euler. In offering this appointment, Frederick The Great proclaimed that the "greatest king in Europe" ought to have the "greatest mathematician in Europe" at his court. In 1787, Lagrange was invited to Paris by Louis XVI and became a good friend of the king and his wife, Marie Antoinette. In 1793, Lagrange headed a commission, which included Laplace and Lavoisier, to devise a new system of weights and measures. Out of this came the metric system. Late in his life he was made a count by Napoleon. Lagrange died on April 10, 1813.

9

Normal Subgroups and Factor Groups

It is tribute to the genius of Galois that he recognized that those subgroups for which the left and right cosets coincide are distinguished ones. Very often in mathematics the crucial problem is to recognize and to discover what are the relevant concepts; once this is accomplished the job may be more than half done.

I. N. Herstein, *Topics in Algebra*

NORMAL SUBGROUPS

As we have seen in Chapter 8, if G is a group and H is a subgroup of G, it is not always true that $aH = Ha$ for all a in G. There are certain situations where this does hold, however, and these cases turn out to be of critical importance in the theory of groups. It was Galois, about 160 years ago, who first recognized that such subgroups were worthy of special attention.

DEFINITION   Normal Subgroup
A subgroup H of a group G is called a *normal* subgroup of G if $aH = Ha$ for all a in G. We denote this by $H \triangleleft G$.

There are several equivalent formulations of the definition of normality. We have chosen the one that is the easiest to use in applications.

However, to *verify* that a subgroup is normal, it is usually better to use the next theorem which is a weaker version of property 6 of the lemma in Chapter 8. It allows us to substitute a condition about two subgroups of G for a condition about two cosets of G.

Theorem 9.1   *Normal Subgroup Test*

A subgroup H of G is normal in G if and only if $xHx^{-1} \subseteq H$ for all x in G.

Many students make the mistake of thinking that H is normal in G means $ah = ha$ for $a \in G$ and $h \in H$. This is not what normality of H means; rather, it means that if $a \in G$ and $h \in H$, then there exists some $h' \in H$ such that $ah = h'a$.

Example 1   Every subgroup of an Abelian group is normal. (In this case, $ah = ha$ for a in the group and h in the subgroup.)   ❑

Example 2   The center $Z(G)$ of a group is always normal. [Again, $ah = ha$ for $a \in G$ and $h \in Z(G)$.]   ❑

Example 3   The alternating group A_n of even permutations is a normal subgroup of S_n. [Note, for example, that for $(12) \in S_n$ and $(123) \in A_n$, we have $(12)(123) \neq (123)(12)$ but $(12)(123) = (132)(12)$ and $(132) \in A_n$.]   ❑

Example 4   The subgroup of rotations in D_n is normal in D_n. (For any rotation r and any reflection f, we have $fr = r^{-1}f$, whereas for any rotations r and r', we have $rr' = r'r$.)   ❑

Example 5   The group $SL(2, \mathbf{R})$ of 2×2 matrices with determinant 1 is a normal subgroup of $GL(2, \mathbf{R})$, the group of 2×2 matrices with nonzero determinant.   ❑

Example 6   Referring to the group table for A_4 given in Table 5.1 on page 94, we may observe that $H = \{\alpha_1, \alpha_2, \alpha_3, \alpha_4\}$ is a normal subgroup of A_4 whereas $K = \{\alpha_1, \alpha_5, \alpha_9\}$ is *not* a normal subgroup of A_4. To see that H is normal, simply note that for any β in A_4, $\beta H \beta^{-1}$ is a subgroup of order 4 and H is the only subgroup of A_4 of order 4. Thus, $\beta H \beta^{-1} = H$. In contrast, $\alpha_2 \alpha_5 \alpha_2^{-1} = \alpha_7$, so that $\alpha_2 K \alpha_2^{-1} \not\subseteq K$.   ❑

FACTOR GROUPS

We have yet to explain why normal subgroups are of special significance. The reason is simple. When the subgroup H of G is normal, then the set of left (or right) cosets of H in G is itself a group—called the *factor* group of G by H (or the *quotient group of G by H*). Quite often, one can obtain information about a group by studying one of its factor groups. This method will be illustrated in the next section of this chapter.

Theorem 9.2 *Factor Groups (O. Hölder, 1889)*

Let G be a group and H a normal subgroup of G. The set $G/H = \{aH \mid a \in G\}$ is a group under the operation $(aH)(bH) = abH.$

Proof. Our first task is to show that the operation is well defined; that is, we must show that the correspondence defined above from $G/H \times G/H$ into G/H is actually a function. To do this, assume that $aH = a'H$ and $bH = b'H$. Then $a' = ah_1$ and $b' = bh_2$ for some h_1, h_2 in H, and therefore $a'b'H = ah_1bh_2H = ah_1bH = ah_1Hb = aHb = abH$. Here we have used property 2 of the lemma from Chapter 8 and the fact that $H \lhd G$. The rest is easy: $eH = H$ is the identity; $a^{-1}H$ is the inverse of aH; and $(aHbH)cH = (ab)HcH = (ab)cH = a(bc)H = aH(bc)H = aH(bHcH)$. This proves that G/H is a group. ∎

Although it is merely a curiosity, we point out that the converse of Theorem 9.2 is also true; that is, if the correspondence $aHbH = abH$ defines a group operation on the set of left cosets of H in G, then H is normal in G.

The next few examples illustrate the factor group concept.

Example 7 Let $4Z = \{0, \pm 4, \pm 8, \ldots\}$. To construct $Z/4Z$, we first must determine the left cosets of $4Z$ in Z. Consider the following four cosets:

$$0 + 4Z = 4Z = \{0, \pm 4, \pm 8, \ldots\},$$
$$1 + 4Z = \{1, 5, 9, \ldots; -3, -7, -11, \ldots\},$$
$$2 + 4Z = \{2, 6, 10, \ldots; -2, -6, -10, \ldots\},$$
$$3 + 4Z = \{3, 7, 11, \ldots; -1, -5, -9, \ldots\}.$$

We claim that there are no others. For if $k \in Z$, then $k = 4q + r$ where $0 \leq r < 4$; and, therefore, $k + 4Z = r + 4q + 4Z = r + 4Z$. Now that we know the elements of the factor group, our next job is to determine the structure of $Z/4Z$. Its Cayley table is

	$0 + 4Z$	$1 + 4Z$	$2 + 4Z$	$3 + 4Z$
$0 + 4Z$	$0 + 4Z$	$1 + 4Z$	$2 + 4Z$	$3 + 4Z$
$1 + 4Z$	$1 + 4Z$	$2 + 4Z$	$3 + 4Z$	$0 + 4Z$
$2 + 4Z$	$2 + 4Z$	$3 + 4Z$	$0 + 4Z$	$1 + 4Z$
$3 + 4Z$	$3 + 4Z$	$0 + 4Z$	$1 + 4Z$	$2 + 4Z$

---

*The notation G/H was first used by C. Jordan.

Clearly, then, $Z/4Z \approx Z_4$. More generally, if for any $n > 0$ we let $nZ = \{0, \pm n, \pm 2n, \pm 3n, \ldots\}$, then Z/nZ is isomorphic to Z_n. ☐

Example 8 Let $G = Z_{18}$ and $H = \langle 6 \rangle = \{0, 6, 12\}$. Then $G/H = \{0 + H, 1 + H, 2 + H, 3 + H, 4 + H, 5 + H\}$. To illustrate how the group elements are combined, consider $(5 + H) + (4 + H)$. This should be one of the six elements listed in the set G/H. Well, $(5 + H) + (4 + H) = 5 + 4 + H = 9 + H = 3 + 6 + H = 3 + H$, since H absorbs all multiples of 6. ☐

A few words of caution about notation are warranted here. When H is a normal subgroup of G, the expression $|aH|$ has two possible interpretations. One could be thinking of aH as a *set* of elements and $|aH|$ as the size of the set; or, as is more often the case, one could be thinking of aH as a group element of the factor group G/H and $|aH|$ as the order of the *element aH* in G/H. In Example 8, for instance, the *set* $3 + H$ has order 3 since $3 + H = \{3, 9, 15\}$. But the *group element* $3 + H$ has order 2 since $(3 + H) + (3 + H) = 6 + H = 0 + H$. As is usually the case when one notation has more than one meaning, the appropriate interpretation will be clear from the context.

Example 9 Let $\mathcal{H} = \{R_0, R_{180}\}$, and consider the factor group of the dihedral group D_4 by $\mathcal{H}$,

$$D_4/\mathcal{H} = \{\mathcal{H}, R_{90}\mathcal{H}, H\mathcal{H}, D\mathcal{H}\}.$$

The multiplication table for $D_4/\mathcal{H}$ is given in Table 9.1. (Notice that even though $H\mathcal{H}R_{90}\mathcal{H} = D'\mathcal{H}$, we have used $D\mathcal{H}$ in Table 9.1 because $D'\mathcal{H} = D\mathcal{H}$, and we use the same symbols in the table as in the heading for the table.)

$D_4/\mathcal{H}$ provides a good opportunity to demonstrate how a factor group of G is related to G itself. Suppose we arrange the heading of the Cayley table for D_4 in such a way that elements from the same coset of $\mathcal{H}$ are in adjacent columns. Then, the multiplication table for D_4 can be blocked off into boxes that are cosets of $\mathcal{H}$, and the substitution that

Table 9.1

	$\mathcal{H}$	$R_{90}\mathcal{H}$	$H\mathcal{H}$	$D\mathcal{H}$
$\mathcal{H}$	$\mathcal{H}$	$R_{90}\mathcal{H}$	$H\mathcal{H}$	$D\mathcal{H}$
$R_{90}\mathcal{H}$	$R_{90}\mathcal{H}$	$\mathcal{H}$	$D\mathcal{H}$	$H\mathcal{H}$
$H\mathcal{H}$	$H\mathcal{H}$	$D\mathcal{H}$	$\mathcal{H}$	$R_{90}\mathcal{H}$
$D\mathcal{H}$	$D\mathcal{H}$	$H\mathcal{H}$	$R_{90}\mathcal{H}$	$\mathcal{H}$

Table 9.2

	R_0	R_{180}	R_{90}	R_{270}	H	V	D	D'
R_0	R_0	R_{180}	R_{90}	R_{270}	H	V	D	D'
R_{180}	R_{180}	R_0	R_{270}	R_{90}	V	H	D'	D
R_{90}	R_{90}	R_{270}	R_{180}	R_0	D'	D	H	V
R_{270}	R_{270}	R_{90}	R_0	R_{180}	D	D'	V	H
H	H	V	D	D'	R_0	R_{180}	R_{90}	R_{270}
V	V	H	D'	D	R_{180}	R_0	R_{270}	R_{90}
D	D	D'	V	H	R_{270}	R_{90}	R_0	R_{180}
D'	D'	D	H	V	R_{90}	R_{270}	R_{180}	R_0

replaces a box containing the element x with the coset $x\mathcal{H}$ yields the Cayley table for $D_4/\mathcal{H}$ (Table 9.2).

Thus, when we pass from D_4 to $D_4/\mathcal{H}$, the box

H	V
V	H

in Table 9.2 becomes the element $H\mathcal{H}$ in Table 9.1. Similarly, the box

D	D'
D'	D

becomes the element $D\mathcal{H}$, and so on.   ❑

In this way, one can see that the formation of a factor group G/H causes a systematic collapsing of the elements of G. In particular, all the elements in the coset of H containing a collapse to the single group element aH in G/H.

Example 10   Consider the group A_4 as represented by Table 5.1 on page 94. (Here i denotes the permutation α_i.) Let $H = \{1, 2, 3, 4\}$. Then the three cosets of H are H, $5H = \{5, 6, 7, 8\}$, and $9H = \{9, 10, 11, 12\}$. (In this case, rearrangement of the headings is unnecessary.) Blocking off the table for A_4 into boxes that are cosets of H and replacing the boxes with 1, 5, and 9 (see Table 9.3) with the cosets $1H$, $5H$, and $9H$, we obtain the Cayley table for G/H given in Table 9.4.

This procedure can be illustrated more vividly with colors. Let's say we had printed the elements of H in green, the elements of $5H$ in red,

Table 9.3

1	2	3	4	5	6	7	8	9	10	11	12
2	1	4	3	6	5	8	7	10	9	12	11
3	4	1	2	7	8	5	6	11	12	9	10
4	3	2	1	8	7	6	5	12	11	10	9
5	8	6	7	9	12	10	11	1	4	2	3
6	7	5	8	10	11	9	12	2	3	1	4
7	6	8	5	11	10	12	9	3	2	4	1
8	5	7	6	12	9	11	10	4	1	3	2
9	11	12	10	1	3	4	2	5	7	8	6
10	12	11	9	2	4	3	1	6	8	7	5
11	9	10	12	3	1	2	4	7	5	6	8
12	10	9	11	4	2	1	3	8	6	5	7

Table 9.4

	$1H$	$5H$	$9H$
$1H$	$1H$	$5H$	$9H$
$5H$	$5H$	$9H$	$1H$
$9H$	$9H$	$1H$	$5H$

group as consisting of the three colors that define a group table isomorphic to G/H.

	Green	Red	Blue
Green	Green	Red	Blue
Red	Red	Blue	Green
Blue	Blue	Green	Red

It is instructive to see what happens if we attempt the same procedure with a group G and a subgroup H that is not normal in G—that is, arrange the heading of the Cayley table so that the elements from the same coset of H are in adjacent columns and attempt to block off the table into boxes that are also cosets of H to produce a Cayley table for the set of cosets. Say, for instance, we were to take G to be A_4 and $H = \{1, 5, 9\}$. The cosets of H would be H, $2H = \{2, 6, 10\}$, $3H = \{3, 7, 11\}$,

$4H = \{4, 8, 12\}$. Then the first three rows of the rearranged Cayley table for A_4 would be

	1	5	9	2	8	11	3	6	12	4	7	10
1	1	5	9	2	8	11	3	6	12	4	7	10
5	5	9	1	8	11	2	6	12	3	7	10	4
9	9	1	5	11	2	8	12	3	6	10	4	7

But already we are in trouble, for blocking these off into 3×3 boxes yields boxes that contain elements of different cosets. Hence, it is impossible to represent an entire box by a single element of the box in the same way we could for boxes made from the cosets of a normal subgroup. Had we printed the rearranged table in four colors with all members of the same coset having the same color, we would see multicolored boxes rather than the uniformly colored boxes produced by a normal subgroup. ❑

In Chapter 11, we will prove that every finite Abelian group is isomorphic to a direct product of cyclic groups. In particular, an Abelian group of order 8 is isomorphic to one of Z_8, $Z_4 \oplus Z_2$, or $Z_2 \oplus Z_2 \oplus Z_2$. In the next two examples, we examine Abelian factor groups of order 8 and determine the isomorphism type of each.

Example 11 Let

$$G = U(32) = \{1, 3, 5, 7, 9, 11, 13, 15, 17, 19, 21, 23, 25, 27, 29, 31\}$$

and $H = U_{16}(32) = \{1, 17\}$. Then G/H is an Abelian group of order $16/2 = 8$. Which of the three Abelian groups of order 8 is it—Z_8, $Z_4 \oplus Z_2$, or $Z_2 \oplus Z_2 \oplus Z_2$? To answer this question, we need only determine the elements of G/H and their orders. Observe that the eight cosets

$$1H = \{1, 17\}, \quad 3H = \{3, 19\}, \quad 5H = \{5, 21\}, \quad 7H = \{7, 23\},$$
$$9H = \{9, 25\}, \quad 11H = \{11, 27\}, \quad 13H = \{13, 29\}, \quad 15H = \{15, 31\}$$

are all distinct, so that they form the factor group G/H. Clearly, $(3H)^2 = 9H \neq H$, and so $3H$ has order at least 4. Thus, G/H is not $Z_2 \oplus Z_2 \oplus Z_2$. On the other hand, direct computations show that both $7H$ and $9H$ have order 2, so that G/H cannot be Z_8 either. This proves that $U(32)/U_{16}(32) \approx Z_4 \oplus Z_2$, which (not so incidentally!) is isomorphic to $U(16)$. ❑

Example 12 Let $G = U(32)$ and $K = \{1, 15\}$. Then $|G/K| = 8$, and we ask which of the three Abelian groups of order 8 is G/K? Since $(3K)^4 = 81K = 17K \neq K$, $|3K| = 8$. Thus, $G/K \approx Z_8$. ❑

It is crucial to understand that when one factors out by a normal subgroup H, what one is really doing is defining every element in H to be the *identity*. Thus, in Example 9, we are making $R_{180}\mathcal{H} = \mathcal{H}$ the identity. Likewise, $R_{270}\mathcal{H} = R_{90}R_{180}\mathcal{H} = R_{90}\mathcal{H}$. Similarly, in Example 7, we are declaring any multiple of 4 to be 0 in the factor group $Z/4Z$. This is why $5 + 4Z = 1 + 4 + 4Z = 1 + 4Z$, and so on. In Example 11, we have $3H = 19H$, since $19 = 3 \cdot 17$ in $U(32)$ and going to the factor group makes 17 the identity. Algebraists often refer to the process of creating the factor group G/H as "killing" H.

APPLICATIONS OF FACTOR GROUPS

Why are factor groups important? Well, when G is finite and $H \neq \{e\}$, G/H is smaller than G, and its structure is usually less complicated than that of G. At the same time, G/H simulates G in many ways. In fact, we may think of a factor group of G as a less complicated approximation of G (similar to using the rational number 3.14 for the irrational number π). What makes factor groups important is that one can often deduce properties of G by examining the less complicated group G/H instead. We illustrate this by giving another proof that A_4 has no subgroup of order 6.

Example 13 A_4 has no subgroup of order 6.
The group A_4 of even permutations on the set $\{1, 2, 3, 4\}$ has no subgroup H of order 6. To see this, suppose that A_4 does have a subgroup H of order 6. By Example 6, we know that $H \triangleleft A_4$. Thus, the factor group A_4/H exists and has order 2. Since the order of an element divides the order of the group, we have $(\alpha H)^2 = \alpha^2 H = H$ for all $\alpha \in A_4$. Thus, $\alpha^2 \in H$ for all α in A_4. Referring to the main diagonal of the group table for A_4 given in Table 5.1 on page 94, however, we observe that A_4 has nine different elements of the form α^2, all of which must belong to H, a subgroup of order 6. This is clearly impossible, so a subgroup of order 6 cannot exist in A_4.* ❑

The next three theorems illustrate how knowledge of a factor group of G reveals information about G itself.

Theorem 9.3   *The G/Z Theorem*
Let G be a group and let $Z(G)$ be the center of G. If $G/Z(G)$ is cyclic, then G is Abelian.

Proof.   Let $gZ(G)$ be a generator of the factor group $G/Z(G)$, and let $a, b \in G$. Then there exist integers i and j such that

$$aZ(G) = (gZ(G))^i = g^i Z(G)$$

How often have I said to you that when you have eliminated the impossible, whatever remains, however improbable, must be the truth. Sir Arthur Conan Doyle.

and

$$bZ(G) = (gZ(G))^j = g^jZ(G).$$

Thus, $a = g^ix$ for some x in $Z(G)$ and $b = g^jy$ for some y in $Z(G)$. It follows then that

$$ab = (g^ix)(g^jy) = g^i(xg^j)y = g^i(g^jx)y$$
$$= (g^ig^j)(xy) = (g^jg^i)(yx) = (g^jy)(g^ix) = ba. \quad \blacksquare$$

A few remarks about Theorem 9.3 are in order. First, our proof shows that a better result is possible: if G/H is cyclic where H is a subgroup of $Z(G)$, then G is Abelian. Second, in practice, it is the contrapositive of the theorem that is most often used—that is, if G is non-Abelian, then $G/Z(G)$ is not cyclic. For example, it follows immediately from this statement and Lagrange's Theorem that a non-Abelian group of order pq, where p and q are primes, must have a trivial center. Third, if $G/Z(G)$ is cyclic, it must be trivial.

Theorem 9.4 $G/Z(G) \approx \text{Inn}(G)$
For any group G, $G/Z(G)$ is isomorphic to $\text{Inn}(G)$.

Proof.   Consider the correspondence from $G/Z(G)$ to $\text{Inn}(G)$ given by $T: gZ(G) \rightarrow \phi_g$ (where, recall, $\phi_g(x) = gxg^{-1}$ for all x in G). First, we show that T is a function. Suppose $gZ(G) = hZ(G)$, so that $h^{-1}g \in Z(G)$. Then, for all x in G, $h^{-1}gx = xh^{-1}g$. Thus, $gxg^{-1} = hxh^{-1}$ for all x in G, and, therefore, $\phi_g = \phi_h$. Reversing this argument shows that T is one-to-one, as well. Clearly, T is onto.

That T is operation-preserving follows directly from the fact that $\phi_g\phi_h = \phi_{gh}$ for all g and h in G. $\quad \blacksquare$

As an application of Theorems 9.3 and 9.4, we may easily determine $\text{Inn}(D_6)$ without looking at $\text{Inn}(D_6)$!

Example 14   We know from Example 11 in Chapter 3 that $|Z(D_6)| = 2$. Thus, $|D_6/Z(D_6)| = 6$. So, by our classification of groups of order 6 (Example 4 in Chapter 8), we know that $\text{Inn}(D_6)$ is isomorphic to D_3 or Z_6. Now, if $\text{Inn}(D_6)$ were cyclic, then, by Theorem 9.4, $D_6/Z(D_6)$ would be also. But then, Theorem 9.3 would tell us that D_6 is Abelian. So, $\text{Inn}(D_6)$ is isomorphic to D_3. $\quad \square$

The next theorem demonstrates one of the most powerful proof techniques available in the theory of finite groups—the combined use of factor groups and induction.

Theorem 9.5   *Existence of Elements of Prime Order*
Let G be a finite Abelian group and let p be a prime that divides the order of G. Then G has an element of order p.

Proof. Clearly, this statement is vacuously true for the case in which G has order 1. We prove the theorem by using the Second Principle of Mathematical Induction on $|G|$. That is, we assume that the statement is true for all Abelian groups with fewer elements than G and use this assumption to show that the statement is true for G as well. Certainly, G has elements of prime order, for if $|x| = m$ and $m = qn$ where q is prime, then $|x^n| = q$. So let x be an element of G of some prime order q, say. If $q = p$, we are finished; so assume that $q \neq p$. Since every subgroup of an Abelian group is normal, we may construct the factor group $G^* = G/\langle x \rangle$. Then G^* is Abelian and p divides $|G^*|$, since $|G^*| = |G|/q$. By induction, then, G^* has an element—call it $y\langle x \rangle$—of order p. Thus the coset $y\langle x \rangle$ raised to the pth power is the identity element $\langle x \rangle$ in G^*. That is, $(y\langle x \rangle)^p = y^p\langle x \rangle = \langle x \rangle$. It follows, then, that $y^p \in \langle x \rangle$, so that $y^p = e$ or y^p has order q. If $y^p = e$, then y is the desired element of order p; if y^p has order q, then y^q has order p. In either case, we have produced an element of order p. ∎

INTERNAL DIRECT PRODUCTS

As we have seen, the external direct product provides a way of putting groups together into a larger group. It would be quite useful to be able to reverse this process—that is, to be able to start with a large group and break it down into a product of smaller groups. It is occasionally possible to do this. To this end, suppose that H and K are subgroups of some group G. We define the set $HK = \{hk \mid h \in H, k \in K\}$.

Example 15 In $U(24) = \{1, 5, 7, 11, 13, 17, 19, 23\}$, let $H = \{1, 17\}$ and $K = \{1, 13\}$. Then, $HK = \{1, 13, 17, 5\}$, since $5 = 17 \cdot 13 \bmod 24$. ☐

Example 16 In S_3, let $H = \{(1), (12)\}$ and $K = \{(1), (13)\}$. Then, $HK = \{(1), (13), (12), (12)(13)\} = \{(1), (13), (12), (132)\}$. ☐

The student should be careful not to assume that the set HK is a subgroup of G; in Example 15 it is, but in Example 16 it is not.

DEFINITION Internal Direct Product of H and K

Let H and K be normal subgroups of a group G. We say that G is the *internal direct product of H and K* and write $G = H \times K$ if

$$G = HK \quad \text{and} \quad H \cap K = \{e\}.$$

The wording of the phrase "internal direct product" is easy to justify. We want to call G the internal direct product of H and K if H and K are subgroups of G, and if G is naturally isomorphic to the external direct product of H and K. One forms the internal direct product by

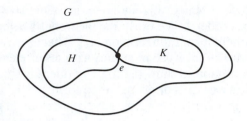

G

H

e

K

Figure 9.1   For the internal direct product, H and K must be subgroups of the same group.

starting with a group G and then proceeding to produce two subgroups H and K within G, such that G is *isomorphic* to the external direct product of H and K. (The definition ensures that this is the case—see Theorem 9.6.) On the other hand, one forms an external direct product by *starting* with any two groups H and K, related or not, and proceeding to produce the larger group $H \oplus K$. The difference between the two products is that the internal direct product can be formed within G itself, using subgroups of G and the operation of G, whereas the external direct product can be formed with totally unrelated groups by creating a new set and a new operation. (See Figures 9.1 and 9.2.)

Perhaps the following analogy with integers will be useful in clarifying the distinction between the two products of groups discussed in the preceding paragraph. Just as one may take any (finite) collection of integers and form their product, one may also take any collection of groups and form their external direct product. Conversely, just as one may start with a particular integer and express it as a product of certain of its divisors, one may be able to start with a particular group and factor it as an internal direct product of certain of its subgroups.

Example 17   In D_6, the dihedral group of order 12, let F denote some reflection and let R_k denote a rotation of k degrees. Then,

$$D_6 = \{R_0, R_{120}, R_{240}, F, R_{120}F, R_{240}F\} \times \{R_0, R_{180}\}.\qquad \square$$

Students should be cautioned about the necessity of having all conditions of the definition of internal direct product satisfied to ensure that $HK \approx H \oplus K$. For example, if we take

$$G = S_3, \quad H = \langle (123) \rangle, \quad \text{and} \quad K = \langle (12) \rangle,$$

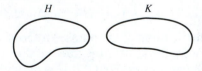

H

K

Figure 9.2   For the external direct product, H and K can be any groups.

then $G = HK$, and $H \cap K = \{e\}$. But, G is *not* isomorphic to $H \oplus K$, since, by Theorem 7.2, $H \oplus K$ is cyclic, whereas S_3 is not. Note that K is not normal.

A group G can also be the internal direct product of a collection of subgroups.

DEFINITION   Internal Direct Product of $H_1 \times H_2 \times \cdots \times H_n$

Let $H_1, H_2, \ldots, H_n$ be a finite collection of normal subgroups of G. We say that G is the *internal direct product* of $H_1, H_2, \ldots, H_n$ and write $G = H_1 \times H_2 \times \cdots \times H_n$, if

1. $G = H_1 H_2 \cdots H_n = \{h_1 h_2 \cdots h_n \mid h_i \in H_i\}$
2. $(H_1 H_2 \cdots H_i) \cap H_{i+1} = \{e\}$ for $i = 1, 2, \ldots, n - 1$

This definition is somewhat more complicated than the one given for two subgroups. The student may wonder about the motivation for it—that is, why should we want the subgroups to be normal and why is it desirable for each subgroup to be disjoint from the product of all the others? The reason is quite simple. We want the internal direct product to be isomorphic to the external direct product. As the next theorem shows, the conditions in the definition of internal direct product were chosen to ensure that the two products are isomorphic.

Theorem 9.6 $H_1 \times H_2 \times \cdots \times H_n \approx H_1 \oplus H_2 \oplus \cdots \oplus H_n$
If a group G is the internal direct product of a finite number of subgroups $H_1, H_2, \ldots, H_n$, then G is isomorphic to the external direct product of $H_1, H_2, \ldots, H_n$.

Proof.   We first show that the normality of the H's together with the second condition of the definition guarantee that h's from different H_i's commute. For if $h_i \in H_i$ and $h_j \in H_j$ with $i \neq j$, then

$$(h_i h_j h_i^{-1}) h_j^{-1} \in H_j h_j^{-1} = H_j$$

and

$$h_i (h_j h_i^{-1} h_j^{-1}) \in h_i H_i = H_i.$$

Thus, $h_i h_j h_i^{-1} h_j^{-1} \in H_i \cap H_j = \{e\}$ (see exercise 2) and therefore $h_i h_j = h_j h_i$. We next claim that each member of G can be expressed uniquely in the form $h_1 h_2 \cdots h_n$, where $h_i \in H_i$. That there is at least one such representation is the content of condition 1 of the definition. To prove uniqueness, suppose that $g = h_1 h_2 \cdots h_n$ and $g = h_1' h_2' \cdots h_n'$, where h_i and h_i' belong to H_i for $i = 1, \ldots, n$. Then, using the fact that the h's from different H_i's commute, we can solve the equation

(*) $h_1 h_2 \cdots h_n = h_1' h_2' \cdots h_n'$

for $h_n' h_n^{-1}$ to obtain

$$h_n' h_n^{-1} = (h_1')^{-1} h_1 (h_2')^{-1} h_2 \cdots (h_{n-1}')^{-1} h_{n-1}.$$

But then,

$$h'_n h_n^{-1} \in H_1 H_2 \cdots H_{n-1} \cap H_n = \{e\},$$

so that $h'_n h_n^{-1} = e$ and, therefore, $h'_n = h_n$. At this point, we can cancel h_n and h'_n from opposite sides of the equal sign in (*) and repeat the preceding argument to obtain $h_{n-1} = h'_{n-1}$. Continuing in this fashion, we eventually have $h_i = h'_i$ for $i = 1, \ldots, n$. With our claim established, we may now define a function ϕ from G to $H_1 \oplus H_2 \oplus \cdots \oplus H_n$ by $\phi(h_1 h_2 \cdots h_n) = (h_1, h_2, \ldots, h_n)$. It is an easy exercise to verify that ϕ is an isomorphism.   ∎

The topic of direct products, like that of mappings, is one in which notation and terminology vary widely. Many authors use $H \times K$ to denote the internal direct product and the external direct product of H and K, making no notational distinction between the two products. A few authors define only the external direct product. Many people reserve the notation $H \oplus K$ for the situation where H and K are Abelian groups under addition and call it the *direct sum* of H and K. In fact, we will adopt this terminology in the section on rings (Part 3), since rings are always Abelian groups under addition.

The U-groups provide a convenient way to illustrate the preceding ideas and clarify the distinction between internal and external direct products. It follows directly from Theorem 7.3 and Theorem 9.6 that if $m = n_1 n_2 \cdots n_k$, where $\gcd(n_i, n_j) = 1$ for $i \neq j$, then,

$$U(m) = U_{m/n_1}(m) \times U_{m/n_2}(m) \times \cdots \times U_{m/n_k}(m)$$
$$\approx U(n_1) \oplus U(n_2) \oplus \cdots \oplus U(n_k).$$

Let us return to the examples given following Theorem 7.3.

$$
\begin{aligned}
U(105) &= U(15 \cdot 7) = U_{15}(105) \times U_7(105) \\
&= \{1, 16, 31, 46, 61, 76\} \times \{1, 8, 22, 29, 43, 64, 71, 92\} \\
&\approx U(7) \oplus U(15), \\
U(105) &= U(5 \cdot 21) = U_5(105) \times U_{21}(105) \\
&= \{1, 11, 16, 26, 31, 41, 46, 61, 71, 76, 86, 101\} \\
&\quad \times \{1, 22, 43, 64\} \approx U(21) \oplus U(5), \\
U(105) &= U(3 \cdot 5 \cdot 7) = U_{35}(105) \times U_{21}(105) \times U_{15}(105) \\
&= \{1, 71\} \times \{1, 22, 43, 64\} \times \{1, 16, 31, 46, 61, 76\} \\
&\approx U(3) \oplus U(5) \oplus U(7).
\end{aligned}
$$

EXERCISES

Error is a hardy plant; it flourisheth in every soil.

Martin Farquhar Tupper, *Proverbial Philosophy*
[1838–1842].

1. Let $H = \{(1), (12)\}$. Is H normal in S_3?
2. Show that if G is the internal direct product of $H_1, H_2, \ldots, H_n$ and $i \neq j$ with $1 \leq i \leq n$, $1 \leq j \leq n$, then $H_i \cap H_j = \{e\}$.
3. Prove that A_n is normal in S_n.
4. Let $H = \left\{ \begin{bmatrix} a & b \\ 0 & d \end{bmatrix} \,\middle|\, a, b, d \in \mathbf{R},\ ad \neq 0 \right\}$. Is H a normal subgroup of $GL(2, \mathbf{R})$?
5. Prove that $SL(2, \mathbf{R})$ is a normal subgroup of $GL(2, \mathbf{R})$.
6. Viewing $\langle 3 \rangle$ and $\langle 12 \rangle$ as subgroups of Z, prove that $\langle 3 \rangle / \langle 12 \rangle$ is isomorphic to Z_4. Similarly, prove that $\langle 8 \rangle / \langle 48 \rangle$ is isomorphic to Z_6. Generalize to arbitrary integers k and n. ↦ # of left cosets
7. Prove that if H has index 2 in G, then H is normal in G. (This exercise is referred to in Chapters 24 and 25.)
8. Let $H = \{(1), (12)(34)\}$ in A_4.
   a. Show that H is not normal in A_4.
   b. Referring to the multiplication table for A_4 in Table 5.1 on page 94, show that, although $\alpha_6 H = \alpha_7 H$ and $\alpha_9 H = \alpha_{11} H$, it is not true that $\alpha_6 \alpha_9 H = \alpha_7 \alpha_{11} H$.

   Explain why this proves that the left cosets of H do not form a group under coset multiplication.
9. Let $G = Z_4 \oplus U(4)$, $H = \langle (2, 3) \rangle$, and $K = \langle (2, 1) \rangle$. Determine the isomorphism class of G/H and G/K. (This shows that $H \approx K$ does not imply $G/H \approx G/K$.)
10. Prove that a factor group of a cyclic group is cyclic.
11. What is the order of element $5 + \langle 6 \rangle$ in the factor group $Z_{18}/\langle 6 \rangle$?
12. What is the order of the element $14 + \langle 8 \rangle$ in the factor group $Z_{24}/\langle 8 \rangle$?
13. What is the order of the element $4U_5(105)$ in the factor group $U(105)/U_5(105)$?
14. Recall that $Z(D_6) = \{R_0, R_{180}\}$. What is the order of the element $R_{60} Z(D_6)$ in the factor group $D_6/Z(D_6)$?
15. Let $G = Z/\langle 20 \rangle$ and $H = \langle 4 \rangle / \langle 20 \rangle$. List the elements of H and G/H.
16. What is the order of the factor group $Z_{60}/\langle 15 \rangle$?
17. What is the order of the factor group $(Z_{10} \oplus U(10))/\langle (2, 9) \rangle$?
18. Construct the Cayley table for $U(20)/U_5(20)$.
19. Is $U(30)/U_5(30)$ isomorphic to $Z_2 \oplus Z_2$ or Z_4?

20. Determine the order of $(Z \oplus Z)/\langle(2, 2)\rangle$. Is the group cyclic?

21. Determine the order of $(Z \oplus Z)/\langle(4, 2)\rangle$. Is the group cyclic?

22. The group $Z_4 \oplus Z_{12}/\langle(2, 2)\rangle$ is isomorphic to one of Z_8, $Z_4 \oplus Z_2$, or $Z_2 \oplus Z_2 \oplus Z_2$. Which one?

23. Let $G = U(32)$ and $H = \{1, 31\}$. The group G/H is isomorphic to one of Z_8, $Z_4 \oplus Z_2$, or $Z_2 \oplus Z_2 \oplus Z_2$. Which one?

24. Let G be the group given by the table in exercise 4 of the Supplementary Exercises for Chapters 1–4 on page 82, and let H be the subgroup $\{e, a^2\}$. Is G/H isomorphic to Z_4 or $Z_2 \oplus Z_2$?

25. Let $G = U(16)$, $H = \{1, 15\}$, and $K = \{1, 9\}$. Are H and K isomorphic? Are G/H and G/K isomorphic?

26. Let $G = Z_4 \oplus Z_4$, $H = \{(0, 0), (2, 0), (0, 2), (2, 2)\}$, and $K = \langle(1, 2)\rangle$. Is G/H isomorphic to Z_4 or $Z_2 \oplus Z_2$? Is G/K isomorphic to Z_4 or $Z_2 \oplus Z_2$?

27. Let $G = GL(2, \mathbf{R})$ and $H = \{A \in G \mid \det A = 3^k, k \in Z\}$. Prove that H is a normal subgroup of G.

28. Express $U(165)$ as an internal direct product of proper subgroups in three different ways.

29. Let $\mathbf{R}^\#$ denote the group of all nonzero real numbers under multiplication. Let $\mathbf{R}^+$ denote the group of positive real numbers under multiplication. Prove that $\mathbf{R}^\#$ is the internal direct product of $\mathbf{R}^+$ and the subgroup $\{1, -1\}$.

30. Prove that D_4 cannot be expressed as an internal direct product of two proper subgroups.

31. Let H and K be subgroups of a group G. If $G = HK$ and $g = hk$, where $h \in H$ and $k \in K$, is there any relationship among $|g|$, $|h|$, and $|k|$? What if $G = H \times K$?

32. In Z, let $H = \langle5\rangle$ and $K = \langle7\rangle$. Prove that $Z = HK$. Does $Z = H \times K$?

33. Let $G = \{3^a 6^b 10^c \mid a, b, c \in Z\}$ under multiplication and $H = \{3^a 6^b 12^c \mid a, b, c \in Z\}$ under multiplication. Prove that $G = \langle3\rangle \times \langle6\rangle \times \langle10\rangle$, whereas $H \neq \langle3\rangle \times \langle6\rangle \times \langle12\rangle$.

34. Determine all subgroups of $\mathbf{R}^\#$ (nonzero reals under multiplication) of index 2.

35. Show, by example, that in a factor group G/H it can happen that $aH = bH$ but $|a| \neq |b|$.

36. Let H be a normal subgroup of G and let a belong to G. If the element aH has order 3 in the group G/H and $|H| = 10$, what are the possibilities for the order of a?

37. Prove that a factor group of an Abelian group is Abelian.

38. An element is called a *square* if it can be expressed in the form b^2 for some b. Suppose that G is an Abelian group and H is a subgroup of G. If every element of H is a square and every element of G/H is a square, prove that every element of G is a square.

39. Observe from the table for A_4 given in Table 5.1 on page 94 that the subgroup given in Example 6 of this chapter is the only subgroup of A_4 of order 4. Why does this imply that this subgroup must be normal in A_4? Generalize this to arbitrary finite groups.

40. Let G be a finite group and H a normal subgroup of G. Prove that the order of the element gH in G/H must divide the order of g in G.

41. Suppose that G is a non-Abelian group of order p^3 (where p is a prime) and $Z(G) \neq \{e\}$. Prove that $|Z(G)| = p$.

42. If $|G| = pq$, where p and q are not necessarily distinct primes, prove that $|Z(G)| = 1$ or pq.

43. Let N be a normal subgroup of G and let H be a subgroup of G. If N is a subgroup of H, prove that H/N is a normal subgroup of G/N if and only if H is a normal subgroup of G.

44. Let $G = \{\pm 1, \pm i, \pm j, \pm k\}$, where $i^2 = j^2 = k^2 = -1, -i = (-1)i,$ $1^2 = (-1)^2 = 1, ij = -ji = k, jk = -kj = i,$ and $ki = -ik = j$.
  a. Construct the Cayley table for G.
  b. Show that $H = \{1, -1\} \triangleleft G$. prove that H is normal
  c. Construct the Cayley table for G/H.

(The rules involving i, j, and k can be remembered by using the circle below.

Going clockwise, the product of two consecutive elements is the third one. The same is true for going counterclockwise, except that we obtain the negative of the third element.) This group is called the *quaternions* and was invented by William Hamilton in 1843. The quaternions are used to describe rotations in three-dimensional space, and they are used in physics. The quaternions can be used to extend the complex numbers in a natural way.

45. In D_4, let $K = \{R_0, D\}$ and let $L = \{R_0, D, D', R_{180}\}$. Show that $K \triangleleft L \triangleleft D_4$, but that K is not normal in D_4. (Normality is not transitive. Compare exercise 4, Supplementary Exercises for Chapters 5–7.)

46. If N is a normal subgroup of G and H is any subgroup of G, prove that NH is a subgroup of G. (This exercise is referred to in Chapter 24.)

47. Show that the intersection of two normal subgroups of G is a normal subgroup of G.

48. If N and M are normal subgroups of G, prove that NM is also a normal subgroup of G.

49. Without looking at inner automorphisms of D_n, determine the number of inner automorphisms of D_n.

50. Let N be a normal subgroup of a group G. If N is cyclic, prove that every subgroup of N is also normal in G. (This exercise is referred to in Chapter 24.)

51. Let H be a normal subgroup of a finite group G. If $\gcd(|x|, |G/H|) = 1$, show that $x \in H$. (This exercise is referred to in Chapter 25.)

52. Let G be a group and let G' be the subgroup of G generated by the set $S = \{x^{-1}y^{-1}xy \mid x, y \in G\}$. (See exercise 3, Supplementary Exercises for Chapters 5–7, for a more complete description of G'.)

    a. Prove that G' is normal in G.

    b. Prove that G/G' is Abelian.

    c. If G/N is Abelian, prove that $N \geq G'$.

    d. Prove that if H is a subgroup of G and $H \geq G'$, then H is normal in G.

53. If N is a normal subgroup of G and $|G/N| = m$, show that $x^m \in N$ for all x in G.

54. Suppose that G has a subgroup of order n. Prove that the intersection of all subgroups of G of order n is a normal subgroup of G.

55. If G is non-Abelian, show that $\text{Aut}(G)$ is not cyclic.

56. Let $|G| = p^n m$, where p is prime and $\gcd(p, m) = 1$. Suppose that H is a normal subgroup of G of order p^n. If K is a subgroup of G of order p^k, show that $K \subseteq H$.

57. Suppose that H is a normal subgroup of a finite group G. If G/H has an element of order n, show that G has an element of order n. Show, by example, that the assumption that G is finite is necessary.

58. Recall that a subgroup N of a group G is called characteristic if $\phi(N) = N$ for all automorphisms ϕ of G. (See exercises 1–4, Supplementary Exercises for Chapters 5–7.) If N is a characteristic subgroup of G, show that N is a normal subgroup of G.

59. In D_4, let $\mathcal{H} = \{R_0, H\}$. Form an operation table for the cosets $\mathcal{H}$, $D\mathcal{H}$, $V\mathcal{H}$, and $D'\mathcal{H}$. Is the result a group table? Does your answer contradict Theorem 9.2?

60. Show that S_4 has a unique subgroup of order 12.

61. If $|G| = 30$ and $|Z(G)| = 5$, what is the structure of $G/Z(G)$?

62. If H is a normal subgroup of G, and $|H| = 2$, prove that H is contained in the center of G.

63. Prove that A_5 cannot have a normal subgroup of order 2.

64. Let G be a finite group and H an odd order subgroup of G of index 2. Show that the product of all the elements of G (taken in any order) cannot belong to H.

65. Let G be a group. If $H = \{g^2 \mid g \in G\}$ is a subgroup of G, prove that it is a normal subgroup of G.

66. Suppose that H is a normal subgroup of G. If $|H| = 4$ and gH has order 3 in G/H, find a subgroup of order 12 in G.

REFERENCE

1. Tony Rothman, "Genius and Biographers: The Fictionalization of Évariste Galois," *The American Mathematical Monthly* 89 (1982): 84–106.

SUGGESTED READINGS

J. A. Gallian, R. S. Johnson, and S. Peng, "On the Quotient Structure of Z^n," *Pi Mu Epsilon Journal,* 9 (1993): 524–526.

The authors determine the structure of the group $(Z \oplus Z)/\langle(a, b)\rangle$ and related groups.

K. R. McLean, "When Isomorphic Groups Are Not the Same," *Mathematical Gazette* 57 (1973): 207–208.

This article gives a simple example showing that two groups may be isomorphic, but behave differently, when they are subgroups of a larger group.

Évariste Galois

Galois at seventeen was making discoveries of epochal significance in the theory of equations, discoveries whose consequences are not yet exhausted after more than a century.

E. T. Bell, *Men of Mathematics*

This French stamp was issued in the 1984 "Celebrity Series" in support of the Red Cross Fund.

ÉVARISTE GALOIS (pronounced GAL-wah) was born on October 25, 1811, near Paris. He took his first mathematics course when he was 15 and quickly mastered the works of Legendre and Lagrange. At 18, Galois wrote his important research on the theory of equations and submitted it to the French Academy of Sciences for publication. The paper was given to Cauchy for refereeing. Cauchy, impressed by the paper, agreed to present it to the academy, but never did. At the age of 19, Galois entered a paper of the highest quality in the competition for the Grand Prize in Mathematics, given by the French Academy of Sciences. The paper was given to Fourier, who died shortly thereafter. Galois's paper was never seen again.

Galois twice failed his entrance examination to l'École Polytechnique. He did not know some basic mathematics, and he did mathematics almost entirely in his head, to the annoyance of the examiner. Legend has it that Galois became so enraged at the stupidity of the examiner that he threw an eraser at him.

Galois spent most of the last year and a half of his life in prison for revolutionary political offenses. While in prison, he attempted suicide, and prophesied his death in a duel. On May 30, 1832, Galois was shot in a duel and died the next day at the age of 20. The life and death of Galois have long been a source of fascination and speculation for mathematics historians. One article [1] argues convincingly that three of the most widely read accounts of Galois's life are highly fictitious.

Among the many concepts introduced by Galois are normal subgroups, isomorphisms, simple groups, finite fields, and Galois theory. His work provided a method for disposing of several famous constructability problems, such as trisecting an arbitrary angle and doubling a cube. Galois's entire collected works fill only 60 pages.

10

Group Homomorphisms

All modern theories of nuclear and electromagnetic
interactions are based on group theory.

Andrew Watson, *New Scientist*

DEFINITION AND EXAMPLES

In this chapter, we consider one of the most fundamental ideas of algebra—homomorphisms. The term *homomorphism* comes from the
Greek words *homo,* "like," and *morphe,* "form." We will see that a
homomorphism is a natural generalization of an isomorphism and that
there is an intimate connection between factor groups of a group and
homomorphisms of a group. The concept of group homomorphisms
was introduced by Camille Jordan in 1870, in his influential book *Traité
des Substitutions.*

DEFINITION   Group Homomorphism

A homomorphism ϕ from a group G to a group $\overline{G}$ is a mapping from G
into $\overline{G}$ that preserves the group operation; that is, $\phi(ab) = \phi(a)\phi(b)$ for
all a, b in G.

Before giving examples and stating numerous properties of homomorphisms, it is convenient to introduce an important subgroup that is
intimately related to the image of a homomorphism.

DEFINITION   Kernel of a Homomorphism

The *kernel* of a homomorphism ϕ from a group G to a group with identity e is the set $\{x \in G \mid \phi(x) = e\}$. The kernel of ϕ is denoted by Ker ϕ.

Example 1   Any isomorphism is a homomorphism that is also onto and one-to-one. The kernel of an isomorphism is the identity.   □

Example 2   Let $G = GL(2, \mathbf{R})$, and let $\mathbf{R}^{\#}$ be the group of nonzero real numbers under multiplication. Then the determinant mapping $A \rightarrow$ det A is a homomorphism from G to $\mathbf{R}^{\#}$. The kernel of the determinant mapping is $SL(2, \mathbf{R})$.   □

Example 3   The mapping from $\mathbf{R}^{\#}$ to $\mathbf{R}^{\#}$ that sends x to the absolute value of x is a homomorphism. The kernel of the absolute value mapping is $\{1, -1\}$.   □

Example 4   Let $\mathbf{R}[x]$ denote the group of all polynomials with real coefficients under addition. For any f in $\mathbf{R}[x]$, let f' denote the derivative of f. Then the mapping $f \rightarrow f'$ is a homomorphism from $\mathbf{R}[x]$ to itself. The kernel of the derivative mapping is the set of all constant polynomials.   □

Example 5   The mapping of ϕ from Z to Z_n, defined by $\phi(m) = r$ where r is the remainder of m when divided by n [that is, $\phi(m) = m$ mod n], is a homomorphism. The kernel of this mapping is $\langle n \rangle$.   □

The natural homomorphism from Z to Z_n given in Example 5 has many applications in number theory. In 1770, Lagrange proved that every positive integer can be written as the sum of four squares (that is, in the form $a^2 + b^2 + c^2 + d^2$). Our next example shows that there are infinitely many integers that are not the sum of three squares.

Example 6   No integer equal to 7 modulo 8 can be written in the form $a^2 + b^2 + c^2$. If this were so, then $7 = a^2$ mod $8 + b^2$ mod $8 + c^2$ mod 8. Now observe that the square of any even integer is 0 or 4 mod 8, whereas the square of any odd integer is 1 mod 8 (see exercise 37 in Chapter 0). But no three numbers chosen from 0, 1, and 4 add up to 7 mod 8.   □

Example 7   The mapping from the group of real numbers under addition to itself given by $x \rightarrow [x]$, the greatest integer less than or equal to x, is *not* a homomorphism, since $[1/2 + 1/2] \neq [1/2] + [1/2]$.   □

When defining a homomorphism from a group in which there are several ways to represent the elements, caution must be exercised to ensure that the correspondence is a function. (The term *well defined* is often used in this context.) For example, since $3(x + y) = 3x + 3y$ in Z_6, one might believe that the correspondence $x + \langle 3 \rangle \rightarrow 3x$ from $Z/\langle 3 \rangle$ to Z_6 is a homomorphism. But it is not a function since $0 + \langle 3 \rangle = 3 + \langle 3 \rangle$ in $Z/\langle 3 \rangle$ but $3 \cdot 0 \neq 3 \cdot 3$ in Z_6.

For students who have had linear algebra, we remark that every linear transformation is a group homomorphism and the nullspace is the same as the kernel. An invertible linear transformation is a group isomorphism.

PROPERTIES OF HOMOMORPHISMS

Theorem 10.1 *Properties of Homomorphisms*
Let ϕ be a homomorphism from a group G to a group $\overline{G}$. Let g be an element of G and H a subgroup of G. Then

1. *ϕ carries the identity of G to the identity of $\overline{G}$.*
2. *$\phi(g^n) = (\phi(g))^n$.*
3. *$\phi(H) = \{\phi(h) \mid h \in H\}$ is a subgroup of $\overline{G}$.*
4. *If H is cyclic, then $\phi(H)$ is cyclic.*
5. *If H is Abelian, then $\phi(H)$ is Abelian.*
6. *If H is normal in G, then $\phi(H)$ is normal in $\phi(G)$.*
7. *If $|g| = n$, then $|\phi(g)|$ divides n.*
8. *If $\phi(g) = g'$, then $\phi^{-1}(g') = \{x \in G \mid \phi(x) = g'\} = g\text{Ker }\phi$.*
9. *If $|H| = n$, then $|\phi(H)|$ divides n.*
10. *If $|\text{Ker }\phi| = n$, then ϕ is an n-to-1 mapping from G onto $\phi(G)$.*
11. *If $\overline{K}$ is a subgroup of $\overline{G}$, then $\phi^{-1}(\overline{K}) = \{k \in G \mid \phi(k) \in \overline{K}\}$ is a subgroup of G.*
12. *If $\overline{K}$ is a normal subgroup of $\overline{G}$, then $\phi^{-1}(\overline{K}) = \{k \in G \mid \phi(k) \in \overline{K}\}$ is a normal subgroup of G.*
13. *If ϕ is onto and $\text{Ker }\phi = \{e\}$, then ϕ is an isomorphism from G to $\overline{G}$.*

Proof. The proofs of these properties of homomorphisms are straightforward and are left as exercises. ∎

A few remarks about Theorem 10.1 are in order here. Students should remember the various properties of the theorem in words. For example, property 4 says that the homomorphic image of a cyclic group is cyclic. Property 6 says that the homomorphic image of a normal subgroup of G is normal in the image of G. Property 10 says that if ϕ is a homomorphism from G to $\overline{G}$, then every element of $\overline{G}$ that gets "hit" by ϕ gets hit the same number of times as does the identity. The set $\phi^{-1}(g')$ defined in property 8 is called the *inverse image of g'* (or *pullback of g'*). Note that the inverse image of an element is a coset of the kernel. Similarly, the set $\phi^{-1}(\overline{K})$ defined in property 11 is called the *inverse image* of $\overline{K}$ (or *pullback* of $\overline{K}$).

Property 8 of Theorem 10.1 is reminiscent of something from linear algebra and differential equations. Recall that if x is a particular solution to a system of linear equations and S is the entire solution set of the corresponding homogeneous system of linear equations, then $x + S$ is the entire solution set of the nonhomogeneous system. In reality, this statement is just a special case of property 8. Properties 1, 8, and 10 of Theorem 10.1 are pictorially represented in Figure 10.1.

The special case of property 12 where $\overline{K} = \{e\}$ is of such importance that we single it out.

Corollary   *Kernels Are Normal*
Let ϕ be a group homomorphism from G to $\overline{G}$. Then Ker ϕ is a normal subgroup of G.

The following example illustrates several properties of Theorem 10.1.

Example 8   Consider the mapping ϕ from $\mathbf{C}^*$ to $\mathbf{C}^*$ given by $x \to x^4$. Since $(xy)^4 = x^4y^4$, ϕ is a homomorphism. Clearly, Ker $\phi = \{x \mid x^4 = 1\} = \{1, -1, i, -i\}$. So, by property 10, we know that ϕ is a 4-to-1 mapping. Now let's find all elements that map to 2, say. Certainly, $\phi(\sqrt[4]{2}) = 2$. Then, by property 8, the set of all elements that map to 2 is $\sqrt[4]{2}$ Ker $\phi = \{\sqrt[4]{2}, -\sqrt[4]{2}, \sqrt[4]{2}i, -\sqrt[4]{2}i\}$.

Finally, we verify a specific instance of properties 4, 7, and 9. Let $H = \langle \cos 30° + i \sin 30° \rangle$. It follows from DeMoivre's Theorem (Example 11 in Chapter 0) that $|H| = 12$, $\phi(H) = \langle \cos 120° + i \sin 120° \rangle$, and $|\phi(H)| = 3$.   □

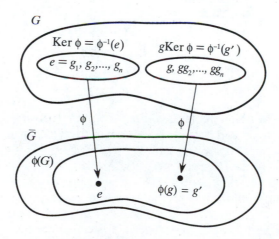

Figure 10.1

The next example illustrates how one can easily determine all homomorphisms from a cyclic group to a cyclic group.

Example 9   We determine all homomorphisms from Z_{12} to Z_{30}. By property 2 of Theorem 10.1, a homomorphism is completely specified by the image of 1. That is, if 1 maps to a, then x maps to xa. Lagrange's Theorem and property 7 of Theorem 10.1 require that $|a|$ divide both 12 and 30. So, $|a| = 1, 2, 3, 6$. Thus, $a = 0, 15, 10, 20, 5$, or 25. This gives us a list of candidates for the homomorphisms. That each of these six possibilities yields an operation-preserving, well-defined function can now be verified by direct calculations. [Note that $\gcd(12, 30) = 6$. This is not a coincidence!]   ❑

Example 10   The mapping from S_n to Z_2 that takes an even permutation to 0 and an odd permutation to 1 is a homomorphism. Figure 10.2 illustrates the telescoping nature of the mapping.   ❑

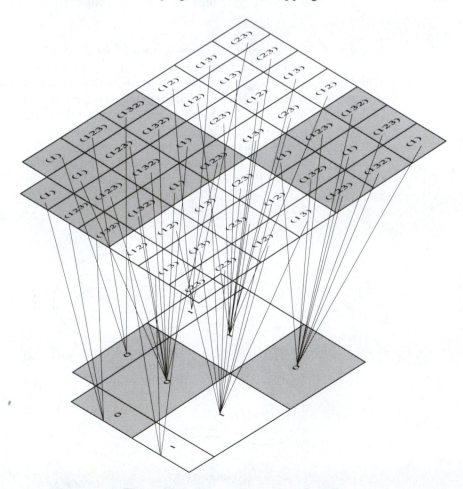

Figure 10.2   Homomorphism from S_3 to Z_2.

THE FIRST ISOMORPHISM THEOREM

In Chapter 9, we showed that for a group G and a normal subgroup H, we could arrange the Cayley table of G into boxes that represented the cosets of H in G, and these boxes then became a Cayley table for G/H. The next theorem shows that for any homomorphism ϕ of G and the normal subgroup Ker ϕ, the same process produces a Cayley table isomorphic to the homomorphic image of G. Thus, homomorphisms, like factor groups, cause a *systematic* collapsing of a group to a simpler but closely related group. This can be likened to viewing a group through the reverse end of a telescope—the general features of the groups are present, but the apparent size is diminished. The important relationship between homomorphisms and factor groups given below is often called the Fundamental Theorem of Group Homomorphisms.

Theorem 10.2 *First Isomorphism Theorem (Jordan, 1870)*

Let ϕ be a group homomorphism from G to $\overline{G}$. Then the mapping from $G/\text{Ker } \phi$ to $\phi(G)$, given by $g\text{Ker } \phi \rightarrow \phi(g)$, is an isomorphism. In symbols, $G/\text{Ker } \phi \approx \phi(G)$.

Proof.   Let us use ψ to denote the correspondence $g\text{Ker } \phi \rightarrow \phi(g)$. First, we show that ψ is well defined (that is, the correspondence is independent of the particular coset representative chosen). Suppose $x\text{Ker } \phi = y\text{Ker } \phi$. Then $y^{-1}x \in \text{Ker } \phi$ and $e = \phi(y^{-1}x) = (\phi(y))^{-1}\phi(x)$. Thus $\phi(x) = \phi(y)$, and ψ is indeed a function. To show that ψ is operation-preserving, observe that $\psi(x\text{Ker } \phi y\text{Ker } \phi) = \psi(xy\text{Ker } \phi) = \phi(xy) = \phi(x)\phi(y) = \psi(x\text{Ker } \phi)\psi(y\text{Ker } \phi)$. Finally, if $\psi(g_1\text{Ker } \phi) = \psi(g_2\text{Ker } \phi)$, then $\phi(g_1) = \phi(g_2)$, so that $g_2^{-1}g_1 \in \text{Ker } \phi$. It follows that ψ is one-to-one. ∎

Example 11   To illustrate Theorem 10.2 and its proof, consider the homomorphism ϕ from D_4 to itself given by

$$R_0 \quad R_{180} \qquad R_{90} \quad R_{270} \qquad H \quad V \qquad D \quad D'$$
$$\searrow \quad \swarrow \qquad\qquad \searrow \quad \swarrow \qquad\qquad \searrow \quad \swarrow \qquad\qquad \searrow \quad \swarrow$$
$$R_0 \qquad\qquad\qquad H \qquad\qquad\qquad R_{180} \qquad\qquad\qquad V$$

Then Ker $\phi = \{R_0, R_{180}\}$, and the mapping ψ in Theorem 10.2 is $R_0\text{Ker } \phi \rightarrow R_0$, $R_{90}\text{Ker } \phi \rightarrow H$, $H\text{Ker } \phi \rightarrow R_{180}$, $D\text{Ker } \phi \rightarrow V$. It is straightforward to verify that the mapping ψ is an isomorphism. ❑

Mathematicians often give a pictorial representation of Theorem 10.2, as follows:

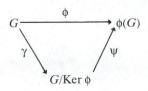

where $\gamma: G \to G/\text{Ker } \phi$ is defined as $\gamma(g) = g\text{Ker } \phi$. The mapping γ is called the *natural mapping* from G to $G/\text{Ker } \phi$. Our proof of Theorem 10.2 shows that $\psi\gamma = \phi$. In this case, one says that the above diagram is *commutative*.

As a consequence of Theorem 10.2, we see that all homomorphic images of G can be determined using G. For we may simply consider the various factor groups of G. For example, we know that the homomorphic image of an Abelian group is Abelian because the factor group of an Abelian group is Abelian. We know that the number of homomorphic images of a cyclic group G of order n is the number of divisors of n, since there is exactly one subgroup of G (and therefore one factor group of G) for each divisor of n. (Be careful: the number of homomorphisms of a cyclic group of order n need not be the same as the number of divisors of n, since different homomorphisms can have the same image.)

An appreciation for Theorem 10.2 can be gained by looking at a few examples.

Example 12 $Z/\langle n \rangle \approx Z_n$
Consider the mapping from Z to Z_n defined in Example 5. Clearly, its kernel is $\langle n \rangle$. So, by Theorem 10.2, $Z/\langle n \rangle \approx Z_n$.   ❑

Example 13   The Wrapping Function
Recall the wrapping function W from trigonometry. The real number line is wrapped around a unit circle in the plane centered at $(0, 0)$ with the number 0 on the number line at the point $(1, 0)$, the positive reals in the counterclockwise direction, and the negative reals in the clockwise direction (see Figure 10.3). The function W assigns to each real number a, the point a radians from $(1, 0)$ on the circle. This mapping is a homomorphism from the group **R** under addition onto the circle group (the group of complex numbers of magnitude 1 under multiplication—see exercise 11, Supplementary Exercises for Chapters 5–7). Indeed, it follows from elementary facts of trigonometry that $W(x) = \cos x + i \sin x$ and $W(x + y) = W(x)W(y)$. Since W is periodic of period 2π, Ker $W = \langle 2\pi \rangle$. So, from the First Isomorphism Theorem, we see that **R**$/\langle 2\pi \rangle$ is isomorphic to the circle group.   ❑

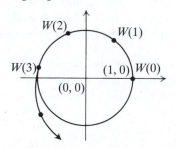

Figure 10.3

Our next example is a theorem that is used repeatedly in Chapters 24 and 25.

Example 14   The N/C Theorem
Let H be a subgroup of a group G. Recall that the normalizer of H in G is $N(H) = \{x \in G \mid xHx^{-1} = H\}$ and the centralizer of H in G is $C(H) = \{x \in G \mid xhx^{-1} = h \text{ for all } h \text{ in } H\}$. Consider the mapping from $N(H)$ to Aut(H) given by $g \to \phi_g$, where ϕ_g is the inner automorphism of H induced by g (that is, $\phi_g(h) = ghg^{-1}$ for all h in H). This mapping is a homomorphism with kernel $C(H)$. So, by Theorem 10.2, $N(H)/C(H)$ is isomorphic to a subgroup of Aut(H).   ☐

The corollary of Theorem 10.1 says that the kernel of every homomorphism of a group is a normal subgroup of the group. We conclude this chapter by verifying that the converse of this statement is also true.

Theorem 10.3   *Normal Subgroups Are Kernels*
Every normal subgroup of a group G is the kernel of a homomorphism of G. In particular, a normal subgroup N is the kernel of the mapping $g \to gN$ from G to G/N.

Proof.   Define $\phi:G \to G/N$ by $\phi(g) = gN$. (This mapping is called the *natural homomorphism* from G to G/N.) Then, $\phi(xy) = (xy)N = xNyN = \phi(x)\phi(y)$.   ■

Examples 12, 13, and 14 illustrate the utility of the First Isomorphism Theorem. But what about homomorphisms in general? Why would one care to study a homomorphism of a group? The answer is that, just as was the case with factor groups of a group, homomorphic images of a group tell us *some* of the properties of the original group. One measure of the likeness of a group and its homomorphic image is the size of the kernel. If the kernel of the homomorphism of group G is the identity, then the image of G tells us everything (group theoretically) about G (the two being isomorphic). On the other hand, if the kernel of the homomorphism is G itself, then the image tells us nothing about G. Between these two extremes, some information about G is preserved and some is lost. The utility of a particular homomorphism lies in its ability to preserve the properties of the group we want, while losing some inessential ones. In this way, we have replaced G by a group less complicated (and therefore easier to study) than G; but, in the process, we have saved enough information to answer questions that we have about G itself. For example, if G is a group of order 60 and G has a homomorphic image of order 12 that is cyclic, then we know from properties 10, 11, and 12 of Theorem 10.1 that G has normal subgroups of orders 5, 10, 15, 20, 30, and 60.
Perhaps the following analogy between homomorphisms and

photography is instructive.* A photograph of a person cannot tell us exactly the person's height, weight, or age. Nevertheless, it *may* be possible to decide from a photograph whether the person is tall or short, heavy or thin, old or young, male or female. In the same way, a homomorphic image of a group gives us *some* information about the group.

In certain branches of group theory, and especially in physics and chemistry, one often wants to know all homomorphic images of a group that are matrix groups over the complex numbers (these are called *group representations*). Here, we may carry our analogy with photography one step further by saying that this is like wanting photographs of a person from many different angles (front view, profile, head-to-toe view, close-up, etc.), as well as x-rays! Just as this composite information from the photographs reveals much about the person, several homomorphic images of a group reveal much about the group.

EXERCISES

The greater the difficulty, the more glory in surmounting it. Skillful pilots gain their reputation from storms and tempests.

Epicurus

1. Let $\mathbf{R}^{\#}$ be the group of nonzero real numbers under multiplication, and let r be a positive integer. Show that the mapping that takes x to x^r is a homomorphism from $\mathbf{R}^{\#}$ to $\mathbf{R}^{\#}$.

2. Let G be the group of all polynomials with real coefficients under addition. For each f in G, let $\int f$ denote the antiderivative of f that passes through the point $(0, 0)$. Show that the mapping $f \rightarrow \int f$ from G to G is a homomorphism. What is the kernel of this mapping? Is this mapping a homomorphism if $\int f$ denotes the antiderivative of f that passes through $(0, 1)$?

3. Prove that the mapping given in Example 2 is a homomorphism.

4. Prove that the mapping given in Example 3 is a homomorphism.

5. Prove that the mapping given in Example 4 is a homomorphism.

6. Let G be a group of permutations. For each σ in G, define

$$\text{sgn}(\sigma) = \begin{cases} +1 & \text{if } \sigma \text{ is an even permutation,} \\ -1 & \text{if } \sigma \text{ is an odd permutation.} \end{cases}$$

Prove that sgn is a homomorphism from G to $\{+1, -1\}$. What is the kernel?

---

*All perception of truth is the detection of an analogy. Henry David Thoreau, *Journal*.

7. Prove that the mapping from $G \oplus H$ to G given by $(g, h) \rightarrow g$ is a homomorphism. What is the kernel? This mapping is called the *projection* of $G \oplus H$ onto G.

8. Let G be a subgroup of some dihedral group. For each x in G, define

$$\phi(x) = \begin{cases} +1 & \text{if } x \text{ is a rotation,} \\ -1 & \text{if } x \text{ is a reflection.} \end{cases}$$

   Prove that ϕ is a homomorphism from G to $\{+1, -1\}$. What is the kernel of ϕ?

9. Prove that $(Z \oplus Z)/(\langle(a, 0)\rangle \oplus \langle(0, b)\rangle)$ is isomorphic to $Z_a \oplus Z_b$.

10. Suppose that k is a divisor of n. Prove that $Z_n/\langle k \rangle \approx Z_k$.

11. Prove that $(A \oplus B)/(A \oplus \{e\}) \approx B$.

12. Explain why the correspondence $x \rightarrow 3x$ from Z_{12} to Z_{10} is not a homomorphism.

13. Suppose that ϕ is a homomorphism from Z_{30} to Z_{30} and Ker $\phi = \{0, 10, 20\}$. If $\phi(23) = 6$, determine all elements that map to 6.

14. Prove that there is no homomorphism from $Z_8 \oplus Z_2$ onto $Z_4 \oplus Z_4$.

15. Prove that there is no homomorphism from $Z_{16} \oplus Z_2$ onto $Z_4 \oplus Z_4$.

16. Can there be a homomorphism from $Z_4 \oplus Z_4$ onto Z_8? Can there be a homomorphism from Z_{16} onto $Z_2 \oplus Z_2$? Explain your answers.

17. Suppose that there is a homomorphism ϕ from Z_{17} to some group and that ϕ is not one-to-one. Determine ϕ.

18. How many homomorphisms are there from Z_{20} onto Z_8? How many are there to Z_8?

19. If ϕ is a homomorphism from Z_{30} onto a group of order 5, determine the kernel of ϕ.

20. Suppose that ϕ is a homomorphism from a finite group G onto $\overline{G}$ and that $\overline{G}$ has an element of order 8. Prove that G has an element of order 8. Generalize.

21. How many homomorphisms are there from Z_{20} onto Z_{10}? How many are there to Z_{10}?

22. Determine all homomorphisms from Z_4 to $Z_2 \oplus Z_2$.

23. Determine all homomorphisms from Z_n to itself.

24. Suppose that ϕ is a homomorphism from S_4 onto Z_2. Determine Ker ϕ. Determine all homomorphisms from S_4 to Z_2.

25. Suppose that there is a homomorphism from a finite group G onto Z_{10}. Prove that G has normal subgroups of index 2 and 5.

26. Suppose that ϕ is a homomorphism from a group G onto $Z_6 \oplus Z_2$ and that the kernel of ϕ has order 5. Explain why G must have normal subgroups of orders 5, 10, 15, 20, 30, and 60.

27. Suppose that ϕ is a homomorphism from $U(30)$ to $U(30)$ and that Ker $\phi = \{1, 11\}$. If $\phi(7) = 7$, find all elements of $U(30)$ that map to 7.

28. Find a homomorphism ϕ from $U(30)$ to $U(30)$ with kernel $\{1, 11\}$ and $\phi(7) = 7$.

29. Suppose that ϕ is a homomorphism from $U(40)$ to $U(40)$ and that Ker $\phi = \{1, 9, 17, 33\}$. If $\phi(11) = 11$, find all elements of $U(40)$ that map to 11.

30. Find a homomorphism ϕ from $U(40)$ to $U(40)$ with kernel $\{1, 9, 17, 33\}$ and $\phi(11) = 11$.

31. Prove that the mapping $\phi: Z \oplus Z \rightarrow Z$ given by $(a, b) \rightarrow a - b$ is a homomorphism. What is the kernel of ϕ? Describe the set $\phi^{-1}(3)$ (that is, all elements that map to 3).

32. Suppose that there is a homomorphism ϕ from $Z \oplus Z$ to a group G such that $\phi((3, 2)) = a$ and $\phi((2, 1)) = b$. Determine $\phi((4, 4))$ in terms of a and b. Assume that the operation of G is addition.

33. Prove that the mapping $x \rightarrow x^6$ from C^* to C^* is a homomorphism. What is the kernel?

34. For each pair of positive integers m and n, we can define a homomorphism from Z to $Z_m \oplus Z_n$ by $x \rightarrow (x \bmod m, x \bmod n)$. What is the kernel when $(m, n) = (3, 4)$? What is the kernel when $(m, n) = (6, 4)$? Generalize.

35. (Second Isomorphism Theorem) If K is a subgroup of G and N is a normal subgroup of G, prove that $K/(K \cap N)$ is isomorphic to KN/N.

36. (Third Isomorphism Theorem) If M and N are normal subgroups of G and $N \leq M$, prove that $(G/N)/(M/N) \approx G/M$.

37. Let $\phi(d)$ denote the Euler phi function of d (see page 74). Show that the number of homomorphisms from Z_n to Z_k is $\Sigma\phi(d)$, where the sum runs over all common divisors d of n and k. [It follows from number theory that this sum is actually $\gcd(n, k)$.]

38. Let k be a divisor of n. Consider the homomorphism from $U(n)$ to $U(k)$ given by $x \rightarrow x \bmod k$. What is the relationship between this homomorphism and the subgroup $U_k(n)$ of $U(n)$?

39. Determine all homomorphic images of D_4 (up to isomorphism).

40. Let N be a normal subgroup of a finite group G. Use the theorems of this chapter to prove that the order of the group element gN in G/N divides the order of g.

41. Suppose that G is a finite group and that Z_{10} is a homomorphic image of G. What can we say about $|G|$?

42. Suppose that Z_{10} and Z_{15} are both homomorphic images of a finite group G. What can be said about $|G|$?

43. Suppose that for each prime p, Z_p is the homomorphic image of a group G. What can we say about $|G|$?

44. (For students who have had linear algebra.) Suppose that x is a particular solution to a system of linear equations and that S is the entire solution set of the corresponding homogeneous system of linear equations.

Explain why property 8 of Theorem 10.1 guarantees that $x + S$ is the entire solution set of the nonhomogeneous system. In particular, describe the relevant groups and the homomorphism between them.

45. Let N be a normal subgroup of a group G. Use property 12 of Theorem 10.1 to prove that every subgroup of G/N has the form H/N, where H is a subgroup of G. (This exercise is referred to in Chapter 24.)

46. Show that a homomorphism defined on a cyclic group is completely determined by its action on a generator of the group.

47. Use the First Isomorphism Theorem to prove Theorem 9.4.

48. Suppose that $\overline{G}$ is the homomorphic image of a finite group G. If $\overline{G}$ has an element of order n, show that G has an element of order n.

49. Let $Z[x]$ be the group of polynomials in x with integer coefficients under addition. Prove that the mapping from $Z[x]$ into Z given by $f(x) \rightarrow f(3)$ is a homomorphism. Give a geometric description of the kernel of this homomorphism. Generalize.

50. If H and K are normal subgroups of G and $H \cap K = \{e\}$, prove that G is isomorphic to a subgroup of $G/H \oplus G/K$.

51. Suppose that H and K are distinct subgroups of G of index 2. Prove that $H \cap K$ is a normal subgroup of G of index 4 and that $G/(H \cap K)$ is not cyclic.

52. Suppose that the number of homomorphisms from G to H is n. How many homomorphisms are there from G to $H \oplus H \oplus \cdots \oplus H$ (s terms)? When H is Abelian, how many homomorphisms are there from $G \oplus G \oplus \cdots \oplus G$ (s terms) to H?

SUGGESTED READING

Loren Larson, "A Theorem About Primes Proved on a Chessboard," *Mathematics Magazine* 50 (1977): 69–74.

This paper gives a chessboard interpretation of several algebraic concepts such as cosets and group homomorphisms. These ideas are used to solve the "*n*-queens" problem and to prove Fermat's Two Square Theorem: every prime $p = 1 \bmod 4$ is the sum of two unique squares.

Camille Jordan

Although these contributions [to analysis and topology] would have been enough to rank Jordan very high among his mathematical contemporaries, it is chiefly as an algebraist that he reached celebrity when he was barely thirty; and during the next forty years he was universally regarded as the undisputed master of group theory.

J. Dieudonné, *Dictionary of Scientific Biography*

CAMILLE JORDAN was born into a well-to-do family on January 5, 1838, in Lyons, France. Like his father, he graduated from the École Polytechnique and became an engineer. Nearly all of his 120 research papers in mathematics were written before his retirement from engineering in 1885. From 1873 until 1912, Jordan taught simultaneously at the École Polytechnique and at the College of France.

In the great French tradition, Jordan was a universal mathematician who published in nearly every branch of mathematics. Among the concepts named after him are the Jordan canonical form in matrix theory, the Jordan curve theorem from topology, the Jordan-Hölder theorem from group theory, and Jordan algebras. His classic book *Traité des Substitutions,* published in 1870, was the first to be devoted solely to group theory and its applications to other branches of mathematics. This book provided the first clear and complete account of the theory invented by Galois to determine which polynomials are solvable by radicals, and it was the first major investigation of infinite groups. In the book, Jordan coined the word *Abelian* to describe commutative groups and, although Galois had introduced the term *group,* it was through the influence of Jordan's book that the term became standard.

Another book that had great influence and set a new standard for rigor was his *Cours d'analyse*. This book gave the first clear definitions of the notions of *volume* and *multiple integral*. It also gave conditions under which a multiple integral can be evaluated by successive integrations. Nearly 100 years after this book appeared, the distinguished mathematician and mathematical historian B. L. van der Waerden wrote "For me, every single chapter of the *Cours d'analyse* is a pleasure to read." Jordan died in Paris on January 22, 1922.

11

Fundamental Theorem of Finite Abelian Groups

By a small sample we may judge of the whole piece.

Miguel de Cervantes, *Don Quixote*

THE FUNDAMENTAL THEOREM

In this chapter, we present a theorem that describes to an algebraist's eyes (that is, up to isomorphism) all finite Abelian groups in a standardized way. Before giving the proof, which is long and difficult, we discuss some consequences of the theorem and its proof. The first proof of the theorem was given by Leopold Kronecker in 1858.

Theorem 11.1 *Fundamental Theorem of Finite Abelian Groups*
Every finite Abelian group is a direct product of cyclic groups of prime-power order. Moreover, the factorization is unique except for rearrangement of the factors.

Since a cyclic group of order n is isomorphic to Z_n, Theorem 11.1 shows that every finite Abelian group G is isomorphic to a group of the form

$$Z_{p_1^{n_1}} \oplus Z_{p_2^{n_2}} \oplus \cdots \oplus Z_{p_k^{n_k}},$$

where the p_i's are not necessarily distinct primes and the prime-powers $p_1^{n_1}$, $p_2^{n_2}$, ... , $p_k^{n_k}$ are uniquely determined by G. Writing a group in this form is called *determining the isomorphism class of G.*

THE ISOMORPHISM CLASSES OF ABELIAN GROUPS

The Fundamental Theorem is extremely powerful. As an application, we can use it as an algorithm for constructing all Abelian groups of any order. Let's look at groups whose orders have the form p^k where p is prime and $k \le 4$.

Order of G	Possible direct products for G
p	Z_p
p^2	Z_{p^2}
	$Z_p \oplus Z_p$
p^3	Z_{p^3}
	$Z_{p^2} \oplus Z_p$
	$Z_p \oplus Z_p \oplus Z_p$
p^4	Z_{p^4}
	$Z_{p^3} \oplus Z_p$
	$Z_{p^2} \oplus Z_{p^2}$
	$Z_{p^2} \oplus Z_p \oplus Z_p$
	$Z_p \oplus Z_p \oplus Z_p \oplus Z_p$

In general, there is one group of order p^k for each set of positive integers whose sum is k (such a set is called a *partition* of k); that is, if k can be written as

$$k = n_1 + n_2 + \cdots + n_t,$$

where each n_i is a positive integer, then

$$Z_{p^{n_1}} \oplus Z_{p^{n_2}} \oplus \cdots \oplus Z_{p^{n_t}}$$

is an Abelian group of order p^k. Furthermore, the uniqueness portion of the Fundamental Theorem guarantees that distinct partitions of k yield distinct isomorphism classes. Thus, for example, $Z_9 \oplus Z_3$ is not isomorphic to $Z_3 \oplus Z_3 \oplus Z_3$. A reliable mnemonic for comparing external direct products is the cancellation property: if A is *finite,* then

$$A \oplus B \approx A \oplus C \quad \text{if and only if} \quad B \approx C \qquad \text{(see [1]).}$$

Thus $Z_4 \oplus Z_4$ is not isomorphic to $Z_4 \oplus Z_2 \oplus Z_2$ because Z_4 is not isomorphic to $Z_2 \oplus Z_2$.

To appreciate fully the potency of the Fundamental Theorem, contrast the ease with which the Abelian groups of order p^k, $k \le 4$, were determined with the corresponding problem for non-Abelian groups.

Even a description of the two non-Abelian groups of order 8 is a challenge (see Chapter 26), and a description of the nine non-Abelian groups of order 16 is well beyond the level of this text.

Now that we know how to construct all the Abelian groups of prime-power order, we move to the problem of constructing all Abelian groups of a certain order n where n has two or more distinct prime divisors. We begin by writing n in prime-power decomposition form $n = p_1^{n_1} p_2^{n_2} \cdots p_k^{n_k}$. Next, individually form all Abelian groups of order $p_1^{n_1}$, then $p_2^{n_2}$, and so on, as described earlier. Finally, form all possible external direct products of these groups. For example, let $n = 1176 = 2^3 \cdot 3 \cdot 7^2$. Then, the complete list of the distinct isomorphism classes of Abelian groups of order 1176 is

$$Z_8 \oplus Z_3 \oplus Z_{49},$$
$$Z_4 \oplus Z_2 \oplus Z_3 \oplus Z_{49},$$
$$Z_2 \oplus Z_2 \oplus Z_2 \oplus Z_3 \oplus Z_{49},$$
$$Z_8 \oplus Z_3 \oplus Z_7 \oplus Z_7,$$
$$Z_4 \oplus Z_2 \oplus Z_3 \oplus Z_7 \oplus Z_7,$$
$$Z_2 \oplus Z_2 \oplus Z_2 \oplus Z_3 \oplus Z_7 \oplus Z_7.$$

If we are given any particular Abelian group G of order 1176, the question we want to answer about G is which of the preceding six isomorphism classes represents the structure of G? We can answer this question by comparing the orders of the elements of G with the orders of the elements in the six direct products, since it can be shown that two Abelian groups are isomorphic if and only if they have the same number of elements of each order. For instance, we could determine whether G has any elements of order 8. If so, then G must be isomorphic to the first or fourth group above, since these are the only ones with elements of order 8. To narrow G down to a single choice, we now need only check whether or not G has an element of order 49, since the first product above has such an element, whereas the fourth one does not.

What if we have some specific Abelian group G of order $p_1^{n_1} p_2^{n_2} \cdots p_k^{n_k}$ where the p_i's are distinct primes? How can G be expressed as an *internal* direct product of cyclic groups of prime-power order? For simplicity, let us say that the group has 2^n elements. First, we must compute the orders of the elements. After this is done, pick an element of maximum order 2^r, call it a_1. Then $\langle a_1 \rangle$ is one of the factors in the desired internal direct product. If $G \neq \langle a_1 \rangle$, choose an element a_2 of maximum order 2^s such that $s \leq n - r$ and none of $a_2, a_2^2, a_2^4, \ldots, a_2^{2^{s-1}}$ is in $\langle a_1 \rangle$. Then $\langle a_2 \rangle$ is a second direct factor. If $n \neq r + s$, select an element a_3 of maximum order 2^t such that $t \leq n - r - s$ and none of $a_3, a_3^2, a_3^4, \ldots, a_3^{2^{t-1}}$ is in $\langle a_1 \rangle \times \langle a_2 \rangle = \{a_1^i a_2^j \mid 0 \leq i < 2^r, 0 \leq j < 2^s\}$. Then $\langle a_3 \rangle$ is another direct factor. We continue in this fashion until our direct product has the same order as G.

A formal presentation of this algorithm for any Abelian group G of prime-power order p^n is as follows:

Greedy Algorithm for an Abelian Group of Order p^n

1. *Compute the orders of the elements of the group G.*
2. *Select an element a_1 of maximum order and define $G_1 = \langle a_1 \rangle$. Set $i = 1$.*
3. *If $|G| = |G_i|$, stop. Otherwise, replace i by $i + 1$.*
4. *Select an element a_i of maximum order p^k such that $p^k \leq |G|/|G_{i-1}|$ and none of $a_i, a_i^p, a_i^{p^2}, \ldots, a_i^{p^{k-1}}$ is in G_{i-1}, and define $G_i = G_{i-1} \times \langle a_i \rangle$.*
5. *Return to step 3.*

In the general case where $|G| = p_1^{n_1} p_2^{n_2} \cdots p_k^{n_k}$, one simply uses the algorithm to build up a direct product of order $p_1^{n_1}$, then another of order $p_2^{n_2}$, and so on. The direct product of all of these pieces is the desired factorization of G. The following example is small enough that we can compute the appropriate internal and external direct products by hand.

Example 1 Let $G = \{1, 8, 12, 14, 18, 21, 27, 31, 34, 38, 44, 47, 51, 53, 57, 64\}$ under multiplication modulo 65. Since G has order 16, we know it is isomorphic to one of

$$Z_{16},$$
$$Z_8 \oplus Z_2,$$
$$Z_4 \oplus Z_4,$$
$$Z_4 \oplus Z_2 \oplus Z_2,$$
$$Z_2 \oplus Z_2 \oplus Z_2 \oplus Z_2.$$

To decide which one, we dirty our hands to calculate the orders of the elements of G.

Element	1	8	12	14	18	21	27	31	34	38	44	47	51	53	57	64
Order	1	4	4	2	4	4	4	4	4	4	4	4	2	4	4	2

From the table of orders we can instantly rule out all but $Z_4 \oplus Z_4$ and $Z_4 \oplus Z_2 \oplus Z_2$ as possibilities. Finally, we observe that this latter group has only eight elements of order 4 (exercise 4) so that $G \approx Z_4 \oplus Z_4$.

Expressing G as an internal direct product is even easier. Pick an element of maximum order, say, the element 8. Then $\langle 8 \rangle$ is a factor in the product. Next, choose a second element, say, a, so that a has order 4 and a and a^2 are not in $\langle 8 \rangle = \{1, 8, 64, 57\}$. Since 12 has this property, we have $G = \langle 8 \rangle \times \langle 12 \rangle$. ☐

Example 1 illustrates how quickly and easily one can write an Abelian group as a direct product given the orders of the elements of the group. But calculating all those orders is certainly not an appealing prospect! The good news is that, in practice, a combination of theory and calculation of the orders of a few elements will usually suffice.

Example 2 Let $G = \{1, 8, 17, 19, 26, 28, 37, 44, 46, 53, 62, 64, 71, 73,$ $82, 89, 91, 98, 107, 109, 116, 118, 127, 134\}$ under multiplication modulo 135. Since G has order 24, it is isomorphic to one of

$$Z_8 \oplus Z_3 \approx Z_{24},$$
$$Z_4 \oplus Z_2 \oplus Z_3 \approx Z_{12} \oplus Z_2,$$
$$Z_2 \oplus Z_2 \oplus Z_2 \oplus Z_3 \approx Z_6 \oplus Z_2 \oplus Z_2.$$

Consider the element 8. Direct calculations show that $8^6 = 109$ and $8^{12} = 1$. (Be sure to mod as you go. For example, $8^3 = 512 = 107 \bmod 135$, so compute 8^4 as $8 \cdot 107$ rather than $8 \cdot 512$.) But now we know G. Why? Clearly, $|8| = 12$ rules out the third group in the list. At the same time, $|109| = 2 = |134|$ (remember, $134 = -1 \bmod 135$) implies that G is not Z_{24} (see Theorem 4.4). Thus, $G \approx Z_{12} \oplus Z_2$, and $G = \langle 8 \rangle \times \langle 134 \rangle$. ◻

Rather than express an Abelian group as a direct product of cyclic groups of prime-power orders, it is often more convenient to combine the cyclic factors of relatively prime order, as we did in Example 2, to obtain a direct product of the form $Z_{n_1} \oplus Z_{n_2} \oplus \cdots \oplus Z_{n_k}$ where n_i divides n_{i-1}. For example, $Z_4 \oplus Z_4 \oplus Z_2 \oplus Z_9 \oplus Z_3 \oplus Z_5$ would be written as $Z_{180} \oplus Z_{12} \oplus Z_2$ (see exercise 10). The algorithm above is easily adapted to accomplish this by replacing step 4 by 4′: select an element a_i of maximum order m such that $m \leq |G|/|G_{i-1}|$ and none of $a_i, a_i^2, \ldots, a_i^{m-1}$ is in G_{i-1}, and define $G_i = G_{i-1} \times \langle a_i \rangle$.

As a consequence of the Fundamental Theorem of Finite Abelian Groups, we have the following corollary.

Corollary   *Existence of Subgroups of Abelian Groups*
If m divides the order of a finite Abelian group G, then G has a subgroup of order m.

It is instructive to verify this corollary for a specific case. Let us say that G is an Abelian group of order 72 and we wish to produce a subgroup of order 12. According to the Fundamental Theorem, G is isomorphic to one of the following six groups:

$$Z_8 \oplus Z_9, \qquad\qquad Z_8 \oplus Z_3 \oplus Z_3,$$
$$Z_4 \oplus Z_2 \oplus Z_9, \qquad Z_4 \oplus Z_2 \oplus Z_3 \oplus Z_3,$$
$$Z_2 \oplus Z_2 \oplus Z_2 \oplus Z_9, \quad Z_2 \oplus Z_2 \oplus Z_2 \oplus Z_3 \oplus Z_3,$$

Obviously, $Z_8 \oplus Z_9 \approx Z_{72}$ and $Z_4 \oplus Z_2 \oplus Z_3 \oplus Z_3 \approx Z_{12} \oplus Z_6$ each has a subgroup of order 12. To construct a subgroup of order 12 in $Z_4 \oplus Z_2 \oplus Z_9$, we simply piece together all of Z_4 and the subgroup of order 3 in Z_9; that is, $\{(a, 0, b) \mid a \in Z_4, b \in \{0, 3, 6\}\}$. A subgroup of order 12 in $Z_8 \oplus Z_3 \oplus Z_3$ is given by $\{(a, b, 0) \mid a \in \{0, 2, 4, 6\},$ $b \in Z_3\}$. An analogous procedure applies to the remaining cases and indeed to any finite Abelian group.

PROOF OF THE FUNDAMENTAL THEOREM

Because of the length and complexity of the proof of the Fundamental Theorem of Finite Abelian Groups, we will break it up into a series of lemmas.

Lemma 1 *Let G be a finite Abelian group of order $p^n m$ where p is a prime that does not divide m. Then $G = H \times K$, where $H = \{x \in G \mid x^{p^n} = e\}$ and $K = \{x \in G \mid x^m = e\}$. Moreover, $|H| = p^n$.*

Proof. It is an easy exercise to prove that H and K are subgroups of G (see exercise 23 in Chapter 3). Because G is Abelian, to prove that $G = H \times K$ we need only prove that $G = HK$ and $H \cap K = \{e\}$. Since we have $\gcd(m, p^n) = 1$, there are integers s and t such that $1 = sm + tp^n$. For any x in G, we have $x = x^1 = x^{sm + tp^n} = x^{sm} x^{tp^n}$ and, by Corollary 3 of Lagrange's Theorem, $x^{sm} \in H$ and $x^{tp^n} \in K$. Thus, $G = HK$. Now suppose that some $x \in H \cap K$. Then $x^{p^n} = e = x^m$ and, by the corollary to Theorem 4.1, $|x|$ divides both p^n and m. Since p does not divide m, we have $|x| = 1$ and, therefore, $x = e$.

To prove the second assertion of the lemma, note that $p^n m = |HK| = |H| |K|/|H \cap K| = |H| |K|$ (see exercise 7 in the Supplementary Exercises for Chapters 5–7). It follows from Theorem 9.5 and the corollary to Theorem 4.1 that p does not divide $|K|$ and therefore $|H| = p^n$. ∎

Given an Abelian group G with $|G| = p_1^{n_1} p_2^{n_2} \cdots p_k^{n_k}$, where the p's are distinct primes, we let $G(p_i)$ denote the set $\{x \in G \mid x^{p_i^{n_i}} = e\}$. It then follows immediately from Lemma 1 and induction that $G = G(p_1) \times G(p_2) \times \cdots \times G(p_k)$ and $|G(p_i)| = p_i^{n_i}$. Hence, we turn our attention to groups of prime-power order.

Lemma 2 *Let G be an Abelian group of prime-power order and let a be an element of maximal order in G. Then G can be written in the form $\langle a \rangle \times K$.*

Proof. We denote $|G|$ by p^n and induct on n. If $n = 1$, then $G = \langle a \rangle \times \langle e \rangle$. Now assume that the statement is true for all Abelian groups of order p^k where $k < n$. Among all the elements of G, choose a of maximal order p^m. Then $x^{p^m} = e$ for all x in G. We may assume that $G \neq \langle a \rangle$, for otherwise there is nothing to prove. Now, among all the elements of G, choose b of smallest order such that $b \notin \langle a \rangle$. We claim that $\langle a \rangle \cap \langle b \rangle = \{e\}$. Clearly, we may establish this claim by showing that $|b| = p$. Since $|b^p| = |b|/p$, we know that $b^p \in \langle a \rangle$ by the manner in which b was chosen. Say, $b^p = a^i$. Notice that $e = b^{p^m} = (b^p)^{p^{m-1}} = (a^i)^{p^{m-1}}$, so $|a^i| \le p^{m-1}$. Thus, a^i is not a generator of $\langle a \rangle$ and, therefore, by Theorem 4.2, $\gcd(p^m, i) \ne 1$. This proves that p divides i, so that we can write $i = pj$. Then $b^p = a^i = a^{pj}$. Consider the element $c = a^{-j}b$. Certainly, c is not in $\langle a \rangle$, for if it were, b would be, too. Also, $c^p = a^{-jp}b^p = a^{-i}b^p = b^{-p}b^p = e$. Thus, we have found an element c of order p such that $c \notin \langle a \rangle$. Since b was chosen to have smallest order so that $b \notin \langle a \rangle$, we conclude that b also has order p, and our claim is verified.

Now consider the factor group $\overline{G} = G/\langle b \rangle$. To simplify the notation, we let $\overline{x}$ denote the coset $x\langle b \rangle$ in $\overline{G}$. If $|\overline{a}| < |a| = p^m$, then $\overline{a}^{p^{m-1}} = \overline{e}$. This means that $(a\langle b \rangle)^{p^{m-1}} = a^{p^{m-1}}\langle b \rangle = \langle b \rangle$, so that $a^{p^{m-1}} \in \langle a \rangle \cap \langle b \rangle = \{e\}$, contradicting the fact that $|a| = p^m$. Thus, $|\overline{a}| = |a| = p^m$, and therefore $\overline{a}$ is an element of maximal order in $\overline{G}$. By induction, we know that $\overline{G}$ can be written in the form $\langle \overline{a} \rangle \times \overline{K}$ for some subgroup $\overline{K}$ of $\overline{G}$. Let K be the pullback of $\overline{K}$ under the natural homomorphism from G to $\overline{G}$ (that is, $K = \{x \in G \mid \overline{x} \in \overline{K}\}$). We claim that $\langle a \rangle \cap K = \{e\}$. For if $x \in \langle a \rangle \cap K$, then $\overline{x} \in \langle \overline{a} \rangle \cap \overline{K} = \{\overline{e}\} = \langle b \rangle$ and $x \in \langle a \rangle \cap \langle b \rangle = \{e\}$. It now follows from an order argument (see exercise 34) that $G = \langle a \rangle K$, and therefore $G = \langle a \rangle \times K$. ∎

Lemma 2 and induction on the order of the group now give the following.

Lemma 3 *A finite Abelian group of prime-power order is a direct product of cyclic groups.*

Let us pause to determine where we are in our effort to prove the Fundamental Theorem of Finite Abelian Groups. The remark following Lemma 1 shows that $G = G(p_1) \times G(p_2) \times \cdots \times G(p_n)$, where each $G(p_i)$ is a group of prime-power order, and Lemma 3 shows that each of these factors is a direct product of cyclic groups. Thus, we have proved that G is a direct product of cyclic groups of prime-power order. All that remains to be proved is the uniqueness of the factors. Certainly the groups $G(p_i)$ are uniquely determined by G since they comprise the elements of G whose orders are powers of p_i. So we must prove that there is only one way (up to isomorphism and rearrangement of factors) to write each $G(p_i)$ as a direct product of cyclic groups.

Lemma 4 *Suppose that G is a finite Abelian group of prime-power order. If $G = H_1 \times H_2 \times \cdots \times H_m$ and $G = K_1 \times K_2 \times \cdots \times K_n$, where the H's and K's are nontrivial cyclic subgroups with $|H_1| \ge |H_2| \ge \cdots \ge |H_m|$ and $|K_1| \ge |K_2| \ge \cdots \ge |K_n|$, then $m = n$ and $|H_i| = |K_i|$ for all i.*

Proof. We proceed by induction on $|G|$. Clearly, the case where $|G| = p$ is true. Now suppose that the statement is true for all Abelian groups of order less than $|G|$. For any Abelian group L, the set $L^p = \{x^p \mid x \in L\}$ is a subgroup of L (see exercise 13, Supplementary Exercises for Chapters 1–4). It follows that $G^p = H_1^p \times H_2^p \times \cdots \times H_{m'}^p$, and $G^p = K_1^p \times K_2^p \times \cdots \times K_{n'}^p$, where m' is the largest integer i such that $|H_i| > p$, and n' is the largest integer j such that $|K_j| > p$. (This ensures that our two direct products for G^p do not have trivial factors.) Since $|G^p| < |G|$, we have, by induction, $m' = n'$ and $|H_i^p| = |K_i^p|$ for $i = 1, \ldots, m'$. Since $|H_i| = p|H_i^p|$, this proves that $|H_i| = |K_i|$ for all $i = 1, \ldots, m'$. All that remains to be proved is that the number of H_i of order p equals the number of K_i of order p; that is, we must prove that $m - m' = n - n'$ (since $n' = m'$). This follows directly from the facts that $|H_1||H_2| \cdots |H_{m'}|p^{m-m'} = |G| = |K_1||K_2| \cdots |K_{n'}|p^{n-n'}$, $|H_i| = |K_i|$, and $m' = n'$. ∎

EXERCISES

You know it ain't easy, you know how hard it can be.

John Lennon and Paul McCartney, *The Ballad of John and Yoko*

1. What is the smallest positive integer n such that there are two nonisomorphic groups of order n?

2. What is the smallest positive integer n such that there are three nonisomorphic Abelian groups of order n?

3. What is the smallest positive integer n such that there are exactly four nonisomorphic Abelian groups of order n?

4. Calculate the number of elements of order 4 in each of Z_{16}, $Z_8 \oplus Z_2$, $Z_4 \oplus Z_4$, and $Z_4 \oplus Z_2 \oplus Z_2$.

5. Prove that any Abelian group of order 45 has an element of order 15. Does every Abelian group of order 45 have an element of order 9?

6. Show that there are two Abelian groups of order 108 that have exactly one subgroup of order 3.

7. Show that there are two Abelian groups of order 108 that have exactly four subgroups of order 3.

8. Show that there are two Abelian groups of order 108 that have exactly 13 subgroups of order 3.

9. Suppose that G is an Abelian group of order 120 and that G has exactly three elements of order 2. Determine the isomorphism class of G.

10. Prove that every finite Abelian group can be expressed as the (external) direct product of cyclic groups of orders $n_1, n_2, \ldots, n_t$, where n_{i+1} divides n_i for $i = 1, 2, \ldots, t - 1$. This exercise is referred to in Chapter 22.

11. Find all Abelian groups (up to isomorphism) of order 360.

12. Suppose that the order of some finite Abelian group is divisible by 10. Prove that the group has a cyclic subgroup of order 10.

13. Show, by example, that if the order of a finite Abelian group is divisible by 4, the group need not have a cyclic subgroup of order 4.

14. On the basis of exercises 12 and 13, draw a general conclusion about the existence of cyclic subgroups of a finite Abelian group.

15. How many Abelian groups (up to isomorphism) are there
    a. of order 6?
    b. of order 15?
    c. of order 42?
    d. of order pq where p and q are distinct primes?
    e. of order pqr where p, q, and r are distinct primes?
    f. Generalize parts a, b, c, d, and e.

16. How does the number (up to isomorphism) of Abelian groups of order n compare with the number (up to isomorphism) of Abelian groups of order m where
    a. $n = 3^2$ and $m = 5^2$?
    b. $n = 2^4$ and $m = 5^4$?
    c. $n = p^r$ and $m = q^r$ where p and q are prime?
    d. $n = p^r$ and $m = p^r q$ where p and q are distinct primes?
    e. $n = p^r$ and $m = p^r q^2$ where p and q are distinct primes?

17. The symmetry group of a nonsquare rectangle is an Abelian group of order 4. Is it isomorphic to Z_4 or $Z_2 \oplus Z_2$?

18. Verify the corollary to the Fundamental Theorem of Finite Abelian Groups in the case that the group has order 1080 and the divisor is 180.

19. The set {1, 9, 16, 22, 29, 53, 74, 79, 81} is a group under multiplication modulo 91. Determine the isomorphism class of this group.

20. Determine the isomorphism class of the Nim group given in exercise 37 in Chapter 2.

21. Characterize those integers n such that the only Abelian groups of order n are cyclic.

22. Characterize those integers n such that any Abelian group of order n belongs to one of exactly four isomorphism classes.

23. Refer to Example 1 in this chapter and explain why it is unnecessary to compute the orders of the last five elements listed to determine the isomorphism class of G.

24. Let $G = \{1, 7, 17, 23, 49, 55, 65, 71\}$ under multiplication modulo 96. Express G as an external and an internal direct product of cyclic groups.

25. Let $G = \{1, 7, 43, 49, 51, 57, 93, 99, 101, 107, 143, 149, 151, 157, 193, 199\}$ under multiplication modulo 200. Express G as an external and an internal direct product of cyclic groups.

26. The set $G = \{1, 4, 11, 14, 16, 19, 26, 29, 31, 34, 41, 44\}$ is a group under multiplication modulo 45. Write G as an external and an internal direct product of cyclic groups of prime-power order.

27. Suppose that G is an Abelian group of order 9. What is the maximum number of elements (excluding the identity) of which one needs to compute the order to determine the isomorphism class of G? What if G has order 18? What about 16?

28. Suppose that G is an Abelian group of order 16, and in computing the orders of its elements, you come across an element of order 8 and two elements of order 2. Explain why no further computations are needed to determine the isomorphism class of G.

29. Let G be an Abelian group of order 16. Suppose that there are elements a and b in G such that $|a| = |b| = 4$ and $a^2 \neq b^2$. Determine the isomorphism class of G.

30. Prove that an Abelian group of order $2^n (n \geq 1)$ must have an odd number of elements of order 2.

31. Without using Lagrange's Theorem, show that an Abelian group of odd order cannot have an element of even order.

32. Suppose that G is an Abelian group with an odd number of elements. Show that the product of all of the elements of G is the identity.

33. Suppose that G is a finite Abelian group. Prove that G has order p^n where p is prime if and only if the order of every element of G is a power of p.

34. Prove the assertion made in Lemma 2 that $G = \langle a \rangle K$.

35. Dirichlet's Theorem says that, for every pair of relatively prime integers a and b, there are infinitely many primes of the form $at + b$. Use Dirichlet's Theorem to prove that every finite Abelian group is isomorphic to a subgroup of a U-group.

36. Suppose that G is a finite Abelian group that has exactly one subgroup for each divisor of $|G|$. Show that G is cyclic.

PROGRAMMING EXERCISES

The purpose of computation is insight, not numbers.

Richard Hamming

1. Write a program that will list the isomorphism classes of all finite Abelian groups of order n. Assume that $n < 1{,}000{,}000$. Run your program for $n = 16, 24, 512, 2048, 441{,}000$, and $999{,}999$.

2. Write a program that will determine how many integers in a given interval are the order of exactly one Abelian group, of exactly two Abelian groups, and so on, up to exactly nine Abelian groups. Run your program for the integers up to 1000.

3. Write a program that will implement the algorithm given in this chapter for expressing an Abelian group as an internal direct product. Run your program for the groups $U(32)$, $U(80)$, and $U(65)$.

REFERENCE

1. R. Hirshon, "On Cancellation in Groups," *American Mathematical Monthly* 76 (1969): 1037–1039.

SUGGESTED READINGS

J. A. Gallian, "Computers in Group Theory," *Mathematics Magazine* 49 (1976): 69–73.
    This paper discusses several computer-related projects in group theory done by undergraduate students.

J. Kane, "Distribution of Orders of Abelian Groups," *Mathematics Magazine* 49 (1976): 132–135.
    In this note, the author determines the percentages of integers k between 1 and n, for sufficiently large n, that have exactly one isomorphism class of Abelian groups of order k, exactly two isomorphism classes of Abelian groups of order k, and so on, up to 13 isomorphism classes.

SUPPLEMENTARY EXERCISES FOR CHAPTERS 8–11

Every prospector drills many a dry hole, pulls out his rig, and moves on.

John L. Hess

1. Suppose that H is a subgroup of G and that each left coset of H in G is some right coset of H in G. Prove that H is normal in G.

2. Use a factor group-induction argument to prove that a finite Abelian group of order n has a subgroup of order m for every divisor m of n.

3. Suppose that N is a subgroup of G of index 2. If $x, y \in G$ and $x \notin N$, $y \notin N$, prove that $xy \in N$.

4. Show that a group of order p^2, where p is prime, can be generated by two elements a and b such that every element can be expressed in the form $a^i b^j$.

5. Suppose that H and K are subgroups of a group G and that some left coset of H equals some left coset of K. Show that $H = K$. Show, by example, that H need not equal K if some left coset of H equals some right coset of K.

6. Show that a group of order 9 is Abelian.

7. Let H be a subgroup of G and let $a, b \in G$. Show that $aH = bH$ if and only if $Ha^{-1} = Hb^{-1}$.

8. Let $H \triangleleft G$. Show that $ab \in H$ implies $ba \in H$. Is this true when H is not normal?

9. Let $\text{diag}(G) = \{(g, g) \mid g \in G\}$. Prove that $\text{diag}(G) \triangleleft G \oplus G$ if and only if G is Abelian. When G is the set of real numbers, describe $\text{diag}(G)$ geometrically. When G is finite, what is the index of $\text{diag}(G)$ in G?

10. Let H be any group of rotations in D_n. Prove that H is normal in D_n.

11. Prove that $\text{Inn}(G) \triangleleft \text{Aut}(G)$.

12. Let G be a group, and let $a \in G$. Show that $\langle a \rangle Z(G)$ is Abelian.

13. The factor group $GL(2, \mathbf{R})/SL(2, \mathbf{R})$ is isomorphic to some very familiar group. What is this group?

14. Let k be a divisor of n. The factor group $(Z/\langle n \rangle)/(\langle k \rangle/\langle n \rangle)$ is isomorphic to some very familiar group. What is this group?

15. Let

$$H = \left\{ \begin{bmatrix} 1 & a & b \\ 0 & 1 & c \\ 0 & 0 & 1 \end{bmatrix} \middle| \ a, b, c \in Q \right\}.$$

Prove that
a. $Z(H)$ is isomorphic to Q under addition.
b. $H/Z(H)$ is isomorphic to $Q \oplus Q$.

c. Are your proofs for parts a and b valid when Q is replaced by **R**? Are they valid when Q is replaced by Z_p?

16. Prove that $D_4/Z(D_4)$ is isomorphic to $Z_2 \oplus Z_2$.

17. Prove that Q/Z under addition is an infinite group in which every element has finite order.

18. Show that the intersection of any collection of normal subgroups of a group is a normal subgroup.

19. Let n be a fixed positive integer and G a group. If $H = \{x \in G \mid |x| = n\}$ is a subgroup of G, prove that it is normal in G. Give an example of a group G and an integer n where H is not a subgroup of G.

20. Suppose that H and K are subgroups of a group and that $|H|$ and $|K|$ are relatively prime. Show that $H \cap K = \{e\}$.

21. How many subgroups of order 2 does $Z_2 \oplus Z_2$ have? How many of order 3 are there in $Z_3 \oplus Z_3$? How many subgroups of order p (p a prime) does $Z_p \oplus Z_p$ have? Prove that your answer is correct.

22. Show that Q/Z has a unique subgroup of order n for each positive integer n.

23. If H and K are normal Abelian subgroups of a group and if $H \cap K = \{e\}$, prove that HK is Abelian.

24. Let G be a group of permutations on the set $\{1, 2, \ldots, n\}$. Recall that $\mathrm{stab}_G(1) = \{\alpha \in G \mid \alpha(1) = 1\}$. If γ sends 1 to k, prove that $\gamma\,\mathrm{stab}_G(1) = \{\beta \in G \mid \beta(1) = k\}$.

25. Let G be an Abelian group and let n be a positive integer. Let $G_n = \{g \mid g^n = e\}$ and $G^n = \{g^n \mid g \in G\}$. Prove that G/G_n is isomorphic to G^n.

26. Prove that the mapping $x \to x^2$ from a finite group to itself is one-to-one if the group has odd order.

27. Suppose that G is a group of permutations on some set. If $|G| = 60$ and $\mathrm{orb}_G(5) = \{1, 5\}$, prove that $\mathrm{stab}_G(5)$ is normal in G.

28. Find a subgroup H and elements a and b of some group such that $aH = bH$ but $Ha \neq Hb$.

29. Let $n = 2m$, where m is odd. How many elements of order 2 does $D_n/Z(D_n)$ have? How many elements are in the subgroup $\langle R_{360/n}\rangle/Z(D_n)$? How do these compare with the number of elements of order 2 in D_m?

30. Suppose that G is isomorphic to $Z_2 \oplus Z_2 \oplus \cdots \oplus Z_2$ and $|G| > 2$. Show that the product of all the elements of G is the identity.

31. Let G be a finite Abelian group of order $2^n m$, where m is odd. If the subgroup of order 2^n is not cyclic, show that the product of all the elements of G is the identity. What is this product if the subgroup of order 2^n is cyclic?

32. Suppose that $G = H \times K$ and that N is a normal subgroup of H. Prove that N is normal in G.

33. Show that there is no homomorphism from $Z_8 \oplus Z_2 \oplus Z_2$ onto $Z_4 \oplus Z_4$.

34. Show that there is no homomorphism from A_4 onto a group of order 2, 4, or 6 but that there is a homomorphism from A_4 onto a group of order 3.

35. Let H be a normal subgroup of S_4 of order 4. Prove that S_4/H is isomorphic to S_3.

PART 3

RINGS

12

Introduction to Rings

Example is the school of mankind, and they will learn at no other.

Edmund Burke, *On a Regicide Peace*

MOTIVATION AND DEFINITION

Many sets are naturally endowed with two binary operations: addition and multiplication. Examples that quickly come to mind are the integers, the integers modulo n, the real numbers, matrices, and polynomials. When considering these sets as groups, we simply used addition and ignored multiplication. In many instances, however, one wishes to take into account both addition and multiplication. One abstract concept that does this is the concept of a ring.* This notion was originated in the mid-19th century by Richard Dedekind, although its first formal abstract definition was not given until Abraham Fraenkel presented it in 1914.

DEFINITION   Ring

A *ring R* is a set with two binary operations, addition (denoted by $a + b$) and multiplication (denoted by ab), such that for all a, b, c in R:

1. $a + b = b + a$.
2. $(a + b) + c = a + (b + c)$.

---

*The term *ring* was coined in 1897 by the German mathematician David Hilbert (1862–1943).

3. There is an element 0 in R such that $a + 0 = a$.
4. There is an element $-a$ in R such that $a + (-a) = 0$.
5. $a(bc) = (ab)c$.
6. $a(b + c) = ab + ac$ and $(b + c)a = ba + ca$.

So, a ring is an Abelian group under addition, also having an associative multiplication that is left and right distributive over addition. Note that multiplication need not be commutative. When it is, we say that the ring is *commutative*. Also, a ring need not have an identity under multiplication. When a ring other than {0} has an identity under multiplication, we say that the ring has a *unity* (or *identity*). A nonzero element of a commutative ring with unity need not have a multiplicative inverse. When it does, we say that it is a *unit* of the ring. Thus, a is a unit if a^{-1} exists.

The following terminology and notation are convenient. If a and b belong to a commutative ring R and a is nonzero, we say that a *divides* b (or that a is a *factor* of b) and write $a \mid b$, if there exists an element c in R such that $b = ac$. If a does not divide b, we write $a \nmid b$.

For an abstraction to be worthy of study, it must have many diverse concrete realizations. The following list of examples shows that the ring concept is pervasive.

EXAMPLES OF RINGS

Example 1   The set Z of integers under ordinary addition and multiplication is a commutative ring with unity 1. The units of Z are 1 and -1. ❑

Example 2   The set $Z_n = \{0, 1, \ldots, n - 1\}$ under addition and multiplication modulo n is a commutative ring with unity 1. The set of units is $U(n)$. ❑

Example 3   The set $Z[x]$ of all polynomials in the variable x with integer coefficients under ordinary addition and multiplication is a commutative ring with unity $f(x) = 1$. ❑

Example 4   The set $M_2(Z)$ of 2×2 matrices with integer entries is a noncommutative ring with unity $\begin{bmatrix} 1 & 0 \\ 0 & 1 \end{bmatrix}$. ❑

Example 5   The set $2Z$ of even integers under ordinary addition and multiplication is a commutative ring without unity. ❑

Example 6   The set of all continuous real-valued functions of a real variable whose graphs pass through the point $(1, 0)$ is a commutative ring without unity under the operations of pointwise addition and multiplication [that is, the operations $(f + g)(a) = f(a) + g(a)$ and $(f \cdot g)(a) = f(a) \cdot g(a)$]. ☐

Example 7 Let $R_1, R_2, \ldots, R_n$ be rings. We can use these to construct a new ring, as follows. Let

$$R_1 \oplus R_2 \oplus \cdots \oplus R_n = \{(a_1, a_2, \ldots, a_n) \mid a_i \in R_i\}$$

and perform componentwise addition and multiplication; that is, define

$$(a_1, a_2, \ldots, a_n) + (b_1, b_2, \ldots, b_n) = (a_1 + b_1, a_2 + b_2, \ldots, a_n + b_n)$$

and

$$(a_1, a_2, \ldots, a_n)(b_1, b_2, \ldots, b_n) = (a_1 b_1, a_2 b_2, \ldots, a_n b_n).$$

This ring is called the *direct sum* of $R_1, R_2, \ldots, R_n$. ☐

All of the preceding examples are quite familiar to you, and the fact that they are rings is obvious. Our last example is included to show that rings, even those with only a handful of elements, can be unusual. It is one of the 11 different rings of order 4 (recall that there are only two groups of order 4).

Example 8 Let $R = \{0, a, b, c\}$. Define addition and multiplication by the Cayley tables:

+	0	a	b	c
0	0	a	b	c
a	a	0	c	b
b	b	c	0	a
c	c	b	a	0

·	0	a	b	c
0	0	0	0	0
a	0	a	b	c
b	0	a	b	c
c	0	0	0	0

Then R is a noncommutative ring without unity. ☐

PROPERTIES OF RINGS

Our first theorem shows how the operations of addition and multiplication intertwine.

Theorem 12.1 *Rules of Multiplication*

Let a, b, and c belong to a ring R. Then

1. $a0 = 0a = 0$.
2. $a(-b) = (-a)b = -(ab)$.
3. $(-a)(-b) = ab$.*
4. $a(b - c) = ab - ac$ and $(b - c)a = ba - ca$.

Minus times minus is plus.
The reason for this we need not discuss.
        W. H. Auden

Furthermore, if R has a unity element 1, then

5. $(-1)a = -a$.
6. $(-1)(-1) = 1$.

Proof.   We will prove rules 1 and 2 and leave the rest as easy exercises. To prove statements such as those in Theorem 12.1, we need only "play off" the distributive property against the fact that R is a group under addition with additive identity 0. Consider rule 1. Clearly,

$$0 + a0 = a0 = a(0 + 0) = a0 + a0.$$

So, by cancellation, $0 = a0$. Similarly, $0a = 0$.
    To prove rule 2, we observe that

$$-(ab) + (ab) = 0 \quad \text{and} \quad a(-b) + ab = a(-b + b) = a0 = 0.$$

So $-(ab) + ab = a(-b) + ab$, and it follows from cancellation that $-(ab) = a(-b)$. The remainder of rule 2 is done analogously.   ■

Recall that in the case of groups, the identity and inverses are unique. What about rings? The same is true for rings, provided they exist. The proofs are identical to the ones given for groups and therefore are omitted.

Theorem 12.2   *Uniqueness of the Unity and Inverses*
If a ring has a unity, it is unique. If a ring element has an inverse, it is unique.

Many students have the mistaken tendency to treat a ring as if it were a group under *multiplication*. It is not. The two most common errors are the assumptions that ring elements have multiplicative inverses—they need not—and that a ring has a multiplicative identity—it need not. For example, if a, b, and c belong to a ring, $a \neq 0$ and $ab = ac$, we *cannot* conclude that $b = c$. Similarly, if $a^2 = a$, we *cannot* conclude that $a = 0$ or 1 (as is the case with real numbers). In the first place, the ring need not have multiplicative cancellation, and in the second place, the ring need not have a multiplicative identity. There is an important class of rings wherein multiplicative identities exist and for which multiplicative cancellation holds. This class is taken up in the next chapter.

SUBRINGS

In our study of groups, subgroups played a crucial role. Subrings, the analogous structures in ring theory, play a much less important role than their counterparts in group theory. Nevertheless, subrings are important.

DEFINITION   Subring

A subset S of a ring R is a *subring of R* if S is itself a ring with the operations of R.

Just as was the case for subgroups, there is a simple test for subrings.

Theorem 12.3   *Subring Test*

A nonempty subset S of a ring R is a subring if S is closed under subtraction and multiplication—that is, if $a - b$ and ab are in S whenever a and b are in S.

Proof.   Since addition in R is commutative and S is closed under subtraction, we know by the One-Step Subgroup Test (Theorem 3.1) that S is an Abelian group under addition. Also, since multiplication in R is associative as well as distributive over addition, the same is true for multiplication in S. Thus, the only condition remaining to be checked is that multiplication is a binary operation on S. But this is exactly what closure means.   ∎

We leave it to the student to confirm that each of the following examples is a subring.

Example 9 $\{0\}$ and R are subrings of any ring R. $\{0\}$ is called the *trivial* subring of R.   ❑

Example 10 $\{0, 2, 4\}$ is a subring of the ring Z_6, the integers modulo 6.   ❑

Example 11   For each positive integer n, the set

$$nZ = \{0, \pm n, \pm 2n, \pm 3n, \dots\}$$

is a subring of the integers Z.   ❑

Example 12   The set of Gaussian integers

$$Z[i] = \{a + bi \mid a, b \in Z\}$$

is a subring of the complex numbers **C**.   ❑

Example 13   Let R be the ring of all real-valued functions of a single real variable under pointwise addition and multiplication. The subset S of R of functions whose graphs pass through the origin forms a subring of R.   ❑

Example 14   The set

$$\left\{ \begin{bmatrix} a & 0 \\ 0 & b \end{bmatrix} \middle| a, b \in Z \right\}$$

of diagonal matrices is a subring of the ring of all 2×2 matrices over Z.   ❑

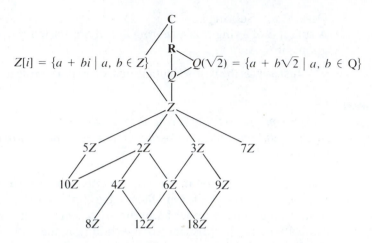

Figure 12.1   Partial subring lattice diagram of **C**.

We can picture the relationship between a ring and its various sub-rings by way of a subring lattice diagram. In such a diagram, any ring is a subring of all the rings that it is connected to by one or more upward lines. Figure 12.1 shows the relationship among some of the rings we have already discussed.

In the next several chapters, we will see that many of the fundamental concepts of group theory can be naturally extended to rings. In particular, we will introduce ring homomorphisms and factor rings.

EXERCISES

There is no substitute for hard work.
Thomas Alva Edison, *Life*

1. The ring in Example 8 is finite and noncommutative. Give another example of a finite, noncommutative ring. Give an example of an infinite noncommutative ring that does not have a unity.
2. The set $\{0, 2, 4\}$ under addition and multiplication modulo 6 has a unity. Find it.
3. Let $Z[\sqrt{2}] = \{a + b\sqrt{2} \mid a, b \in Z\}$. Prove that $Z[\sqrt{2}]$ is a ring under the ordinary addition and multiplication of real numbers.
4. Show that for fixed elements a and b in a ring, the equation $ax = b$ can have more than one solution. How does this compare with groups?

5. Prove that a ring can have at most one unity.

6. Find an integer n that shows that the rings Z_n need not have the following properties that the ring of integers has.
   a. $a^2 = a$ implies $a = 0$ or $a = 1$.
   b. $ab = 0$ implies $a = 0$ or $b = 0$.
   c. $ab = ac$ and $a \neq 0$ imply $b = c$.
   Is the n you found prime?

7. Show that the three properties listed in exercise 6 are valid for Z_p, where p is prime.

8. Show that the ring given in Example 8 has two left identities (that is, an element r such that $rx = x$ for all x in the ring) but no right identity.

9. Show that a ring that is cyclic under addition is commutative.

10. Verify that examples 9 through 14 in this chapter are as stated.

11. Prove parts 3 through 6 of Theorem 12.1.

12. Give an example of a noncommutative ring that has exactly 16 elements.

13. Describe all the subrings of the ring of integers.

14. Show that if m and n are integers and a and b are elements from a ring, then $(ma)(nb) = (mn)(ab)$. (This exercise is referred to in Chapter 15.)

15. Show that if n is an integer and a is an element from a ring, then $n(-a) = -(na)$.

16. Let a and b belong to a ring R and let m be an integer. Prove that $m(ab) = (ma)b = a(mb)$.

17. Prove that the intersection of any collection of subrings of a ring R is a subring of R.

18. Let a belong to a ring R. Let $S = \{x \in R \mid ax = 0\}$. Show that S is a subring of R.

19. Let R be a ring. The *center of R* is the set $\{x \in R \mid ax = xa$ for all a in $R\}$. Prove that the center of a ring is a subring.

20. Describe the elements of $M_2(Z)$ (see Example 4) that have multiplicative inverses.

21. Suppose that $R_1, R_2, \ldots, R_n$ are rings that contain nonzero elements. Show that $R_1 \oplus R_2 \oplus \cdots \oplus R_n$ has a unity if and only if each R_i has a unity.

22. Let R be a commutative ring with unity, and let $U(R)$ denote the set of units of R. Prove that $U(R)$ is a group under the multiplication of R. (This group is called the *group of units of R*.)

23. Determine $U(Z[i])$ (see Example 12).

24. If $R_1, R_2, \ldots, R_n$ are commutative rings with unity, show that $U(R_1 \oplus R_2 \oplus \cdots \oplus R_n) = U(R_1) \oplus U(R_2) \oplus \cdots \oplus U(R_n)$.

25. Determine $U(Z[x])$. (This exercise is referred to in Chapter 17.)

26. Determine $U(\mathbf{R}[x])$.

27. Show that a unit of a ring divides every element of the ring.

28. In Z_6, show that $4 \mid 2$; in Z_8, show that $3 \mid 7$; in Z_{15}, show that $9 \mid 12$.

29. Suppose that a and b belong to a commutative ring R. If a is a unit of R and $b^2 = 0$, show that $a + b$ is a unit of R.

30. Suppose that there is an integer $n > 1$ such that $a^n = a$ for all elements a of some ring. If m is a positive integer and $a^m = 0$ for some a, show that $a = 0$.

31. Give an example of ring elements a and b with the properties that $ab = 0$ but $ba \neq 0$.

32. In a ring in which $x^3 = x$ for all x, show that $ab = 0$ implies $ba = 0$.

33. For $p = 3, 7,$ and 11, find all solutions of $a^2 + b^2 = 0$ in Z_p. Show that for $p = 5, 13,$ and 17, there are nontrivial solutions (that is, $a \neq 0, b \neq 0$) of $a^2 + b^2 = 0$. Make a conjecture about the existence of nontrivial solutions of this equation in Z_p (p a prime) and the form of p.

34. Let m and n be positive integers and let k be the least common multiple of m and n. Show that $mZ \cap nZ = kZ$.

35. Explain why every subgroup of Z_n under addition is also a subring of Z_n.

36. Is Z_6 a subring of Z_{12}?

37. Suppose that R is a ring with unity 1 and a is an element of R such that $a^2 = 1$. Let $S = \{ara \mid r \in R\}$. Prove that S is a subring of R. Does S contain 1?

38. Let $M_2(Z)$ be the ring of all 2×2 matrices over the integers and let $R = \left\{ \begin{bmatrix} a & a + b \\ a + b & b \end{bmatrix} \middle| a, b \in Z \right\}$. Prove or disprove that R is a subring of $M_2(Z)$.

39. Let $M_2(Z)$ be the ring of all 2×2 matrices over the integers and let $R = \left\{ \begin{bmatrix} a & a - b \\ a - b & b \end{bmatrix} \middle| a, b \in Z \right\}$. Prove or disprove that R is a subring of $M_2(Z)$.

40. Let $R = \left\{ \begin{bmatrix} a & a \\ b & b \end{bmatrix} \middle| a, b \in Z \right\}$. Prove or disprove that R is a subring of $M_2(Z)$.

41. Let $R = Z \oplus Z \oplus Z$ and $S = \{(a, b, c) \in R \mid a + b = c\}$. Prove or disprove that S is a subring of R.

42. Does the unity of a subring have to be the same as the unity of the whole ring? If yes, prove it; if no, give an example. What is the analogous situation for groups?

43. Show that $2Z \cup 3Z$ is not a subring of Z.

44. Let R be a ring with unity e. Show that $S = \{ne \mid n \in Z\}$ is a subring of R. (This exercise is referred to in Chapter 15.)

45. Determine the smallest subring of Q that contains $\frac{1}{2}$.

46. Determine the smallest subring of Q that contains $\frac{2}{3}$.

47. Let R be a ring. Prove that $a^2 - b^2 = (a + b)(a - b)$ for all a, b in R if and only if R is commutative.

48. Suppose that R is a ring and that $a^2 = a$ for all a in R. Show that R is commutative. [A ring in which $a^2 = a$ for all a is called a *Boolean* ring, in honor of the English mathematician George Boole (1815–1864).]

49. Suppose that there is a positive even integer n such that $a^n = a$ for all elements a of some ring. Show that $-a = a$ for all a in the ring.

50. Give an example of a subset of a ring that is a subgroup under addition but not a subring.

PROGRAMMING EXERCISES

Theory is the general; experiments are the soldiers.

Leonardo da Vinci

1. Write a program that will find all solutions of the equation $a^2 + b^2 = 0$ in Z_p. Run your program for all odd primes up to 37. Make a conjecture about the existence of nontrivial solutions in Z_p (p a prime) and the form of p.

2. Let $Z_n[i] = \{a + bi \mid a, b \in Z_n, i^2 = -1\}$ (the Gaussian integers modulo n). Write a program that will find the group of units of this ring and the order of each element of the group. Run your program for $n = 3, 7, 11$, and 23. Is the group of units cyclic for these cases? Try to guess a formula for the order of the group of units of $Z_n[i]$ as a function of n when n is a prime and $n = 3$ modulo 4. Run your program for $n = 9$. Is this group cyclic? Does your formula predict the correct order for the group in this case? Run your program for $n = 5, 13$, and 17. Is the group cyclic for these cases? Try to guess a formula for the order of the group of units of $Z_n[i]$ as a function of n when n is a prime and $n = 1$ modulo 4. Run your program for $n = 25$. Is this group cyclic? Does your formula predict the correct order of the group in this case?

SUGGESTED READINGS

D. B. Erickson, "Orders for Finite Noncommutative Rings," *American Mathematical Monthly* 73 (1966): 376–377.

   In this elementary paper, it is shown that there exists a noncommutative ring of order $m > 1$ if and only if m is divisible by the square of a prime.

Colin R. Fletcher, "Rings of Small Order," *The Mathematical Gazette* 64 (1980): 9–22.

   This article gives a complete list of the 24 rings of order less than 8. The only hard case is the 11 rings of order 4.

I. N. Herstein

A whole generation of textbooks and an entire generation of mathematicians, myself included, have been profoundly influenced by that text [Herstein's *Topics in Algebra*].

Georgia Benkart

I. N. HERSTEIN was born on March 28, 1923, in Poland. His family moved to Canada when he was seven. He grew up in a poor and tough environment, on which he commented that in his neighborhood you either became a gangster or a college professor. During his school years he played football, hockey, golf, tennis, and pool. During this time he worked as a steeple-jack and as a "barber" at a fair. Herstein received a B.S. degree from the University of Manitoba, an M.A. from the University of Toronto, and, in 1948, a Ph.D. degree from Indiana University under the supervision of Max Zorn. Before permanently settling at the University of Chicago in 1962, he held positions at the University of Kansas, Ohio State University, the University of Pennsylvania, and Cornell University.

Herstein wrote more than 100 research papers and a dozen books. Although his principal interest was noncommutative ring theory, he also wrote papers on finite groups, linear algebra, and mathematical economics. His textbook *Topics in Algebra,* first published in 1964, dominated the field for 20 years and has become a classic. Herstein had great influence through his teaching and his collaboration with colleagues. He had 30 Ph.D. students, and traveled and lectured widely. His nonmathematical interests included languages and art. He spoke Italian, Hebrew, Polish, and Portuguese.

Herstein died on February 9, 1988, after a long battle with cancer.

13

Integral Domains

The interplay between generality and individuality, deduction and construction, logic and imagination—this is the profound essence of live mathematics. Any one or another of these aspects of mathematics can be at the center of a given achievement. In a far reaching development all of them will be involved. Generally speaking, such a development will start from the "concrete" ground, then discard ballast by abstraction and rise to the lofty layers of thin air where navigation and observation are easy; after this flight comes the crucial test of landing and reaching specific goals in the newly surveyed low plains of individual "reality." In brief, the flight into abstract generality must start from and return again to the concrete and specific.

Richard Courant

DEFINITION AND EXAMPLES

To a certain degree, the notion of a ring was invented in an attempt to put the algebraic properties of the integers into an abstract setting. A ring is not the appropriate abstraction of the integers, however, for too much is lost in the process. Besides the two obvious properties of commutativity and existence of a unity, there is one other essential feature of the integers that rings in general do not enjoy—the cancellation property. In this chapter, we introduce integral domains—a particular class of rings that have all three of these properties. Integral domains play a prominent role in number theory and algebraic geometry.

DEFINITION   Zero-Divisors
A nonzero element a in a commutative ring R is called a *zero-divisor* if there is a nonzero element b in R such that $ab = 0$.

DEFINITION   Integral Domain
A commutative ring with a unity is said to be an *integral domain* if it has no zero-divisors.

Thus, in an integral domain, a product is 0 only when one of the factors is 0; that is, $ab = 0$ only when $a = 0$ or $b = 0$. The following examples show that many familiar rings are integral domains and some familiar rings are not. For each example, the student should verify the assertion made.

Example 1   The ring of integers is an integral domain.   ❏

Example 2   The ring of Gaussian integers $Z[i] = \{a + bi \mid a, b \in Z\}$ is an integral domain.   ❏

Example 3   The ring $Z[x]$ of polynomials with integer coefficients is an integral domain.   ❏

Example 4   The ring $Z[\sqrt{2}] = \{a + b\sqrt{2} \mid a, b \in Z\}$ is an integral domain.   ❏

Example 5   The ring Z_p of integers modulo a prime p is an integral domain.   ❏

Example 6   The ring Z_n of integers modulo n is *not* an integral domain when n is not prime.   ❏

Example 7   The ring $M_2(Z)$ of 2×2 matrices over the integers is *not* an integral domain.   ❏

What makes integral domains particularly appealing is that they have an important multiplicative group-theoretic property, in spite of the fact that the nonzero elements need not form a group under multiplication. This property is cancellation.

Theorem 13.1   *Cancellation*
Let a, b, and c belong to an integral domain. If $a \neq 0$ and $ab = ac$, then $b = c$.

Proof. From $ab = ac$ we have $a(b - c) = 0$. Since $a \neq 0$, we must have $b - c = 0$.   ∎

Many authors prefer to define integral domains by the cancellation property—that is, as commutative rings with unity in which the cancellation property holds. This definition is equivalent to ours.

FIELDS

In many applications, a particular kind of integral domain called a *field* is necessary.

DEFINITION   Field
A commutative ring with a unity is called a *field* if every nonzero element is a unit.

It is often helpful to think of ab^{-1} as a divided by b. With this in mind, a field can be thought of as simply an algebraic system that is closed under addition, subtraction, multiplication, and division (except by 0). We have had numerous examples of fields: the complex numbers, the real numbers, the rational numbers. The abstract theory of fields was initiated by Heinrich Weber in 1893. Groups, rings, and fields are the three main branches of abstract algebra. Theorem 13.2 says that, in the finite case, fields and integral domains are the same.

Theorem 13.2   *Finite Integral Domains Are Fields*
A finite integral domain is a field.

Proof.   Let D be a finite integral domain with unity 1. Let a be any nonzero element of D. We must show that a is a unit. If $a = 1$, a is its own inverse, so we may assume that $a \neq 1$. Now consider the following sequence of elements of D: $a, a^2, a^3, \ldots$. Since D is finite, there must be two positive integers i and j such that $i > j$ and $a^i = a^j$. Then, by cancellation, $a^{i-j} = 1$. Since $a \neq 1$, we know that $i - j > 1$, and we have shown that a^{i-j-1} is the inverse of a.   ∎

Corollary Z_p *Is a Field*
For every prime p, Z_p, the ring of integers modulo p, is a field.

Proof.   According to Theorem 13.2, we need only prove that Z_p has no zero-divisors. So, suppose that $a, b \in Z_p$ and $ab = 0$. Then $ab = pk$ for some integer k. But, then, by Euclid's Lemma (see Chapter 0), p divides a or p divides b. Thus, in Z_p, $a = 0$ or $b = 0$.   ∎

Putting the above corollary together with Example 6, we see that Z_n is a field if and only if n is prime. In a later chapter, we will describe how all finite fields can be constructed. For now, we give one example of a finite field not of the form Z_p.

Table 13.1   Multiplication Table for $Z_3[i]$

	1	2	i	$1 + i$	$2 + i$	$2i$	$1 + 2i$	$2 + 2i$
1	1	2	i	$1 + i$	$2 + i$	$2i$	$1 + 2i$	$2 + 2i$
2	2	1	$2i$	$2 + 2i$	$1 + 2i$	i	$2 + i$	$1 + i$
i	i	$2i$	2	$2 + i$	$2 + 2i$	1	$1 + i$	$1 + 2i$
$1 + i$	$1 + i$	$2 + 2i$	$2 + i$	$2i$	1	$1 + 2i$	2	i
$2 + i$	$2 + i$	$1 + 2i$	$2 + 2i$	1	i	$1 + i$	$2i$	2
$2i$	$2i$	i	1	$1 + 2i$	$1 + i$	2	$2 + 2i$	$2 + i$
$1 + 2i$	$1 + 2i$	$2 + i$	$1 + i$	2	$2i$	$2 + 2i$	i	1
$2 + 2i$	$2 + 2i$	$1 + i$	$1 + 2i$	i	2	$2 + i$	1	$2i$

Example 8   Field with Nine Elements
Let

$$Z_3[i] = \{a + bi \mid a, b \in Z_3\}$$
$$= \{0, 1, 2, i, 1 + i, 2 + i, 2i, 1 + 2i, 2 + 2i\},$$

where $i^2 = -1$. This is the ring of Gaussian integers modulo 3. Elements are added and multiplied as in the complex numbers, except that the coefficients are reduced modulo 3. In particular, $-1 = 2$. Table 13.1 is the multiplication table for the nonzero elements of $Z_3[i]$.   ❑

Example 9   Let $Q[\sqrt{2}] = \{a + b\sqrt{2} \mid a, b \in Q\}$. It is easy to see that $Q[\sqrt{2}]$ is a ring. Viewed as an element of **R**, the multiplication inverse of any nonzero element of the form $a + b\sqrt{2}$ is simply $1/(a + b\sqrt{2})$. To verify that $Q[\sqrt{2}]$ is a field, we must show that $1/(a + b\sqrt{2})$ can be written in the form $c + d\sqrt{2}$. In high school algebra, this process is called "rationalizing the denominator." Specifically,

$$\frac{1}{a + b\sqrt{2}} = \frac{1}{a + b\sqrt{2}} \frac{a - b\sqrt{2}}{a - b\sqrt{2}} = \frac{a}{a^2 - 2b^2} - \frac{b}{a^2 - 2b^2}\sqrt{2}. \quad ❑$$

CHARACTERISTIC OF A RING

Note that for any element x in $Z_3[i]$, we have $3x = x + x + x = 0$, since addition is done modulo 3. Similarly, in the subring $\{0, 3, 6, 9\}$ of Z_{12} we have $4x = x + x + x + x = 0$ for all x. This observation motivates the following definition.

DEFINITION   Characteristic of a Ring
The *characteristic* of a ring R is the least positive integer n such that $nx = 0$ for all x in R. If no such integer exists, we say that R has characteristic 0.

Thus, the ring of integers has characteristic 0, and Z_n has characteristic n. An infinite ring can have nonzero characteristic. Indeed, the ring $Z_2[x]$ of all polynomials with coefficients in Z_2 has characteristic 2. (Addition and multiplication are done as for polynomials with ordinary integer coefficients except that the coefficients are reduced modulo 2.) When a ring has a unity, the task of determining the characteristic is simplified by Theorem 13.3

Theorem 13.3 *Characteristic of a Ring with Unity*

Let R be a ring with unity 1. If 1 has infinite order under addition, then the characteristic of R is 0. If 1 has order n under addition, then the characteristic of R is n.

Proof. If 1 has infinite order, then there is no positive integer n such that $n \cdot 1 = 0$, so R has characteristic 0. Now suppose that 1 had additive order n. Then $n \cdot 1 = 0$, and n is the least positive integer with this property. So for any x in R we have

$$nx = n(1x) = (n \cdot 1)x = 0x = 0.$$

Thus, R has characteristic n. ∎

In the case of an integral domain, the possibilities for the characteristic are severely limited.

Theorem 13.4 *Characteristic of an Integral Domain*

The characteristic of an integral domain is 0 or prime.

Proof. By Theorem 13.3, it suffices to show that if the additive order of 1 is finite, it must be prime. Suppose that 1 has order n and that $n = st$, where $1 < s, t < n$. Then

$$0 = n \cdot 1 = (st) \cdot 1 = (s \cdot 1)(t \cdot 1).$$

So, $s \cdot 1 = 0$ or $t \cdot 1 = 0$. But this contradicts the fact that 1 has order n. Hence, n cannot be factored as a product of two smaller integers. ∎

We conclude this chapter with a brief discussion of polynomials with coefficients from a ring—a topic we will consider in detail in later chapters. The existence of zero-divisors in a ring causes unusual results when one is finding roots of polynomials with coefficients in the ring. Consider, for example, the equation $x^2 - 4x + 3 = 0$. In the integers, we could find all solutions by factoring

$$x^2 - 4x + 3 = (x - 3)(x - 1) = 0$$

and setting each factor equal to 0. But notice that when we say we can find *all* solutions in this manner, we are using the fact that the only way for a product to equal 0 is for one of the factors to be 0—that is, we are using the fact that Z is an integral domain. In Z_{12}, there are many pairs of nonzero elements whose products are 0: $2 \cdot 6 = 0$, $3 \cdot 4 = 0$,

Table 13.2   Summary of Rings and Their Properties

Ring	Form of Element	Unity	Commutative	Integral Domain	Field	Characteristic
Z	k	1	yes	yes	no	0
Z_n, n composite	k	1	yes	no	no	n
Z_p, p prime	k	1	yes	yes	yes	p
$Z[x]$	$a_n x^n + \cdots + a_1 x + a_0$	$f(x) = 1$	yes	yes	no	0
nZ, $n > 1$	nk	none	yes	no	no	0
$M_2(Z)$	$\begin{bmatrix} a & b \\ c & d \end{bmatrix}$	$\begin{bmatrix} 1 & 0 \\ 0 & 1 \end{bmatrix}$	no	no	no	0
$M_2(2Z)$	$\begin{bmatrix} 2a & 2b \\ 2c & 2d \end{bmatrix}$	none	no	no	no	0
$Z[i]$	$a + bi$	1	yes	yes	no	0
$Z_3[i]$	$a + bi; a, b \in Z_3$	1	yes	yes	yes	3
$Z[\sqrt{2}]$	$a + b\sqrt{2}; a, b \in Z$	1	yes	yes	no	0
$Q[\sqrt{2}]$	$a + b\sqrt{2}; a, b \in Q$	1	yes	yes	yes	0
$Z \oplus Z$	(a, b)	$(1, 1)$	yes	no	no	0

$4 \cdot 6 = 0$, $6 \cdot 8 = 0$, and so on. So, how do we find *all* solutions of $x^2 - 4x + 3 = 0$ in Z_{12}? The easiest way is simply to try every element! Upon doing so, we find four solutions: $x = 1$, $x = 3$, $x = 7$, and $x = 9$. Observe that we can find all solutions of $x^2 - 4x + 3 = 0$ over Z_{11} or Z_{13}, say, by setting the two factors $x - 3$ and $x - 1$ equal to 0. Of course, the reason why this works for these rings is that they are integral domains. Perhaps this will convince you that integral domains are particularly advantageous rings. Table 13.2 gives a summary of some of the rings we have introduced and their properties.

EXERCISES

It looked absolutely impossible. But it so happens that you go on worrying away at a problem in science and it seems to get tired, and lies down and lets you catch it.

William Lawrence Bragg.*

    1. Verify that Examples 1 through 7 are as claimed.

    2. Which of Examples 1 through 5 are fields?

---

*Bragg, at age 24, won the Nobel Prize for the invention of x-ray crystallography. He remains the youngest person ever to receive the Nobel Prize.

3. Show that a commutative ring with the cancellation property (under multiplication) has no zero-divisors.

4. List all zero-divisors in Z_{20}. Can you see a relationship between the zero-divisors of Z_{20} and the units of Z_{20}?

5. Show that every nonzero element of Z_n is a unit or a zero-divisor.

6. Find a nonzero element in a ring that is neither a zero-divisor nor a unit.

7. Let R be a finite commutative ring with unity. Prove that every nonzero element of R is either a zero-divisor or a unit. What happens if we drop the "finite" condition on R?

8. Describe all zero-divisors and units of $Z \oplus Q \oplus Z$.

9. Let d be an integer. Prove that $Z[\sqrt{d}] = \{a + b\sqrt{d} \mid a, b \in Z\}$ is an integral domain. (This exercise is referred to in Chapter 18.)

10. Show that a field is an integral domain.

11. Give an example of a commutative ring without zero-divisors that is not an integral domain.

12. Find two elements a and b in a ring such that both a and b are zero-divisors, $a + b \neq 0$, and $a + b$ is not a zero-divisor.

13. Let a belong to a ring R with unity and $a^n = 0$ for some positive integer n. (Such an element is called *nilpotent*.) Prove that $1 - a$ has a multiplicative inverse in R. [*Hint:* Consider $(1 - a)(1 + a + a^2 + \cdots + a^{n-1})$.]

14. Show that the nilpotent elements of a commutative ring form a subring.

15. Show that 0 is the only nilpotent element in an integral domain.

16. A ring element a is called an *idempotent* if $a^2 = a$. Prove that the only idempotents in an integral domain are 0 and 1.

17. Prove that the set of idempotents of a commutative ring is closed under multiplication.

18. Find a zero-divisor and a nonzero idempotent in $Z_5[i] = \{a + bi \mid a, b \in Z_5, i^2 = -1\}$.

19. Determine the isomorphism class of the group of units of $Z_5[i]$.

20. Find all units, zero-divisors, idempotents, and nilpotent elements in $Z_3 \oplus Z_6$.

21. Determine all elements of a ring that are both units and idempotents.

22. Let R be the set of all real-valued functions defined for all real numbers under function addition and multiplication.
    a. Determine all zero-divisors of R.
    b. Determine all nilpotent elements of R.
    c. Show that every nonzero element is a zero-divisor or a unit.

23. (Subfield Test) Let F be a field and let K be a subset of F with at least two elements. Prove that K is a subfield of F if, for any a, b ($b \neq 0$) in K, $a - b$ and ab^{-1} belong to K.

24. Let d be a positive integer. Prove that $Q[\sqrt{d}] = \{a + b\sqrt{d} \mid a, b \in Q\}$ is a field.

25. Let R be a ring with unity 1. If the product of any pair of nonzero elements of R is nonzero, prove that $ab = 1$ implies $ba = 1$.

26. Let $R = \{0, 2, 4, 6, 8\}$ under addition and multiplication modulo 10. Prove that R is a field.

27. Formulate the appropriate definition of a subdomain (that is, a "sub" integral domain). Let D be an integral domain with unity 1. Show that $P = \{n1 \mid n \in Z\}$ (that is, all integral multiples of 1) is a subdomain of D. Show that P is contained in every subdomain of D. What can we say about the order of P?

28. Prove that there is no integral domain with exactly six elements. Can your argument be adapted to show that there is no integral domain with exactly four elements? What about 15 elements? Use these observations to guess a general result about the number of elements in a finite integral domain.

29. Is $Z \oplus Z$ an integral domain? Explain.

30. Determine all elements of an integral domain that are their own inverses under multiplication.

31. Suppose that a and b belong to an integral domain.
    a. If $a^5 = b^5$ and $a^3 = b^3$, prove that $a = b$.
    b. If $a^m = b^m$ and $a^n = b^n$, where m and n are positive integers that are relatively prime, prove that $a = b$.

32. Find an example of an integral domain and distinct positive integers m and n such that $a^m = b^m$ and $a^n = b^n$, but $a \neq b$.

33. Verify Table 13.1. Look for shortcuts. (For example, the sixth row can be obtained from the third one by doubling.)

34. Construct a multiplication table for $Z_2[i]$, the ring of Gaussian integers modulo 2. Is this ring a field? Is it an integral domain?

35. The nonzero elements of $Z_3[i]$ form an Abelian group of order 8 under multiplication. Is it isomorphic to Z_8, $Z_4 \oplus Z_2$, or $Z_2 \oplus Z_2 \oplus Z_2$?

36. Define $Z_n[i] = \{a + bi \mid a, b \in Z_n\}$, where $i^2 = -1$ in $Z_n[i]$. For some primes p, $Z_p[i]$ is a field, and for others it is not. In particular, when $p = 2, 5$, or 13, $Z_p[i]$ is *not* a field; but when $p = 3, 7$, or 11, $Z_p[i]$ *is* a field. From these examples, try to guess a condition that determines whether or not $Z_p[i]$ is a field. (*Hint:* Consider p modulo 4.)

37. Show that a finite commutative ring with no zero-divisors has a unity.

38. Suppose that a and b belong to a commutative ring and ab is a zero-divisor. Show that either a or b is a zero-divisor.

39. Suppose that R is a commutative ring without zero-divisors. Show that all the nonzero elements of R have the same additive order.

40. Suppose that R is a commutative ring without zero-divisors. Show that the characteristic of R is 0 or prime.

41. Show that any finite field has order p^n where p is a prime. (*Hint:* Use facts about finite Abelian groups.)

42. Let x and y belong to an integral domain of prime characteristic p.
    a. Show that $(x + y)^p = x^p + y^p$.
    b. Show that, for all positive integers n, $(x + y)^{p^n} = x^{p^n} + y^{p^n}$.
    c. Find elements x and y in a ring of characteristic 4 such that $(x + y)^4 \neq x^4 + y^4$.

43. Exhibit a finite field that contains two nonzero elements a and b such that $a^2 + b^2 = 0$.

44. Give an example of an infinite integral domain that has characteristic 3.

45. Let R be a ring and let $M_2(R)$ be the ring of 2×2 matrices with entries from R. Explain why these two rings have the same characteristic.

46. Let R be a ring with m elements. Show that the characteristic of R divides m.

47. Explain why a finite ring must have a nonzero characteristic.

48. Find all solutions of $x^2 - x + 2 = 0$ over $Z_3[i]$. (See Example 8.)

49. Consider the equation $x^2 - 5x + 6 = 0$.
    a. How many solutions does this equation have in Z_7?
    b. Find all solutions of this equation in Z_8.
    c. Find all solutions of this equation in Z_{12}.
    d. Find all solutions of this equation in Z_{14}.

50. Find the characteristic of $Z_4 \oplus 4Z$.

51. Suppose that R is an integral domain in which $20 \cdot 1 = 0$ and $12 \cdot 1 = 0$. (Recall that $n \cdot 1$ means the sum $1 + 1 + \cdots + 1$ with n terms.) What is the characteristic of R?

52. In a commutative ring of characteristic 2, prove that the idempotents form a subring.

53. Describe the smallest subfield of the field of real numbers that contains $\sqrt{2}$.

54. Let F be a finite field with n elements. Prove that $x^{n-1} = 1$ for all nonzero x in F.

55. Let F be a field of prime characteristic p. Prove that $K = \{x \in F \mid x^p = x\}$ is a subfield of F.

56. Suppose that a and b belong to a field of order 8 and that $a^2 + ab + b^2 = 0$. Prove that $a = 0$ and $b = 0$. Do the same when the field has order 2^n with n odd.

57. Let F be a field of characteristic 2 with more than two elements. Show that $(x + y)^3 \neq x^3 + y^3$ for some x and y in F.

58. Suppose that F is a field with characteristic not 2, and that the nonzero elements of F form a cyclic group under multiplication. Prove that F is finite.

59. Suppose that D is an integral domain and that ϕ is a nonconstant function from D to the nonnegative integers such that $\phi(xy) = \phi(x)\phi(y)$. If x is a unit in D, show that $\phi(x) = 1$.

60. Let F be a field of order 32. Show that the only subfields of F are F itself and $\{0, 1\}$.

PROGRAMMING EXERCISES

"Data! data! data!" he cried impatiently. "I can't make bricks without clay."

Sherlock Holmes

1. Write a program that will list all the idempotents (see exercise 16 for definition) in Z_n. Run your program for $1 \leq n \leq 200$. Use this output to make conjectures about the number of idempotents in Z_n as a function of n. For example, how many idempotents are there when n is a prime-power? What about when n is divisible by two distinct primes?

2. Write a program that will list all the nilpotent elements (see exercise 13 for definition) in Z_n. Run your program for $1 \leq n \leq 200$. Use this output to make conjectures about nilpotent elements in Z_n.

3. Write a program that will determine all the roots of the equation $x^3 - 2x^2 + x - 2 = (x - 2)(x^2 + 1)$ over Z_n. Run your program for $n = 5$, 9, 25, 50, and 100. How many roots does this polynomial have over Z?

SUGGESTED READINGS

N. A. Khan, "The Characteristic of a Ring," *American Mathematical Monthly* 70 (1963): 736.

> Here it is shown that a ring has nonzero characteristic n if and only if n is the maximum of the orders of the elements of R.

K. Robin McLean, "Groups in Modular Arithmetic," *The Mathematical Gazette* 62 (1978): 94–104.

> This article explores the interplay between various groups of integers under multiplication modulo n and the ring Z_n. It shows how to construct groups of integers in which the identity is not obvious; for example, 1977 is the identity of the group {1977, 5931} under multiplication modulo 7908.

Nathan Jacobson

Here, as in so many other parts of algebra, Jake's influence has been very noticeable.

Paul M. Cohn

NATHAN JACOBSON was born on September 8, 1910, in Warsaw, Poland. After arriving in the United States in 1917, Jacobson grew up in Alabama, Mississippi, and Georgia, where his father owned small clothing stores. He received a B.A. degree from the University of Alabama in 1930 and a Ph.D. from Princeton in 1934. After brief periods as a professor at Bryn Mawr, the University of Chicago, the University of North Carolina, and Johns Hopkins, Jacobson accepted a position at Yale, where he remained until his retirement in 1981.

Jacobson's principal contributions to algebra have been in the fields of rings, Lie algebras, and Jordan algebras. In particular, he developed structure theories for these systems. He is the author of nine books and numerous articles.

Jacobson has held visiting positions in France, India, Italy, Israel, China, Australia, and Switzerland. Among his many honors are the presidency of the American Mathematical Society, memberships in the National Academy of Sciences and the American Academy of Arts and Sciences, a Guggenheim Fellowship, and an honorary degree from the University of Chicago.

14

Ideals and Factor Rings

The secret of science is to ask the right questions, and it is the choice of problem more than anything else that marks the man of genius in the scientific world.

Sir Henry Tizard in C. P. Snow, *A Postscript to Science and Government*

IDEALS

Normal subgroups play a special role in group theory—they permit us to construct factor groups. In this chapter, we introduce the analogous concepts for rings—ideals and factor rings.

DEFINITION   Ideal

A subring A of a ring R is called a (two-sided) *ideal* of R if for every $r \in R$ and every $a \in A$ both ra and ar are in A.

So, a subring A of a ring R is an ideal of R if A "absorbs" elements from R—that is, if $rA \subseteq A$ and $Ar \subseteq A$ for all $r \in R$.

An ideal A of R is called a *proper* ideal of R if A is a proper subset of R. In practice, one identifies ideals with the following test, which is an immediate consequence of the definition of ideal, and the subring test given in Chapter 12.

Theorem 14.1 *Ideal Test*

A nonempty subset A of a ring R is an ideal of R if

1. $a - b \in A$ whenever $a, b \in A$.
2. ra and ar are in A whenever $a \in A$ and $r \in R$.

Example 1 For any ring R, $\{0\}$ and R are ideals of R. The ideal $\{0\}$ is called the *trivial* ideal. ❑

Example 2 For any positive integer n, the set $nZ = \{0, \pm n, \pm 2n, \ldots\}$ is an ideal of Z. ❑

Example 3 Let R be a commutative ring with unity and let $a \in R$. The set $\langle a \rangle = \{ra \mid r \in R\}$ is an ideal of R called the *principal ideal generated by a*. (Notice that $\langle a \rangle$ is also the notation we used for the cyclic subgroup generated by a. However, the intended meaning will always be clear from the context.) The assumption that R is commutative is necessary in this example (see exercise 29 of the Supplementary Exercises for Chapters 12–14). ❑

Example 4 Let $\mathbf{R}[x]$ denote the set of all polynomials with real coefficients and let A denote the subset of all polynomials with constant term 0. Then A is an ideal of $\mathbf{R}[x]$. ❑

Example 5 Let R be the ring of all real-valued functions of a real variable. The subset S of all differentiable functions is a subring of R but not an ideal of R. ❑

FACTOR RINGS

Let R be a ring and let A be an ideal of R. Since R is a group under addition and A is a normal subgroup of R, we may form the factor group $R/A = \{r + A \mid r \in R\}$. The natural question at this point is how may we form a ring of this group of cosets? The addition is already taken care of, and, by analogy with groups of cosets, we define the product of two cosets of $a + A$ and $b + A$ as $ab + A$. The next theorem shows that this definition works as long as A is an ideal, and not just a subring, of R.

Theorem 14.2 *Existence of Factor Rings*

Let R be a ring and let A be a subring of R. The set of cosets $\{r + A \mid r \in R\}$ is a ring under the operations $(s + A) + (t + A) = s + t + A$ and $(s + A)(t + A) = st + A$ if and only if A is an ideal of R.

Proof. We know that the set of cosets forms a group under addition, and it is trivial to check that the multiplication is associative and that multiplication is distributive over addition once we know that multiplication is indeed a binary operation on the set of cosets. Hence, the proof boils down to showing that multiplication is well defined if and only if A is an ideal of R. To do this, let us suppose that A is an ideal and let $s + A = s' + A$ and $t + A = t' + A$. Then we must show that $st + A = s't' + A$. Well, by definition, $s = s' + a$ and $t = t' + b$, where a and b belong to A. Then,

$$st = (s' + a)(t' + b) = s't' + at' + s'b + ab,$$

and, so,

$$st + A = s't' + at' + s'b + ab + A = s't' + A,$$

since A absorbs the last three summands of the middle expression. Thus, multiplication is well defined when A is an ideal.

On the other hand, suppose that A is not an ideal of R. Then there exist elements $a \in A$ and $r \in R$ such that $ar \notin A$ or $ra \notin A$. For convenience, say, $ar \notin A$. Consider the elements $a + A = 0 + A$ and $r + A$. Clearly, $(a + A)(r + A) = ar + A$ but $(0 + A)(r + A) = A$. Since $ar + A \neq A$, the multiplication is not well defined and the set of cosets is not a ring. ∎

Let's look at a few factor rings.

Example 6 $Z/4Z = \{0 + 4Z, 1 + 4Z, 2 + 4Z, 3 + 4Z\}$. To see how to add and multiply, consider $2 + 4Z$ and $3 + 4Z$.

$$(2 + 4Z) + (3 + 4Z) = 5 + 4Z = 1 + 4 + 4Z = 1 + 4Z,$$
$$(2 + 4Z)(3 + 4Z) = 6 + 4Z = 2 + 4 + 4Z = 2 + 4Z.$$

One can readily see that the two operations are essentially modulo 4 arithmetic. ❑

Example 7 $2Z/6Z = \{0 + 6Z, 2 + 6Z, 4 + 6Z\}$. Here the operations are essentially modulo 6 arithmetic. ❑

Here is a noncommutative example of an ideal and factor ring.

Example 8   Let $R = \left\{ \begin{bmatrix} a_1 & a_2 \\ a_3 & a_4 \end{bmatrix} \middle| a_i \in Z \right\}$ and let I be the subset of R consisting of matrices with even entries. It is easy to show that I is indeed an ideal of R (exercise 16). Consider the factor ring R/I. The interesting question about this ring is its size. Just what is it? We claim it is 16; in fact, $R/I = \left\{ \begin{bmatrix} r_1 & r_2 \\ r_3 & r_4 \end{bmatrix} + I \middle| r_i \in \{0, 1\} \right\}$. An example illustrates the

typical situation. Which of the 16 elements is $\begin{bmatrix} 7 & 8 \\ 5 & -3 \end{bmatrix} + I$?

Well, observe that $\begin{bmatrix} 7 & 8 \\ 5 & -3 \end{bmatrix} + I = \begin{bmatrix} 1 & 0 \\ 1 & 1 \end{bmatrix} + \begin{bmatrix} 6 & 8 \\ 4 & -4 \end{bmatrix} + I =$

$\begin{bmatrix} 1 & 0 \\ 1 & 1 \end{bmatrix} + I$, since an ideal absorbs its own elements. The general case is left to the reader. $\qquad \square$

Example 9 Let $\mathbf{R}[x]$ denote the ring of polynomials with real coefficients and let $\langle x^2 + 1 \rangle$ denote the principal ideal generated by $x^2 + 1$; that is,

$$\langle x^2 + 1 \rangle = \{f(x)(x^2 + 1) \mid f(x) \in \mathbf{R}[x]\}.$$

Then

$$\mathbf{R}[x]/\langle x^2 + 1 \rangle = \{g(x) + \langle x^2 + 1 \rangle\}$$
$$= \{ax + b + \langle x^2 + 1 \rangle \mid a, b \in \mathbf{R}\}.$$

To see this last equality, note that if $g(x)$ is any member of $\mathbf{R}[x]$, then we may write $g(x)$ in the form of $q(x)(x^2 + 1) + r(x)$, where $q(x)$ is the quotient and $r(x)$ is the remainder upon dividing $g(x)$ by $x^2 + 1$. In particular, $r(x) = 0$ or degree $r(x) < 2$, so that $r(x) = ax + b$ for some a and b in $\mathbf{R}$. Thus,

$$g(x) + \langle x^2 + 1 \rangle = q(x)(x^2 + 1) + r(x) + \langle x^2 + 1 \rangle$$
$$= r(x) + \langle x^2 + 1 \rangle,$$

since the ideal $\langle x^2 + 1 \rangle$ absorbs the term $q(x)(x^2 + 1)$.

How is multiplication done? Since

$$x^2 + 1 + \langle x^2 + 1 \rangle = 0 + \langle x^2 + 1 \rangle,$$

one should think of $x^2 + 1$ as 0 or, equivalently, as $x^2 = -1$. So, for example,

$$(x + 3 + \langle x^2 + 1 \rangle) \cdot (2x + 5 + \langle x^2 + 1 \rangle)$$
$$= 2x^2 + 11x + 15 + \langle x^2 + 1 \rangle = 11x + 13 + \langle x^2 + 1 \rangle.$$

In view of the fact that the elements of this ring have the form $ax + b + \langle x^2 + 1 \rangle$, where $x^2 + \langle x^2 + 1 \rangle = -1 + \langle x^2 + 1 \rangle$, it is perhaps not surprising that this ring turns out to be algebraically the same ring as the ring of complex numbers. This observation was first made by Cauchy in 1847. $\qquad \square$

Example 9 illustrates one of the most important applications of factor rings—the construction of rings with highly desirable properties. In particular, we shall show how one may use factor rings to construct integral domains and fields.

PRIME IDEALS AND MAXIMAL IDEALS

DEFINITION Prime Ideal, Maximal Ideal

A proper ideal A of a commutative ring R is said to be a *prime ideal* of R if $a, b \in R$ and $ab \in A$ implies $a \in A$ or $b \in A$. A proper ideal A of R

is said to be a *maximal ideal* of R if, whenever B is an ideal of R and $A \subseteq B \subseteq R$, then $B = A$ or $B = R$.

So, the only ideal that properly contains a maximal ideal is the entire ring. The motivation for the definition of a prime ideal comes from the integers.

Example 10   Let n be a positive integer. Then, in the ring of integers, the ideal nZ is prime if and only if n is prime.   □

Example 11   The lattice of ideals of Z_{36} (Figure 14.1) shows that only $\langle 2 \rangle$ and $\langle 3 \rangle$ are maximal ideals.   □

Example 12   The ideal $\langle x^2 + 1 \rangle$ is maximal in $\mathbf{R}[x]$. To see this, assume that A is an ideal of $\mathbf{R}[x]$ that properly contains $\langle x^2 + 1 \rangle$. We will prove that $A = \mathbf{R}[x]$ by showing that A contains some nonzero real number c. [This is the constant polynomial $h(x) = c$ for all x.] Then $1 = (1/c)c \in A$ and therefore, by exercise 14, $A = \mathbf{R}[x]$. To this end, let $f(x) \in A$, but $f(x) \notin \langle x^2 + 1 \rangle$. Then

$$f(x) = q(x)(x^2 + 1) + r(x),$$

where $r(x) \neq 0$ and degree $r(x) < 2$. It follows that $r(x) = ax + b$, where not both a and b are 0, and

$$ax + b = r(x) = f(x) - q(x)(x^2 + 1) \in A.$$

Thus,

$$a^2x^2 - b^2 = (ax + b)(ax - b) \in A \quad \text{and} \quad a^2(x^2 + 1) \in A,$$

also. So,

$$a^2 + b^2 = (a^2x^2 + a^2) - (a^2x^2 - b^2) \in A.$$ □

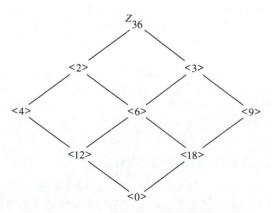

Figure 14.1

Example 13 The ideal $\langle x^2 + 1 \rangle$ is not prime in $Z_2[x]$ since it contains $(x + 1)^2 = x^2 + 2x + 1 = x^2 + 1$ but does not contain $x + 1$. ☐

Theorem 14.3 *R/A Is an Integral Domain if and Only if A Is Prime*

Let R be a commutative ring with unity and let A be an ideal of R. Then R/A is an integral domain if and only if A is prime.

Proof. Suppose that R/A is an integral domain and $ab \in A$. Then $(a + A)(b + A) = ab + A = A$, the zero element of the ring R/A. So, either $a + A = A$ or $b + A = A$; that is, either $a \in A$ or $b \in A$. Hence, A is prime.

To prove the other half of the theorem, we first observe that R/A is a commutative ring with unity for any proper ideal A. Thus, our task is simply to show that when A is prime, R/A has no zero-divisors. So, suppose that A is prime and $(a + A)(b + A) = 0 + A = A$. Then $ab \in A$ and, therefore, $a \in A$ or $b \in A$. Thus, one of $a + A$ or $b + A$ is the zero coset in R/A. ∎

For maximal ideals, we can do even better.

Theorem 14.4 *R/A Is a Field if and Only if A Is Maximal*

Let R be a commutative ring with unity and let A be an ideal of R. Then R/A is a field if and only if A is maximal.

Proof. Suppose that R/A is a field and B is an ideal of R that properly contains A. Let $b \in B$ but $b \notin A$. Then $b + A$ is a nonzero element of R/A and, therefore, there exists an element $c + A$ such that $(b + A)(c + A) = 1 + A$, the multiplicative identity of R/A. Since $b \in B$, we have $bc \in B$. Because

$$1 + A = (b + A)(c + A) = bc + A,$$

we have $1 - bc \in A \subset B$. So, $1 = (1 - bc) + bc \in B$. By exercise 14, $B = R$. This proves that A is maximal.

Now suppose that A is maximal and let $b \in R$ but $b \notin A$. It suffices to show that $b + A$ has a multiplicative inverse. (All other properties for a field follow trivially.) Consider $B = \{br + a \mid r \in R, a \in A\}$. This is an ideal of R that properly contains A (exercise 19). Since A is maximal, we must have $B = R$. Thus, $1 \in B$, say, $1 = bc + a'$, where $a' \in A$. Then

$$1 + A = bc + a' + A = bc + A = (b + A)(c + A).$$ ∎

EXERCISES

Problems worthy
of attack
prove their worth
by hitting back.

Piet Hein, "Problems," *Grooks* **(1966)**

1. Show that the set of rational numbers is a subring of the reals but is not an ideal.
2. Find a subring of $Z \oplus Z$ that is not an ideal of $Z \oplus Z$.
3. Let $S = \{a + bi \mid a, b \in Z, b \text{ is even}\}$. Show that S is a subring of $Z[i]$, but not an ideal of $Z[i]$.
4. Find all maximal ideals in

    a. Z_8, **b.** Z_{10} **c.** Z_{12} **d.** Z_n.
5. Let a belong to a commutative ring R. Show that $aR = \{ar \mid r \in R\}$ is an ideal of R. If R is the ring of even integers, list the elements of $4R$.
6. If n is a positive integer, show that $\langle n \rangle = nZ$ is a prime ideal of Z if and only if n is prime.
7. Prove that the intersection of any set of ideals of a ring is an ideal.
8. If A and B are ideals of a ring, show that the *sum* of A and B, $A + B = \{a + b \mid a \in A, b \in B\}$, is an ideal.
9. In the ring of integers, find a positive integer a such that
    a. $\langle a \rangle = \langle 2 \rangle + \langle 3 \rangle$,
    b. $\langle a \rangle = \langle 3 \rangle + \langle 6 \rangle$,
    c. $\langle a \rangle = \langle m \rangle + \langle n \rangle$.
10. If A and B are ideals of a ring, show that the *product* of A and B, $AB = \{a_1 b_1 + a_2 b_2 + \cdots + a_n b_n \mid a_i \in A, b_i \in B, n \text{ a positive integer}\}$, is an ideal.
11. Find a positive integer a such that
    a. $\langle a \rangle = \langle 3 \rangle \langle 4 \rangle$,
    b. $\langle a \rangle = \langle 6 \rangle \langle 8 \rangle$,
    c. $\langle a \rangle = \langle m \rangle \langle n \rangle$.
12. Let A and B be ideals of a ring. Prove that $AB \subseteq A \cap B$.
13. If A and B are ideals of a commutative ring R with unity and $A + B = R$, show that $A \cap B = AB$.
14. If A is an ideal of a ring R and 1 belongs to A, prove that $A = R$. (This exercise is referred to in this chapter.)
15. If an ideal I of a ring R contains a unit, show that $I = R$.
16. Let R and I be as described in Example 8. Prove that I is an ideal of R.
17. Verify the claim made in Example 8 about the size of R/I.
18. Prove that the ideal $\langle x^2 + 1 \rangle$ is prime in $Z[x]$ but not maximal in $Z[x]$.

19. Show that the set B in the latter half of the proof of Theorem 14.4 is an ideal of R. (This exercise is referred to in this chapter.)

20. Construct a multiplication table for the ring $3Z/9Z$.

21. If R is a commutative ring with unity and A is an ideal of R, show that R/A is a commutative ring with unity.

22. Show that $\mathbf{R}[x]/\langle x^2 + 1 \rangle$ is a field.

23. Let R be a commutative ring with unity. Show that every maximal ideal of R is prime.

24. Show that $A = \{(3x, y) \mid x, y \in Z\}$ is a maximal ideal of $Z \oplus Z$.

25. Let R be the ring of continuous functions from $\mathbf{R}$ to $\mathbf{R}$. Show that $A = \{f \in R \mid f(0) = 0\}$ is a maximal ideal of R.

26. Let $R = Z_8 \oplus Z_{30}$. Find all maximal ideals of R, and, for each maximal ideal I, identify the size of the field R/I.

27. How many elements are in $Z[i]/\langle 3 + i \rangle$? Give reasons for your answer.

28. In $Z[x]$, the ring of polynomials with integer coefficients, let $I = \{f \in Z[x] \mid f(0) = 0\}$. Prove that I is not a maximal ideal.

29. In $Z \oplus Z$, let $I = \{(a, 0) \mid a \in Z\}$. Show that I is a prime ideal but not a maximal ideal.

30. Let R be a ring and I an ideal of R. Prove that the factor ring R/I is commutative if and only if $rs - sr \in I$ for all r and s in R.

31. In $Z[x]$, let $I = \{f \in Z[x] \mid f(0)$ is an even integer$\}$. Prove that I is a prime ideal of $Z[x]$. Is it maximal?

32. Prove that $I = \langle 2 + 2i \rangle$ is not a prime ideal of $Z[i]$. How many elements are in $Z[i]/I$? What is the characteristic of $Z[i]/I$?

33. In $Z_5[x]$, let $I = \langle x^2 + x + 2 \rangle$. Find the multiplicative inverse of $2x + 3 + I$ in $Z_5[x]/I$.

34. An integral domain D is called a *principal ideal domain* if every ideal of D has the form $\langle a \rangle = \{ad \mid d \in D\}$ for some a in D. Show that Z is a principal ideal domain.

35. Let R be a ring and let p be a fixed prime. Show that $I_p = \{r \in R \mid$ additive order of r is a power of $p\}$ is an ideal of R.

36. Let a and b belong to a commutative ring R. Prove that $\{x \in R \mid ax \in bR\}$ is an ideal.

37. Let R be a commutative ring and let A be any subset of R. Show that the *annihilator* of A, $\text{Ann}(A) = \{r \in R \mid ra = 0$ for all a in $A\}$, is an ideal.

38. Let R be a commutative ring and let A be an ideal of R. Show that the *nil radical* of A, $\sqrt{A} = \{r \in R \mid r^n \in A$ for some positive integer n (n depends on r)$\}$, is an ideal of $\mathbf{R}$. ($\sqrt{\langle 0 \rangle}$ is called the *nil radical* of R.)

39. Let $R = Z_{27}$. Find
    a. $\sqrt{0}$,   b. $\sqrt{\langle 3 \rangle}$,   c. $\sqrt{\langle 9 \rangle}$.

40. Let $R = Z_{36}$. Find
    a. $\sqrt{\langle 0 \rangle}$,   b. $\sqrt{\langle 4 \rangle}$,   c. $\sqrt{\langle 6 \rangle}$.

41. Let R be a commutative ring. Show that $R/\sqrt{\langle 0 \rangle}$ has no nonzero nilpotent elements.

42. Let A be an ideal of a commutative ring. Prove that $\sqrt{\sqrt{A}} = \sqrt{A}$.

43. Let $Z_2[x]$ be the ring of all polynomials with coefficients in Z_2 (that is, coefficients are 0 or 1, and addition and multiplication of coefficients are done modulo 2). Show that $Z_2[x]/\langle x^2 + x + 1 \rangle$ is a field.

44. List the elements of the field given in exercise 43, and make an addition and multiplication table for the field.

45. Show that $Z_3[x]/\langle x^2 + x + 1 \rangle$ is not a field.

46. Let R be a commutative ring without unity, and let $a \in R$. Describe the smallest ideal of R that contains a.

47. Let R be the ring of continuous functions from $\mathbf{R}$ to $\mathbf{R}$. Let $A = \{f \in R \mid f(0)$ is an even integer$\}$. Show that A is a subring of R, but not an ideal of R.

48. Show that $Z[i]/\langle 1 - i \rangle$ is a field. How many elements does this field have?

49. If R is a principal ideal domain and I is an ideal of R, prove that every ideal of R/I is principal (see exercise 34).

50. How many elements are in $Z_5[i]/\langle 1 + i \rangle$?

51. Let R be a ring with unity that has the property that $a^2 = a$ for all a in R. Let I be a prime ideal in R. Show that $|R/I| = 2$.

Richard Dedekind

Richard Dedekind was not only a mathematician, but one of the wholly great in the history of mathematics, now and in the past, the last hero of a great epoch, the last pupil of Gauss, for four decades himself a classic, from whose works not only we, but our teachers and the teachers of our teachers, have drawn.

Edmund Landau, Commemorative Address to the Royal Society of Göttingen

This stamp was issued by East Germany in 1981 to commemorate the 150th anniversary of Dedekind's birth. Notice that it features the representation of an ideal as the product of powers of prime ideals.

RICHARD DEDEKIND was born on October 6, 1831, in Brunswick, Germany, the birthplace of Gauss. Dedekind was the youngest of four children of a law professor. His early interests were in chemistry and physics, but he obtained a doctor's degree in mathematics at the age of 21 under Gauss at the University of Göttingen. Dedekind continued his studies at Göttingen for a few years, and in 1854 he began to lecture there.

Dedekind spent the years 1858–1862 as a professor in Zürich. Then he accepted a position at an institute in Brunswick where he had once been a student. Although this school was less than university level, Dedekind remained there for the next 50 years. He died in Brunswick in 1916.

During his career, Dedekind made numerous fundamental contributions to mathematics. His treatment of irrational numbers, "Dedekind cuts," put analysis on a firm, logical foundation. His work on unique factorization led to the modern theory of algebraic numbers. He was a pioneer in the theory of rings and fields. The notion of ideal as well as the term itself are due to Dedekind. Mathematics historian Morris Kline has called him "the effective founder of abstract algebra."

Emmy Noether

In the judgment of the most competent living mathematicians, Fräulein Noether was the most significant creative mathematical genius thus far produced since the higher education of women began. In the realm of algebra, in which the most gifted mathematicians have been busy for centuries, she discovered methods which have proved of enormous importance in the development of the present-day younger generation of mathematicians.

Albert Einstein, *The New York Times*

EMMY NOETHER was born on March 23, 1882, in Germany. When she entered the University of Erlangen, she was one of only two women among the 1000 students. Noether completed her doctorate in 1907.

In 1916, Noether went to Göttingen and, under the influence of David Hilbert and Felix Klein, became interested in general relativity. While there, she made a major contribution to physics with her theorem that, whenever there is a symmetry in nature, there is also a conservation law, and vice versa. Hilbert tried unsuccessfully to obtain a faculty appointment at Göttingen for Noether, saying, "I do not see that the sex of the candidate is an argument against her admission as Privatdozent. After all, we are a university and not a bathing establishment."

It was not until she was 38 that Noether's true genius revealed itself. Over the next 13 years, she used an axiomatic method to develop a general theory of ideals and noncommutative algebras. With this abstract theory, Noether was able to weld together many important concepts. Her approach was even more important than the individual results. Hermann Weyl said of Noether, "She originated above all a new and epoch-making style of thinking in algebra."

Noether was not good at lecturing, but she was an inspiring teacher. Her students were known as "the Noether boys," and many turned out to be important mathematicians. Weyl once said, "In my Göttingen years, 1930–1933, she was without doubt the strongest center of mathematical activity there, considering both the fertility of her scientific research program and her influence upon a large circle of pupils."

With the rise of Hitler in 1933, Noether, a Jew, fled to the United States and took a position at Bryn Mawr College. She died suddenly on April 14, 1935, following an operation.

SUPPLEMENTARY EXERCISES FOR CHAPTERS 12–14

If at first you do succeed—try to hide your astonishment.

Harry F. Banks

1. Find all idempotent elements in Z_{10}, Z_{20}, and Z_{30}. (Recall that a is idempotent if $a^2 = a$.)

2. If m and n are relatively prime integers greater than 1, prove that Z_{mn} has at least two idempotents besides 0 and 1.

3. Suppose that R is a ring in which $a^2 = 0$ implies $a = 0$. Show that R has no nonzero nilpotent elements.

4. Let R be a commutative ring with more than one element. Prove that if for every nonzero element a of R we have $aR = R$, then R is a field.

5. Let A, B, and C be ideals of a ring R. If $AB \subseteq C$ and C is a prime ideal of R, show that $A \subseteq C$ or $B \subseteq C$.

6. Show, by example, that the intersection of two prime ideals need not be a prime ideal.

7. Show that the only ideals of a field F are $\{0\}$ and F.

8. Determine all factor rings of Z.

9. Let p be a prime. Show that $A = \{(px, y) \mid x, y \in Z\}$ is a maximal ideal of $Z \oplus Z$.

10. Let R be a commutative ring with unity. Suppose that a is a unit and b is nilpotent. Show that $a + b$ is a unit. (*Hint:* See exercise 29 in Chapter 12.)

11. Let A, B, and C be subrings of a ring R. If $A \subseteq B \cup C$, show that $A \subseteq B$ or $A \subseteq C$.

12. For any element a in a ring R, define $\langle a \rangle$ to be the smallest ideal of R that contains a. If R is a commutative ring with unity, show that $\langle a \rangle = aR = \{ar \mid r \in R\}$. Show, by example, that if R is commutative but does not have a unity, then $\langle a \rangle$ and aR may be different.

13. Let R be a ring with unity. Show that $\langle a \rangle = \{s_1 a t_1 + s_2 a t_2 + \cdots + s_n a t_n \mid s_i, t_i \in R$ and n is a positive integer$\}$.

14. Let p be a prime. Show that $Z_p[x]$ has characteristic p.

15. Let A and B be ideals of a ring R. If $A \cap B = \{0\}$, show that $ab = 0$ when $a \in A$ and $b \in B$.

16. Show that the direct sum of two integral domains is not an integral domain.

17. Consider the ring $R = \{0, 2, 4, 6, 8, 10\}$ under addition and multiplication modulo 12. What is the characteristic of R?

18. What is the characteristic of $Z_m \oplus Z_n$?

19. Let R be a commutative ring with unity. Suppose that the only ideals of R are $\{0\}$ and R. Show that R is a field.

20. Suppose that I is an ideal of J and that J is an ideal of R. Prove that if I has a unity, then I is an ideal of R. (Be careful not to assume that the unity of I is the unity of R. It need not be—see exercise 2 in Chapter 12.)

21. Recall that an idempotent element b in a ring is one with the property that $b^2 = b$. Find a nontrivial idempotent (that is, not 0 and not 1) in $Q[x]/\langle x^4 + x^2 \rangle$.

22. In a principal ideal domain, show that every nontrivial prime ideal is a maximal ideal.

23. Find an example of a commutative ring R with unity such that $a, b \in R$, $a \neq b$, and $a^n = b^n$ and $a^m = b^m$, where n and m are positive integers that are relatively prime. (Compare with exercise 31, part b, in Chapter 13.)

24. Let $Q(\sqrt[3]{2})$ denote the smallest subfield of $\mathbf{R}$ that contains Q and $\sqrt[3]{2}$. Describe the elements of $Q(\sqrt[3]{2})$.

25. Let R be an integral domain with nonzero characteristic. If A is a proper ideal of R, show that R/A has the same characteristic as R.

26. Let F be a field of order p^n. Determine the group isomorphism class of F under the operation addition.

27. If R is a finite commutative ring with unity, prove that every prime ideal of R is a maximal ideal of R.

28. Let R be a noncommutative ring and $C(R)$ the center of R (see exercise 19 in Chapter 12). Prove that the additive group of $R/C(R)$ is not cyclic.

29. Let

$$R = \left\{ \begin{bmatrix} a & b \\ c & d \end{bmatrix} \,\middle|\, a, b, c, d \in Z_2 \right\}$$

with ordinary matrix addition and multiplication modulo 2. Show that

$$\left\{ \begin{bmatrix} 1 & 0 \\ 0 & 0 \end{bmatrix} r \,\middle|\, r \in R \right\}$$

is not an ideal of R. (Hence, in exercise 5 in Chapter 14, the commutativity assumption is necessary.)

30. If R is an integral domain and A is a proper ideal of R, must R/A be an integral domain?

31. Let $A = \{a + bi \mid a, b \in Z, a \equiv b \bmod 2\}$. Show that A is a maximal ideal of $Z[i]$. How many elements does $Z[i]/A$ have?

32. Suppose that R is a commutative ring with unity such that for each a in R there is a positive integer n greater than 1 (n depends on a) so that $a^n = a$. Prove that every prime ideal of R is a maximal ideal of R.

33. State a "finite subfield test." That is, state conditions that guarantee that a finite subset of a field is a subfield.

34. Let F be a finite field with more than two elements. Prove that the sum of all of the elements of F is 0.

35. Let n be a positive integer. Show that the ring Z_n cannot contain a proper subring that is a field.

36. Suppose that R is a ring with no zero-divisors and that R contains a non-zero element b such that $b^2 = b$. Show that b is a unity for R.

37. Find the characteristic of $Z[i]/\langle 2 + i \rangle$.

38. Show that the characteristic of $Z[i]/\langle a + bi \rangle$ divides $a^2 + b^2$.

15

Ring Homomorphisms

If there is one central idea which is common to all aspects of modern algebra it is the notion of homomorphism.

I. N. Herstein, *Topics in Algebra*

DEFINITION AND EXAMPLES

In our work with groups, we saw that one way to discover information about a group is to examine its interaction with other groups by way of homomorphisms. It should not be surprising to learn that this concept extends to rings with equally profitable results.

Just as a group homomorphism preserves the group operation, a ring homomorphism preserves the ring operations.

DEFINITIONS Ring Homomorphism, Ring Isomorphism
A *ring homomorphism* ϕ from a ring R to a ring S is a mapping from R to S that preserves the two ring operations; that is,

$$\phi(a + b) = \phi(a) + \phi(b) \quad \text{and}$$
$$\phi(ab) = \phi(a)\phi(b) \qquad \text{for all } a, b \text{ in } R.$$

A ring homomorphism that is both one-to-one and onto is called a *ring isomorphism*.

As is the case for groups, in the preceding definition the operations on the left of the equal signs are those of R, whereas the operations on the right of the equal signs are those of S.

As with group theory also, the roles of isomorphisms and homo-morphisms are entirely distinct. An isomorphism is used to show that two rings are algebraically identical; a homomorphism is used to sim-plify a ring while retaining certain of its features.

A schematic representation of a ring homomorphism is given in Figure 15.1. The dashed arrows indicate the results of performing the ring operations.

The following examples illustrate ring homomorphisms. The reader should supply the missing details.

Example 1   For any positive integer n, the mapping $k \rightarrow k \bmod n$ is a ring homomorphism from Z onto Z_n. This mapping is called the *natu-ral homomorphism* from Z to Z_n.   ❏

Example 2   The mapping $a + bi \rightarrow a - bi$ is a ring isomorphism from complex numbers onto the complex numbers.   ❏

Example 3   Let $\mathbf{R}[x]$ denote the ring of all polynomials with real coef-ficients. The mapping $f(x) \rightarrow f(1)$ is a ring homomorphism from $\mathbf{R}[x]$ onto $\mathbf{R}$.   ❏

Example 4   The mapping $\phi: x \rightarrow 5x$ from Z_4 to Z_{10} is a ring homo-morphism. Although showing that $\phi(x + y) = \phi(x) + \phi(y)$ appears to be accomplished by the simple statement that $5(x + y) = 5x + 5y$, we must bear in mind that the addition on the left is done modulo 4 whereas the addition on the right and the multiplication on both sides are done modulo 10. An analogous difficulty arises with showing that ϕ preserves multiplication. So, to verify that ϕ preserves both operations, we write $x + y = 4q_1 + r_1$ and $xy = 4q_2 + r_2$ where $0 \leq r_1 < 4$ and $0 \leq r_2 < 4$. Then $\phi(x + y) = \phi(r_1) = 5r_1 = 5(x + y - 4q_1) = 5x + 5y - 20q_1 = 5x + 5y$ in Z_{10}. Similarly, using the additional fact that $5 \cdot 5 = 5$ in Z_{10}, we have $\phi(xy) = \phi(r_2) = 5r_2 = 5(xy - 4q_2) = 5xy - 20q_2 = (5 \cdot 5)xy = 5x5y = \phi(x)\phi(y)$ in Z_{10}.   ❏

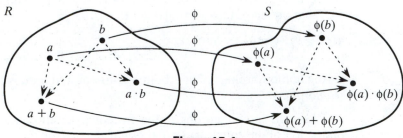

Figure 15.1

Example 5 We determine all ring homomorphisms from Z_{12} to Z_{30}. By Example 9 in Chapter 10, the only group homomorphisms from Z_{12} to Z_{30} are $x \rightarrow ax$, where $a = 0, 15, 10, 20, 5$, or 25. But, since $1 \cdot 1 = 1$ in Z_{12}, we must have $a \cdot a = a$ in Z_{30}. This requirement rules out 20 and 5 as possibilities for a. Finally, simple calculations show that each of the remaining four choices does yield a ring homomorphism.   ☐

Example 6 Let R be a commutative ring of characteristic 2. Then the mapping $a \rightarrow a^2$ is a ring homomorphism from R to R.   ☐

Example 7 Although $2Z$, the group of even integers under addition, is group-isomorphic to the group Z under addition, the ring $2Z$ is not ring-isomorphic to the ring Z. (Quick! What does Z have that $2Z$ doesn't?)   ☐

Our next three examples are applications of the natural homomorphism given in Example 1 to number theory.

Example 8 Test for Divisibility by 9
An integer n with decimal representation $a_k a_{k-1} \cdots a_0$ is divisible by 9 if and only if $a_k + a_{k-1} \cdots + a_0$ is divisible by 9. To verify this, observe that $n = a_k 10^k + a_{k-1} 10^{k-1} + \cdots + a_0$. Then, letting α denote the natural homomorphism from Z to Z_9 [in particular, $\alpha(10) = 1$], we note that n is divisible by 9 if and only if

$$0 = \alpha(n) = \alpha(a_k)(\alpha(10))^k + \alpha(a_{k-1})(\alpha(10))^{k-1} + \cdots + \alpha(a_0)$$
$$= \alpha(a_k) + \alpha(a_{k-1}) + \cdots + \alpha(a_0)$$
$$= \alpha(a_k + a_{k-1} + \cdots + a_0).$$

But $\alpha(a_k + a_{k-1} + \cdots + a_0) = 0$ is equivalent to $a_k + a_{k-1} + \cdots + a_0$ being divisible by 9.   ☐

Example 9 Consider the sequence 3, 7, 11, 15, ... Is it possible that one of these integers is the sum of two squares? If so, the equation $3 + 4k = a^2 + b^2$ is satisfied for some integers k, a, and b. Then, applying the natural homomorphism from Z to Z_4 to both sides of this equation, we see that the equation $3 = x^2 + y^2$ has a solution in Z_4. But, by direct substitution, one may verify that this equation has no solution. Thus, no integer in the sequence is the sum of two squares.   ☐

Example 10 We determine the positive integer solutions to the equation $x^3 + 15xy = 2192$. Let α be the natural homomorphism from Z to Z_3 and let β be the natural homomorphism from Z to Z_5. If x and y are integer solutions to the equation, then $\alpha(x^3 + 15xy) = \alpha(2192)$ and $\beta(x^3 + 15xy) = \beta(2192)$. It follows then that $(\alpha(x))^3 = 2$ and $(\beta(x))^3 = 2$. Solving these two equations in Z_3 and Z_5, respectively, we see that

$\alpha(x) = 2$ and $\beta(x) = 3$. Thus, $x \in \{2, 5, 8, \ldots\} \cap \{3, 8, 13, \ldots\}$. This means that x has the form $8 + 15k$. If $k \geq 1$, then $x^3 > 2192$, so we conclude that $x = 8$. Substituting this into the original equation now gives $y = 14$. □

PROPERTIES OF RING HOMOMORPHISMS

Theorem 15.1 *Properties of Ring Homomorphisms*

Let ϕ be a homomorphism from a ring R to a ring S. Let A be a subring of R and B an ideal of S.

1. *For any $r \in R$ and any positve integer n, $\phi(nr) = n\phi(r)$ and $\phi(r^n) = (\phi(r))^n$.*
2. *$\phi(A) = \{\phi(a) \mid a \in A\}$ is a subring of S.*
3. *If A is an ideal and ϕ is onto S, then $\phi(A)$ is an ideal.*
4. *$\phi^{-1}(B) = \{r \in R \mid \phi(r) \in B\}$ is an ideal of R.*
5. *If R is commutative, then $\phi(R)$ is commutative.*
6. *If R has a unity 1, $S \neq \{0\}$, and ϕ is onto, then $\phi(1)$ is the unity of S.*
7. *ϕ is an isomorphism if and only if ϕ is onto and Ker $\phi = \{r \in R \mid \phi(r) = 0\} = \{0\}$.*
8. *If ϕ is an isomorphism from R onto S, then ϕ^{-1} is an isomorphism from S onto R.*

Proof. The proofs of these properties are straightforward and are left as exercises. ∎

As with groups, the student should learn in words the various properties of Theorem 15.1 in addition to the symbols. Property 2 says that the homomorphic image of a subring is a subring. Property 4 says that the pullback of an ideal is an ideal, and so on.

The next three theorems parallel results we had for groups. The proofs are nearly identical to the group theory counterparts and are left as exercises.

Theorem 15.2 *Kernels Are Ideals*

Let ϕ be a homomorphism from a ring R to a ring S. Then Ker $\phi = \{r \in R \mid \phi(r) = 0\}$ is an ideal of R.

Theorem 15.3 *First Isomorphism Theorem for Rings*

Let ϕ be a ring homomorphism from R to S. Then the mapping from $R/\text{Ker } \phi$ to $\phi(R)$, given by $r + \text{Ker } \phi \to \phi(r)$ is an isomorphism. In symbols, $R/\text{Ker } \phi \approx \phi(R)$.

Theorem 15.4 *Ideals Are Kernels*

Every ideal of a ring R is the kernel of a ring homomorphism of R. In particular, an ideal A is the kernel of the mapping $r \to r + A$ from R to R/A.

The homomorphism from R to R/A given in Theorem 15.4 is called the *natural homomorphism* from R to R/A. Theorem 15.3 is often referred to as the Fundamental Theorem of Ring Homomorphisms.

Theorem 15.5 *Homomorphism from Z to a Ring with Unity*
Let R be a ring with unity e. The mapping $\phi: Z \to R$ given by $n \to ne$ is a ring homomorphism.

Proof. Let $m, n \in Z$. To show that addition is preserved, we consider three cases. First suppose that both m and n are nonnegative. Then

$$\phi(m + n) = (m + n)e = \underbrace{e + e + \cdots + e}_{(m + n) \text{ summands}}$$

$$= \underbrace{(e + e + \cdots + e)}_{m \text{ summands}} + \underbrace{(e + e + \cdots + e)}_{n \text{ summands}}$$

$$= me + ne = \phi(m) + \phi(n).$$

Next, suppose that both m and n are negative. Then,

$$\phi(m + n) = (m + n)e = (-m - n)(-e)$$
$$= (-m)(-e) + (-n)(-e) = me + ne = \phi(m) + \phi(n).$$

Third, suppose that one of m and n is nonnegative and the other is negative, say $m \geq 0, n < 0$.
Then,

$$\phi(m + n) = (m + n)e$$
$$= \underbrace{(e + e + \cdots + e)}_{m \text{ summands}} - \underbrace{(e + e + \cdots + e)}_{-n \text{ summands}}$$
$$= me + (-n)(-e) = me + ne = \phi(m) + \phi(n).$$

So, ϕ preserves addition.

The multiplication can be handled in a single case with the aid of exercise 14 in Chapter 12, which says that $(ma)(nb) = (mn)(ab)$ for all integers m and n. Thus, $\phi(mn) = (mn)e = (mn)(ee) = (me)(ne) = \phi(m)\phi(n)$. So, ϕ preserves multiplication as well. ∎

Corollary 1 *A Ring with Unity Contains Z_n or Z*
If R is a ring with unity and the characteristic of R is $n > 0$, then R contains a subring isomorphic to Z_n. If the characteristic R is 0, then R contains a subring isomorphic to Z.

Proof. Let e be the unity element of R, and recall that, by exercise 44 in Chapter 12, $S = \{ke \mid k \in Z\}$ is a subring of R. Now Theorem 15.5 shows that the mapping ϕ from Z onto S given by $\phi(k) = ke$ is a homomorphism, and by the First Isomorphism Theorem for rings, we have $Z/\text{Ker } \phi \approx S$. But, clearly, $\text{Ker } \phi = \langle n \rangle$, where n is the additive order of e and, by Theorem 13.3, n is the characteristic of R. So, when R has characteristic n, $S \approx Z/\langle n \rangle \approx Z_n$. When R has characteristic 0, $S \approx Z/\langle 0 \rangle \approx Z$. ∎

Corollary 2 Z_m *Is a Homomorphic Image of Z*
For any positive integer m, the mapping of $\phi: Z \to Z_m$ given by $x \to x$ mod m is a ring homomorphism.

Proof.   This follows directly from the statement of Theorem 15.5, since in the ring Z_m, the integer x mod m is $x1$. [For example, in Z_3, if $x = 5$, we have $5(1) = 1 + 1 + 1 + 1 + 1 = 2$.] ∎

Corollary 3   *A Field Contains Z_p or Q (Steinitz, 1910)*
If F is a field of characteristic p, then F contains a subfield isomorphic to Z_p. If F is a field of characteristic 0, then F contains a subfield isomorphic to the rational numbers.

Proof.   By Corollary 1, F contains a subring isomorphic to Z_p if F has characteristic p, and F has a subring S isomorphic to Z if F has characteristic 0. In the latter case, let

$$T = \{ab^{-1} \mid a, b \in S, b \neq 0\}.$$

Then T is isomorphic to the rationals (exercise 49). ∎

Since the intersection of all subfields of a field is itself a subfield, every field has a smallest subfield (that is, a subfield that is contained in every subfield). This subfield is called the *prime subfield* of the field. It follows from Corollary 3 that the prime subfield of a field of characteristic p is isomorphic to Z_p, whereas the prime subfield of a field of characteristic 0 is isomorphic to Q. (See exercise 53.)

THE FIELD OF QUOTIENTS

Although the integral domain Z is not a field, it is at least contained in a field—the field of rational numbers. And notice that the field of rational numbers is nothing more than quotients of integers. Can we mimic the construction of the rationals from the integers for other integral domains? Yes.

Theorem 15.6   *Field of Quotients*
Let D be an integral domain. Then there exists a field F (called the field of quotients of D) that contains a subring isomorphic to D.

Proof.   (Throughout this proof, you should keep in mind that we are using the construction of the rationals from the integers as a model for our construction of the field of quotients of D.)

Let S be the set of all formal symbols of the form a/b, where a, $b \in D$ and $b \neq 0$. Define an equivalence relation $\equiv$ on S by $a/b \equiv c/d$ if $ad = bc$ (just as we have $1/2 = 3/6$ for rationals). Now, let F be the set of equivalence classes of S under the relation $\equiv$. We define addition and multiplication on F by

$$[a/b] + [c/d] = [(ad + bc)/bd] \quad \text{and} \quad [a/b] \cdot [c/d] = [ac/bd].$$

(Notice that here we need the fact that D is an integral domain to ensure that multiplication is closed; that is, $bd \neq 0$ whenever $b \neq 0$ and $d \neq 0$.)

Since there are many representations of any particular element of F (just as in the rationals, we have $1/2 = 3/6 = 4/8$), we must show that these two operations are well defined. To do this, suppose that $[a/b] = [a'/b']$ and $[c/d] = [c'/d']$, so that $ab' = a'b$ and $cd' = c'd$. It then follows that

$$
\begin{aligned}
(ad + bc)b'd' &= adb'd' + bcb'd' = (ab')dd' + (cd')bb' \\
&= (a'b)dd' + (c'd)bb' = a'd'bd + b'c'bd \\
&= (a'd' + b'c')bd.
\end{aligned}
$$

Thus, by definition, we have

$$[(ad + bc)/bd] = [(a'd' + b'c')/b'd'],$$

and, therefore, addition is well defined. We leave the verification that multiplication is well defined as an exercise. That F is a field is straightforward. Let 1 denote the unity of D. Then $[0/1]$ is the additive identity of F. The additive inverse of $[a/b]$ is $[-a/b]$; the multiplicative inverse of a nonzero element $[a/b]$ is $[b/a]$. The remaining field properties can be checked easily.

Finally, the mapping $\phi: D \to F$ given by $x \to [x/1]$ is a ring isomorphism from D to $\phi(D)$. ∎

In practice, it is customary to identify the equivalence class $[a/b]$ with the symbol a/b. This notation is less cumbersome and should not cause confusion. Again, the rational numbers provide a model, since we traditionally use $1/2$ in place of $[1/2]$, and so on. Just remember that a/b and c/d represent the same equivalence class if and only if $ad = cb$.

Example 11 Let $D = Z[x]$. Then the field of quotients of D is $\{f(x)/g(x) \mid f(x), g(x) \in D$, where $g(x)$ is not the zero polynomial$\}$. This set is usually called the set of *rational functions*. ❑

When F is a field, the field of quotients of $F[x]$ is traditionally denoted by $F(x)$. Thus, $Q(x) = \{f(x)/g(x) \mid f(x), g(x) \in Q[x]$, where $g(x)$ is not the zero polynomial$\}$.

EXERCISES

We can work it out.

Title of song by John Lennon and Paul McCartney, single, December 1965.

1. Show that the correspondence $x \to 5x$ from Z_5 to Z_{10} does not preserve addition.

2. Show that the correspondence $x \to 3x$ from Z_4 to Z_{12} does not preserve multiplication.

3. **a.** Is the ring $2Z$ isomorphic to the ring $3Z$?
   b. Is the ring $2Z$ isomorphic to the ring $4Z$?

4. Let $Z_3[i] = \{a + bi \mid a, b \in Z_3\}$ (see Example 8 in Chapter 13). Show that $Z_3[i]$ is isomorphic to $Z_3[x]/\langle x^2 + 1 \rangle$ as fields.

5. Let

$$S = \left\{ \begin{bmatrix} a & b \\ -b & a \end{bmatrix} \;\middle|\; a, b \in \mathbf{R} \right\}.$$

Show that $\phi : \mathbf{C} \to S$ given by

$$\phi(a + bi) = \begin{bmatrix} a & b \\ -b & a \end{bmatrix}$$

is a ring isomorphism.

6. Let $Z[\sqrt{2}] = \{a + b\sqrt{2} \mid a, b \in Z\}$. Let

$$H = \left\{ \begin{bmatrix} a & 2b \\ b & a \end{bmatrix} \;\middle|\; a, b \in Z \right\}.$$

Show that $Z[\sqrt{2}]$ and H are isomorphic as rings.

7. Consider the mapping from $M_2(Z)$ into Z given by $\begin{bmatrix} a & b \\ c & d \end{bmatrix} \to a$. Prove or disprove that this is a ring homomorphism.

8. Let $R = \left\{ \begin{bmatrix} a & b \\ 0 & c \end{bmatrix} \;\middle|\; a, b, c \in Z \right\}$. Prove or disprove that the mapping $\begin{bmatrix} a & b \\ 0 & c \end{bmatrix} \to a$ is a ring homomorphism.

9. Is the mapping from Z_5 to Z_{30} given by $x \to 6x$ a ring homomorphism? Note that the image of the unity is the unity of the image but not the unity of Z_{30}.

10. Is the mapping from Z_{10} to Z_{10} given by $x \to 2x$ a ring homomorphism?

11. Describe the kernel of the homomorphism given in Example 3.

12. Let ϕ be a ring homomorphism from Z_m to Z_n. Prove that $\phi(1)$ is an idempotent of Z_n. (Recall that an idempotent is an element a with the property that $a^2 = a$.)

13. In Z, let $A = \langle 2 \rangle$ and $B = \langle 8 \rangle$. Show that the group A/B is isomorphic to the group Z_4 but that the ring A/B is not isomorphic to the ring Z_4.

14. Determine all ring homomorphisms from Z_6 to Z_6. Determine all ring homomorphisms from Z_{20} to Z_{30}.

15. Determine all ring homomorphisms from Z to Z.

16. Determine all ring homomorphisms from Q to Q.

17. Prove that the sequence 2, 10, 18, 26, . . . contains no cube.

18. Show that $(Z \oplus Z)/(\langle (a, 0) \rangle \oplus \langle (0, b) \rangle)$ is ring-isomorphic to $Z_a \oplus Z_b$.

19. Determine all ring homomorphisms from $Z \oplus Z$ to Z.

20. Prove that the sum of the squares of three consecutive integers cannot be a square.

21. Let m be a positive integer and let n be an integer obtained from m by rearranging the digits of m in some way. (For example, 72345 is a rearrangement of 35274.) Show that $m - n$ is divisible by 9.

22. (Test for divisibility by 11) Let n be an integer with decimal representation $a_k a_{k-1} \cdots a_1 a_0$. Prove that n is divisible by 11 if and only if $a_0 - a_1 + a_2 - \cdots (-1)^k a_k$ is divisible by 11.

23. Show that the number 7,176,825,942,116,027,211 is divisible by 9 but not divisible by 11.

24. Show that the number 9,897,654,527,609,877 is divisible by 99.

25. (Test for divisibility by 3) Let n be an integer with decimal representation $a_k a_{k-1} \cdots a_1 a_0$. Prove that n is divisible by 3 if and only if $a_k + a_{k-1} + \cdots + a_1 + a_0$ is divisible by 3.

26. (Test for divisibility by 4) Let n be an integer with decimal representation $a_k a_{k-1} \cdots a_1 a_0$. Prove that n is divisible by 4 if and only if $a_1 a_0$ is divisible by 4.

27. In your head, determine $(2 \cdot 10^{75} + 2)^{100} \bmod 3$ and $(10^{100} + 1)^{99} \bmod 3$.

28. Suppose ϕ is a ring homomorphism from $Z \oplus Z$ into $Z \oplus Z$. What are the possibilities for $\phi((1, 0))$?

29. Let R and S be commutative rings with unity. If ϕ is a homomorphism from R onto S and the characteristic of R is nonzero, prove that the characteristic of S divides the characteristic of R.

30. Let R be a commutative ring of prime characteristic p. Show that the *Frobenius* map $x \to x^p$ is a ring homomorphism from R to R.

31. Is there a ring homomorphism from the reals to some ring whose kernel is the integers?

32. Show that a homomorphism from a field onto a ring with more than one element must be an isomorphism.

33. Let ϕ be a ring homomorphism from a commutative ring R onto a commutative ring S and let A be an ideal of S.
    a. If A is prime in S, show that $\phi^{-1}(A) = \{x \in R \mid \phi(x) \in A\}$ is prime in R.
    b. If A is maximal in S, show that $\phi^{-1}(A)$ is maximal in R.

34. Show that the homomorphic image of a principal ideal ring is a principal ideal ring.

35. Let R and S be rings.
    a. Show that the mapping from $R \oplus S$ onto R given by $(a, b) \rightarrow a$ is a ring homomorphism.
    b. Show that the mapping from R to $R \oplus S$ given by $a \rightarrow (a, 0)$ is a one-to-one ring homomorphism.
    c. Show that $R \oplus S$ is ring-isomorphic to $S \oplus R$.

36. Show that if m and n are distinct positive integers, then mZ is not ring-isomorphic to nZ.

37. Prove or disprove that the field of real numbers is ring-isomorphic to the field of complex numbers.

38. Show that the only ring-automorphism of the real numbers is the identity mapping.

39. Determine all ring homomorphisms from **R** to **R**.

40. Show that the operation of multiplication defined in the proof of Theorem 15.6 is well defined.

41. Suppose that n divides m and that a is an idempotent of Z_n (that is, $a^2 = a$). Show that the mapping $x \rightarrow ax$ is a ring homomorphism from Z_m to Z_n. Show that the same correspondence need not yield a well-defined function if n does not divide m.

42. Let $Q[\sqrt{2}] = \{a + b\sqrt{2} \mid a, b \in Q\}$ and $Q[\sqrt{5}] = \{a + b\sqrt{5} \mid a, b \in Q\}$. Show that these two rings are not isomorphic.

43. Let $Z[i] = \{a + bi \mid a, b \in Z\}$. Show that the field of quotients of $Z[i]$ is ring-isomorphic to $Q[i] = \{r + si \mid r, s \in Q\}$. (This exercise is referred to in Chapter 18).

44. Let F be a field. Show that the field of quotients of F is isomorphic to F.

45. Let D be an integral domain and let F be the field of quotients of D. Show that if E is any field that contains D, then E contains a subfield isomorphic to F. (Thus, the field of quotients of an integral domain D is the smallest field containing D.)

46. Explain why a commutative ring with unity that is not an integral domain cannot be contained in a field.

47. Give an example of a ring without unity that is contained in a field.

48. Show that the relation $\equiv$ defined in the proof of Theorem 15.6 is an equivalence relation.

49. Prove that the set T in the proof of Corollary 3 of Theorem 15.5 is ring-isomorphic to the field of rational numbers.

50. Let $f(x) \in R[x]$. If $a + bi$ is a complex zero of $f(x)$ (here $i = \sqrt{-1}$), show that $a - bi$ is a zero of $f(x)$. (This exercise is referred to in Chapter 32.)

51. Suppose that $\phi: R \rightarrow S$ is a ring homomorphism and that the image of ϕ is not $\{0\}$. If R has a unity and S is an integral domain, show that ϕ carries the unity of R to the unity of S. Give an example to show that the previous statement need not be true if S is not an integral domain.

52. Show that the mapping of $\phi: D \rightarrow F$ in the proof of Theorem 15.6 is a ring homomorphism.

53. Show that the prime subfield of a field of characteristic p is ring-isomorphic to Z_p and that the prime subfield of a field of characteristic 0 is ring-isomorphic to Q.

54. Let n be a positive integer. Show that there is a ring-isomorphism from Z_2 to a subring of Z_{2n} if and only if n is odd.

SUGGESTED READINGS

J. A. Gallian and D. S. Jungreis, "Homomorphisms from $Z_m[i]$ into $Z_n[i]$ and $Z_m[\rho]$ into $Z_n[\rho]$ where $i^2 + 1 = 0$ and $\rho^2 + \rho + 1 = 0$," *The American Mathematical Monthly* 95 (1988): 247–249.

This article gives formulas for counting the homomorphisms mentioned in the title.

J. A. Gallian and J. Van Buskirk, "The Number of Homomorphisms from Z_m into Z_n," *American Mathematical Monthly* 91 (1984): 196–197.

In this note, formulas are given for the number of group homomorphisms from Z_m into Z_n and the number of ring homomorphisms from Z_m into Z_n.

Lillian Kinkade and Joyce Wagner, "When Polynomial Rings are Principal Ideal Rings," *Journal of Undergraduate Mathematics* 23 (1991) 59–62.

In this article written by undergraduates, it is shown that $R[x]$ is a principal ideal ring if and only if $R \approx R_1 \oplus R_2 \oplus \cdots \oplus R_n$, where each R_i is a field.

W. C. Waterhouse, "Rings with Cyclic Additive Group," *American Mathematical Monthly* 71 (1964): 449–450.

In this brief note, it is proved that, up to isomorphism, the number of rings whose additive group is cyclic of order m is the number of divisors of m.

16

Polynomial Rings

Very early in our mathematical education—in fact in junior high school or early in high school itself—we are introduced to polynomials. For a seemingly endless amount of time we are drilled, to the point of utter boredom, in factoring them, multiplying them, dividing them, simplifying them. Facility in factoring a quadratic becomes confused with genuine mathematical talent.

I. N. Herstein, *Topics in Algebra*

NOTATION AND TERMINOLOGY

One of the mathematical concepts that students are most familiar with and most comfortable with is that of a polynomial. In high school, students study polynomials with integer coefficients, rational coefficients, real coefficients, and, perhaps, even complex coefficients. In earlier chapters of this book, we introduced something that was probably new—polynomials with coefficients from Z_n. Notice that all of these sets of polynomials are rings and, in each case, the set of coefficients is also a ring. In this chapter, we abstract all of these examples into one.

DEFINITION Ring of Polynomials over R

Let R be a commutative ring. The set of formal symbols

$$R[x] = \{a_n x^n + a_{n-1} x^{n-1} + \cdots + a_1 x + a_0 \mid a_i \in R, \quad n \text{ is a nonnegative integer}\}$$

is called the *ring of polynomials over R in the indeterminate x*. Two elements

$$a_n x^n + a_{n-1} x^{n-1} + \cdots + a_1 x + a_0$$

and

$$b_m x^m + b_{m-1} x^{m-1} + \cdots + b_1 x + b_0$$

of $R[x]$ are considered equal if and only if $a_i = b_i$ for all nonnegative integers i. (Define $a_i = 0$ when $i > n$ and $b_i = 0$ when $i > m$.)

In this definition, the symbols x, x^2, ... , x^n do not represent "unknown" elements or variables from the ring R. Rather, their purpose is to serve as convenient placeholders that separate the ring elements $a_n, a_{n-1}, \ldots , a_0$. We could have avoided the x's by defining a polynomial as an infinite sequence $a_0, a_1, a_2, \ldots , a_n, 0, 0, 0, \ldots$, but our method takes advantage of the student's experience in manipulating polynomials where x does represent a variable. The disadvantage of our method is that one must be careful not to confuse a polynomial with the function determined by a polynomial. For example, in $Z_3[x]$, the polynomials $f(x) = x^3 + 2x$ and $g(x) = x^5 + 2x$ determine the same function from Z_3 to Z_3 since $f(a) = g(a)$ for all a in Z_3.* But $f(x)$ and $g(x)$ are different elements of $Z_3[x]$. Also, in the ring $Z_n[x]$, be careful to reduce only the coefficients and not the exponents modulo n. For example, in $Z_3[x]$, $5x = 2x$, but $x^5 \neq x^2$.

To make $R[x]$ into a ring, we define addition and multiplication in the usual way.

DEFINITION Addition and Multiplication in $R[x]$

Let R be a commutative ring and let

$$f(x) = a_n x^n + a_{n-1} x^{n-1} + \cdots + a_1 x + a_0$$

and

$$g(x) = b_m x^m + b_{m-1} x^{m-1} + \cdots + b_1 x + b_0$$

belong to $R[x]$. Then

$$f(x) + g(x) = (a_s + b_s)x^s + (a_{s-1} + b_{s-1})x^{s-1}$$
$$+ \cdots + (a_1 + b_1)x + a_0 + b_0,$$

where $a_i = 0$ for $i > n$ and $b_i = 0$ for $i > m$. Also,

$$f(x)g(x) = c_{m+n}x^{m+n} + c_{m+n-1}x^{m+n-1} + \cdots + c_1 x + c_0,$$

where

$$c_k = a_k b_0 + a_{k-1} b_1 + \cdots + a_1 b_{k-1} + a_0 b_k$$

for $k = 0, \ldots , m + n$.

---

*In general, given $f(x)$ in $R[x]$ and a in R, $f(a)$ means substitute a for x in the formula for $f(x)$. This substitution is a homomorphism from $R[x]$ to R.

Although the definition of multiplication might appear complicated, it is just a formalization of the familiar process of using the distributive property and collecting like terms. So, just multiply polynomials over a commutative ring R in the same way that polynomials are always multiplied. Here is an example.

Consider $f(x) = 2x^3 + x^2 + 2x + 2$ and $g(x) = 2x^2 + 2x + 1$ in $Z_3[x]$. Then, in our preceding notation, $a_5 = 0$, $a_4 = 0$, $a_3 = 2$, $a_2 = 1$, $a_1 = 2$, $a_0 = 2$ and $b_5 = 0$, $b_4 = 0$, $b_3 = 0$, $b_2 = 2$, $b_0 = 1$. Now, using the definitions and remembering that addition and multiplication of the coefficients are done modulo 3, we have

$$f(x) + g(x) = (2 + 0)x^3 + (1 + 2)x^2 + (2 + 2)x + (2 + 1)$$
$$= 2x^3 + 0x^2 + 1x + 0$$

and

$$\begin{aligned}
f(x) \cdot g(x) = {} & (0 \cdot 1 + 0 \cdot 2 + 2 \cdot 2 + 1 \cdot 0 + 2 \cdot 0 + 2 \cdot 0)x^5 \\
& + (0 \cdot 1 + 2 \cdot 2 + 1 \cdot 2 + 2 \cdot 0 + 2 \cdot 0)x^4 \\
& + (2 \cdot 1 + 1 \cdot 2 + 2 \cdot 2 + 2 \cdot 0)x^3 \\
& + (1 \cdot 1 + 2 \cdot 2 + 2 \cdot 2)x^2 + (2 \cdot 1 + 2 \cdot 2)x + 2 \cdot 1 \\
= {} & x^5 + 0x^4 + 2x^3 + 0x^2 + 0x + 2.
\end{aligned}$$

Our definitions for addition and multiplication of polynomials were formulated so that $R[x]$ is commutative and associative, and so that multiplication is distributive over addition. After all, $R[x]$ is called the "ring of polynomials over R." We leave the verification that $R[x]$ is a ring as an exercise.

It is time to introduce some terminology for polynomials. If

$$f(x) = a_n x^n + a_{n-1} x^{n-1} + \cdots + a_1 x + a_0,$$

where $a_n \neq 0$, we say that $f(x)$ has *degree n;* the term a_n is called the *leading coefficient* of $f(x)$ and, if the leading coefficient is the multiplicative identity element of R, we say that $f(x)$ is a *monic* polynomial. The polynomial $f(x) = 0$ has no degree. Polynomials of the form $f(x) = a_0$ are called *constant*. We often write $\deg f(x) = n$ to indicate that $f(x)$ has degree n. In keeping with our experience with polynomials with real coefficients, we adopt the following notational conventions: we may insert or delete terms of the form $0x^k$; $1x^k$ will be denoted by x^k; $+(-a_k)x^k$ will be denoted by $-a_k x^k$.

Very often properties of R carry over to $R[x]$. Our first theorem is a case in point.

Theorem 16.1   D an Integral Domain Implies D[x] Is
   If D is an integral domain, then D[x] is an integral domain.

Proof.   Since we already know that $D[x]$ is a ring, all we need to show is that $D[x]$ is commutative with a unity and has no zero-divisors. Clearly, $D[x]$ is commutative whenever D is. If 1 is the unity element of D, it is easy to check that $f(x) = 1$ is the unity element of $D[x]$. Finally, suppose that

$$f(x) = a_n x^n + a_{n-1} x^{n-1} + \cdots + a_0$$

and

$$g(x) = b_m x^m + b_{m-1} x^{m-1} + \cdots + b_0,$$

where $a_n \neq 0$ and $b_m \neq 0$. Then, by definition, $f(x)g(x)$ has leading coefficient $a_n b_m$ and, since D is an integral domain, $a_n b_m \neq 0$. ∎

THE DIVISION ALGORITHM AND CONSEQUENCES

One of the properties of integers that we have used repeatedly is the division algorithm: if a and b are integers and $b \neq 0$, then there exist unique integers q and r such that $a = bq + r$, where $0 \leq r < |b|$. The next theorem is the analogous statement for polynomials over a field.

Theorem 16.2 *Division Algorithm for $F[x]$*

Let F be a field and let $f(x)$ and $g(x) \in F[x]$ with $g(x) \neq 0$. Then there exist unique polynomials $q(x)$ and $r(x)$ in $F[x]$ such that $f(x) = g(x)q(x) + r(x)$ and either $r(x) = 0$ or $\deg r(x) < \deg g(x)$.

Proof. We begin by showing the existence of $q(x)$ and $r(x)$. If $f(x) = 0$ or $\deg f(x) < \deg g(x)$, we simply put $q(x) = 0$ and $r(x) = f(x)$. So, we may assume that $n = \deg f(x) \geq \deg g(x) = m$ and let $f(x) = a_n x^n + \cdots + a_0$ and $g(x) = b_m x^m + \cdots + b_0$. The idea behind this proof is to begin just as if you were going to "long divide" $g(x)$ into $f(x)$, then use the Second Principle of Induction on $\deg f(x)$ to finish up. Thus, resorting to long division, we let $f_1(x) = f(x) - a_n b_m^{-1} x^{n-m} g(x)$.* Then, $f_1(x) = 0$ or $\deg f_1(x) < \deg f(x)$; so, by our induction hypothesis, there exist $q_1(x)$ and $r_1(x)$ in $F[x]$ such that $f_1(x) = g(x)q_1(x) + r_1(x)$, where $r_1(x) = 0$ or $\deg r_1(x) < \deg g(x)$. [Technically, we should get the induction started by proving the case $\deg f(x) = 0$, but this is trivial.]

---

*For example,

$$
\begin{array}{r}
3/2 x^2 \\
2x^2 + 2 \overline{)3x^4 + x + 1} \\
\underline{3x^4 + 3x^2} \\
-3x^2 + x + 1
\end{array}
$$

So, $\qquad -3x^2 + x + 1 = 3x^4 + x + 1 - 3/2 x^2 (2x^2 + 2)$

In general,

$$
\begin{array}{r}
a_n b_m^{-1} x^{n-m} \\
b_m x^m + \cdots \overline{)a_n x^n + \cdots} \\
\underline{a_n x^n + \cdots} \\
f_1(x)
\end{array}
$$

So, $\qquad f_1(x) = (a_n x^n + \cdots) - a_n b_m^{-1} x^{n-m} (b_m x^m + \cdots)$

Thus,

$$
\begin{aligned}
f(x) &= a_n b_m^{-1} x^{n-m} g(x) + f_1(x) \\
&= a_n b_m^{-1} x^{n-m} g(x) + q_1(x)g(x) + r_1(x) \\
&= [a_n b_m^{-1} x^{n-m} + q_1(x)]g(x) + r_1(x).
\end{aligned}
$$

So, the polynomials $q(x) = a_n b_m^{-1} x^{n-m} + q_1(x)$ and $r(x) = r_1(x)$ have the desired properties.

To prove uniqueness, suppose that $f(x) = g(x)q(x) + r(x)$ and $f(x) = g(x)\bar{q}(x) + \bar{r}(x)$, where $r(x) = 0$ or $\deg r(x) < \deg g(x)$ and $\bar{r}(x) = 0$ or $\deg \bar{r}(x) < \deg g(x)$. Then, subtracting these two equations, we obtain

$$
0 = g(x)[q(x) - \bar{q}(x)] + [r(x) - \bar{r}(x)]
$$

or

$$
\bar{r}(x) - r(x) = g(x)[q(x) - \bar{q}(x)].
$$

Thus, $\bar{r}(x) - r(x)$ is 0, or the degree of $[\bar{r}(x) - r(x)]$ is at least that of $g(x)$. Since the latter is clearly impossible, we have $\bar{r}(x) = r(x)$ and $q(x) = \bar{q}(x)$ as well. ∎

The polynomials $q(x)$ and $r(x)$ in the division algorithm are called the *quotient* and *remainder* in the division of $f(x)$ by $g(x)$. When the ring of coefficients of a polynomial ring is a field, we can use the long division process to determine the quotient and remainder.

Example 1   To find the quotient and remainder upon dividing $f(x) = 3x^4 + x^3 + 2x^2 + 1$ by $g(x) = x^2 + 4x + 2$, where $f(x)$ and $g(x)$ belong to $Z_5[x]$, we may proceed by long division, provided we keep in mind that addition and multiplication are done modulo 5. Thus,

```
                         3x² + 4x
          x² + 4x + 2 )3x⁴ +   x³ + 2x²        + 1
                       3x⁴ + 2x³ +   x²
                       ────────────────────
                             4x³ +   x²        + 1
                             4x³ +   x² + 3x
                             ──────────────────
                                          2x + 1
```

So, $3x^2 + 4x$ is the quotient and $2x + 1$ is the remainder. Also,

$$
3x^4 + x^3 + 2x^2 + 1 = (x^2 + 4x + 2)(3x^2 + 4x) + 2x + 1. \quad \square
$$

Let D be an integral domain. If $f(x)$ and $g(x) \in D[x]$, we say that $g(x)$ *divides* $f(x)$ in $D[x]$ [and write $g(x) \mid f(x)$] if there exists an $h(x) \in D[x]$ such that $f(x) = g(x)h(x)$. In this case, we also call $g(x)$ a *factor* of $f(x)$. An element a is a *zero* (or a *root*) of a polynomial $f(x)$ if $f(a) = 0$. [Recall that $f(a)$ means substitute a for x in the expression for $f(x)$.] When F is a field, $a \in F$, and $f(x) \in F[x]$, we say that a is a *zero of mul-*

tiplicity k $(k \geq 1)$ if $(x - a)^k$ is a factor of $f(x)$ but $(x - a)^{k+1}$ is not a factor of $f(x)$. With these definitions, we may now give several important corollaries of the division algorithm. No doubt you have seen these for the special case where F is the field of real numbers.

Corollary 1 *The Remainder Theorem*
Let F be a field, $a \in F$, and $f(x) \in F[x]$. Then $f(a)$ is the remainder in the division of $f(x)$ by $x - a$.

Proof.   The proof of Corollary 1 is left as an exercise. ∎

Corollary 2 *The Factor Theorem*
Let F be a field, $a \in F$, and $f(x) \in F[x]$. Then a is a zero of $f(x)$ if and only if $x - a$ is a factor of $f(x)$.

Proof.   The proof of Corollary 2 is left as an exercise. ∎

Corollary 3 *Polynomials of Degree n Have at Most n Zeros*
A polynomial of degree n over a field has at most n zeros counting multiplicity.

Proof.   We proceed by induction on n. Clearly, a polynomial of degree 1 over a field has exactly one zero. Now suppose that $f(x)$ is a polynomial of degree n over a field and a is a zero of $f(x)$ of multiplicity k. Then, $f(x) = (x - a)^k q(x)$ and $q(a) \neq 0$; and, since $n = \deg f(x) = \deg(x - a)^k q(x) = k + \deg q(x)$, we have $k \leq n$ (see exercise 16). If $f(x)$ has no zeros other than a, we are done. On the other hand, if $b \neq a$ and b is a zero of $f(x)$, then $0 = f(b) = (b - a)^k q(b)$, so that b is also a zero of $q(x)$. By the Second Principle of Mathematical Induction, we know that $q(x)$ has at most $\deg q(x) = n - k$ zeros, counting multiplicity. Thus, $f(x)$ has at most $k + n - k = n$ zeros, counting multiplicity. ∎

We remark that Corollary 3 is not true for arbitrary polynomial rings. For example, the polynomial $x^2 + 3x + 2$ has four zeros in Z_6. Lagrange was the first to prove Corollary 3 for polynomials in $Z_p[x]$.

Example 2   The Complex Zeros of $x^n - 1$
We find all complex zeros of $x^n - 1$. Let $\omega = \cos(360°/n) + i \sin(360°/n)$. It follows from DeMoivre's Theorem (see Example 11 in Chapter 0) that $\omega^n = 1$ and $\omega^k \neq 1$ for $1 \leq k < n$. Thus, each of $1, \omega, \omega^2, \ldots, \omega^{n-1}$ is a zero of $x^n = 1$ and, by Corollary 3, there are no others. ∎

The complex number ω in Example 2 is called a *primitive nth root of unity*.
We conclude this chapter with an important theoretical application of the division algorithm, but first an important definition.

DEFINITION   Principal Ideal Domain (PID)
A *principal ideal domain* is an integral domain R in which every ideal has the form $\langle a \rangle = \{ra \mid r \in R\}$ for some a in R.

Theorem 16.3 $F[x]$ *Is a Principal Ideal Domain*
Let F be a field. Then $F[x]$ is a principal ideal domain.

Proof.   By Theorem 16.1, we know that $F[x]$ is an integral domain. Now, let I be an ideal in $F[x]$. If $I = \{0\}$, then $I = \langle 0 \rangle$. If $I \neq \{0\}$, then among all the elements of I, let $g(x)$ be one of minimum degree. We will show that $I = \langle g(x) \rangle$. Since $g(x) \in I$, we have $\langle g(x) \rangle \subseteq I$. Now let $f(x) \in I$. Then, by the division algorithm, we may write $f(x) = g(x)q(x) + r(x)$, where $r(x) = 0$ or $\deg r(x) < \deg g(x)$. Since $r(x) = f(x) - g(x)q(x) \in I$, the minimality of $\deg g(x)$ implies that the latter condition cannot hold. So, $r(x) = 0$ and, therefore, $f(x) \in \langle g(x) \rangle$. This shows that $I \subseteq \langle g(x) \rangle$.   ∎

The proof of Theorem 16.3 also establishes the following.

Theorem 16.4   *Criterion for $I = \langle g(x) \rangle$*
Let F be a field, I an ideal in $F[x]$, and $g(x)$ an element of $F[x]$. Then, $I = \langle g(x) \rangle$ if and only if $g(x)$ is a nonzero polynomial of minimum degree in I.

As an application of the First Isomorphism Theorem for Rings (Theorem 15.3) and Theorem 16.4, we verify the remark we made in Example 9 in Chapter 14 that the ring $\mathbf{R}[x]/\langle x^2 + 1 \rangle$ is isomorphic to the ring of complex numbers.

Example 3   Consider the homomorphism ϕ from $\mathbf{R}[x]$ onto $\mathbf{C}$ given by $f(x) \rightarrow f(i)$ (that is, evaluate every polynomial in $\mathbf{R}[x]$ at i). Then $x^2 + 1 \in \operatorname{Ker} \phi$ and is clearly a polynomial of minimum degree in $\operatorname{Ker} \phi$. Thus, $\operatorname{Ker} \phi = \langle x^2 + 1 \rangle$ and $\mathbf{R}[x]/\langle x^2 + 1 \rangle$ is isomorphic to $\mathbf{C}$.   □

EXERCISES

The difference between a text without problems and a text with problems is like the difference between learning to read a language and learning to speak it.

                                    Freeman Dyson, *Disturbing the Universe*

1. Let $f(x) = 4x^3 + 2x^2 + x + 3$ and $g(x) = 3x^4 + 3x^3 + 3x^2 + x + 4$, where $f(x), g(x) \in Z_5[x]$. Compute $f(x) + g(x)$ and $f(x) \cdot g(x)$.
2. Show that $x^2 + 3x + 2$ has four zeros in Z_6. (This exercise is referred to in this chapter.)
3. In $Z_3[x]$, show that $x^4 + x$ and $x^2 + x$ determine the same function from Z_3 to Z_3.

4. Prove Corollary 1 of Theorem 16.2.
5. Prove Corollary 2 of Theorem 16.2.
6. List all the polynomials of degree 2 in $Z_2[x]$.
7. If R is a commutative ring, show that the characteristic of $R[x]$ is the same as the characteristic of R.
8. If $\phi:R \rightarrow S$ is a ring homomorphism, define $\bar{\phi}:R[x] \rightarrow S[x]$ by $(a_n x^n + \cdots + a_0 \rightarrow \phi(a_n)x^n + \cdots + \phi(a_0)$. Show that $\bar{\phi}$ is a ring homomorphism.
9. If the rings R and S are isomorphic, show that $R[x]$ and $S[x]$ are isomorphic.
10. Let R be a commutative ring. Show that $R[x]$ has a subring isomorphic to R.
11. Let $f(x) = x^3 + 2x + 4$ and $g(x) = 3x + 2$ in $Z_5[x]$. Determine the quotient and remainder upon dividing $f(x)$ by $g(x)$.
12. Let $f(x) = 5x^4 + 3x^3 + 1$ and $g(x) = 3x^2 + 2x + 1$ in $Z_7[x]$. Determine the quotient and remainder upon dividing $f(x)$ by $g(x)$.
13. Show that the polynomial $2x + 1$ in $Z_4[x]$ has a multiplicative inverse in $Z_4[x]$.
14. Are there any nonconstant polynomials in $Z[x]$ that have multiplicative inverses? Explain your answer.
15. Let p be a prime. Are there any nonconstant polynomials in $Z_p[x]$ that have multiplicative inverses? Explain your answer.
16. (Degree Rule) Let D be an integral domain and $f(x), g(x) \in D[x]$. Prove that $\deg(f(x) \cdot g(x)) = \deg f(x) + \deg g(x)$.
17. Prove that the ideal $\langle x \rangle$ in $Z[x]$ is prime but not maximal.
18. Prove that the ideal $\langle x \rangle$ in $Q[x]$ is maximal.
19. Let F be an infinite field and $f(x) \in F[x]$. If $f(a) = 0$ for infinitely many elements a of F, show that $f(x) = 0$.
20. Let F be an infinite field and $f(x), g(x) \in F[x]$. If $f(a) = g(a)$ for infinitely many elements a of F, show that $f(x) = g(x)$.
21. Let F be a field and let $p(x) \in F[x]$. If $f(x), g(x) \in F[x]$ and $\deg f(x) < \deg p(x)$ and $\deg g(x) < \deg p(x)$, show that $f(x) + \langle p(x) \rangle = g(x) + \langle p(x) \rangle$ implies $f(x) = g(x)$. (This exercise is referred to in Chapter 20.)
22. Prove that $Z[x]$ is not a principal ideal domain. (Compare this with Theorem 16.3.)
23. Find a polynomial with integer coefficients that has $1/2$ and $-1/3$ as zeros.
24. Let $f(x) \in \mathbf{R}[x]$. Suppose that $f(a) = 0$ but $f'(a) \neq 0$, where $f'(x)$ is the derivative of $f(x)$. Show that a is a zero of $f(x)$ of multiplicity 1.
25. Show the Corollary 2 of Theorem 16.2 is true over any commutative ring.
26. Show that Corollary 3 of Theorem 16.2 is true for polynomials over integral domains.

27. Let F be a field and let

$$I = \{a_n x^n + a_{n-1} x^{n-1} + \cdots + a_0 \mid a_n, a_{n-1}, \ldots, a_0 \in F \quad \text{and}$$
$$a_n + a_{n-1} + \cdots + a_0 = 0\}.$$

Show that I is an ideal of $F[x]$ and find a generator for I.

28. Let m be a fixed positive integer. For any integer a, let $\bar{a}$ denote a and m. Show that the mapping of $\phi: Z[x] \rightarrow Z_m[x]$ given by

$$\phi(a_n x^n + a_{n-1} x^{n-1} + \cdots + a_0) = \bar{a}_n x^n + \bar{a}_{n-1} x^{n-1} + \cdots + \bar{a}_0$$

is a ring homomorphism. (This exercise is referred to in Chapter 17.)

29. Let F be a field and let $f(x) = a_n x^n + a_{n-1} x^{n-1} + \cdots + a_0 \in F[x]$. Prove that $x - 1$ is a factor of $f(x)$ if and only if $a_n + a_{n-1} + \cdots + a_0 = 0$.

30. Find infinitely many polynomials $f(x)$ in $Z_3[x]$ such that $f(a) = 0$ for all a in Z_3.

31. For every prime p, show that

$$x^{p-1} - 1 = (x - 1)(x - 2) \cdots [x - (p - 1)]$$

in $Z_p[x]$.

32. (Wilson's Theorem) For every integer $n > 1$, prove that $(n - 1)! = n - 1 \bmod n$ if and only if n is prime.

33. For every prime p, show that $(p - 2)! = 1 \bmod p$.

34. Find the remainder upon dividing 98! by 101.

35. Prove that $(50!)^2 = -1 \bmod 101$.

36. If I is an ideal of a ring R, prove that $I[x]$ is an ideal of $R[x]$.

37. Give an example of a commutative ring R with unity and a maximal ideal I of R such that $I[x]$ is not a maximal ideal of $R[x]$.

38. Let R be a commutative ring with unity. If I is a prime ideal of R, prove that $I[x]$ is a prime ideal of $R[x]$.

39. Prove that $Q[x]/\langle x^2 - 2 \rangle$ is isomorphic to $Q[\sqrt{2}] = \{a + b\sqrt{2} \mid a, b \in Q\}$.

40. Let F be a field, and let $f(x)$ and $g(x)$ belong to $F[x]$. If there is no polynomial of positive degree in $F[x]$ that divides both $f(x)$ and $g(x)$ [in this case, $f(x)$ and $g(x)$ are said to be *relatively prime*], prove that there exist polynomials $h(x)$ and $k(x)$ in $F[x]$ with the property that $f(x)h(x) + g(x)k(x) = 1$. (This exercise is referred to in Chapter 20.)

41. Let $f(x) \in R[x]$. If $f(a) = 0$ and $f'(a) = 0$ [$f'(a)$ is the derivative of $f(x)$ at a], show that $(x - a)^2$ divides $f(x)$.

42. Let F be a field and let $I = \{f(x) \in F[x] \mid f(a) = 0 \text{ for all } a \text{ in } F\}$. Prove that I is an ideal in $F[x]$. Prove that I is infinite when F is finite and $I = \{0\}$ when F is infinite.

17

Factorization of Polynomials

The value of a principle is the number of things it will explain.

Ralph Waldo Emerson

REDUCIBILITY TESTS

In high school, students spend much time factoring polynomials and finding their roots. In this chapter, we consider the same problems in a more abstract setting.

To discuss factorization of polynomials, we must first introduce the polynomial analog of a prime integer.

DEFINITION   Irreducible Polynomial, Reducible Polynomial

Let D be an integral domain. A polynomial $f(x)$ from $D[x]$ that is neither the zero polynomial nor a unit in $D[x]$ is said to be *irreducible over D* if, whenever $f(x)$ is expressed as a product $f(x) = g(x)h(x)$, with $g(x)$ and $h(x)$ from $D[x]$, then $g(x)$ or $h(x)$ is a unit in $D[x]$. A nonzero, nonunit element of $D[x]$ that is not irreducible over D is called *reducible over D*.

In the case that an integral domain is a field F, it is more convenient, although equivalent, to define a nonconstant $f(x) \in F[x]$ to be irreducible if and only if $f(x)$ cannot be expressed as a product of two polynomials of lower degree.

Example 1   The polynomial $f(x) = 2x^2 + 4$ is irreducible over Q but reducible over Z.   ❏

Example 2   The polynomial $f(x) = 2x^2 + 4$ is irreducible over **R** but reducible over **C**.   ❏

Example 3   The polynomial $x^2 - 2$ is irreducible over Q but reducible over **R**.   ❏

Example 4   The polynomial $x^2 + 1$ is irreducible over Z_3 but reducible over Z_5.   ❏

In general, it is a difficult problem to decide whether or not a particular polynomial is reducible over an integral domain, but there are special cases when it is easy. Our first theorem is a case in point. It applies to the three preceding examples.

Theorem 17.1   *Reducibility Test for Degrees 2 and 3*
Let F be a field. If $f(x) \in F[x]$ and $\deg f(x) = 2$ or 3, then $f(x)$ is reducible over F if and only if $f(x)$ has a zero in F.

Proof.   Suppose that $f(x) = g(x)h(x)$, where both $g(x)$ and $h(x)$ belong to $F[x]$ and have degree less than that of $f(x)$. Since $\deg f(x) = \deg g(x) + \deg h(x)$ and $\deg f(x) = 2$ or 3, at least one of $g(x)$ and $h(x)$ has degree 1. Say, $g(x) = ax + b$. Then, clearly, $-a^{-1}b$ is a zero of $g(x)$ and therefore a zero of $f(x)$ as well.

Conversely, suppose that $f(a) = 0$, where $a \in F$. Then, by the Factor Theorem, we know that $x - a$ is a factor of $f(x)$ and, therefore, $f(x)$ is reducible over F.   ∎

Theorem 17.1 is particularly easy to use when the field is Z_p, because, in this case, we can check for reducibility of $f(x)$ by simply testing to see if $f(x) = 0$ for $x = 0, 1, \ldots, p - 1$.

Note that polynomials of degree larger than 3 may be reducible over a field, even though they do not have zeros in the field. For example, in $Q[x]$, the polynomial $x^4 + 2x^2 + 1$ is equal to $(x^2 + 1)^2$, but has no zeros in Q.

Our next three tests deal with polynomials with integer coefficients. To simplify the proof of the first of these, we introduce some terminology and isolate a portion of the argument in the form of a lemma.

DEFINITION   Content of Polynomial, Primitive Polynomial
The *content* of a nonzero polynomial $a_n x^n + a_{n-1}x^{n-1} + \cdots + a_0$, where the a's are integers, is the greatest common divisor of the integers $a_n, a_{n-1}, \ldots, a_0$. A *primitive polynomial* is an element of $Z[x]$ with content 1.

Gauss's Lemma   *The product of two primitive polynomials is primitive.*

Proof. [2] Let $f(x)$ and $g(x)$ be primitive polynomials, and suppose that $f(x)g(x)$ is not primitive. Let p be a prime divisor of the content of $f(x)g(x)$, and let $\bar{f}(x)$, $\bar{g}(x)$, and $\overline{f(x)g(x)}$ be the polynomials obtained from $f(x)$, $g(x)$, and $f(x)g(x)$ by reducing the coefficients modulo p. Then, $\bar{f}(x)$ and $\bar{g}(x)$ belong to the integral domain $Z_p[x]$ and $\bar{f}(x)\bar{g}(x) = \overline{f(x)g(x)} = 0$, the zero element of $Z_p[x]$ (see exercise 28 in Chapter 16). Thus, $\bar{f}(x) = 0$ or $\bar{g}(x) = 0$. This means that either p divides every coefficient of $f(x)$ or p divides every coefficient of $g(x)$. Hence, either $f(x)$ is not primitive or $g(x)$ is not primitive. This contradiction completes the proof. ∎

Remember that the question of reducibility depends on which ring of coefficients one permits. Thus, $x^2 - 2$ is irreducible over Z but reducible over $Q[\sqrt{2}]$. In Chapter 20, we will prove that every polynomial of degree greater than 1 with coefficients from an integral domain is reducible over some field. Theorem 17.2 shows that in the case of the integers, this field must be larger than the field of rational numbers.

Theorem 17.2   *Over Q Implies over Z*
Let $f(x) \in Z[x]$. If $f(x)$ is reducible over Q, then it is reducible over Z.

Proof.   Suppose that $f(x) = g(x)h(x)$, where $g(x)$ and $h(x) \in Q[x]$. Clearly, we may assume that $f(x)$ is primitive because we can divide both $f(x)$ and $g(x)h(x)$ by the content of $f(x)$. Let a be the least common multiple of the denominators of the coefficients of $g(x)$, and b the least common multiple of the denominators of the coefficients of $h(x)$. Then $abf(x) = ag(x) \cdot bh(x)$, where $ag(x)$ and $bh(x) \in Z[x]$. Let c_1 be the content of $ag(x)$ and let c_2 be the content of $bh(x)$. Then $ag(x) = c_1 g_1(x)$ and $bh(x) = c_2 h_1(x)$, where both $g_1(x)$ and $h_1(x)$ are primitive and $abf(x) = c_1 c_2 g_1(x)h_1(x)$. Since $f(x)$ is primitive, the content of $abf(x)$ is ab. Also, since the product of two primitive polynomials is primitive, it follows that the content of $c_1 c_2 g_1(x)h_1(x)$ is $c_1 c_2$. Thus, $ab = c_1 c_2$ and $f(x) = g_1(x)h_1(x)$, where $g_1(x)$ and $h_1(x) \in Z[x]$. ∎

IRREDUCIBILITY TESTS

Theorem 17.1 reduces the question of irreducibility of a polynomial of degree 2 or 3 to one of finding a zero. The next theorem often allows us to simplify the problem even further.

Theorem 17.3   *Mod p Irreducibility Test*
Let p be a prime and suppose that $f(x) \in Z[x]$ with $\deg f(x) \geq 1$. Let $\bar{f}(x)$ be the polynomial in $Z_p[x]$ obtained from $f(x)$ by reducing all the coefficients of $f(x)$ modulo p. If $\bar{f}(x)$ is irreducible over Z_p and $\deg \bar{f}(x) = \deg f(x)$, then $f(x)$ is irreducible over Q.

Proof.   It follows from the proof of Theorem 17.2 that if $f(x)$ is reducible over Q, then $f(x) = g(x)h(x)$ with $g(x), h(x) \in Z[x]$ and both $g(x)$ and $h(x)$ have degree less than that of $f(x)$. Let $\bar{f}(x), \bar{g}(x)$, and $\bar{h}(x)$ be the polynomials obtained from $f(x)$, $g(x)$, and $h(x)$ by reducing all the coefficients modulo p. Since $\deg f(x) = \deg \bar{f}(x)$, we have $\deg \bar{g}(x) \leq \deg g(x) < \deg \bar{f}(x)$ and $\deg \bar{h}(x) \leq \deg h(x) < \deg \bar{f}(x)$. But, $\bar{f}(x) = \bar{g}(x)\bar{h}(x)$, and this contradicts our assumption that $\bar{f}(x)$ is irreducible over Z_p. ∎

Example 5   Let $f(x) = 21x^3 - 3x^2 + 2x + 9$. Then, over $Z_2[x]$, we have $\bar{f}(x) = x^3 + x^2 + 1$ and, since $\bar{f}(0) = 1$ and $\bar{f}(1) = 1$, we see that $\bar{f}(x)$ is irreducible over Z_2. Thus, $f(x)$ is irreducible over Q. Notice that, over Z_3, $\bar{f}(x) = 2x$ is irreducible, but we may *not* apply Theorem 17.3 to conclude that $f(x)$ is irreducible over Q. ❑

   Be cautious not to use the converse of Theorem 17.3. If $\bar{f}(x) \in Z[x]$ and $\bar{f}(x)$ is reducible over Z_p for some p, $f(x)$ may still be irreducible over Q. For example, consider $f(x) = 21x^3 - 3x^2 + 2x + 8$. Then, over Z_2, $\bar{f}(x) = x^3 + x^2 = x^2(x + 1)$. But over Z_5, $\bar{f}(x)$ has no roots and therefore is irreducible over Z_5. So, $f(x)$ is irreducible over Q. Note that this example shows that the Mod p Irreducibility Test may fail for some p and work for others. To conclude that a particular $f(x)$ in $Z[x]$ is irreducible over Q, all we need to do is find a single p for which the corresponding polynomial $\bar{f}(x)$ in Z_p is irreducible. However, this is not always possible since $f(x) = x^4 + 1$ is irreducible over Q but reducible over Z_p for *every* prime p. (See exercise 29.)
   The Mod p Irreducibility Test can also be helpful in checking for irreducibility of polynomials of degree greater than 3 and polynomials with rational coefficients.

Example 6   Let $f(x) = (3/7)x^4 - (2/7)x^2 + (9/35)x + 3/5$. We will show that $f(x)$ is irreducible over Q. First, let $h(x) = 35f(x) = 15x^4 - 10x^2 + 9x + 21$. Then $f(x)$ is irreducible over Q if and only if $h(x)$ is irreducible over Z. Next, applying the Mod 2 Irreducibility Test to $h(x)$, we get $\bar{h}(x) = x^4 + x + 1$. Clearly, $\bar{h}(x)$ has no roots in Z_2. Furthermore, $\bar{h}(x)$ has no quadratic factor in $Z_2[x]$ either. [For if so, the factor would have to be either $x^2 + x + 1$ or $x^2 + 1$. Long division shows that $x^2 + x + 1$ is not a factor, and $x^2 + 1$ cannot be a factor because it has a root whereas $\bar{h}(x)$ does not.] Thus $\bar{h}(x)$ is irreducible over $Z_2[x]$. This guarantees that $h(x)$ is irreducible over Q. ❑

Example 7   Let $f(x) = x^5 + 2x + 4$. Obviously, neither Theorem 17.1 nor the Mod 2 Irreducibility Test helps here. Let's try mod 3. Substitution of 0, 1, and 2 into $\bar{f}(x)$ does not yield 0, so there are no linear factors. But $\bar{f}(x)$ may have a quadratic factor. If so, we may assume it has the form $x^2 + ax + b$ (see exercise 5). This gives nine possibilities to check. We can immediately rule out each of the nine that has a zero over Z_3, since $\bar{f}(x)$ does not have one. This leaves only $x^2 + 1$, $x^2 + x + 2$, and $x^2 + 2x + 2$ to check. These are eliminated by long division. So, since $\bar{f}(x)$ is irreducible over Z_3, $f(x)$ is irreducible over Q. (Why is it unnecessary to check for cubic or fourth-degree factors?)   □

Another important irreducibility test is the following one, credited to Ferdinand Eisenstein (1823–1852), a student of Gauss. The corollary was first proved by Gauss by a different method.

Theorem 17.4   *Eisenstein's Criterion (1850)*
   Let

$$f(x) = a_n x^n + a_{n-1} x^{n-1} + \cdots + a_0 \in Z[x].$$

If there is a prime p such that $p \nmid a_n$, $p \mid a_{n-1}, \ldots, p \mid a_0$ and $p^2 \nmid a_0$, then $f(x)$ is irreducible over Q.

Proof.   If $f(x)$ is reducible over Q, we know by Theorem 17.2 that there exist elements $g(x)$ and $h(x)$ in $Z[x]$ such that $f(x) = g(x)h(x)$ and $1 \le$ deg $g(x)$, deg $h(x) < n$. Say, $g(x) = b_r x^r + \cdots + b_0$ and $h(x) = c_s x^s + \cdots + c_0$. Then, since $p \mid a_0$, $p^2 \nmid a_0$, and $a_0 = b_0 c_0$, it follows that p divides one of b_0 and c_0 but not the other. Let us say $p \mid b_0$ and $p \nmid c_0$. Also, since $p \nmid a_n = b_r c_s$, we know that $p \nmid b_r$. So, there is a least integer t such that $p \nmid b_t$. Now, consider $a_t = b_t c_0 + b_{t-1} c_1 + \cdots + b_0 c_t$. By assumption, p divides a_t and, by choice of t, every summand on the right after the first one is divisible by p. Clearly, this forces p to divide $b_t c_0$ as well. This is impossible, however, since p is prime and p divides neither b_t nor c_0.   ∎

Corollary   *Irreducibility of pth Cyclotomic Polynomial*
   For any prime p, the pth cyclotomic polynomial

$$\Phi_p(x) = \frac{x^p - 1}{x - 1} = x^{p-1} + x^{p-2} + \cdots + x + 1$$

is irreducible over Q.

Proof.   Let

$$f(x) = \Phi_p(x + 1) = \frac{(x + 1)^p - 1}{(x + 1) - 1} = x^{p-1} + p x^{p-2} + \cdots + p.$$

Then, since every coefficient except that of x^{p-1} is divisible by p, by Eisenstein's criterion, $f(x)$ is irreducible over Q. So, if $\Phi_p(x) = g(x)h(x)$ were a nontrivial factorization of $\Phi_p(x)$ over Q, then $f(x) = \Phi_p(x + 1) = g(x + 1)h(x + 1)$ would be a nontrivial factorization of $f(x)$ over Q. Since this is impossible, we conclude that $\Phi_p(x)$ is irreducible over Q.   ∎

Example 8   The polynomial $3x^5 + 15x^4 - 20x^3 + 10x + 20$ is irreducible over Q because $5 \nmid 3$ and $25 \nmid 20$ but 5 does divide $15, -20, 10$, and 20.   ❑

The principal reason for our interest in irreducible polynomials stems from the fact that there is an intimate connection between them, maximal ideals, and fields. This connection is revealed in the next theorem and its first corollary.

Theorem 17.5 $p(x)$ *Irreducible if and only if* $\langle p(x) \rangle$ *Is Maximal*

Let F be a field and let $p(x) \in F[x]$. Then $\langle p(x) \rangle$ is a maximal ideal in $F[x]$ if and only if $p(x)$ is irreducible over F.

Proof.   Suppose first that $\langle p(x) \rangle$ is a maximal ideal in $F[x]$. Clearly, $p(x)$ is neither the zero polynomial nor a unit in $F[x]$, because neither 0 nor $F[x]$ is a maximal ideal in $F[x]$. If $p(x) = g(x)h(x)$ is a factorization of $p(x)$ over F, then $\langle p(x) \rangle \subseteq \langle g(x) \rangle \subseteq F[x]$. Thus, $\langle p(x) \rangle = \langle g(x) \rangle$ or $F[x] = \langle g(x) \rangle$. In the first case, we must have deg $p(x) = $ deg $g(x)$. In the second case, it follows that deg $g(x) = 0$ and, consequently, deg $h(x) = $ deg $p(x)$. Thus, $p(x)$ cannot be written as a product of two polynomials in $F[x]$ of lower degree.

Now, suppose that $p(x)$ is irreducible over F. Let I be any ideal of $F[x]$ such that $\langle p(x) \rangle \subseteq I \subseteq F[x]$. Because $F[x]$ is a principal ideal domain, we know that $I = \langle g(x) \rangle$ for some $g(x)$ in $F[x]$. So, $p(x) \in \langle g(x) \rangle$ and, therefore, $p(x) = g(x)h(x)$, where $h(x) \in F[x]$. Since $p(x)$ is irreducible over F, it follows that either $g(x)$ is a constant or $h(x)$ is a constant. In the first case, we have $I = F[x]$; in the second case, we have $\langle p(x) \rangle = \langle g(x) \rangle = I$. So, $\langle p(x) \rangle$ is maximal in $F[x]$.   ∎

Corollary 1 $F[x]/\langle p(x) \rangle$ *Is a Field*

Let F be a field and $p(x)$ an irreducible polynomial over F. Then $F[x]/\langle p(x) \rangle$ is a field.

Proof.   This follows directly from Theorems 17.5 and 14.4.   ∎

The next corollary is a polynomial analog of Euclid's Lemma for primes (see Chapter 0).

Corollary 2 $p(x) \mid a(x)b(x)$ *Implies* $p(x) \mid a(x)$ *or* $p(x) \mid b(x)$

Let F be a field and let $p(x), a(x), b(x) \in F[x]$. If $p(x)$ is irreducible over F and $p(x) \mid a(x)b(x)$, then $p(x) \mid a(x)$ or $p(x) \mid b(x)$.

Proof.   Since $p(x)$ is irreducible, $F[x]/\langle p(x) \rangle$ is a field and, therefore, an integral domain. Let $\bar{a}(x)$ and $\bar{b}(x)$ be the images of $a(x)$ and $b(x)$ under the natural homomorphism from $F[x]$ to $F[x]/\langle p(x) \rangle$. Since $p(x) \mid a(x)b(x)$, we have $\bar{a}(x)\bar{b}(x) = 0$, the zero element of $F[x]/\langle p(x) \rangle$. Thus, $\bar{a}(x) = \bar{0}$ or $\bar{b}(x) = \bar{0}$, and it follows that $p(x) \mid a(x)$ or $p(x) \mid b(x)$.   ∎

The next two examples put the theory to work.

Example 9   We construct a field with eight elements. By Theorem 17.1 and Corollary 1 of Theorem 17.5, it suffices to find a cubic polynomial over Z_2 that has no zero in Z_2. By inspection, $x^3 + x + 1$ fills the bill. Thus, $Z_2[x]/\langle x^3 + x + 1\rangle = \{ax^2 + bx + c + \langle x^3 + x + 1\rangle \mid a, b, c \in Z_2\}$ is a field with eight elements. For practice, let us do a few calculations in this field. Since the sum of two polynomials of the form $ax^2 + bx + c$ is another one of the same form, addition is easy. Thus,

$$(x^2 + x + 1 + \langle x^3 + x + 1\rangle) + (x^2 + 1 + \langle x^3 + x + 1\rangle)$$
$$= x + \langle x^3 + x + 1\rangle.$$

On the other hand, multiplication of two coset representatives need not yield one of the original eight coset representatives:

$$(x^2 + x + 1 + \langle x^3 + x + 1\rangle) \cdot (x^2 + 1 + \langle x^3 + x + 1\rangle)$$
$$= x^4 + x^3 + x + 1 + \langle x^3 + x + 1\rangle = x^4 + \langle x^3 + x + 1\rangle$$

(since the ideal absorbs the last three terms). How do we express this in the form $ax^2 + bx + c + \langle x^3 + x + 1\rangle$? One way is to long divide by $x^3 + x + 1$ to obtain the remainder of $x^2 + x$ (just as one reduces $12 + \langle 5\rangle$ to $2 + \langle 5\rangle$ by dividing 12 by 5 to obtain the remainder 2). Another way is to observe that $x^3 + x + 1 + \langle x^3 + x + 1\rangle = 0 + \langle x^3 + x + 1\rangle$ implies $x^3 + \langle x^3 + x + 1\rangle = x + 1 + \langle x^3 + x + 1\rangle$. Thus, we may multiply both sides by x to obtain

$$x^4 + \langle x^3 + x + 1\rangle = x^2 + x + \langle x^3 + x + 1\rangle.$$

A partial multiplication table for this field is given in Table 17.1. To simplify the notation, we indicate a coset by its representative only. (Complete the table yourself. Keep in mind that x^3 can be replaced by $x + 1$ and x^4 by $x^2 + x$.)   ❑

Example 10   Since $x^2 + 1$ has no root in Z_3, it is irreducible. Thus, $Z_3[x]/\langle x^2 + 1\rangle$ is a field. Analogous to Example 9 in Chapter 14, $Z_3[x]/\langle x^2 + 1\rangle = \{ax + b + \langle x^2 + 1\rangle \mid a, b \in Z_3\}$. Thus, this field has 9 elements. A multiplication table for this field can be obtained from Table 13.1 by replacing i by x. (Why does this work?)   ❑

Table 17.1   A Partial Multiplication Table for Example 9

	1	x	$x + 1$	x^2	$x^2 + 1$	$x^2 + x$	$x^2 + x + 1$
1	1	x	$x + 1$	x^2	$x^2 + 1$	$x^2 + x$	$x^2 + x + 1$
x	x	x^2	$x^2 + x$	$x + 1$	1	$x^2 + x + 1$	$x^2 + 1$
$x + 1$	$x + 1$	$x^2 + x$	$x^2 + 1$	$x^2 + x + 1$	x^2	1	x
x^2	x^2	$x + 1$	$x^2 + x + 1$	$x^2 + x$	x	$x^2 + 1$	1
$x^2 + 1$	$x^2 + 1$	1	x^2	x	$x^2 + x + 1$	$x + 1$	$x^2 + x$

Unique Factorization in $Z[x]$

As a further application of the ideas presented in this chapter, we next prove that $Z[x]$ has an important factorization property. In Chapter 18, we will study this property in greater depth. The first proof of Theorem 17.6 was given by Gauss. In reading this theorem and its proof, keep in mind that the units in $Z[x]$ are precisely $f(x) = 1$ and $f(x) = -1$ (see exercise 25 in Chapter 12), the irreducible polynomials of degree 0 over Z are precisely those of the form $f(x) = p$ and $f(x) = -p$ where p is a prime, and every nonconstant polynomial from $Z[x]$ that is irreducible over Z is primitive (see exercise 2).

Theorem 17.6   *Unique Factorization in $Z[x]$*

Every polynomial in $Z[x]$ that is not the zero polynomial or a unit in $Z[x]$ can be written in the form $b_1 b_2 \cdots b_s p_1(x) p_2(x) \cdots p_m(x)$, where the b's are irreducible polynomials of degree 0, and the $p_i(x)$'s are irreducible polynomials of positive degree. Furthermore, if

$$b_1 b_2 \cdots b_s p_1(x) p_2(x) \cdots p_m(x) = c_1 c_2 \cdots c_t q_1(x) q_2(x) \cdots q_n(x),$$

where the b's and c's are irreducible polynomials of degree 0, and the $p(x)$'s and $q(x)$'s are irreducible polynomials of positive degree, then $s = t$, $m = n$, and, after renumbering the c's and $q(x)$'s, we have $b_i = \pm c_i$ for $i = 1, \ldots, s$; and $p_i(x) = \pm q_i(x)$ for $i = 1, \ldots, m$.

Proof.   Let $f(x)$ be a nonzero, nonunit polynomial from $Z[x]$. If $\deg f(x) = 0$, then $f(x)$ is constant and the result follows from the Fundamental Theorem of Arithmetic. If $\deg f(x) > 0$, let b denote the content of $f(x)$, and let $b_1 b_2 \cdots b_s$ be the factorization of b as a product of primes. Then, $f(x) = b_1 b_2 \cdots b_s f_1(x)$, where $f_1(x)$ belongs to $Z[x]$, is primitive and has positive degree. Thus, to prove the existence portion of the theorem, it suffices to show that a primitive polynomial $f(x)$ of positive degree can be written as a product of irreducible polynomials of positive degree. We proceed by induction of $\deg f(x)$. If $\deg f(x) = 1$, then $f(x)$ is already irreducible and we are done. Now suppose that every primitive polynomial of degree less than $\deg f(x)$ can be written as a product of irreducibles of positive degree. If $f(x)$ is irreducible, there is nothing to prove. Otherwise, $f(x) = g(x)h(x)$, where both $g(x)$ and $h(x)$ are primitive and have degree less than that of $f(x)$. Thus, by induction, both $g(x)$ and $h(x)$ can be written as a product of irreducibles of positive degree. Clearly, then, $f(x)$ is also such a product.

To prove the uniqueness portion of the theorem, suppose that $f(x) = b_1 b_2 \cdots b_s p_1(x) p_2(x) \cdots p_m(x) = c_1 c_2 \cdots c_t q_1(x) q_2(x) \cdots q_n(x)$, where the b's and c's are irreducible polynomials of degree 0, and the $p(x)$'s and $q(x)$'s are irreducible polynomials of positive degree. Let $b = b_1 b_2 \cdots b_s$ and $c = c_1 c_2 \cdots c_t$. Since the $p(x)$'s and $q(x)$'s are primitive, it follows from Gauss's Lemma that $p_1(x) p_2(x) \cdots p_m(x)$

and $q_1(x)q_2(x) \cdots q_n(x)$ are primitive. Hence, both b and c must equal plus-or-minus the content of $f(x)$ and, therefore, are equal in absolute value. It then follows from the Fundamental Theorem of Arithmetic that $s = t$ and, after renumbering, $b_i = \pm c_i$ for $i = 1, 2, \ldots, s$. Thus, by cancelling the constant terms in the two factorizations for $f(x)$, we have $p_1(x)p_2(x) \cdots p_m(x) = \pm q_1(x) q_2(x) \cdots q_n(x)$. Now, viewing the $p(x)$'s and $q(x)$'s as elements of $Q[x]$ and noting that $p_1(x)$ divides $q_1(x) \cdots q_n(x)$, it follows from Corollary 2 of Theorem 17.5 and induction (see exercise 28) that $p_1(x) \mid q_i(x)$ for some i. By renumbering, we may assume $i = 1$. Then, since $q_1(x)$ is irreducible, we have $q_1(x) = (r/s)p_1(x)$, where $r, s \in Z$. However, because both $q_1(x)$ and $p_1(x)$ are primitive, we must have $r/s = \pm 1$. So, $q_1(x) = \pm p_1(x)$. Also, after cancelling, we have $p_2(x) \cdots p_m(x) = \pm q_2(x) \cdots q_n(x)$. Now, we may repeat the above argument with $p_2(x)$ in place of $p_1(x)$. If $m < n$, after m such steps we would have 1 on the left and a nonconstant polynomial on the right. Clearly, this is impossible. On the other hand, if $m > n$, after n steps we would have ± 1 on the right and a nonconstant polynomial on the left—another impossibility. So, $m = n$ and $p_i(x) = \pm q_i(x)$ after suitable renumbering of the $q(x)$'s. ∎

WEIRD DICE: AN APPLICATION OF UNIQUE FACTORIZATION

Example 11 Consider an ordinary pair of dice whose faces are labelled 1 through 6. The probability of rolling a sum of 7 is 6/36, the probability of rolling a sum of 6 is 5/36, and so on. In a 1978 issue of *Scientific American*[1], Martin Gardner remarked that if one were to label the six faces of one cube with integers 1, 2, 2, 3, 3, 4 and the six faces of another cube with the integers 1, 3, 4, 5, 6, 8, then the probability of obtaining any particular sum with these dice (called *Sicherman dice*) is the same as the probability of rolling that sum with ordinary dice (that is, 6/36 for a 7, 5/36 for a 6, and so on). See Figure 17.1. In this example, we show

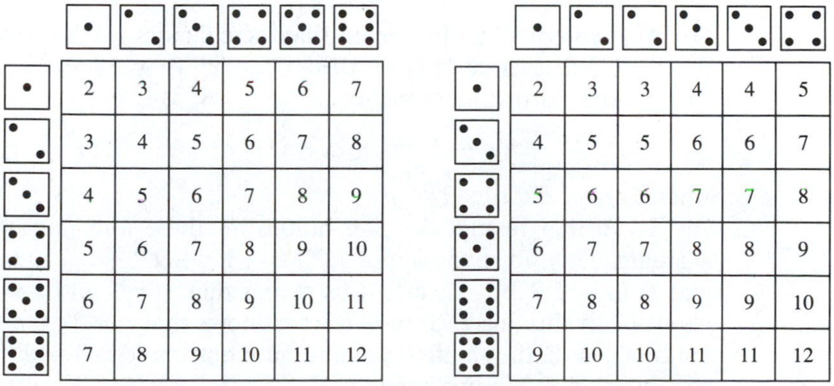

	2	3	4	5	6	7
	3	4	5	6	7	8
	4	5	6	7	8	9
	5	6	7	8	9	10
	6	7	8	9	10	11
	7	8	9	10	11	12

	2	3	3	4	4	5
	4	5	5	6	6	7
	5	6	6	7	7	8
	6	7	7	8	8	9
	7	8	8	9	9	10
	9	10	10	11	11	12

Figure 17.1

how the Sicherman labels can be derived, and that they are the only possible such labels besides 1 through 6. To do so, we utilize that fact that $Z[x]$ has the unique factorization property.

To begin with, let us ask ourselves how we may obtain a sum of 6, say, with an ordinary pair of dice. Well, there are five possibilities for the two faces: (5, 1), (4, 2), (3, 3), (2, 4), and (1, 5). Next we consider the product of the two polynomials created by using the ordinary dice labels as exponents:

$$(x^6 + x^5 + x^4 + x^3 + x^2 + x)(x^6 + x^5 + x^4 + x^3 + x^2 + x).$$

Observe that we pick up the term x^6 in this product in precisely the following ways: $x^5 \cdot x^1$, $x^4 \cdot x^2$, $x^3 \cdot x^3$, $x^2 \cdot x^4$, $x^1 \cdot x^5$. Notice the correspondence between pairs of labels whose sums are 6 and pairs of terms whose products are x^6. This correspondence is one-to-one, and it is valid for all sums and all dice—including the Sicherman dice and any other dice that yield the desired probabilities. So, let $a_1, a_2, a_3, a_4, a_5, a_6$ and $b_1, b_2, b_3, b_4, b_5, b_6$ be any two lists of positive integer labels for a pair of cubes with the property that the probability of rolling any particular sum with these dice (let us call them *weird dice*) is the same as the probability of rolling that sum with ordinary dice labeled 1 through 6. Using our observation about products of polynomials, this means that

$$
\begin{aligned}
(x^6 + x^5 &+ x^4 + x^3 + x^2 + x)(x^6 + x^5 + x^4 + x^3 + x^2 + x) \\
&= (x^{a_1} + x^{a_2} + x^{a_3} + x^{a_4} + x^{a_5} + x^{a_6}) \cdot \\
&\quad (x^{b_1} + x^{b_2} + x^{b_3} + x^{b_4} + x^{b_5} + x^{b_6}).
\end{aligned}
\tag{1}
$$

Now all we have to do is solve this equation for the a's and b's. Here is where unique factorization in $Z[x]$ comes in. The polynomial $x^6 + x^5 + x^4 + x^3 + x^2 + x$ factors uniquely into irreducibles as

$$x(x + 1)(x^2 + x + 1)(x^2 - x + 1)$$

so that the left-hand side of equation (1) has the irreducible factorization

$$x^2(x + 1)^2(x^2 + x + 1)^2(x^2 - x + 1)^2.$$

So, by Theorem 17.6, this means that these factors are the only possible irreducible factors of $P(x) = x^{a_1} + x^{a_2} + x^{a_3} + x^{a_4} + x^{a_5} + x^{a_6}$. Thus, $P(x)$ has the form

$$x^q(x + 1)^r(x^2 + x + 1)^t(x^2 - x + 1)^u,$$

where $0 \leq q, r, t, u \leq 2$.

To further restrict the possibilities for these four parameters, we evaluate $P(1)$ in two ways. $P(1) = 1^{a_1} + 1^{a_2} + \cdots + 1^{a_6} = 6$ and $P(1) = 1^q 2^r 3^t 1^u$. Clearly, this means that $r = 1$ and $t = 1$. What about q? Evaluating $P(0)$ in two ways shows that $q \neq 0$. On the other hand, if $q = 2$, the smallest possible sum one could roll with the corresponding labels for dice would be 3. Since this violates our assumption, we have now reduced our list of possibilities for q, r, t, and u to $q = 1$,

$r = 1, t = 1$, and $u = 0, 1, 2$. Let's consider each of these possibilities in turn.

When $u = 0$, $P(x) = x^4 + x^3 + x^3 + x^2 + x^2 + x$, so the die labels are 4, 3, 3, 2, 2, 1—a Sicherman die.

When $u = 1$, $P(x) = x^6 + x^5 + x^4 + x^3 + x^2 + x$, so the die labels are 6, 5, 4, 3, 2, 1—an ordinary die.

When $u = 2$, $P(x) = x^8 + x^6 + x^5 + x^4 + x^3 + x$, so the die labels are 8, 6, 5, 4, 3, 1—the other Sicherman die.

This proves that the Sicherman dice do give the same probabilities as ordinary dice *and* that they are the *only* other pair of dice that have this property.   ∎

EXERCISES

It is a great nuisance that knowledge can only be acquired by hard work.

W. Somerset Maugham

1. Suppose that D is an integral domain and F is a field containing D. If $f(x) \in D[x]$ and $f(x)$ is irreducible over F but reducible over D, what can you say about the factorization of $f(x)$ over D?

2. Show that a nonconstant polynomial from $Z[x]$ that is irreducible over Z is primitive.

3. Show that Theorem 17.1 is false if F is Z_6 instead of a field.

4. Suppose that $f(x) = x^n + a_{n-1}x^{n-1} + \cdots + a_0 \in Z[x]$. If r is rational and $x - r$ divides $f(x)$, show that r is an integer.

5. Let F be a field and let a be a nonzero element of F.
   a. If $af(x)$ is irreducible over F, prove that $f(x)$ is irreducible over F.
   b. If $f(ax)$ is irreducible over F, prove that $f(x)$ is irreducible over F.
   c. If $f(x + a)$ is irreducible over F, prove that $f(x)$ is irreducible over F.
   d. Use part c to prove that $8x^3 - 6x + 1$ is irreducible over Q.

6. Show that $x^4 + 1$ is irreducible over Q but reducible over **R**.

7. Construct a field of order 25.

8. Construct a field of order 27.

9. Show that $x^3 + x^2 + x + 1$ is reducible over Q. Does this fact contradict the corollary to Theorem 17.4?

10. Determine which of the polynomials below are irreducible over Q.
    a. $x^5 + 9x^4 + 12x^2 + 6$
    b. $x^4 + x + 1$
    c. $x^4 + 3x^2 + 3$
    d. $x^5 + 5x^2 + 1$
    e. $5/2x^5 + 9/2x^4 + 15x^3 + 3/7x^2 + 6x + 3/14$

11. Show that $x^2 + x + 4$ is irreducible over Z_{11}.

12. Suppose that $f(x) \in Z_p[x]$ and is irreducible over Z_p, where p is a prime. If $\deg f(x) = n$, prove that $Z_p[x]/\langle f(x) \rangle$ is a field with p^n elements.

13. Let $f(x) = x^3 + 6 \in Z_7[x]$. Write $f(x)$ as a product of irreducible polynomials over Z_7.

14. Let $f(x) = x^3 + x^2 + x + 1 \in Z_2[x]$. Write $f(x)$ as a product of irreducible polynomials over Z_2.

15. Let p be a prime.
   a. Show that the number of reducible polynomials over Z_p of the form $x^2 + ax + b$ is $p(p + 1)/2$.
   b. Determine the number of reducible quadratic polynomials over Z_p.

16. Let p be a prime.
   a. Determine the number of irreducible polynomials over Z_p of the form $x^2 + ax + b$.
   b. Determine the number of irreducible quadratic polynomials over Z_p.

17. Show that for every prime p there exists a field of order p^2.

18. Show that the field given in Example 10 in this chapter is isomorphic to the field given in Example 8 in Chapter 13.

19. Prove that, for every positive integer n, there are infinitely many polynomials of degree n in $Z[x]$ that are irreducible over Q.

20. Let $f(x) \in Z_p[x]$. Prove that if $f(x)$ has no factor of the form $x^2 + ax + b$, then it has no quadratic factor over Z_p.

21. Find all monic irreducible polynomials of degree 2 over Z_3.

22. Given that π is not the zero of a polynomial with rational coefficients, prove that π^2 cannot be written in the form $a\pi + b$ where a and b are rational.

23. Find all the zeros and their multiplicity of $x^5 + 4x^4 + 4x^3 - x^2 - 4x + 1$ over Z_5.

24. Find all zeros of $f(x) = 2x^2 + 2x + 1$ over Z_5 by substitution. Find all zeros of $f(x)$ by using the quadratic formula $(-b \pm \sqrt{b^2 - 4ac})(2a)^{-1}$. Do your answers agree? Should they? Find all zeros of $g(x) = 2x^2 + x + 3$ over Z_5 by substitution. Try the quadratic formula on $g(x)$. Why doesn't it work? State necessary and sufficient conditions for the quadratic formula to yield the zeros of a quadratic from $Z_p[x]$ where p is a prime greater than 2.

25. (Rational Root Theorem) Let

$$f(x) = a_n x^n + a_{n-1} x^{n-1} + \cdots + a_0 \in Z[x]$$

and $a_n \neq 0$. Prove that if r and s are relatively prime integers and $f(r/s) = 0$, then $r \mid a_0$ and $s \mid a_n$.

26. Let F be a field and $f(x) \in F[x]$. Show that, as far as deciding upon the irreducibility of $f(x)$ over F is concerned, we may assume that $f(x)$ is monic. (This assumption is useful when one uses a computer to check for irreducibility.)

27. Explain how the Mod p Irreducibility Test (Theorem 17.3) can be used to test members of $Q[x]$ for irreducibility.

28. Let F be a field and let $p(x), a_1(x), a_2(x), \ldots, a_k(x) \in F[x]$ where $p(x)$ is irreducible over F. If $p(x) \mid a_1(x)a_2(x) \cdots a_k(x)$, show that $p(x)$ divides some $a_i(x)$. (This exercise is referred to in the proof of Theorem 17.6.)

29. Show that $x^4 + 1$ is reducible over Z_p for every prime p.

30. Let F be a field and let $p(x)$ be irreducible over F. Show that $\{a + \langle p(x) \rangle \mid a \in F\}$ is a subfield of $F[x]/\langle p(x) \rangle$ isomorphic to F. (This exercise is referred to in Chapter 20.)

31. Let F be a field and let $p(x)$ be irreducible over F. If E is a field that contains F and there is an element a in E such that $p(a) = 0$, show that the mapping $\phi{:}F[x] \to E$ given by $f(x) \to f(a)$ is a ring homomorphism with kernel $\langle p(x) \rangle$. (This exercise is referred to in Chapter 20.)

32. Prove that the irreducible factorization of $x^6 + x^5 + x^4 + x^3 + x^2 + x$ over Z is $x(x + 1)(x^2 + x + 1)(x^2 - x + 1)$.

33. If p is a prime, prove that $x^{p-1} - x^{p-2} + x^{p-3} - \cdots - x + 1$ is irreducible over Q.

34. Carry out the analysis given in Example 11 for a pair of tetrahedrons instead of a pair of cubes. (Define ordinary tetrahedron dice as the ones labeled 1 through 4.)

35. Suppose in Example 11 that we begin with n ($n > 2$) ordinary dice each labeled 1 through 6, instead of just two. Show that the only possible labels that produce the same probabilities as n ordinary dice are the labels 1 through 6 and the Sicherman labels.

36. Show that one two-sided die labeled with 1 and 4 and another eighteen-sided die labeled with 1, 2, 2, 3, 3, 3, 4, 4, 4, 5, 5, 5, 6, 6, 6, 7, 7, 8 yield the same probabilities as an ordinary pair of cubes labeled 1 through 6. Carry out an analysis similar to that given in Example 11 to derive these labels.

37. In the game of Monopoly, would the probabilities of landing on various properties be different if the game were played with Sicherman dice instead of ordinary dice? Why?

38. (Algorithm for Factoring $x^n - 1$) *Assume the following:*
    1. $\Phi_1(x) = x - 1$.
    2. For any prime p, define $\Phi_p(x) = x^{p-1} + x^{p-2} + \cdots + x + 1$.
    3. For $k > 1$, p a prime, and m positive, $\Phi_{mp}^k(x) = \Phi_{mp}(x^{p^{k-1}})$.
    4. For $n \geq 3$ odd, $\Phi_{2n}(x) = \Phi_n(-x)$.
    5. For p a prime and $p \nmid m$, $\Phi_{mp}(x) = \Phi_m(x^p)/\Phi_m(x)$. [For example,

$$\Phi_{15}(x) = \Phi_3(x^5)/\Phi_3(x) = \Phi_5(x^3)/\Phi_5(x)$$
$$= x^8 - x^7 + x^5 - x^4 + x^3 - x + 1.]$$

    6. $x^n - 1 = \Pi_{d \mid n} \Phi_d(x)$ is the irreducible factorization of $x^n - 1$ over Z. [For example,

$$x^{15} - 1 = \Phi_1(x)\ \Phi_3(x)\ \Phi_5(x)\ \Phi_{15}(x) = (x - 1)(x^2 + x + 1)\ .$$
$$(x^4 + x^3 + x^2 + x + 1)(x^8 - x^7 + x^5 - x^4 + x^3 - x + 1).]$$

Use this algorithm to find the irreducible factorization of each of the following:

a. $x^4 + x^3 + x^2 + x = \dfrac{x(x^4 - 1)}{x - 1}$

b. $x^8 + x^7 + \cdots + x = \dfrac{x(x^8 - 1)}{x - 1}$

c. $x^{20} + x^{19} + \cdots + x = \dfrac{x(x^{20} - 1)}{x - 1}$

PROGRAMMING EXERCISES

The experiment serves two purposes, often independent one from the other: it allows the observation of new facts, hitherto either unsuspected, or not yet well defined; and it determines whether a working hypothesis fits the world of observable facts.

René J. Dubos

1. Write a program that will test any polynomial in $Z_p[x]$ for zeros. Make reasonable assumptions about the size of p and the degree of the polynomial.

2. Write a program that will implement the "Mod p Irreducibility Test." Assume that the degree of the polynomial is at most 5 and that p is at most 11. Test your program with the examples given in this chapter and the polynomials given in exercise 10.

3. Write a program that will implement the Rational Root Theorem (see exercise 25) for polynomials with integer coefficients. Make reasonable assumptions about the sizes of the coefficients and the degree of the polynomial.

4. Program the algorithm given in exercise 38 to find the irreducible factorization over Z of all polynomials of the form $x^n - 1$, where n is between 2 and 100. On the basis of this information, make a conjecture about the coefficients of the irreducible factors of $x^n - 1$ for all n. Test your conjecture for $n = 105$.

5. Write a program that will implement Eisenstein's Criterion. Randomly generate polynomials of degree at most 5 with integer coefficients between -100 and 100. Keep track of how often Eisenstein's Criterion applies to these polynomials. If one of the generated polynomials $f(x)$ does not satisfy Eisenstein's Criterion, apply it to the polynomials $f(x + k)$ for $k = -10, -9, \ldots, -1, 1, \ldots, 10$. [From exercise 5c of this chapter, if $f(x + k)$ is irreducible, so is $f(x)$.] Keep track of how often Eisenstein's Criterion applies to $f(x)$ or $f(x + k)$. Does this give significantly better results than Eisenstein's Criterion alone?

REFERENCES

1. Martin Gardner, "Mathematical Games," *Scientific American* 238/2 (1978): 19–32.
2. Richard Singer, "Some Applications of a Morphism," *American Mathematical Monthly* 76 (1969): 1131–1132.

SUGGESTED READINGS

Duane Broline, "Renumbering the Faces of Dice," *Mathematics Magazine* 52 (1979): 312–315.
   In this article, the author extends the analysis we carried out in Example 11 to dice in the shape of Platonic solids.

J. A. Gallian and D. J. Rusin, "Cyclotomic Polynomials and Nonstandard Dice," *Discrete Mathematics* 27 (1979): 245–259.
   Here Example 11 is generalized to the case of n dice each with m labels for all n and m greater than 1.

M. A. Lee, "Some Irreducible Polynomials Which Are Reducible mod p for All p," *American Mathematical Monthly* 76 (1969): 1125.
   This brief note gives a class of polynomials that are reducible mod p for all primes p, but are irreducible over the integers.

Carl Friedrich Gauss

He [Gauss] lives everywhere in mathematics.

E. T. Bell, *Men of Mathematics*

This stamp was issued by East Germany in 1977. It commemorates Gauss's construction of a regular 17-sided polygon with a straightedge and compass.

CARL FRIEDRICH GAUSS, considered by many to be the greatest mathematician who has ever lived, was born in Brunswick, Germany, on April 30, 1777. By the age of three, he was able to perform long computations in his head; at 10, he studied algebra and analysis. While still a teenager, he made many fundamental discoveries. Among these were the method of ''least squares'' for handling statistical data, a proof that a 17-sided regular polygon can be constructed with a straightedge and compass (this result was the first of its kind since discoveries by the Greeks 2000 years earlier), and his quadratic reciprocity theorem. Gauss obtained his Ph.D. in 1799 from the University of Helmstedt, under the supervision of Pfaff. In his dissertation, he proved the Fundamental Theorem of Algebra.

In 1801, Gauss published his monumental book on number theory. *Disquisitiones Arithmeticae,* summarizing previous work in a systematic way and introducing many fundamental ideas of his own, including the notion of modular arithmetic. This book won Gauss great fame among mathematicians.

In 1801, Ceres (an asteroid) was observed by astronomers on three occasions before they lost track of it. In what seemed to be an almost superhuman feat, Gauss used these three observations to calculate the orbit of Ceres. In carrying out this work, he showed that the variation inherent in experimentally derived data follows a bell-shaped curve, now called the Gaussian distribution. Gauss also used the method of least squares in this problem. This achievement established Gauss's reputation as a scientific genius before he was 25 years old.

In 1807, Gauss became professor of astronomy and director of the new observatory at the University of Göttingen. During the decades to come, Gauss continued to make important contributions not only in nearly all branches of mathematics, but also in astronomy, mechanics, optics, geodesy, and magnetism. Gauss also invented, with the physicist Wilhelm Weber, the first practical telegraph.

The acceptance of complex numbers among mathematicians was brought about by Gauss's use of them. Gauss coined the term *complex number* and popularized the notation i for $\sqrt{-1}$. He proved that the ring $Z[i]$ is a unique factorization domain and a Euclidean domain.

Throughout his life, Gauss largely ignored the work of his contemporaries and, in fact, made enemies of many of them. Young mathematicians who sought encouragement from him were usually rebuffed. Despite this fact, Gauss had many outstanding students, including Eisenstein, Riemann, Kummer, Dirichlet, and Dedekind.

Gauss died in Göttingen at the age of 77 on February 23, 1855. At Brunswick, there is a statue of him. Appropriately, the base is in the shape of a 17-point star. In 1989, Germany issued a bank note (shown below) depicting Gauss and the Gaussian distribution.

18

Divisibility in Integral Domains

Give me a fruitful error anytime, full of seeds, bursting with its
own corrections. You can keep your sterile truth for yourself.

Vilfredo Pareto

IRREDUCIBLES, PRIMES

In the previous two chapters, we focused on factoring polynomials over
the integers or a field. Several of those results—unique factorization in
$Z[x]$ and the division algorithm for $F[x]$, for instance—are natural
counterparts to theorems about the integers. In this chapter and the
next, we examine factoring in a more abstract setting.

DEFINITION Associates, Irreducibles, Primes

Elements a and b of an integral domain D are called *associates* if $a = ub$, where u is a unit of D. A nonzero element a of an integral domain
D is called an *irreducible* if a is not a unit and, whenever $b, c \in D$ with
$a = bc$, then b or c is a unit. A nonzero element a of an integral domain
D is called *prime* if a is not a unit and $a \mid bc$ implies $a \mid b$ or $a \mid c$.

Roughly speaking, an irreducible is an element that can be factored
only in a trivial way. Notice that an element a is prime if and only if $\langle a \rangle$
is a prime ideal.

Relating the above definitions to the integers may seem a bit confusing, since in Chapter 0 we defined a positive integer to be prime if it

satisfies our definition of an irreducible, and we proved that a prime integer satisfies the definition of a prime in an integral domain (Euclid's Lemma). The source of the confusion is that, in the case of the integers, the concepts of irreducible and prime are equivalent but, in general, as we will soon see, they are not.

The distinction between primes and irreducibles is best illustrated by integral domains of the form $Z[\sqrt{d}] = \{a + b\sqrt{d} \mid a, b \in Z\}$, where d is not 1 and is not divisible by the square of a prime. (These rings are of fundamental importance in number theory.) To analyze these rings, we need a convenient method of determining their units, irreducibles, and primes. To do this, we define a function N, called the *norm*, from $Z[\sqrt{d}]$ into the nonnegative integers by $N(a + b\sqrt{d}) = |a^2 - db^2|$. We leave it to the reader to verify the following four properties: $N(x) = 0$ if and only if $x = 0$; $N(xy) = N(x)N(y)$ for all x and y; x is a unit if and only if $N(x) = 1$; and, if $N(x)$ is prime, then x is irreducible in $Z[\sqrt{d}]$.

Example 1 We exhibit an irreducible in $Z[\sqrt{-3}]$ that is not prime. Here, $N(a + b\sqrt{-3}) = a^2 + 3b^2$. Consider $1 + \sqrt{-3}$. Suppose that we can factor this as xy, where neither x nor y is a unit. Then $N(xy) = N(x)N(y) = N(1 + \sqrt{-3}) = 4$, and it follows that $N(x) = 2$. But there are no integers a and b that satisfy $a^2 + 3b^2 = 2$. Thus, x or y is a unit and $1 + \sqrt{-3}$ is an irreducible. To verify that it is not prime, we observe that $(1 + \sqrt{-3})(1 - \sqrt{-3}) = 4 = 2 \cdot 2$, so that $1 + \sqrt{-3}$ divides $2 \cdot 2$. On the other hand, for integers a and b to exist so that $2 = (1 + \sqrt{-3})(a + b\sqrt{-3}) = (a - 3b) + (a + b)\sqrt{-3}$, we must have $a - 3b = 2$ and $a + b = 0$, which is impossible. □

Example 1 raises the question of whether or not there is an integral domain containing a prime that is not an irreducible. The answer: no.

Theorem 18.1 *Prime Implies Irreducible*
In an integral domain, every prime is an irreducible.

Proof. Suppose that a is a prime in an integral domain and $a = bc$. We must show that b or c is a unit. By definition of prime, we know that $a \mid b$ or $a \mid c$. Say, $at = b$. Then $b \cdot 1 = b = at = (bc)t = b(ct)$ and, by cancellation, $1 = ct$. Thus, c is a unit. ∎

Recall that a principal ideal domain is an integral domain in which every ideal has the form $\langle a \rangle$. The next theorem reveals a circumstance in which primes and irreducibles are equivalent.

Theorem 18.2 *PID Implies Irreducible Equals Prime*
In a principal ideal domain, an element is an irreducible if and only if it is a prime.

Proof.   Theorem 18.1 shows that primes are irreducibles. To prove the converse, let a be an irreducible element of a principal ideal domain D and suppose that $a \mid bc$. We must show that $a \mid b$ or $a \mid c$. Consider the ideal $I = \{ax + by \mid x, y \in D\}$ and let $\langle d \rangle = I$. Since $a \in I$, we can write $a = dr$, and because a is irreducible, d is a unit or r is a unit. If d is a unit, then $I = D$ and we may write $1 = ax + by$. Then $c = acx + bcy$, and since a divides both terms on the right, a also divides c.

On the other hand, if r is a unit, then $\langle a \rangle = \langle d \rangle = I$, and, because $b \in I$, there is an element t in D such that $at = b$. Thus, a divides b. ∎

It is an easy consequence of the respective division algorithms for Z and $F[x]$, where F is a field, that Z and $F[x]$ are principal ideal domains (see exercise 34 in Chapter 14 and Theorem 16.3). Our next example shows, however, that one of the most familiar rings is not a principal ideal domain.

Example 2   We show that $Z[x]$ is not a principal ideal domain. Consider the ideal $I = \{f(x) \in Z[x] \mid f(0) \text{ is even}\}$. We claim that I is not of the form $\langle h(x) \rangle$. If this were so, there would be $f(x)$ and $g(x)$ in $Z[x]$ such that $2 = h(x)f(x)$ and $x = h(x)g(x)$, since both 2 and x belong to I. By the degree rule (exercise 16 in Chapter 16), $0 = \deg 2 = \deg h(x) + \deg f(x)$, so that $h(x)$ is a constant polynomial. To determine which constant, we observe that $2 = h(1)f(1)$. Thus, $h(1) = \pm 1$ or ± 2. Since 1 is not in I, we must have $h(x) = \pm 2$. But then $x = \pm 2g(x)$, which is nonsense. □

We have previously proved that the integral domains Z and $Z[x]$ have important factorization properties: every integer greater than 1 can be uniquely factored as a product of irreducibles (that is, primes), and every nonzero, nonunit polynomial can be uniquely factored as a product of irreducible polynomials. It is natural to ask whether all integral domains have this property. The question of unique factorization in integral domains first arose with the efforts to solve a famous problem in number theory that goes by the misnomer Fermat's Last Theorem.

HISTORICAL DISCUSSION OF FERMAT'S LAST THEOREM

There are infinitely many nonzero integers x, y, z that satisfy the equation $x^2 + y^2 = z^2$. But what about the equation $x^3 + y^3 = z^3$ or, more generally, $x^n + y^n = z^n$, where n is an integer greater than 2 and x, y, z are nonzero integers? Well, no one has ever found a single solution of this equation, and for over three centuries many have tried to prove there is none. The tremendous effort put forth by the likes of Euler, Legendre, Abel, Gauss, Dirichlet, Cauchy, Kummer, Kronecker, and Hilbert to prove that there are no solutions has greatly influenced the development of ring theory.

About a thousand years ago, Arab mathematicians gave an incorrect proof that there were no solutions when $n = 3$. The problem lay dormant until 1637, when the French mathematician Pierre de Fermat (1601–1665) wrote in the margin of a book, ". . . it is impossible to separate a cube into two cubes, a fourth power into two fourth powers, or, generally, any power above the second into two powers of the same degree: I have discovered a truly marvelous demonstration [of this general theorem] which this margin is too narrow to contain."

Because Fermat gave no proof, many mathematicians have tried to prove the result. The case where $n = 3$ was done by Euler in 1770, although his proof was incomplete. The case where $n = 4$ is elementary and was done by Fermat himself (see [1]). The case where $n = 5$ was done in 1825 by Dirichlet, who had just turned 20, and by Legendre, who was past 70. Since the validity of the case for a particular integer implies the validity of all multiples of that integer, the next case of interest was $n = 7$. This case resisted the efforts of the best mathematicians until it was done by Gabriel Lamé in 1839. In 1847, Lamé stirred excitement by announcing that he had completely solved the problem. His approach was to factor the expression $x^p + y^p$, where p is an odd prime, into

$$(x + y)(x + \alpha y) \cdot \cdot \cdot (x + \alpha^{p-1} y),$$

where α is the complex number $\cos(2\pi/p) + i \sin(2\pi/p)$. Thus, his factorization took place in the ring $Z[\alpha] = \{a_0 + a_1\alpha + \cdot \cdot \cdot + a_{p-1}\alpha^{p-1} \mid a_i \in Z\}$. But Lamé made the mistake of assuming that, in such a ring, factorization into the product of irreducibles is unique. In fact, three years earlier, Ernst Eduard Kummer had proved that this is not always the case. Undaunted by the failure of unique factorization, Kummer began developing a theory to "save" factorization by creating a new type of number. Within a few weeks of Lamé's announcement, Kummer had shown that Fermat's Last Theorem is true for all primes of a special type (see [1]). This proved that the theorem was true for all exponents less than 100, prime or not, except for 37, 59, 67, and 74. Kummer's work has led to the theory of ideals as we know it today.

Over the centuries, many proposed proofs have not held up under scrutiny. The famous number theorist Edmund Landau received so many of these that he had a form printed with "On page _____, lines _____ to _____, you will find there is a mistake." Martin Gardner, "Mathematical Games" columnist of *Scientific American,* had postcards printed to decline requests from readers asking him to examine their proofs.

Recent discoveries have tied Fermat's Last Theorem closely to modern mathematical theories, giving hope that these theories might eventually lead to a proof. In March 1988, newspapers and scientific publications worldwide carried news of a proof by Yoichi Miyaoka (see Figure 18.1). Within weeks, however, Miyaoka's proof was shown to be invalid. As this edition went to press in June 1993, excitement spread through the mathematics community with the announcement that Andrew Wiles of Princeton University had proved Fermat's Last Theorem (see Figure 18.2). The Princeton Mathematics department chairperson was quoted as saying, "When we heard it, people started walking on air."

Proof of math theory creates academic stir

Los Angeles Times

Mathematicians around the world were buzzing with excitement Monday over the prospect that Fermat's Last Theorem, one of the oldest and most famous conjectures in mathematics, has finally been proved.

A Japanese mathematician may be working in Germany, has proposed what may be the greatest mathematician Pierre de Fermat in 1637. Experts who have examined the French mathematician Pierre de Fermat in 1637. Experts who have examined Miyaoka at the Max Planck Institute, find no mistakes in it, but they are not sure that the conjecture has been proved.

"It's still not definitive," said one American mathematician. But nobody by telephone from Bonn. "I wouldn't say that it's wrong, but I think there's a mistake somewhere."

Fermat's Last Theorem, in its simplest form "x to the nth power plus y to the nth power equals z to the nth power" for any integer greater than 2 has no solutions. When n equals 2, the equation has many solutions. An example from Pythagoras is 3² + 4² = 5² of a right triangle, the square of the length of the hypotenuse equals...

was found...
many, many ...
Final word d...
...t a few more da...

Solving the Puzzle

About 350 years ago, a French amateur mathematician named Pierre de Fermat scratched a devilishly tricky problem in the margin of a Greek mathematical text. Then he added, "I have discovered a truly remarkable proof [of the theorem], which this margin is too small to contain." Did he really have the answer?

The attempts of generations of scientists to find out have made Fermat's Last Theorem the El Dorado of math problems. Now, at long last, an assistant professor at Tokyo Metropolitan University seems to have broken the code. Last month at Bonn's Max Planck Insti-

tute, Yoichi Miyaoka, 38, sketched out his answer on a blackboard for fellow mathematicians.

Since before Euclid's time it has been known that in the equation $A^2 + B^2 = C^2$, if A and B are whole numbers, then C can also be a whole number—for example, $5^2 + 12^2 = 13^2$. Fermat postulated that if the same equation is taken to a power higher than 2, such as $A^3 + B^3 = C^3$, then C can never be a whole number. Miyaoka has apparently found out why by using an esoteric branch of mathematics called arithmetic geometry. Scientists are now awaiting the first draft of his manuscript. If it checks out, the Frenchman's infuriating puzzle will finally be solved.

Pierre de Fermat

THE GRANGER COLLECTION

Fermat's last theorem: A promising approach

The end of a centuries-long search for a proof of Fermat's last theorem, one of the most famous unsolved problems in mathematics, may at last be in sight. A Japanese mathematician, Yoichi Miyaoka of the Tokyo Metropolitan University, has proposed a proof for a key link in a chain of reasoning that establishes the theorem's truth. If Miyaoka's proof survives the mathematical community's intense scrutiny, then Fermat's conjecture (as it ought to be called until a proof is firmly established) can truly be called a theorem.

Miyaoka's method builds on work done by several Russian mathematicians and links important ideas in three mathematical fields: number theory, algebra and geometry. Though highly technical, his argument fills fewer than a dozen manuscript pages – short for such a significant mathematical proof. Miyaoka recently presented a sketch of his ideas at a seminar at the Max Planck Institute for Mathematics in Bonn, West Germany.

"It looks very nice," mathematician Don B. Zagier of the Max Planck Institute told SCIENCE NEWS. "There are many nice

ideas, but it's very subtle, and there could easily be a mistake. It'll certainly take days, if not weeks, until the proof is completely checked."

Fermat's conjecture is related to a statement by the ancient Greek mathematician Diophantus, who observed that there are positive integers, x, y and z, that satisfy the equation $x^2 + y^2 = z^2$. For example, if $x = 3$ and $y = 4$, then $z = 5$. In fact, this equation has an infinite number of such solutions.

In the 17th century, French amateur mathematician Pierre de Fermat, while reading a book by Diophantus, scribbled a note in a margin proposing that there are no positive-integer solutions to the equation $x^n + y^n = z^n$, when n is greater than 2. In other words, when $n = 3$, no set of positive integers satisfies the equation $x^3 + y^3 = z^3$, and so on. Then, in a tantalizing sentence that was to haunt mathematicians for centuries to come, Fermat added that although he had a wonderful proof for the theorem, he didn't have enough room to write it out.

Later mathematicians found proofs for a number of special cases, and a com-

showed that it was true for all exponents...

But despite the efforts of mathematicians, a proof for the general case remained elusive (SN: 6/20/87, p.397).

In 1983, Gerd Faltings, now at Princeton (N.J.) University, opened up a new direction in the search for a proof. As one consequence of his proof of the Mordell conjecture (SN: 7/23/83, p.58), he showed that if there are any solutions to Fermat's equations, then there are only a finite number of them for each value of n. However, that was still far from the assertion that there are no such solutions.

Some of the key ideas for Faltings' proof came from the work of Russian mathematician S. Arakelov, who was looking for connections between prime numbers, curves and geometrical surfaces. Both Arakelov and Faltings found that analogs of certain classical theorems already well established for geometrical surfaces could apply to curves and provide information about statements, such as Fermat's last theorem, that involve only integers.

About a year ago, A.N. Parshin of the Steklov Institute in Moscow, following

Figure 18.1

Figure 18.2

In view of the fact that so many eminent mathematicians were unable to prove Fermat's Last Theorem, despite the availability of the vastly powerful theories, it seems highly improbable that Fermat had a correct proof. Most likely, he made the error that his successors made by assuming that the properties of integers, such as unique factorization, carry over to integral domains in general.

UNIQUE FACTORIZATION DOMAINS

We now have the necessary terminology to formalize the idea of unique factorization.

DEFINITION Unique Factorization Domain (UFD)

An integral domain D is a *unique factorization domain* if

1. every nonzero element of D that is not a unit can be written as a product of irreducibles of D, and
2. the factorization into irreducibles is unique up to associates and the order in which the factors appear.

Another way to formulate part 2 of this definition is the following. If $p_1^{n_1} p_2^{n_2} \cdots p_r^{n_r}$ and $q_1^{m_1} q_2^{m_2} \cdots q_s^{m_s}$ are two factorizations of some element as a product of irreducibles, where no two of the p_i's are associates and no two of the q_j's, are associates, then $r = s$, and each $p_i^{n_i}$ is an associate of one and only one $q_j^{m_j}$.

Of course, the Fundamental Theorem of Arithmetic tells us that the ring of integers is a unique factorization domain and Theorem 17.6 says that $Z[x]$ is a unique factorization domain. In fact, as we shall soon see, most of the integral domains we have encountered are unique factorization domains.

Before proving our next theorem, we need the ascending chain condition for ideals.

Lemma   *Ascending Chain Condition for a PID*
In a principal ideal domain, any strictly increasing chain of ideals, $I_1 \subset I_2 \subset \cdots$ must be finite in length.

Proof.   Let $I_1 \subset I_2 \subset \cdots$ be a chain of strictly increasing ideals in an integral domain D, and let I be the union of all the ideals in this chain. We leave it as an exercise to verify that I is an ideal of D.

Then, since D is a principal ideal domain, there is an element a in D such that $I = \langle a \rangle$. Because $a \in I$ and $I = \cup I_k$, a belongs to some member of the chain, say $a \in I_n$. Clearly, then, for any member I_i of the chain, we have $I_i \subseteq I = \langle a \rangle \subseteq I_n$, so that I_n must be the last member of the chain. ∎

Theorem 18.3   *PID Implies UFD*
Every principal ideal domain is a unique factorization domain.

Proof.   Let D be a principal ideal domain and let a_0 be any nonzero nonunit in D. We will show that a_0 is a product of irreducibles (the product might consist of only one factor). We begin by showing that a_0 has at least one irreducible factor. If a_0 is irreducible, we are done. Thus, we may assume that $a_0 = b_1 a_1$, where neither b_1 nor a_1 is a unit and a_1 is nonzero. If a_1 is not irreducible, then we can write $a_1 = b_2 a_2$, where neither b_2 nor a_2 is a unit and a_2 is nonzero. Continuing in this fashion, we obtain a sequence $b_1, b_2, \ldots$ of elements that are not units in D and a sequence $a_0, a_1, a_2, \ldots$ of nonzero elements of D with $a_n = b_{n+1} a_{n+1}$ for each n. Hence, $\langle a_0 \rangle \subset \langle a_1 \rangle \subset \cdots$ is a strictly increasing chain of ideals (see exercise 3), which, by the preceding lemma, must be finite, say, $\langle a_0 \rangle \subset \langle a_1 \rangle \subset \cdots \subset \langle a_r \rangle$. In particular, a_r is an irreducible factor of a_0. This argument shows that every nonzero nonunit in D has at least one irreducible factor.

Now write $a_0 = p_1 c_1$, where p_1 is irreducible and c_1 is not a unit. If c_1 is not irreducible, then we can write $c_1 = p_2 c_2$, where p_2 is irreducible and c_2 is not a unit. Continuing in this fashion, we obtain, as before, a strictly increasing sequence $\langle a_0 \rangle \subset \langle c_1 \rangle \subset \langle c_2 \rangle \subset \cdots$, which must end in a finite number of steps. Let us say that the sequence ends with

$\langle c_s \rangle$. Then c_s is irreducible and $a_0 = p_1 p_2 \cdots p_s c_s$, where each p_i is also irreducible. This completes the proof that every nonzero nonunit of a principal ideal domain is a product of irreducibles.

It remains to be shown that the factorization is unique up to associates and the order in which the factors appear. To do this, suppose that some element a of D can be written

$$a = p_1 p_2 \cdots p_r = q_1 q_2 \cdots q_s,$$

where the p's and q's are irreducible and repetition is permitted. We induct on r. If $r = 1$, then a is irreducible and, clearly, $s = 1$ and $p_1 = q_1$. So we may assume that any element that can be expressed as a product of fewer than r irreducible factors can be done so in only one way (up to order and associates). Since p_1 divides $q_1 q_2 \cdots q_s$, it must divide some q_i (see exercise 24), say, $p_1 \mid q_1$. Then, $q_1 = u p_1$, where u is a unit of D. Thus,

$$u a = u p_1 p_2 \cdots p_r = q_1 (u q_2) \cdots q_s$$

and, by cancellation,

$$p_2 \cdots p_r = (u q_2) \cdots q_s.$$

The induction hypothesis now tells us that these two factorizations are identical up to associates and the order in which the factors appear. Hence, the same is true about the two factorizations of a. ■

In the existence portion of the proof of Theorem 18.3, the only way we used the fact that the integral domain D is a principal ideal domain was to say that D has the property that there is no infinite, strictly increasing chain of ideals in D. An integral domain with this property is called a *Noetherian domain,* in honor of Emmy Noether, who inaugurated the use of chain conditions in algebra. Noetherian domains are of the utmost importance in algebraic geometry. One reason for this is that, for many important rings R, the polynomial ring $R[x]$ is a Noetherian domain but not a principal ideal domain. One such example is $Z[x]$. In particular, $Z[x]$ shows that a UFD need not be a PID.

As an immediate corollary of Theorem 18.3, we have the following fact.

Corollary $F[x]$ *Is a UFD*
Let F be a field. Then $F[x]$ is a unique factorization domain.

Proof. By Theorem 16.3, $F[x]$ is a principal ideal domain. So, $F[x]$ is a unique factorization domain, as well. ■

As an application of the preceding corollary, we give an elegant proof, due to Richard Singer, of Eisenstein's Criterion (Theorem 17.4).

Example 3 Let

$$f(x) = a_n x^n + a_{n-1} x^{n-1} + \cdots + a_0 \in Z[x],$$

and suppose that p is prime such that

$$p \nmid a_n, p \mid a_{n-1}, \ldots, p \mid a_0 \text{ and } p^2 \nmid a_0.$$

We will prove that $f(x)$ is irreducible over Q. If $f(x)$ is reducible over Q, we know by Theorem 17.2 that there exist elements $g(x)$ and $h(x)$ in $Z[x]$ such that $f(x) = g(x)h(x)$ and $1 \leq \deg g(x), \deg h(x) < n$. Let $\bar{f}(x), \bar{g}(x),$ and $\bar{h}(x)$ be the polynomials in $Z_p[x]$ obtained from $f(x), g(x),$ and $h(x)$ by reducing all coefficients modulo p. Then, since p divides all the coefficients of $f(x)$ except a_n, we have $\bar{a}_n x^n = \bar{f}(x) = \bar{g}(x)\bar{h}(x)$. Since Z_p is a field, $Z_p[x]$ is a unique factorization domain. Thus, $x \mid \bar{g}(x)$ and $x \mid \bar{h}(x)$. So, $\bar{g}(0) = \bar{h}(0) = 0$ and, therefore, $p \mid g(0)$ and $p \mid h(0)$. But, then, $p^2 \mid g(0)h(0) = f(0) = a_0$, which is a contradiction.   ∎

EUCLIDEAN DOMAINS

Another important kind of integral domain is a Euclidean domain.

DEFINITION   Euclidean Domain

An integral domain D is called a *Euclidean domain* if there is a function d from the nonzero elements of D to the nonnegative integers such that

1. $d(a) \leq d(ab)$ for all nonzero a, b in D; and
2. if $a, b \in D, b \neq 0$, then there exist elements q and r in D such that $a = bq + r$, where $r = 0$ or $d(r) < d(b)$.

Example 4   The ring Z is a Euclidean domain with $d(a) = |a|$ (the absolute value of a).   ∎

Example 5   Let F be a field. Then $F[x]$ is a Euclidean domain with $d(f(x)) = \deg f(x)$ (see Theorem 16.2).   ∎

Examples 4 and 5 illustrate just one of many similarities between the rings Z and $F[x]$. Additional similarities are summarized in Table 18.1.

Example 6   The ring of Gaussian integers

$$Z[i] = \{a + bi \mid a, b \in Z\}$$

is a Euclidean domain with $d(a + bi) = a^2 + b^2$. Unlike the previous two examples, the function d does not obviously satisfy the necessary conditions. That $d(x) \leq d(xy)$ for $x, y \in Z[i]$ follows directly from the fact that $d(xy) = d(x)d(y)$ (exercise 5). If $x, y \in Z[i]$ and $y \neq 0$, then $xy^{-1} \in Q[i]$, the field of quotients of $Z[i]$ (exercise 43 in Chapter 15). Say, $xy^{-1} = s + ti$, where $s, t \in Q$. Now let m be the integer nearest s, and let n be the integer nearest t. (These integers may not be uniquely

Table 18.1   Similarities Between Z and $F[x]$

Z		$F[x]$
Form of elements: $a_n10^n + a_{n-1}10^{n-1} + \cdots + a_110 + a_0$	$\leftrightarrow$	Form of elements: $a_nx^n + a_{n-1}x^{n-1} + \cdots + a_1x + a_0$
Euclidean domain: $d(a) = \|a\|$	$\leftrightarrow$	Euclidean domain: $d(f(x)) = \deg f(x)$
Units: a is a unit if and only if $\|a\| = 1$		Units: $f(x)$ is a unit if and only if $\deg f(x) = 0$
Division algorithm: For $a, b \in Z$, $b \neq 0$, there exist $q, r \in Z$ such that $a = bq + r$, $0 \leq r < \|b\|$	$\leftrightarrow$	Division algorithm: For $f(x), g(x) \in F[x]$, $g(x) \neq 0$, there exist $q(x), r(x) \in F[x]$ such that $f(x) = g(x)q(x) + r(x)$, $0 \leq \deg r(x) < \deg g(x)$ or $r(x) = 0$
PID: Every nonzero ideal $I = \langle a \rangle$, where $a \neq 0$ and $\|a\|$ is minimum	$\leftrightarrow$	PID: Every nonzero ideal $I = \langle f(x) \rangle$, where $\deg f(x)$ is minimum
Prime: No nontrivial factors	$\leftrightarrow$	Irreducible: No nontrivial factors
UFD: Every element is a "unique" product of primes	$\leftrightarrow$	UFD: Every element is a "unique" product of irreducibles

determined, but that does not matter.) Thus, $\|m - s\| \leq 1/2$ and $\|n - t\| \leq 1/2$. Then

$$xy^{-1} = s + ti = (m - m + s) + (n - n + t)i$$
$$= (m + ni) + [(s - m) + (t - n)i].$$

So,

$$x = (m + ni)y + [(s - m) + (t - n)i]y.$$

We claim that the division condition of the definition of a Euclidean domain is satisfied with $q = m + ni$ and

$$r = [(s - m) + (t - n)i]y.$$

Clearly, q belongs to $Z[i]$, and since $r = x - qy$, so does r. Finally,

$$d(r) = d([(s - m) + (t - n)i])d(y)$$
$$= [(s - m)^2 + (t - n)^2]d(y)$$
$$\leq \left(\frac{1}{4} + \frac{1}{4}\right)d(y) < d(y). \qquad \square$$

Theorem 18.4   ED (Euclidean Domain) Implies PID
Every Euclidean domain is a principal ideal domain.

Proof. Let D be a Euclidean domain and I a nonzero ideal of D. Among all the nonzero elements of I, let a be such that $d(a)$ is minimum. Then $I = \langle a \rangle$. For, if $b \in I$, there are elements q and r such that $b = aq + r$, where $r = 0$ or $d(r) < d(a)$. But $r = b - aq \in I$, so $d(r)$ cannot be less than $d(a)$. Thus, $r = 0$ and $b \in \langle a \rangle$. Finally, the zero ideal is $\langle 0 \rangle$.   ∎

Although it is not easy to verify, we remark that there are principal ideal domains that are not Euclidean domains. The first such example was given by T. Motzkin in 1949. A more accessible account of Motzkin's result can be found in [3].

As an immediate consequence of Theorems 18.3 and 18.4, we have the following important result.

Corollary   *ED Implies UFD*
Every Euclidean domain is a unique factorization domain.

We may summarize our theorems and remarks as follows:

$$\text{ED} \Rightarrow \text{PID} \Rightarrow \text{UFD}$$
$$\text{UFD} \not\Rightarrow \text{PID} \not\Rightarrow \text{ED}$$

(You can remember these implications by listing the types alphabetically.)

In Chapter 17, we proved that $Z[x]$ is a unique factorization domain. Since Z is a unique factorization domain, the next theorem is a broad generalization of this fact. The proof is similar to that of the special case, and we therefore omit it.

Theorem 18.5   *D a UFD Implies D[x] a UFD*
If D is a unique factorization domain, then $D[x]$ is a unique factorization domain.

We conclude this chapter with an example of an integral domain that is not a unique factorization domain.

Example 7   The ring $Z[\sqrt{-5}] = \{a + b\sqrt{-5} \mid a, b \in Z\}$ is an integral domain but not a unique factorization domain. It is straightforward that $Z[\sqrt{-5}]$ is an integral domain (see exercise 9 in Chapter 13). To verify that unique factorization does not hold, we mimic the method used in Example 1 with $N(a + b\sqrt{-5}) = a^2 + 5b^2$. Since $N(xy) = N(x)N(y)$ and $N(x) = 1$ if and only if x is a unit (see exercise 14), it follows that the only units of $Z[\sqrt{-5}]$ are ± 1.

Now consider the following factorizations:

$$46 = 2 \cdot 23,$$
$$46 = (1 + 3\sqrt{-5})(1 - 3\sqrt{-5}).$$

We claim that each of these four factors is irreducible over $Z[\sqrt{-5}]$. Suppose that, say, $2 = xy$, where $x, y \in Z[\sqrt{-5}]$ and neither is a unit.

Then $4 = N(2) = N(x)N(y)$ and, therefore, $N(x) = N(y) = 2$, which is impossible. Likewise, if $23 = xy$ were a nontrivial factorization, then $N(x) = 23$. Thus, there would be integers a and b such that $a^2 + 5b^2 = 23$. Clearly, no such integers exist. The same argument applies to $1 \pm 3\sqrt{-5}$.

□

In light of Examples 6 and 7, one can't help but wonder for which $d < 0$ is $Z[\sqrt{d}]$ a unique factorization domain. The answer is only when $d = -1$ or -2 (see [2, p. 297]). The case where $d = -1$ was first proved, naturally enough, by Gauss.

EXERCISES

We are on the verge: Today our program proved Fermat's next-to-last theorem.
Epigrams in Programming, ACM SIGPLAN, September 1982

1. In an integral domain, show that the product of an irreducible and a unit is an irreducible.
2. Show that the union of a chain $I_1 \subset I_2 \subset \cdots$ of ideals of a ring R is an ideal of R.
3. Suppose that a and b belong to an integral domain, $b \neq 0$, and a is not a unit. Show that $\langle ab \rangle$ is a proper subset of $\langle b \rangle$. (This exercise is referred to in the proof of Theorem 18.3.)
4. Let D be an integral domain. Define $a \sim b$ if a and b are associates. Show that this defines an equivalence relation on D.
5. In the notation of Example 6, show that $d(xy) = d(x)d(y)$.
6. Let D be a Euclidean domain and d the associated function. Prove that u is a unit in D if and only if $d(u) = d(1)$.
7. Let D be a Euclidean domain and d the associated function. Show that if a and b are associates in D then $d(a) = d(b)$.
8. Let D be a principal ideal domain. Show that every proper ideal of D is contained in a maximal ideal of D.
9. In $Z[\sqrt{-5}]$, show that 21 does not factor uniquely as a product of irreducibles.
10. Show that $1 - i$ is an irreducible in $Z[i]$.
11. Show that $Z[\sqrt{-6}]$ is not a unique factorization domain. (*Hint:* Factor 10 in two ways.) Why does this show that $Z[\sqrt{-6}]$ is not a principal ideal domain?
12. Give an example of a unique factorization domain with a subdomain that does not have a unique factorization.
13. In $Z[i]$, show that 3 is irreducible but 2 and 5 are not.

14. For the ring $Z[\sqrt{d}] = \{a + b\sqrt{d} \mid a, b \in Z\}$ where $d \neq 1$ and d is not divisible by the square of a prime, prove that the norm $N(a + b\sqrt{d}) = |a^2 - db^2|$ satisfies the four assertions made preceding Example 1.

15. In an integral domain, show that a and b are associates if and only if $\langle a \rangle = \langle b \rangle$.

16. Prove that 7 is irreducible in $Z[\sqrt{6}]$, even though $N(7)$ is not prime. (Thus, the converse of the fourth part of exercise 14 is not true.)

17. Prove that if p is a prime in Z that can be written in the form $a^2 + b^2$, then $a + bi$ is irreducible in $Z[i]$. Find three primes that have this property and the corresponding irreducibles.

18. Prove that $Z[\sqrt{-3}]$ is not a principal ideal domain.

19. In $Z[\sqrt{-5}]$, prove that $1 + 3\sqrt{-5}$ is irreducible but not prime.

20. In $Z[\sqrt{5}]$, prove that both 2 and $1 + \sqrt{5}$ are irreducible but not prime.

21. Let d be an integer less than -1 that is not divisible by the square of a prime. Prove that the only units of $Z[\sqrt{d}]$ are $+1$ and -1.

22. If a and b belong to $Z[\sqrt{d}]$, where d is not divisible by the square of a prime and ab is a unit, prove that a and b are units.

23. Prove or disprove that if D is a principal ideal domain, then $D[x]$ is a principal ideal domain.

24. Let p be a prime in an integral domain. If $p \mid a_1 a_2 \cdots a_n$, prove that p divides some a_i. (This exercise is referred to in this chapter.)

25. Determine the units in $Z[i]$.

26. Show that $3x^2 + 4x + 3 \in Z_5[x]$ factors as $(3x + 2)(x + 4)$ and $(4x + 1)(2x + 3)$. Explain why this does not contradict the corollary of Theorem 18.3.

27. Prove that $Z[\sqrt{5}]$ is not a unique factorization domain.

28. Let D be a principal ideal domain and let $p \in D$. Prove that $\langle p \rangle$ is a maximal ideal in D if and only if p is irreducible.

29. Let D be a principal ideal domain and p an irreducible element of D. Prove that $D/\langle p \rangle$ is a field.

30. Show that an integral domain with the property that every strictly decreasing chain of ideals $I_1 \supset I_2 \supset \cdots$ must be finite in length is a field.

31. An ideal A of a commutative ring R with unity is said to be *finitely generated* if there exist elements $a_1, a_2, \ldots, a_n$ of A such that every element of A can be written in the form $r_1 a_1 + r_2 a_2 + \cdots + r_n a_n$, where $r_1, r_2, \ldots, r_n$ are in R. An integral domain R is said to satisfy the *ascending chain condition* if every strictly increasing chain of ideals $I_1 \subset I_2 \subset \cdots$ must be finite in length. Show that an integral domain R satisfies the ascending chain condition if and only if every ideal of R is finitely generated.

32. Prove or disprove that a subdomain of a Euclidean domain is a Euclidean domain.

33. Show that for any nontrivial ideal I of $Z[i]$, $Z[i]/I$ is finite.

REFERENCES

1. H. M. Edwards, *Fermat's Last Theorem: A Genetic Introduction to Algebraic Number Theory,* New York: Springer-Verlag, 1977.
2. H. M. Stark, *An Introduction to Number Theory,* Chicago, Ill.: Markham, 1970.
3. J. C. Wilson, "A Principal Ideal Ring That Is Not a Euclidean Ring," *Mathematics Magazine* 46 (1973): 74–78.

SUGGESTED READINGS

Oscar Campoli, "A Principal Ideal Domain That Is Not a Euclidean Domain," *The American Mathematical Monthly* 95 (1988) 868–871.

The author shows that $\{a + b\theta \mid a, b \in Z, \theta = (1 + \sqrt{-19})/2\}$ is a PID that is not an ED.

H. M. Edwards, "Fermat's Last Theorem," *Scientific American* 239/4 (1978): 104–122.

This well-written article traces the history of the efforts to prove Fermat's Last Theorem.

Steven Galovich, "Unique Factorization Rings with Zero Divisors," *Mathematics Magazine* 51 (1978): 277–283.

Here the concept of unique factorization is formulated for any commutative ring with unity. Rather complete structure theorems for unique factorization rings with zero-divisors are given.

Gina Kolata, "At Last, Shout of 'Eureka!' in Age-Old Math Mystery," *The New York Times,* June 24, 1993.

This front-page article reports on Andrew Wiles's announced proof of Fermat's Last Theorem.

C. Krauthhammer, "The Joy of Math, or Fermat's Revenge," *Time,* April 18, 1988; 92.

The demise of Miyaoka's proof of Fermat's Last Theorem is charmingly lamented.

Sahib Singh, "Non-Euclidean Domains: An Example," *Mathematics Magazine* 49 (1976): 243.

This note gives a short proof that $Z[\sqrt{-n}] = \{a + b\sqrt{-n} \mid a, b \in Z\}$ is an integral domain that is not Euclidean when $n > 2$ and $-n \equiv 2$ or 3 modulo 4.

Stan Wagon, "Fermat's Last Theorem," *The Mathematical Intelligencer* 8 (1986): 59–61.

This article gives an overview of the work done on Fermat's Last Theorem.

Ernst Eduard Kummer

Modern arithmetic—after
Gauss—began with Kummer.

E. T. Bell, *Men of Mathematics*

ERNST EDUARD KUMMER was born on January 29, 1810, in Sorau, Germany. He entered the University of Halle at the age of 18 to study theology, but within three years, he had obtained a Ph.D. degree in mathematics. The next 10 years of his life were spent doing research and teaching at the high-school level. One student greatly influenced by Kummer was Leopold Kronecker, who also became one of the 19th century's leading mathematicians (see Chapter 20 for a biography of Kronecker). Kummer received a professorship at the University of Breslau in 1842 and, 13 years later, moved on to the University of Berlin as Dirichlet's successor. In 1857, the French Academy of Sciences issued the following statement:

> Report on the competition for the grand prize in mathematical sciences. Already set in the competition for 1853 and prorogued to 1856. The committee, having found no work which seemed to it worthy of the prize among those submitted to it in competition, proposed to the Academy to award it to M. Kummer, for his beautiful researches on complex numbers composed of roots of unity and integers. The Academy adopted this proposal.

Throughout his career, Kummer was a popular teacher because of the clarity of his lectures and his charm and sense of humor. At Berlin, he directed the dissertation of 39 students, several of whom became well-known mathematicians.

Although Kummer did his best work on unique factorization domains and Fermat's Last Theorem, he also made outstanding contributions to analysis, geometry, and physics. He died on May 14, 1893, at the age of 83.

Sophie Germain

One of the very few women to overcome the prejudice and discrimination which have tended to exclude women from the pursuit of higher mathematics up to the present time was Sophie Germain.

SOPHIE GERMAIN was born in Paris on April 1, 1776. Although unable to attend a university because of discrimination against her sex, Germain educated herself by reading the works of Newton and Euler in Latin and the lecture notes of Lagrange. In 1804, Germain wrote to Gauss about her work in number theory but used the pseudonym Monsieur LeBlanc because she feared Gauss would not take seriously the efforts of a woman. Gauss gave Germain's results high praise and a few years later, upon learning her true identity, wrote to her [1, p. 61]:

> But how to describe to you my admiration and astonishment at seeing my esteemed correspondent Mr. LeBlanc metamorphose himself into this illustrious personage who gives such a brilliant example of what I would find it difficult to believe. A taste for the abstract sciences in general and above all the mysteries of numbers is excessively rare: it is not a subject which strikes everyone; the enchanting charms of this sublime science reveal themselves only to those who have the courage to go deeply into it. But when a person of the sex which, according to our customs and prejudices, must encounter infinitely more difficulties than men to familiarize herself with these thorny researches, succeeds nevertheless in surmounting these obstacles and penetrating the most obscure parts of

them, then without doubt she must have the noblest courage, quite extraordinary talents, and a superior genius.

Germain is best known for her result on Fermat's Last Theorem. She proved that $x^n + y^n = z^n$ has no positive integer solutions if x, y, and z are relatively prime to each other and to n where $n < 100$.

Sophie Germain died on June 27, 1831, in Paris. A school and a street in Paris are named in her honor.

SUPPLEMENTARY EXERCISES FOR CHAPTERS 15–18

The intelligence is proved not by ease of learning, but by understanding what we learn.

Joseph Whitney

1. Suppose that F is a field and there is a ring homomorphism from Z onto F. Show that F is isomorphic to Z_p for some prime p.

2. Let $Q[\sqrt{2}] = \{r + s\sqrt{2} \mid r, s \in Q\}$. Determine all ring automorphisms of $Q[\sqrt{2}]$.

3. (Second Isomorphism Theorem for Rings) Let A be a subring of R and let B be an ideal of R. Show that $A \cap B$ is an ideal of A and that $A/(A \cap B)$ is isomorphic to $(A + B)/B$. (Recall $A + B = \{a + b \mid a \in A, b \in B\}$.)

4. (Third Isomorphism Theorem for Rings) Let A and B be ideals of a ring R with $B \subseteq A$. Show that A/B is an ideal of R/B and $(R/B)(A/B)$ is isomorphic to R/A.

5. Let $f(x)$ and $g(x)$ be irreducible polynomials over a field F. If $f(x)$ and $g(x)$ are not associates, prove that $F[x]/\langle f(x)g(x)\rangle$ is isomorphic to $F[x]/\langle f(x)\rangle \oplus F[x]/\langle g(x)\rangle$.

6. (Chinese Remainder Theorem for Rings) If R is a commutative ring and I and J are two proper ideals with $I + J = R$, prove that $R/(I \cap J)$ is isomorphic to $R/I \oplus R/J$. Explain why exercise 5 is a special case of this theorem.

7. Prove that the set of all polynomials all of whose coefficients are even is a prime ideal in $Z[x]$.

8. Let $R = Z[\sqrt{-5}]$ and let $I = \{a + b\sqrt{-5} \mid a, b \in Z, a - b \text{ is even}\}$. Show that I is a maximal ideal of R.

9. Let R be a ring with unity and let a be a unit in R. Show that the mapping from R into itself given by $x \to axa^{-1}$ is a ring automorphism.

10. Let $a + b\sqrt{-5}$ belong to $Z[\sqrt{-5}]$ with $b \neq 0$. Show that 2 does not belong to $\langle a + b\sqrt{-5}\rangle$.

11. Show that $Z[i]/\langle 2 + i\rangle$ is a field. How many elements does it have?

12. Is the homomorphic image of a principal ideal domain a principal ideal domain?

13. In $Z[\sqrt{2}] = \{a + b\sqrt{2} \mid a, b \in Z\}$, show that every element of the form $(3 + 2\sqrt{2})^n$ is a unit.

14. Let p be a prime. Show that there is exactly one ring homomorphism from Z_m to Z_{p^k} if p^k does not divide m, and exactly two ring homomorphisms from Z_m to Z_{p^k} if p^k does divide m.

15. Recall that a is an idempotent if $a^2 = a$. Show that if $1 + k$ is an idempotent in Z_n, then $n - k$ is an idempotent in Z_n.

16. Show that Z_n (where $n > 1$) always has an even number of idempotents. (The number is 2^d, where d is the number of distinct prime divisors of n.)

17. Show that if p is prime, the only idempotents in Z_{p^k} are 0 and 1.

18. Prove that if both k and $k + 1$ are idempotents in Z_n and $k \neq 0$, then $n = 2k$.

19. Prove that $x^4 + 15x^3 + 7$ is irreducible over Q.

20. For any integers m and n, prove that the polynomial $x^3 + (5m + 1)x + 5n + 1$ is irreducible over Z.

21. Prove that $\langle \sqrt{2} \rangle$ is a maximal ideal in $Z[\sqrt{2}]$.

22. Prove that $Z[\sqrt{-2}]$ and $Z[\sqrt{2}]$ are unique factorization domains. (*Hint:* Mimic Example 6 in Chapter 18.)

23. Is $\langle 3 \rangle$ a maximal ideal in $Z[i]$?

24. Express both 13 and $5 + i$ as products of irreducibles from $Z[i]$.

25. Let $R = \{a/b \mid a, b \in Z, 3 \nmid b\}$. Prove that R is an integral domain. Find its field of quotients.

26. Give an example of a ring that contains a subring isomorphic to Z and a subring isomorphic to Z_3.

27. Show that $Z[i]/\langle 3 \rangle$ is not ring-isomorphic to $Z_3 \oplus Z_3$.

28. For any $n > 1$, prove that $R = \left\{ \begin{bmatrix} a & 0 \\ 0 & b \end{bmatrix} \mid a, b \in Z_n \right\}$ is ring-isomorphic to $Z_n \oplus Z_n$.

PART 4

FIELDS

19

Vector Spaces

It is important to appreciate at the outset that the idea of a vector space is the algebraic abstraction and generalization of the cartesian coordinate system introduced into the euclidean plane—that is, a generalization of analytic geometry.

Richard A. Dean, *Elements of Abstract Algebra*

DEFINITION AND EXAMPLES

Abstract algebra has three basic components: groups, rings, and fields. Thus far we have covered groups and rings in some detail and we have touched on the notion of a field. To explore fields more deeply, we need some rudiments of vector space theory that are covered in a linear algebra course. In this chapter, we provide a concise review of this material.

DEFINITION   Vector Space
A set V is said to be a *vector space* over a field F if V is an Abelian group under addition (denoted by $+$) and, if for each $a \in F$ and $v \in V$, there is an element av in V such that the following conditions hold for all a, b in F and all u, v in V.

1. $a(v + u) = av + au$
2. $(a + b)v = av + bv$
3. $a(bv) = (ab)v$
4. $1v = v$

The members of a vector space are called *vectors*. The members of the field are called *scalars*. The operation that combines a scalar a and a vector v to form the vector av is called *scalar multiplication*. In general, we will denote vectors by letters from the end of the alphabet, such as u, v, w, and scalars by letters from the beginning of the alphabet, such as a, b, c.

Example 1   The set $\mathbf{R}^n = \{(a_1, a_2, \ldots, a_n) \mid a_i \in \mathbf{R}\}$ is a vector space over $\mathbf{R}$. Here the operations are the obvious ones.

$$(a_1, a_2, \ldots, a_n) + (b_1, b_2, \ldots, b_n) = (a_1 + b_1, a_2 + b_2, \ldots, a_n + b_n)$$

and

$$b(a_1, a_2, \ldots, a_n) = (ba_1, ba_2, \ldots, ba_n).$$ □

Example 2   The set $M_2(Q)$ of 2×2 matrices with entries from Q is a vector space over Q. The operations are

$$\begin{bmatrix} a_1 & a_2 \\ a_3 & a_4 \end{bmatrix} + \begin{bmatrix} b_1 & b_2 \\ b_3 & b_4 \end{bmatrix} = \begin{bmatrix} a_1 + b_1 & a_2 + b_2 \\ a_3 + b_3 & a_4 + b_4 \end{bmatrix}$$

and

$$b\begin{bmatrix} a_1 & a_2 \\ a_3 & a_4 \end{bmatrix} = \begin{bmatrix} ba_1 & ba_2 \\ ba_3 & ba_4 \end{bmatrix}.$$ □

Example 3   The set $Z_p[x]$ of polynomials with coeffficients from Z_p is a vector space over Z_p, where p is a prime.   □

Example 4   The set of complex numbers $\mathbf{C} = \{a + bi \mid a, b \in \mathbf{R}, i^2 = -1\}$ is a vector space over $\mathbf{R}$. The operations are the usual addition and multiplication of complex numbers.   □

The next example is a generalization of Example 4. Although it appears rather trivial, it is of the utmost importance in the theory of fields.

Example 5   Let E be a field and let F be a subfield of E. Then E is a vector space over F. The operations are the operations of E.   □

SUBSPACES

Of course, there is a natural analog of subgroup and subring.

DEFINITION Subspace
Let V be a vector space over a field F and let U be a subset of V. We say that U is a *subspace* of V if U is also a vector space over F under the operations of V.

Example 6　The set $\{a_2x^2 + a_1x + a_0 \mid a_0, a_1, a_2 \in \mathbf{R}\}$ is a subspace of the set of all polynomials with real coefficients.　□

Example 7　Let V be a vector space over F and let $v_1, v_2, \ldots, v_n$ be (not necessarily distinct) elements of V. Then the subset

$$\langle v_1, v_2, \ldots, v_n \rangle = \{a_1v_1 + a_2v_2 + \cdots + a_nv_n \mid a_1, a_2, \ldots, a_n \in F\}$$

is called the *subspace of V spanned by* $v_1, v_2, \ldots, v_n$. Any summand of the form $a_1v_1 + a_2v_2 + \cdots + a_nv_n$ is called a *linear combination of* $v_1, v_2, \ldots, v_n$. If $\langle v_1, v_2, \ldots, v_n \rangle = V$, we say that $\{v_1, v_2, \ldots, v_n\}$ spans V.　□

LINEAR INDEPENDENCE

The next definition is the heart of the theory.

DEFINITION　Linearly Dependent, Linearly Independent
A set of vectors $\{v_1, v_2, \ldots, v_n\}$ is said to be *linearly dependent* over the field F if there are elements $a_1, a_2, \ldots, a_n$ from F, not all zero, such that $a_1v_1 + a_2v_2 + \cdots + a_nv_n = 0$. A set of vectors that is not linearly dependent over F is called *linearly independent over F*.

Example 8　In $\mathbf{R}^3$ the vectors $(1, 0, 0), (1, 0, 1)$, and $(1, 1, 1)$ are linearly independent over $\mathbf{R}$. To verify this, assume that there are real numbers a, b, and c such that $a(1, 0, 0) + b(1, 0, 1) + c(1, 1, 1) = (0, 0, 0)$. Then $(a + b + c, c, b + c) = (0, 0, 0)$. From this we see that $a = b = c = 0$.　□

Certain kinds of linearly independent sets play a crucial role in the theory of vector spaces.

DEFINITION　Basis
Let V be a vector space over F. A subset B of V is called a *basis* for V if B is linearly independent over F and B spans V.

The motivation for this definition is twofold. First, if B is a basis for a vector space V, then every member of V is a unique linear combination of the elements of B (see exercise 24). Second, with every vector space spanned by finitely many vectors, we can use the notion of basis to associate a unique integer that tells us much about the vector space. (In fact, this integer and the field completely determine the vector space up to isomorphism—see exercise 31.)

Example 9　The set $V = \left\{ \begin{bmatrix} a & a+b \\ a+b & b \end{bmatrix} \middle| a, b \in \mathbf{R} \right\}$ is a vector space over $\mathbf{R}$ (see exercise 17). We claim that the set $B = \left\{ \begin{bmatrix} 1 & 1 \\ 1 & 0 \end{bmatrix}, \begin{bmatrix} 0 & 1 \\ 1 & 1 \end{bmatrix} \right\}$ is a basis for V over $\mathbf{R}$. To prove that the set B is linearly independent, suppose that there are real numbers a and b such that

$$a\begin{bmatrix} 1 & 1 \\ 1 & 0 \end{bmatrix} + b\begin{bmatrix} 0 & 1 \\ 1 & 1 \end{bmatrix} = \begin{bmatrix} 0 & 0 \\ 0 & 0 \end{bmatrix}.$$

But this gives $\begin{bmatrix} a & a+b \\ a+b & b \end{bmatrix} = \begin{bmatrix} 0 & 0 \\ 0 & 0 \end{bmatrix}$, so that $a = b = 0$. On the other hand, since every member of V has the form

$$\begin{bmatrix} a & a+b \\ a+b & b \end{bmatrix} = a\begin{bmatrix} 1 & 1 \\ 1 & 0 \end{bmatrix} + b\begin{bmatrix} 0 & 1 \\ 1 & 1 \end{bmatrix},$$

we see that B spans V. ❑

We now come to the main result of this chapter.

Theorem 19.1   *Invariance of Basis Size*
If $\{u_1, u_2, \ldots, u_m\}$ and $\{w_1, w_2, \ldots, w_n\}$ are both bases of a vector space V, then $m = n$.

Proof.   Suppose that $m \neq n$. To be specific, let us say that $m < n$. Consider the set $\{w_1, u_1, u_2, \ldots, u_m\}$. Since the u's span V, we know that w_1 is a linear combination of the u's, say, $w_1 = a_1u_1 + a_2u_2 + \cdots + a_mu_m$. Clearly, not all the a's are 0. For convenience, say, $a_1 \neq 0$. Then $\{w_1, u_2, \ldots, u_m\}$ spans V (see exercise 22). Next consider the set $\{w_1, w_2, u_2, \ldots, u_m\}$. This time, w_2 is a linear combination of $w_1, u_2, \ldots, u_m$, say, $w_2 = b_1w_1 + b_2u_2 + \cdots + b_mu_m$. Then at least one of $b_2, \ldots, b_m$ is nonzero, for otherwise the w's are not linearly independent. Let us say $b_2 \neq 0$. Then $w_1, w_2, u_3, \ldots, u_m$ span V. Continuing in this fashion, we see that $\{w_1, w_2, \ldots, w_m\}$ spans V. But then w_{m+1} is a linear combination of $w_1, w_2, \ldots, w_m$ and, therefore, the set $\{w_1, \ldots, w_n\}$ is not linearly independent. This contradiction finishes the proof. ∎

Theorem 19.1 shows that any two finite bases for a vector space have the same size. Of course, not all vector spaces have finite bases. However, there is no vector space that has a finite basis and an infinite basis (see exercise 26).

DEFINITION   Dimension
A vector space that has a basis consisting of n elements is said to have *dimension n*. For completeness, the trivial vector space $\{0\}$ is said to be spanned by the empty set and to have dimension 0.

Although it requires a bit of set theory that is beyond the scope of this text, it can be shown that every vector space has a basis. A vector space that has a finite basis is called *finite dimensional;* otherwise it is called *infinite dimensional*.

EXERCISES

> The good Lord made us with two ends—one to sit on and one to think with. How well you succeed in life depends on which one you use.
>
> **Isaac Dworetsky**

1. Verify that each of the sets in Examples 1–4 satisfies the axioms for a vector space. Find a basis for each of the vector spaces in Examples 1–4.

2. (Subspace Test) Prove that a nonempty subset U of a vector space V over a field F is a subspace of V if, for every u and u' in U and every a in F, $u - u' \in U$ and $au \in U$.

3. Verify that the set in Example 6 is a subspace. Find a basis for this subspace. Is $\{x^2 + x + 1, x + 5, 3\}$ a basis?

4. Verify that the set $\langle v_1, v_2, \ldots, v_n \rangle$ defined in Example 7 is a subspace.

5. Determine whether or not the set $\{(2, -1, 0), (1, 2, 5), (7, -1, 5)\}$ is linearly independent over **R**.

6. Determine whether or not the set

$$\left\{ \begin{bmatrix} 2 & 1 \\ 1 & 0 \end{bmatrix}, \begin{bmatrix} 0 & 1 \\ 1 & 2 \end{bmatrix}, \begin{bmatrix} 1 & 1 \\ 1 & 1 \end{bmatrix} \right\}$$

   is linearly independent over Z_5.

7. If $\{u, v, w\}$ is a linearly independent subset of a vector space, show that $\{u, u + v, u + v + w\}$ is also linearly independent.

8. If $\{v_1, v_2, \ldots, v_n\}$ is a linearly dependent set of vectors, prove that one of these vectors is a linear combination of the others.

9. (Every spanning collection contains a basis.) If $v_1, v_2, \ldots, v_n$ span a vector space V, prove that some subset of the v's is a basis for V.

10. (Every independent set is contained in a basis.) Let V be a finite dimensional vector space and let $\{v_1, v_2, \ldots, v_n\}$ be a linearly independent subset of V. Show that there are vectors $w_1, w_2, \ldots, w_m$ such that $\{v_1, v_2, \ldots, v_n, w_1, \ldots, w_m\}$ is a basis for V.

11. If V is a vector space over F of dimension 5 and U and W are subspaces of V dimension 3, prove that $U \cap W \neq \{0\}$. Generalize.

12. Show that the solution set to the system of equations of the form

$$a_{11}x_1 + \cdots + a_{1n}x_n = 0$$
$$a_{21}x_1 + \cdots + a_{2n}x_n = 0$$
$$\vdots \qquad\qquad \vdots$$
$$a_{m1}x_1 + \cdots + a_{mn}x_n = 0,$$

   where the a's are real, is a subspace of $\mathbf{R}^n$.

13. Let V be the set of all polynomials over Q of degree 2 together with the zero polynomial. Is V a vector space over Q?

14. Let $V = \mathbf{R} \oplus \mathbf{R} \oplus \mathbf{R}$ and $W = \{(a, b, c) \in V \mid a^2 + b^2 = c^2\}$. Is W a subspace of V? If so, what is its dimension?

15. Let $V = \mathbf{R} \oplus \mathbf{R} \oplus \mathbf{R}$ and $W = \{(a, b, c) \in V \mid a + b = c\}$. Is W a subspace of V? If so, what is its dimension?

16. Let $V = \left\{ \begin{bmatrix} a & b \\ b & c \end{bmatrix} \;\middle|\; a, b, c \in Q \right\}$. Prove that V is a vector space over Q and find a basis for V over Q.

17. Verify that the set V in Example 9 is a vector space over $\mathbf{R}$.

18. Let $P = \{(a, b, c) \mid a, b, c \in \mathbf{R}, a = 2b + 3c\}$. Prove that P is a subspace of $\mathbf{R}^3$. Find a basis for P. Give a geometric description of P.

19. Let U and W be subspaces of a vector space V. Show that $U \cap W$ is a subspace of V and that $U + W = \{u + w \mid u \in U, w \in W\}$ is a subspace of V.

20. If U is a proper subspace of a finite dimensional vector space V, show that the dimension of U is less than the dimension of V.

21. If V is a vector space of dimension n over the field Z_p, how many elements are in V?

22. Referring to the proof of Theorem 19.1, prove that $\{w_1, u_2, \ldots, u_m\}$ spans V.

23. Let $S = \{(a, b, c, d) \mid a, b, c, d \in \mathbf{R}, a = c, d = a + b\}$. What is the dimension of S?

24. Let B be a subset of a vector space V. Show that B is a basis for V if and only if every member of V is a unique linear combination of the elements of B. (This exercise is referred to in this chapter and in Chapter 20.)

25. Let $u = (2, 3, 1)$, $v = (1, 3, 0)$, and $w = (2, -3, 3)$. Since $\frac{1}{2}u - \frac{2}{3}v - \frac{1}{6}w = (0, 0, 0)$, can we conclude that u, v, and w are linearly dependent over Z_7?

26. If a vector space has one basis that contains infinitely many elements, prove that every basis contains infinitely many elements.

27. Define the vector space analog of group homomorphism and ring homomorphism. Such a mapping is called a *linear transformation*. Define the vector space analog of group isomorphism and ring isomorphism.

28. Let T be a linear transformation from V to W. Prove that the image of V under T is a subspace of W.

29. Let T be a linear transformation of a vector space V. Prove that the *kernel* of $T = \{v \in V \mid T(v) = 0\}$ is a subspace of V.

30. Let T be a linear transformation of V onto W. If $\{v_1, v_2, \ldots, v_n\}$ spans V, show that $\{T(v_1), T(v_2), \ldots, T(v_n)\}$ spans W.

31. If V is a vector space over F of dimension n, prove that V is isomorphic as a vector space to $F^n = \{(a_1, a_2, \ldots, a_n) \mid a_i \in F\}$. (This exercise is referred to in this chapter and in Chapter 21.)

Emil Artin

For Artin, to be a mathematician meant to participate in a great common effort, to continue work begun thousands of years ago, to shed new light on old discoveries, to seek new ways to prepare the developments of the future. Whatever standards we use, he was a great mathematician.

Richard Brauer, *Bulletin of the American Mathematical Society*

EMIL ARTIN was one of the leading mathematicians of the 20th century and a major contributor to linear algebra and abstract algebra. Artin was born on March 3, 1898, in Vienna, Austria, and grew up in what was recently known as Czecho-slovakia. After serving in the Austrian army during World War I, Artin enrolled at the University of Leipzig where he received a Ph.D. in 1921. From 1923 until he emigrated to America in 1937, he was a professor at the University of Hamburg. After one year at Notre Dame, Artin went to Indiana University. In 1946, he moved to Princeton, where he stayed until 1958. The last four years of his career were spent where it began, at Hamburg.

Artin's mathematics is both deep and broad. He made contributions to number theory, group theory, ring theory (in fact, there is a class of rings named after him), field theory, Galois theory, geometric algebra, algebraic topology, and the theory of braids—a field he invented. Artin received the American Mathematical Society's Cole Prize in number theory, and he solved one of the 23 famous problems posed by the eminent mathematician David Hilbert in 1900. Besides mathematics, Artin had a deep interest in chemistry, astronomy, biology, and old music. He played the flute, the harpsichord, and the clavichord.

Artin was an outstanding teacher of mathematics at all levels, from freshman calculus to seminars for colleagues. Many of his Ph.D. students have become leading mathematicians. Through his research, teaching, and books, Artin exerted great influence among his contemporaries. He died of a heart attack, at the age of 64, in 1962.

20

Extension Fields

THE FUNDAMENTAL THEOREM OF FIELD THEORY

In our work on rings, we came across a number of fields both finite and
infinite. Indeed, we saw that $Z_3[x]/\langle x^2 + 1 \rangle$ is a field of order 9 whereas
$\mathbf{R}[x]/\langle x^2 + 1 \rangle$ is a field isomorphic to the complex numbers. In the next
three chapters, we take up, in a systematic way, the subject of fields.

DEFINITION   Extension Field
A field E is an *extension field* of a field F if $F \subseteq E$ and the operations of
F are those of E restricted to F.

Cauchy's observation in 1847 that $\mathbf{R}[x]/\langle x^2 + 1 \rangle$ is a field that contains a zero of $x^2 + 1$ prepared the way for the following sweeping generalization of that fact.

Theorem 20.1 *Fundamental Theorem of Field Theory (Kronecker's Theorem,
1887)*
*Let F be a field and $f(x)$ a nonconstant polynomial in $F[x]$. Then there
is an extension field E of F in which $f(x)$ has a zero.*

Proof. Since $F[x]$ is a unique factorization domain, $f(x)$ has an irreducible factor, say, $p(x)$. Clearly, it suffices to construct an extension field E of F in which $p(x)$ has a zero. Our candidate for E is $F[x]/\langle p(x)\rangle$. We already know that this is a field from Corollary 1 of Theorem 17.5. Also, since the mapping of $\phi:F \rightarrow E$ given by $\phi(a) = a + \langle p(x)\rangle$ is one-to-one and preserves both operations, E has a subfield isomorphic to F. We may think of E as containing F if we simply identify the coset $a + \langle p(x)\rangle$ with its unique coset representative a [that is, think of $a + \langle p(x)\rangle$ as just a and vice versa; see exercise 30 in Chapter 17].

Finally, to show that $p(x)$ has a zero in E, write

$$p(x) = a_n x^n + a_{n-1} x^{n-1} + \cdots + a_0.$$

Then, in E, $x + \langle p(x)\rangle$ is a zero of $p(x)$. For

$$
\begin{aligned}
p(x + \langle p(x)\rangle) &= a_n(x + \langle p(x)\rangle)^n + a_{n-1}(x + \langle p(x)\rangle)^{n-1} + \cdots + a_0 \\
&= a_n(x^n + \langle p(x)\rangle) + a_{n-1}(x^{n-1} + \langle p(x)\rangle) + \cdots + a_0 \\
&= a_n x^n + a_{n-1} x^{n-1} + \cdots + a_0 + \langle p(x)\rangle \\
&= p(x) + \langle p(x)\rangle = 0 + \langle p(x)\rangle. \qquad \blacksquare
\end{aligned}
$$

Example 1 Let $f(x) = x^2 + 1 \in Q[x]$. Then, in $E = Q[x]/\langle x^2 + 1\rangle$, we have

$$
\begin{aligned}
f(x + \langle x^2 + 1\rangle) &= (x + \langle x^2 + 1\rangle)^2 + 1 \\
&= x^2 + \langle x^2 + 1\rangle + 1 \\
&= x^2 + 1 + \langle x^2 + 1\rangle \\
&= 0 + \langle x^2 + 1\rangle.
\end{aligned}
$$

Of course, the polynomial $x^2 + 1$ has the complex number $\sqrt{-1}$ as a zero, but the point we wish to emphasize here is that we have constructed a field that contains the rational numbers and a zero for the polynomial $x^2 + 1$ by using only the rational numbers. No knowledge of complex numbers is necessary. Our method utilizes only the field we are given. $\quad\square$

Example 2 Let $f(x) = x^5 + 2x^2 + 2x + 2 \in Z_3[x]$. Then, the irreducible factorization of $f(x)$ over Z_3 is $(x^2 + 1)(x^3 + 2x + 2)$. So, to find an extension E of Z_3 in which $f(x)$ has a zero, we may take $E = Z_3[x]/\langle x^2 + 1\rangle$, a field with nine elements, or $E = Z_3[x]/\langle x^3 + 2x + 2\rangle$, a field with 27 elements. $\quad\square$

Since every integral domain is contained in its field of quotients (Theorem 15.6), we see that every nonconstant polynomial with coefficients from an integral domain always has a zero in some field containing the ring of coefficients. The next example shows that this is not true for commutative rings in general.

Example 3   Let $f(x) = 2x + 1 \in Z_4[x]$. Then $f(x)$ has no zero in any ring containing Z_4 as a subring. For if β were a zero in such a ring, then $0 = 2\beta + 1$ and therefore $0 = 2(2\beta + 1) = 2(2\beta) + 2 = (2 \cdot 2)\beta + 2 = 0 \cdot \beta + 2 = 2$. But $0 \neq 2$ in Z_4.   ❑

SPLITTING FIELDS

To motivate the next definition and theorem, let's return to Example 1 for a moment. For notational convenience, in $Q[x]/\langle x^2 + 1\rangle$, let $\alpha = x + \langle x^2 + 1\rangle$. Then, since α and $-\alpha$ are both zeros of $x^2 + 1$, it should be the case that $x^2 + 1 = (x - \alpha)(x + \alpha)$. Let's check this out. First note that

$$(x - \alpha)(x + \alpha) = x^2 - \alpha^2 = x^2 - (x^2 + \langle x^2 + 1\rangle).$$

At the same time,

$$x^2 + \langle x^2 + 1\rangle = -1 + \langle x^2 + 1\rangle$$

and we have agreed to identify -1 and $-1 + \langle x^2 + 1\rangle$, so

$$(x - \alpha)(x + \alpha) = x^2 - (-1) = x^2 + 1.$$

This shows that $x^2 + 1$ can be written as a product of linear factors in some extension of Q. That was easy and you might argue coincidental. The polynomial given in Example 2 presents a greater challenge. Is there an extension of Z_3 in which that polynomial factors as a product of linear factors? Yes, there is. But first a definition.

DEFINITION   Splitting Field
Let E be an extension field of F and let $f(x) \in F[x]$. We say that $f(x)$ *splits* in E if $f(x)$ can be factored as a product of linear factors in $E[x]$. We call E a *splitting field for $f(x)$ over F* if $f(x)$ splits in E but in no proper subfield of E.

    Note that a splitting field of a polynomial over a field depends not only on the polynomial but the field as well. Indeed, a splitting field of $f(x)$ over F is just a smallest extension field of F in which $f(x)$ splits. The next example illustrates how a splitting field of a polynomial $f(x)$ over field F depends on F.

Example 4   Consider the polynomial $f(x) = x^2 + 1 \in Q[x]$. Since $x^2 + 1 = (x + \sqrt{-1})(x - \sqrt{-1})$, we see that $f(x)$ splits in $\mathbf{C}$, but a splitting field over Q is $Q(i) = \{r + si \mid r, s \in Q\}$. A splitting field for $x^2 + 1$ over $\mathbf{R}$ is $\mathbf{C}$. Likewise, $x^2 - 2 \in Q[x]$ splits in $\mathbf{R}$, but a splitting field over Q is $Q(\sqrt{2}) = \{r + s\sqrt{2} \mid r, s \in Q\}$.   ❑

    There is a useful analogy between the definition of splitting field and the definition of an irreducible polynomial. Just as it makes no sense to say "$f(x)$ is irreducible," it makes no sense to say "E is a splitting field for $f(x)$." In each case, the underlying field must be specified; that is, one must say "$f(x)$ is irreducible over F" and "E is a splitting field for $f(x)$ over F."

The following notation is convenient. Let F be a field and let $a_1, a_2, \ldots, a_n$ be elements of some extension E of F. We use $F(a_1, a_2, \ldots, a_n)$ to denote the smallest subfield of E that contains F and the set $\{a_1, a_2, \ldots, a_n\}$. It is an easy exercise to show that $F(a_1, a_2, \ldots, a_n)$ is the intersection of all subfields of E that contain F and the set $\{a_1, a_2, \ldots, a_n\}$.

Notice that if $f(x) \in F[x]$ and $f(x)$ factors as

$$b(x - a_1)(x - a_2) \cdots (x - a_n)$$

over some extension E of F, then $F(a_1, \ldots, a_n)$ is the splitting field for $f(x)$ over F.

This notation appears to be inconsistent with the notation that we used in earlier chapters. For example, we denoted the set $\{a + b\sqrt{2} \mid a, b \in Z\}$ by $Z[\sqrt{2}]$ and the set $\{a + b\sqrt{2} \mid a, b \in Q\}$ by $Q[\sqrt{2}]$. The difference is that $Z[\sqrt{2}]$ is merely a ring whereas $Q[\sqrt{2}]$ is a field. In general, parentheses are used when one wishes to indicate that the set is a field, although no harm would be done by using, say, $Q[\sqrt{2}]$ to denote $\{a + b\sqrt{2} \mid a, b \in Q\}$ if we were concerned with its ring properties only. After all, notation is just a convenient way to convey information. Using parentheses rather than brackets simply conveys a bit more information about the set.

Theorem 20.2   *Existence of Splitting Fields*

Let F be a field and let $f(x)$ be a nonconstant element of $F[x]$. Then there exists a splitting field E for $f(x)$ over F.

Proof.   We proceed by induction on $\deg f(x)$. If $\deg f(x) = 1$, then $f(x)$ is already linear and $E = F$. Now suppose that the statement is true for all fields and all polynomials of degree less than that of $f(x)$. By Theorem 20.1, there is an extension E of F in which $f(x)$ has a zero, say, a_1. Then we may write $f(x) = (x - a_1)g(x)$, where $g(x) \in E[x]$. Since $\deg g(x) < \deg f(x)$, by induction, there is a field K that contains E and all the zeros of $g(x)$, say, $a_2, \ldots, a_n$. Clearly, then, the splitting field for $f(x)$ over F is $F(a_1, a_2, \ldots, a_n)$. ∎

Example 5   Consider

$$f(x) = x^4 - x^2 - 2 = (x^2 - 2)(x^2 + 1)$$

over Q. Obviously, the zeros of $f(x)$ are $\pm\sqrt{2}$ and $\pm i$. So a splitting field for $f(x)$ over Q is

$$Q(\sqrt{2}, i) = Q(\sqrt{2})(i) = \{\alpha + \beta i \mid \alpha, \beta \in Q(\sqrt{2})\}$$
$$= \{(a + b\sqrt{2}) + (c + d\sqrt{2})i \mid a, b, c, d \in Q\}. \quad \square$$

Example 6   Consider $f(x) = x^2 + x + 2$ over Z_3. Then $Z_3(i) = \{a + bi \mid a, b \in Z_3\}$ (see Example 8 in Chapter 13) is a splitting field for $f(x)$ over Z_3 because

$$f(x) = [x - (1 + i)][x - (1 - i)].$$

At the same time, we know by the proof of Kronecker's Theorem that the element $x + \langle x^2 + x + 2 \rangle$ of

$$F = Z_3[x]/\langle x^2 + x + 2 \rangle$$

is a zero of $f(x)$. Since $f(x)$ has degree 2, it follows from the Factor Theorem (Corollary 2 of Theorem 16.2) that the other zero of $f(x)$ must also be in F. Thus, $f(x)$ splits in F, and because F has only nine elements, it is obvious that F is also a splitting field of $f(x)$ over Z_3. But how do we factor $f(x)$ in F? Factoring $f(x)$ in F is confusing because we are using the symbol x in two distinct ways. It is used as a placeholder to write the polynomial $f(x)$, and it is used to create the coset representatives of the elements of F. This confusion can be avoided by simply identifying the coset $1 + \langle x^2 + x + 2 \rangle$ with the element 1 in Z_3 and denoting the coset $x + \langle x^2 + x + 2 \rangle$ by β. To obtain the factorization of $f(x)$ in F, we observe that

$$(x - \beta)[x - (2\beta + 2)] = x^2 + (-\beta - 2\beta - 2)x + 2(\beta^2 + \beta)$$
$$= x^2 + x + 2 = f(x).$$

Here we have used the facts that F has characteristic 3 and $\beta^2 + \beta = 1$. Thus, we have found two splitting fields for $x^2 + x + 2$ over Z_3, one of the form $F(a)$ and one of the form $F[x]/\langle p(x) \rangle$ [where $F = Z_3$ and $p(x) = x^2 + x + 2$].   ☐

The next theorem shows how the fields $F(a)$ and $F[x]/\langle p(x) \rangle$ are related in the case where $p(x)$ is irreducible over F and a is a zero of $p(x)$ in some extension of F.

Theorem 20.3 $F(a) \approx F[x]/\langle p(x) \rangle$
Let F be a field and let $p(x) \in F[x]$ be irreducible over F. If a is a zero of $p(x)$ in some extension E of F, then $F(a)$ is isomorphic to $F[x]/\langle p(x) \rangle$. Furthermore, if $\deg p(x) = n$, then every member of $F(a)$ can be uniquely expressed in the form

$$c_{n-1}a^{n-1} + c_{n-2}a^{n-2} + \cdots + c_1a + c_0,$$

where $c_0, c_1, \ldots, c_{n-1} \in F$.

Proof.   Consider the function ϕ from $F[x]$ to $F(a)$ given by $\phi(f(x)) = f(a)$. Clearly, ϕ is a ring homomorphism. We claim that $\text{Ker } \phi = \langle p(x) \rangle$. (This is exercise 31 in Chapter 17.) Since $p(a) = 0$, we have $\langle p(x) \rangle \subseteq \text{Ker } \phi$. On the other hand, we know by Theorem 17.5 that $\langle p(x) \rangle$ is a maximal ideal in $F[x]$. So, because $\text{Ker } \phi \neq F[x]$ [it does not contain the constant polynomial $f(x) = 1$], we have Ker $\phi = \langle p(x) \rangle$. At this point, it follows from the First Isomorphism Theorem for Rings and Corollary 1 of Theorem 17.5 that $\phi(F[x])$ is a subfield of $F(a)$. Noting that $\phi(F[x])$ contains both F and a and recalling that $F(a)$ is the smallest such field, we have $F[x]/\langle p(x) \rangle \approx \phi(F[x]) = F(a)$.

The final assertion of the theorem follows from the fact that every element of $F[x]/\langle p(x) \rangle$ can be expressed uniquely in the form

$$c_{n-1}x^{n-1} + \cdots + c_0 + \langle p(x) \rangle,$$

where $c_0, \ldots, c_{n-1} \in F$ (see exercise 21 in Chapter 16) and the natural isomorphism from $F[x]/\langle p(x) \rangle$ to $F(a)$ carries $c_k x^k + \langle p(x) \rangle$ to $c_k a^k$.   ∎

Recall that a basis for an n-dimensional vector space over a field F is a set of n vectors $v_1, v_2, \ldots, v_n$ with the property that every member of the vector space can be expressed uniquely in the form $a_1 v_1 + a_2 v_2 + \cdots + a_n v_n$, where the a's belong to F (exercise 24 in Chapter 19). So, in the language of vector spaces, the latter portion of Theorem 20.3 says that if a is a zero of an irreducible polynomial over F of degree n, then the set $\{1, a, \ldots, a^{n-1}\}$ is a basis for $F(a)$ over F.

Theorem 20.3 often provides a convenient way of describing the elements of a field.

Example 7   Consider the irreducible polynomial $f(x) = x^6 - 2$ over Q. Since $\sqrt[6]{2}$ is a zero of $f(x)$, we know from Theorem 20.3 that the set $\{1, 2^{1/6}, 2^{2/6}, 2^{3/6}, 2^{4/6}, 2^{5/6}\}$ is a basis for $Q(\sqrt[6]{2})$ over Q. Thus,

$$Q(\sqrt[6]{2}) = \{a_0 + a_1 2^{1/6} + a_2 2^{2/6} + a_3 2^{3/6} + a_4 2^{4/6} + a_5 2^{5/6} \mid a_i \in Q\}.$$

This field is isomorphic to $Q[x]/\langle x^6 - 2 \rangle$.   □

Notice that Theorem 20.3 does not apply to $Q(\pi)$ since π is not the zero of a polynomial in $Q[x]$. [This important fact was first proved by Ferdinand Lindemann (1852–1939) in 1882.]

In Example 6, we produced two splitting fields for the polynomial $x^2 + x + 2$ over Z_3. Likewise, it is an easy exercise to show that both $Q[x]/\langle x^2 + 1 \rangle$ and $Q(i) = \{r + si \mid r, s \in Q\}$ are splitting fields of the polynomial $x^2 + 1$ over Q. But are these different-looking splitting fields algebraically different? Not really. We conclude our discussion of splitting fields by proving that splitting fields are unique up to isomorphism. To make it easier to apply induction, we will prove a more general result.

We begin by observing first that any field isomorphism ϕ from F to F' has a natural extension from $F[x]$ to $F'[x]$ given by $c_n x^n + c_{n-1}x^{n-1} + \cdots + c_1 x + c_0 \rightarrow \phi(c_n)x^n + \phi(c_{n-1})x^{n-1} + \cdots + \phi(c_1)x + \phi(c_0)$. Since this mapping agrees with ϕ on F, it is convenient and natural to use ϕ to denote this mapping as well.

Lemma   *Let F be a field, let $p(x) \in F[x]$ be irreducible over F, and let a be a zero of $p(x)$ in some extension of F. If ϕ is a field isomorphism from F to F' and b is a zero of $\phi(p(x))$ in some extension of F', then there is an isomorphism from $F(a)$ to $F'(b)$ that agrees with ϕ on F and carries a to b.*

Proof.   First observe that since $p(x)$ is irreducible over F, $\phi(p(x))$ is irreducible over $F'[x]$. It is straightforward to check that the mapping from $F[x]/\langle p(x)\rangle$ to $F'[x]/\langle \phi(p(x))\rangle$ given by

$$f(x) + \langle p(x)\rangle \rightarrow \phi(f(x)) + \langle \phi(p(x))\rangle$$

is a field isomorphism. By a slight abuse of notation, we denote this mapping by ϕ also. (If you object, put a bar over the ϕ.) From the proof of Theorem 20.3, we know that there is an isomorphism α from $F(a)$ to $F[x]/\langle p(x)\rangle$ that is the identity on F and carries a to $x + \langle p(x)\rangle$. Similarly, there is an isomorphism β from $F'[x]/\langle \phi(p(x))\rangle$ to $F'(b)$ that is the identity on F' and carries $x + \langle \phi(p(x))\rangle$ to b. Thus, $\beta\phi\alpha$ is the desired mapping. See Figure 20.1.   ■

Theorem 20.4

Let ϕ be an isomorphism from a field F to a field F' and let $f(x) \in F[x]$. If E is a splitting field for $f(x)$ over F and E' is a splitting field for $\phi(f(x))$ over F', then there is an isomorphism from E to E' that agrees with ϕ on F.

Proof.   We induct on $\deg f(x)$. If $\deg f(x) = 1$, then $E = F$ and $E' = F'$, so that ϕ itself is the desired mapping. If $\deg f(x) > 1$, let $p(x)$ be an irreducible factor of $f(x)$, let a be a zero of $p(x)$ in E, and let b be a zero of $\phi(p(x))$ in E'. By the preceding lemma, there is an isomorphism α from $F(a)$ to $F'(b)$ that agrees with ϕ on F and carries a to b. Now write $f(x) = (x - a)g(x)$, where $g(x) \in F(a)[x]$. Then E is a splitting field for $g(x)$ over $F(a)$ and E' is a splitting field for $\alpha(g(x))$ over $F'(b)$. Since $\deg g(x) < \deg f(x)$, there is an isomorphism from E to E' that agrees with α on $F(a)$ and therefore with ϕ on F.   ■

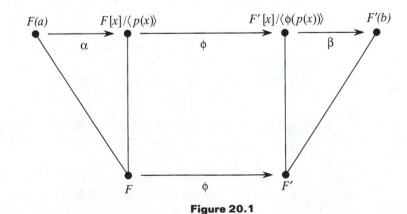

Figure 20.1

Corollary *Splitting Fields Are Unique*

Let F be a field and let $f(x) \in F[x]$. Then any two splitting fields of $f(x)$ over F are isomorphic.

Proof. Suppose that E and E' are splitting fields of $f(x)$ over F. The result follows immediately from Theorem 20.4 by letting ϕ be the identity from F to F. ∎

In light of the above corollary, we may refer to "the" splitting field of a polynomial without ambiguity.

Even though $x^6 - 2$ has a zero in $Q(\sqrt[6]{2})$, it does not split in $Q(\sqrt[6]{2})$. The splitting field is easy to obtain, however.

Example 8 The Splitting Field of $x^n - a$ over Q

Let a be a positive rational number and let ω be a primitive nth root of unity (see Example 2 in Chapter 16). Then each of

$$a^{1/n}, \; \omega a^{1/n}, \; \omega^2 a^{1/n}, \ldots, \; \omega^{n-1} a^{1/n}$$

is a zero of $x^n - a$ in $Q(\sqrt[n]{a}, \omega)$. ☐

ZEROS OF AN IRREDUCIBLE POLYNOMIAL

Now that we know that every nonconstant polynomial over a field splits in some extension, we ask whether irreducible polynomials must split in some special way. Yes, they do. To discover how, we borrow something whose origins are in calculus.

DEFINITION Let $f(x) = a_n x^n + a_{n-1} x^{n-1} + \cdots + a_1 x + a_0$ belong to $F[x]$. The *derivative* of $f(x)$, denoted by $f'(x)$, is the polynomial $na_n x^{n-1} + (n - 1)a_{n-1} x^{n-2} + \cdots + a_1$ in $F[x]$.

Notice that our definition does not involve the notion of a limit. The standard rules for handling sums and products of functions in calculus carry over to arbitrary fields as well.

Lemma Let $f(x)$ and $g(x) \in F[x]$ and let $a \in F$. Then

1. $(f(x) + g(x))' = f'(x) + g'(x)$
2. $(af(x))' = af'(x)$
3. $(f(x)g(x))' = f(x)g'(x) + g(x)f'(x)$.

Proof. Parts 1 and 2 follow from straightforward applications of the definition. Using part 1 and induction on deg $f(x)$, part 3 reduces to the special case that $f(x) = a_n x^n$. This also follows directly from the definition. ∎

Before addressing the question of the nature of the zeros of an irreducible polynomial, we establish a general result concerning zeros of multiplicity greater than 1. Such zeros are called *multiple* zeros.

Theorem 20.5   *Criterion for Multiple Zeros*

> *A polynomial $f(x)$ over a field F has a multiple zero in some extension E if and only if $f(x)$ and $f'(x)$ have a common factor of positive degree in $F[x]$.*

Proof.   If a is a multiple zero of $f(x)$ in some extension E, then there is a $g(x)$ in $E[x]$ such that $f(x) = (x - a)^2 g(x)$. Since $f'(x) = (x - a)^2 g'(x) + 2(x - a)g(x)$, we see that $f'(a) = 0$. Thus $x - a$ is a factor of both $f(x)$ and $f'(x)$ in the extension E of F. Now if $f(x)$ and $f'(x)$ have no common divisor of positive degree in $F[x]$, there are polynomials $h(x)$ and $k(x)$ in $F[x]$ such that $f(x)h(x) + f'(x)k(x) = 1$ (see exercise 40 in Chapter 16). Viewing $f(x)h(x) + f'(x)k(x)$ as an element of $E[x]$, we see also that $x - a$ is a factor of 1. Since this is nonsense, $f(x)$ and $f'(x)$ must have a common divisor of positive degree in $F[x]$.

Conversely, suppose that $f(x)$ and $f'(x)$ have a common factor of positive degree, but $f(x)$ has no multiple zeros. We will obtain a contradiction. Let a be a zero of the common factor. Then a is a zero of $f(x)$ and $f'(x)$. Since a is a zero of $f(x)$ of multiplicity 1, there is a polynomial $q(x)$ such that $f(x) = (x - a)q(x)$ where $q(a) \neq 0$. Then $f'(x) = (x - a)q'(x) + q(x)$ and $0 = f'(a) = q(a)$, which is a contradiction. ∎

Theorem 20.6   *Zeros of an Irreducible*

> *Let $f(x)$ be an irreducible polynomial over a field F. If F has characteristic 0, then $f(x)$ has no multiple zeros. If F has characteristic $p \neq 0$, then $f(x)$ has a multiple zero only if it is of the form $f(x) = g(x^p)$ for some $g(x)$ in $F[x]$.*

Proof.   If $f(x)$ has a multiple zero, then, by Theorem 20.5, $f(x)$ and $f'(x)$ have a common divisor of positive degree in $F[x]$. Since the only divisor of positive degree of $f(x)$ in $F[x]$ is $f(x)$ itself (up to associates), we see that $f(x)$ divides $f'(x)$. Because a polynomial over a field cannot divide a polynomial of smaller degree, we must have $f'(x) = 0$.

Now what does it mean to say that $f'(x) = 0$? If we write $f(x) = a_n x^n + a_{n-1} x^{n-1} + \cdots + a_1 x + a_0$, then $f'(x) = na_n x^{n-1} + (n - 1)a_{n-1} x^{n-2} + \cdots + a_1$. Thus, $f'(x) = 0$ only when $ka_k = 0$ for $k = 1, \ldots, n$.

So, when char $F = 0$, we have $f(x) = a_0$ and when char $F = p \neq 0$, we have $a_k = 0$ when p does not divide k. Thus, the only powers of x that appear in the sum $a_n x^n + \cdots + a_1 x + a_0$ are those of the form $x^{pi} = (x^p)^i$. It follows that $f(x) = g(x^p)$ for some $g(x) \in F[x]$. [For example, if $f(x) = x^{4p} + 3x^{2p} + x^p$, then $g(x) = x^4 + 3x^2 + x$.] ∎

EXERCISES

I have yet to see any problem, however complicated, which, when you looked at it in the right way, did not become still more complicated.

Paul Anderson, *New Scientist*

1. Describe the elements of $Q(\sqrt[3]{5})$.
2. Show that $Q(\sqrt{2}, \sqrt{3}) = Q(\sqrt{2} + \sqrt{3})$.
3. Find the splitting field of $x^3 - 1$ over Q. Express your answer in the form $Q(a)$.
4. Find the splitting field of $x^4 + 1$ over Q.
5. Find the splitting field of

$$x^4 + x^2 + 1 = (x^2 + x + 1)(x^2 - x + 1)$$

   over Q.
6. Let $a, b \in \mathbf{R}$ with $b \neq 0$. Show that $\mathbf{R}(a + bi) = \mathbf{C}$.
7. Find a polynomial $p(x)$ in $Q[x]$ so that $Q(\sqrt{1 + \sqrt{5}})$ is isomorphic to $Q[x]/\langle p(x) \rangle$.
8. Let $F = Z_2$ and let $f(x) = x^3 + x + 1 \in F[x]$. Suppose that a is a zero of $f(x)$ in some extension of F. How many elements does $F(a)$ have? Express each member of $F(a)$ in terms of a. Write out a complete multiplication table for $F(a)$.
9. Let $F(a)$ be the field described in exercise 8. Express each of a^5, a^{-2}, and a^{100} in the form $c_2 a^2 + c_1 a + c_0$.
10. Let $F(a)$ be the field described in exercise 8. Show that $a + 1$, $a^2 + 1$, and $a^2 + a + 1$ are all zeros of $x^3 + x + 1$.
11. Describe the elements in $Q(\pi)$.
12. Let $F = Q(\pi^3)$. Find a basis for $F(\pi)$ over F.
13. Show that $Q(\sqrt{2})$ is not isomorphic to $Q(\sqrt{3})$.
14. Find all automorphisms of $Q(\sqrt[3]{5})$.
15. Let F be a field of characteristic p and let $f(x) = x^p - a \in F[x]$. Show that $f(x)$ is irreducible over F or $f(x)$ splits in F.
16. Let F be a field and let $f(x) \in F[x]$ be irreducible over F. Suppose that a and b are zeros of $f(x)$ in some extension of F. Show that there is an isomorphism from $F(a)$ onto $F(b)$ that acts as the identity on F and carries a to b.
17. Find a, b, c in Q so that

$$(1 + \sqrt[3]{4})/(2 - \sqrt[3]{2}) = a + b\sqrt[3]{2} + c\sqrt[3]{4}.$$

   Note that such a, b, c exist since

$$(1 + \sqrt[3]{4})/(2 - \sqrt[3]{2}) \in Q(\sqrt[3]{2}) = \{a + b\sqrt[3]{2} + c\sqrt[3]{4} \mid a, b, c \in Q\}.$$

18. Express $(3 + 4\sqrt{2})^{-1}$ in the form $a + b\sqrt{2}$, where $a, b \in Q$.

19. Show that $Q(4 - i) = Q(1 + i)$, where $i = \sqrt{-1}$.

20. Let F be a field, and let a and b belong to F with $a \neq 0$. If c belongs to some extension of F, prove that $F(c) = F(ac + b)$. (F "absorbs" its own elements.)

21. Let $f(x) \in F[x]$ and let $a \in F$. Show that $f(x)$ and $f(x + a)$ have the same splitting field over F.

22. Recall that two polynomials $f(x)$ and $g(x)$ from $F[x]$ are said to be relatively prime if there is no polynomial of positive degree in $F[x]$ that divides both $f(x)$ and $g(x)$. Show that if $f(x)$ and $g(x)$ are relatively prime in $F[x]$, they are relatively prime in $K[x]$, where K is any extension of F.

23. Determine all of the subfields of $Q(\sqrt{2})$.

24. Let E be an extension of F and let a and b belong to E. Prove that $F(a, b) = F(a)(b) = F(b)(a)$.

25. Write $x^3 + 2x + 1$ as a product of linear polynomials over some field extension of Z_3.

26. Express $x^8 - x$ as a product of irreducibles over Z_2.

27. Prove or disprove that $Q(\sqrt{3})$ and $Q(\sqrt{-3})$ are field-isomorphic.

Leopold Kronecker

But the worst of it is that Kronecker uses his authority to proclaim that all those who up to now have labored to establish the theory of functions are sinners before the Lord ... such a verdict from a man whose eminent talent and distinguished performance in mathematical research I admire as sincerely and with as much pleasure as all his colleagues [is humiliating].

Karl Weierstrass in a letter to Sonja Kowalewski

LEOPOLD KRONECKER, the son of a businessman, was born on December 7, 1823, in Leignitz, Prussia. As a schoolboy, he excelled in Greek, Latin, Hebrew, philosophy, swimming, and gymnastics, and he received special instruction from the great algebraist Kummer. Kronecker entered the University of Berlin in 1841 and completed his Ph.D. dissertation in 1845 on the units in a certain ring.

Kronecker devoted the years 1845–1853 to business affairs, relegating mathematics to a hobby. Thereafter, being well-off financially, he spent most of his time doing research in algebra and number theory. Kronecker was one of the early advocates of the abstract approach to algebra. He innovatively applied rings and fields in his investigations of algebraic numbers, established the Fundamental Theorem of Finite Abelian Groups, and was the first mathematician to master Galois's theory of fields.

Early in his career, Kronecker embarked on a crusade-like attack on the mathematical analysis of Karl Weierstrass and his disciples. Kronecker advocated constructive methods for all proofs and definitions. He believed that all mathematics should be based on relationships among integers. He went so far as to say to Lindermann, who proved that π is transcendental, that irrational numbers do not exist.

313

His most famous remark on the matter is "God made the integers, all the rest is the work of man." And Kronecker believed that all other numbers, being the work of man, were to be avoided.

Henri Poincaré once remarked that Kronecker was able to produce fine work in number theory and algebra only by temporarily forgetting his own philosophy. Although Kronecker's mathematical philosophy gained no supporters among his contemporaries, his beliefs about constructive proofs and definitions were advocated many years later by the so-called intuitionists. Kronecker died on December 29, 1891, at the age of 68.

21

Algebraic Extensions

Banach once told me, "Good mathematicians see analogies between theorems or theories, the very best ones see analogies between analogies."

S. M. Ulam, *Adventures of a Mathematician*

CHARACTERIZATION OF EXTENSIONS

In Chapter 20, we saw that every element in the field $Q(\sqrt{2})$ has the particularly simple form $a + b\sqrt{2}$, where a and b are rational. On the other hand, the elements of $Q(\pi)$ have the more complicated form

$$(a_n\pi^n + a_{n-1}\pi^{n-1} + \cdots + a_1\pi + a_0)/(b_m\pi^m + b_{m-1}\pi^{m-1} + \cdots + b_1\pi + b_0),$$

where the a's and b's are rational. The fields of the first type have a great deal of structure. This structure is the subject of this chapter.

DEFINITION   Types of Extensions

Let E be an extension field of a field F and let $a \in E$. We call a *algebraic over F* if a is the zero of some nonzero polynomial in $F[x]$. If a is not algebraic over F, it is called *transcendental over F*. An extension E of F is called an *algebraic* extension of F if every element of E is algebraic over F. If E is not an algebraic extension of F, it is called a *transcendental* extension of F. An extension of F of the form $F(a)$ is called a *simple* extension of F.

315

Leonhard Euler used the term *transcendental* for numbers that are not algebraic because "they transcended the power of algebraic methods." Although Euler made this distinction in 1744, it wasn't until 1844 that the existence of transcendental numbers over Q was proved by Joseph Liouville. Charles Hermite proved that e is transcendental over Q in 1873, and Lindemann showed that π is transcendental over Q in 1882. To this day, it is not known whether $\pi + e$ is transcendental over Q. With a precise definition of "almost all," it can be shown that almost all real numbers are transcendental over Q.

Theorem 21.1 shows why we make the distinction between elements that are algebraic over a field and elements that are transcendental over a field. Recall that $F(x)$ is the field of quotients of $F[x]$; that is,

$$F(x) = \{f(x)/g(x) \mid f(x), g(x) \in F[x], g(x) \neq 0\}.$$

Theorem 21.1   *Characterization of Extensions*

Let E be an extension field of the field F and let $a \in E$. If a is trancendental over F, then $F(a) \approx F(x)$. If a is algebraic over F, then $F(a) \approx F[x]/\langle p(x) \rangle$, where $p(x)$ is a polynomial in $F[x]$ of minimum degree such that $p(a) = 0$. Moreover, $p(x)$ is irreducible over F.

Proof.   Consider the homomorphism $\phi:F[x] \to F(a)$ given by $f(x) \to f(a)$. If a is transcendental over F, then Ker $\phi = \{0\}$, and so we may extend ϕ to an isomorphism $\bar{\phi}:F(x) \to F(a)$ by defining $\bar{\phi}(f(x)/g(x)) = f(a)/g(a)$.

If a is algebraic over F, then Ker $\phi \neq \{0\}$; and, by Theorem 16.4, there is a polynomial $p(x)$ in $F[x]$ such that Ker $\phi = \langle p(x) \rangle$ and $p(x)$ has minimum degree among all nonzero elements of Ker ϕ. Thus, $p(a) = 0$ and, since $p(x)$ is a polynomial of minimum degree with this property, it is irreducible over F.   ∎

The proof of Theorem 21.1 can readily be adapted to yield the next two results also. The details are left to the reader (see exercise 1).

Theorem 21.2   *Uniqueness Property*

If a is algebraic over a field F, then there is a unique monic irreducible polynomial $p(x)$ in $F[x]$ such that $p(a) = 0$.

The polynomial with the property specified in Theorem 21.2 is called the *minimal polynomial for a over F*.

Theorem 21.3   *Divisibility Property*

Let a be algebraic over F, and let $p(x)$ be the minimal polynomial for a over F. If $f(x) \in F[x]$ and $f(a) = 0$, then $p(x)$ divides $f(x)$ in $F[x]$.

If E is an extension field of F, we may view E as a vector space over F (that is, the elements of E are the vectors and the elements of F are the scalars). We are then able to use such notions as dimension and basis in our discussion.

FINITE EXTENSIONS

DEFINITION Degree of an Extension

Let E be a field extension of a field F. We say that E has a *degree n over F* and write $[E:F] = n$, if E has dimension n as a vector space over F. If $[E:F]$ is finite, E is called a *finite extension* of F; otherwise, we say that E is an *infinite extension* of F.

Figure 21.1 illustrates a convenient method of depicting the degree of a field extension over a field.

Example 1 The field of complex numbers has degree 2 over the reals since $\{1, i\}$ is a basis. The field of complex numbers is an infinite extension of the rationals. ▫

Example 2 If a is algebraic over F and its minimal polynomial over F has degree n, then, by Theorem 20.3, we know that $\{1, a, \ldots, a^{n-1}\}$ is a basis for $F(a)$ over F; and, therefore, $[F(a):F] = n$. In this case, we say that a has *degree n over F*. ▫

Theorem 21.4 *Finite Implies Algebraic*

If E is a finite extension of F, then E is an algebraic extension of F.

Proof. Suppose that $[E:F] = n$ and $a \in E$. Then the set $\{1, a, \ldots, a^n\}$ is linearly dependent over F; that is, there are elements $c_0, c_1, \ldots, c_n$ in F, not all zero, such that

$$c_n a^n + c_{n-1} a^{n-1} + \cdots + c_1 a + c_0 = 0.$$

Clearly, then, a is a zero of the nonzero polynomial

$$f(x) = c_n x^n + c_{n-1} x^{n-1} + \cdots + c_1 x + c_0. \qquad \blacksquare$$

The converse of Theorem 21.4 is not true. For otherwise, the degrees of the elements of every algebraic extension of E over F would be bounded. But $Q(\sqrt{2}, \sqrt[3]{2}, \sqrt[4]{2}, \ldots)$ is an algebraic extension of Q that contains elements of every degree over Q.

The next theorem is the field theory counterpart of Lagrange's Theorem for finite groups. Like all counting theorems, it has far-reaching consequences.

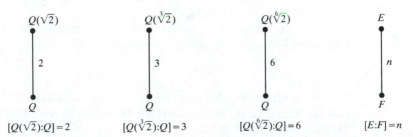

Figure 21.1

Theorem 21.5 $[K:F] = [K:E][E:F]$

> *Let K be a finite extension field of the field E and let E be a finite extension field of the field F. Then K is a finite extension field of F and $[K:F] = [K:E][E:F]$.*

Proof.   Let $X = \{x_1, x_2, \ldots, x_n\}$ be a basis for K over E, and let $Y = \{y_1, y_2, \ldots, y_m\}$ be a basis for E over F. It suffices to prove that

$$YX = \{ y_j x_i \mid 1 \leq j \leq m, \ 1 \leq i \leq n\}$$

is a basis for K over F. To do this, let $a \in K$. Then there are elements $b_1, b_2, \ldots, b_n \in E$ such that

$$a = b_1 x_1 + b_2 x_2 + \cdots + b_n x_n.$$

And, for each $i = 1, \ldots, n$, there are elements $c_{i1}, c_{i2}, \ldots, c_{im} \in F$ such that

$$b_i = c_{i1} y_1 + c_{i2} y_2 + \cdots + c_{im} y_m.$$

Thus,

$$a = \sum_{i=1}^{n} b_i x_i = \sum_{i=1}^{n} \left(\sum_{j=1}^{m} c_{ij} y_j \right) x_i = \sum_{i,j} c_{ij}(y_j x_i).$$

This proves that YX spans K over F.

Now suppose there are elements c_{ij} in F such that

$$0 = \sum_{i,j} c_{ij}(y_j x_i) = \sum_{i} \sum_{j} (c_{ij} y_j) x_i.$$

Then, since each $c_{ij} y_j \in E$ and X is a basis for K over E, we have

$$\sum_{j} c_{ij} y_j = 0$$

for each i. But each $c_{ij} \in F$ and Y is a basis for E over F, so each $c_{ij} = 0$. This proves that the set YX is linearly independent over F.   ∎

Using the fact that for any field extension L of a field J, $[L:J] = n$ if and only if L is isomorphic to J^n as vector spaces (see exercise 28), we may give a concise conceptual proof of Theorem 21.5, as follows. Let $[K:E] = n$ and $[E:F] = m$. Then $K \approx E^n$ and $E \approx F^m$, so that $K \approx E^n \approx (F^m)^n \approx F^{mn}$. Thus, $[K:F] = mn$.

The content of Theorem 21.5 can be pictured as in Figure 21.2. Examples 3, 4, and 5 show how Theorem 21.5 is often utilized.

Example 3   Since $\{1, \sqrt{3}\}$ is a basis for $Q(\sqrt{3}, \sqrt{5})$ over $Q(\sqrt{5})$ and $\{1, \sqrt{5}\}$ is a basis for $Q(\sqrt{5})$ over Q, the proof of Theorem 21.5 shows that $\{1, \sqrt{3}, \sqrt{5}, \sqrt{15}\}$ is a basis for $Q(\sqrt{3}, \sqrt{5})$ over Q. (See Figure 21.3.)   □

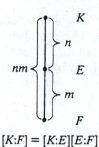

$$[K:F] = [K:E][E:F]$$

Figure 21.2

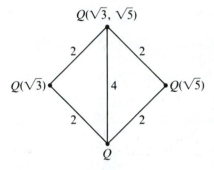

Figure 21.3

Example 4 Consider $Q(\sqrt{2}, \sqrt[3]{2})$. Then $[Q(\sqrt{2}, \sqrt[3]{2}):Q] = 6$. For, clearly, $Q(\sqrt{2}, \sqrt[3]{2}) \subseteq Q(\sqrt[6]{2})$, so that

$$[Q(\sqrt{2}, \sqrt[3]{2}):Q] \leq [Q(\sqrt[6]{2}):Q] = 6.$$

On the other hand, since

$$Q(\sqrt{2}) \subseteq Q(\sqrt{2},\sqrt[3]{2}) \quad \text{and} \quad Q(\sqrt[3]{2}) \subseteq Q(\sqrt{2}, \sqrt[3]{2}),$$

we have $2 = [Q(\sqrt{2}):Q]$ divides $[Q(\sqrt{2}, \sqrt[3]{2}):Q]$ and $3 = [Q(\sqrt[3]{2}):Q]$ divides $[Q(\sqrt{2}, \sqrt[3]{2}):Q]$. It now follows that $Q(\sqrt{2}, \sqrt[3]{2}) = Q(\sqrt[6]{2})$. (See Figure 21.4.) ❑

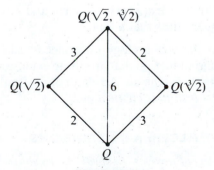

Figure 21.4

Example 5 Consider $Q(\sqrt{3}, \sqrt{5})$. We claim that $Q(\sqrt{3}, \sqrt{5}) = Q(\sqrt{3} + \sqrt{5})$. The inclusion $Q(\sqrt{3} + \sqrt{5}) \subseteq Q(\sqrt{3}, \sqrt{5})$ is clear. Now note that, since $(\sqrt{3} + \sqrt{5})^3 = 18\sqrt{3} + 14\sqrt{5}$ and $-14\sqrt{3} - 14\sqrt{5}$ both belong to $Q(\sqrt{3} + \sqrt{5})$, so does their sum $4\sqrt{3}$. Therefore, $1/4(4\sqrt{3}) = \sqrt{3} \in Q(\sqrt{3} + \sqrt{5})$. Of course, $\sqrt{5} = \sqrt{3} + \sqrt{5} - \sqrt{3} \in Q(\sqrt{3} + \sqrt{5})$ as well. Thus, $Q(\sqrt{3}, \sqrt{5}) \subseteq Q(\sqrt{3} + \sqrt{5})$. □

The preceding two examples show that an extension obtained by adjoining two elements to a field can sometimes be obtained by adjoining a single element to the field. Our next theorem shows that, under certain conditions, this can always be done.

Theorem 21.6 *Primitive Element Theorem (Steinitz, 1910)*
If F is a field of characteristic 0, and a and b are algebraic over F, then there is an element c in F(a, b) such that F(a, b) = F(c).

Proof. Let $p(x)$ and $q(x)$ be the minimum polynomials over F for a and b, respectively. In some extension K of F, let $a = a_1, a_2, \ldots, a_m$ and $b = b_1, b_2, \ldots, b_n$ be the distinct zeros of $p(x)$ and $q(x)$, respectively. Among the infinitely many elements of F, choose an element d not equal to $(a_i - a)/(b - b_j)$ for all $i \geq 1$ and all $j > 1$. In particular, $a_i \neq a + d(b - b_j)$ for $j > 1$.

We shall show that $c = a + db$ has the property that $F(a, b) = F(c)$. Certainly, $F(c) \subseteq F(a, b)$. To verify that $F(a, b) \subseteq F(c)$, it suffices to prove that $b \in F(c)$, for then b, c, and d belong to $F(c)$ and $a = c - bd$. Consider the polynomials $q(x)$ and $r(x) = p(c - dx)$ over $F(c)$. Since both $q(b) = 0$ and $r(b) = p(c - db) = p(a) = 0$, both $q(x)$ and $r(x)$ are divisible by the minimum polynomial $s(x)$ for b over $F(c)$ (see Theorem 21.3). Because $s(x) \in F(c)[x]$, we may complete the proof by proving that $s(x) = x - b$. Since $s(x)$ is a common divisor of $q(x)$ and $r(x)$, the only possible zeros of $s(x)$ in K are the zeros of $q(x)$ that are also zeros of $r(x)$. But $r(b_j) = p(c - db_j) = p(a + db - db_j) = p(a + d(b - b_j))$ and d was chosen so that $a + d(b - b_j) \neq a_i$ for $j > 1$. It follows that b is the only zero of $s(x)$ in $K[x]$ and, therefore, $s(x) = (x - b)^u$. Since $s(x)$ is irreducible and F has characteristic 0, Theorem 20.6 guarantees that $u = 1$. ∎

In the terminology introduced earlier, it follows from Theorem 21.6 and induction that any finite extension of a field of characteristic 0 is a simple extension. An element a with the property that $E = F(a)$ is called a *primitive element* of E.

PROPERTIES OF ALGEBRAIC EXTENSIONS

Theorem 21.7 *Algebraic over Algebraic Is Algebraic*
If K is an algebraic extension of E and E is an algebraic extension of F, then K is an algebraic extension of F.

Proof.   Let $a \in K$. It suffices to show that a belongs to some finite extension of F. Since a is algebraic over E, we know that a is the zero of some irreducible polynomial in $E[x]$, say, $p(x) = b_n x^n + \cdots + b_0$. Now we construct a tower of field extensions of F, as follows:

$$F_0 = F(b_0),$$
$$F_1 = F_0(b_1), \ldots, F_n = F_{n-1}(b_n).$$

In particular,

$$F_n = F(b_0, b_1, \ldots, b_n),$$

so that $p(x) \in F_n[x]$. Thus, $[F_n(a):F_n] = n$; and, because each b_i is algebraic over F, we know that each $[F_{i+1}:F_i]$ is finite. So,

$$[F_n(a):F] = [F_n(a):F_n][F_n:F_{n-1}] \cdots [F_1:F_0][F_0:F]$$

is finite. (See Figure 21.5.)   ■

Corollary *Subfield of Algebraic Elements*
Let E be an extension field of the field F. Then the set of all elements of E that are algebraic over F is a subfield of E.

Proof.   Suppose that $a, b \in E$ are algebraic over F and $b \neq 0$. To show that $a + b$, $a - b$, ab, and a/b are algebraic over F, it suffices to show that $[F(a, b):F]$ is finite, since each of these four elements belongs to $F(a, b)$. But note that

$$[F(a, b):F] = [F(a, b):F(b)][F(b):F].$$

Also, since a is algebraic over F, it is certainly algebraic over $F(b)$. Thus, both $[F(a, b):F(b)]$ and $[F(b):F]$ are finite.   ■

For any extension E of a field F, the subfield of E of the elements that are algebraic over F is called the *algebraic closure of F over E*.

One might wonder if there is such a thing as a maximal algebraic extension of a field F—that is, whether there is an algebraic extension

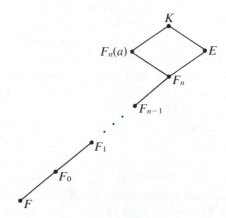

Figure 21.5

E of F that has no proper algebraic extensions. For such an E to exist, it is necessary that every polynomial in $E[x]$ splits in E. Otherwise, it follows from Kronecker's Theorem that E would have a proper algebraic extension. This condition is also sufficient. If every member of $E[x]$ splits in E, and K is an algebraic extension of E, then every member of K is a zero of some element of $E[x]$. But the zeros of elements of $E[x]$ are in E. A field that has no proper algebraic extension is called *algebraically closed*. In 1910, Ernst Steinitz proved that every field has a unique (up to isomorphism) algebraic extension that is algebraically closed. A proof of this result requires a sophisticated set theory background. The reader can find a proof in [1, p. 344–345].

In 1799, Gauss, at the age of 22, proved that $\mathbf{C}$ is algebraically closed. This fact was considered so important at the time that it was called "The Fundamental Theorem of Algebra." Over a 50-year period, Gauss found three additional proofs of the Fundamental Theorem. Today more than 100 proofs exist. In view of the ascendancy of abstract algebra in this century, a more appropriate phrase for Gauss's result would be "The Fundamental Theorem of Classical Algebra."

EXERCISES

The more we do, the more we can do.
William Hazlitt

1. Prove Theorem 21.2 and Theorem 21.3.
2. Prove that $Q(\sqrt{2}, \sqrt[3]{2}, \sqrt[4]{2}, \ldots)$ is an algebraic extension of Q but not a finite extension of Q.
3. Let E be the algebraic closure of F. Show that every polynomial in $F[x]$ splits in E.
4. Let E be an algebraic extension of F. If every polynomial in $F[x]$ splits in E, show that E is algebraically closed.
5. Suppose that F is a field and every irreducible polynomial in $F[x]$ is linear. Show that F is algebraically closed.
6. Suppose that $f(x)$ and $g(x)$ are irreducible over F and deg $f(x)$ and deg $g(x)$ are relatively prime. If a is a zero of $f(x)$ in some extension of F, show that $g(x)$ is irreducible over $F(a)$.
7. Let a and b belong to Q with $b \neq 0$. Show that $Q(\sqrt{a}) = Q(\sqrt{b})$ if and only if there exists some $c \in Q$ such that $a = bc^2$.
8. Find the degree and a basis for $Q(\sqrt{3} + \sqrt{5})$ over $Q(\sqrt{15})$. Find the degree and a basis for $Q(\sqrt{2}, \sqrt[3]{2}, \sqrt[4]{2})$ over Q.
9. Suppose that E is an extension of F of prime degree. Show that, for every a in E, $F(a) = F$ or $F(a) = E$.

10. Let a be a complex number that is algebraic over Q and let $p(x)$ denote the minimum polynomial for a over Q. Show that $\sqrt{a}$ is algebraic over Q and determine the minimum polynomial for $\sqrt{a}$ over Q.

11. Suppose that E is an extension of F and $a, b \in E$. If a is algebraic over F of degree m, and b is algebraic over F of degree n, where m and n are relatively prime, show that $[F(a, b):F] = mn$.

12. Find an example of a field F and elements a and b from some extension field such that $F(a, b) \neq F(a)$, $F(a, b) \neq F(b)$, and $[F(a, b):F] < [F(a):F][F(b):F]$.

13. Let K be a field extension of F and let $a \in K$. Show that $[F(a):F(a^3)] \leq 3$. Find examples to illustrate that $[F(a):F(a^3)]$ can be 1, 2, or 3.

14. Find the minimal polynomial for $\sqrt{-3} + \sqrt{2}$ over Q.

15. Let K be an extension of F. Suppose that E_1 and E_2 are contained in K and are extensions of F. If $[E_1:F]$ and $[E_2:F]$ are both prime, show that $E_1 = E_2$ or $E_1 \cap E_2 = F$.

16. Find the minimal polynomial for $\sqrt[3]{2} + \sqrt[3]{4}$ over Q.

17. Let E be a finite extension of $\mathbf{R}$. Use the fact that $\mathbf{C}$ is algebraically closed to prove that $E = \mathbf{C}$ or $E = \mathbf{R}$.

18. Suppose that $[E:Q] = 2$. Show that there is an integer d such that $E = Q(\sqrt{d})$ and d is not divisible by the square of any prime.

19. Suppose that $p(x) \in F[x]$ and E is a finite extension of F. If $p(x)$ is irreducible over F and deg $p(x)$ and $[E:F]$ are relatively prime, show that $p(x)$ is irreducible over E.

20. Let E be a field extension of F. Show that $[E:F]$ is finite if and only if $E = F(a_1, a_2, \ldots, a_n)$, where $a_1, a_2, \ldots, a_n$ are algebraic over F.

21. If α and β are transcendental over Q, show that either $\alpha\beta$ or $\alpha + \beta$ is also transcendental over Q.

22. Let $f(x) \in F[x]$. If a belongs to some extension of F and $f(a)$ is algebraic over F, prove that a is algebraic over F.

23. Let $f(x) = ax^2 + bx + c \in Q[x]$. Find a primitive element for the splitting field for $f(x)$ over Q.

24. Find the splitting field for $x^4 - x^2 - 2$ over Z_3.

25. Let $f(x) \in F[x]$. If deg $f(x) = 2$ and a is a zero of $f(x)$ in some extension of F, prove that $F(a)$ is the splitting field for $f(x)$ over F.

26. Find the splitting field for $x^3 + x + 1$ over Z_2. Express $x^3 + x + 1$ as a product of linear factors over the splitting field.

27. If F is a field and the multiplicative group of nonzero elements of F is cyclic, prove that F is finite.

28. Prove that, if K is a field extension of F, then $[K:F] = n$ if and only if K is isomorphic to F^n as vector spaces. (See exercise 27 in Chapter 19 for the appropriate definition.)

29. Let a be a complex number that is algebraic over Q and let r be a rational number. Show that a^r is algebraic over Q.

30. Let a be a positive real number and n a positive integer greater than 1. Prove or disprove that $[Q(a^{1/n}):Q] = n$.

31. Let a and b belong to some extension of F and let b be algebraic over F. Prove that $[F(a, b):F(a)] \leq [F(a, b):F]$.

REFERENCE

1. E. A. Walker, *Introduction to Abstract Algebra,* New York: Random House, 1987.

SUGGESTED READINGS

Charles J. Parry and David Perin, "Equivalence of Extension Fields,"*Mathematics Magazine* 50 (1977):36–42.

> Let K be a field, a and b nonzero elements of K, and n a positive integer relatively prime to the characteristic of K. This paper gives necessary and sufficient conditions for $K(a^{1/n}) = K(b^{1/n})$.

R. L. Roth, "On Extensions of Q by Square Roots," *American Mathematical Monthly* 78 (1971):392–393.

> In this paper, it is proved that if $p_1, p_2, \ldots, p_n$ are distinct primes, then $[Q(\sqrt{p_1}, \sqrt{p_2}, \ldots, \sqrt{p_n}):Q] = 2^n$.

Paul B. Yale, "Automorphisms of the Complex Numbers," *Mathematics Magazine* 39 (1966): 135–141.

> This award-winning expository paper is devoted to various results on automorphisms of the complex numbers.

Irving Kaplansky

He got to the top of the heap by being a first-rate doer and expositor of algebra.

Paul R. Halmos, *I Have a Photographic Memory*

IRVING KAPLANSKY was born on March 22, 1917, in Toronto, Canada, a few years after his parents emigrated from Poland. As a young boy, he demonstrated a talent for music and took piano lessons for 11 years. Although his parents thought he would pursue a career in music, Kaplansky knew early on that mathematics was what he wanted to do. (To this day, however, music remains a hobby.) As an undergraduate at the University of Toronto, Kaplansky was a member of the winning team of the first William Lowell Putnam competition, a mathematical contest for United States and Canadian college students. Kaplansky received a B.A. degree from Toronto in 1938 and an M.A. in 1939. In 1939 he entered Harvard University for his doctorate as the first recipient of a Putnam Fellowship. After receiving his Ph.D. from Harvard in 1941, Kaplansky stayed on as Benjamin Peirce instructor until 1944. After one year at Columbia University, he went to the University of Chicago, where he remained until his retirement in 1984. He then became the director of the Mathematical Sciences Research Institute at the University of California, Berkeley.

Kaplansky is the author of numerous books and research papers. His interests are broad, including areas such as ring theory, group theory, field theory, Galois theory, ergodic theory, algebras, metric spaces, number theory, statistics, and probability. In much of his mathematics, there is a recurring theme that might be described as "algebra with an infinite flavor." He is also regarded as a first-rate mathematical stylist.

Among the many honors received by Kaplansky are a Guggenheim Fellowship, election to both the National Academy of Sciences and the American Academy of Arts and Sciences, election to the presidency of the American Mathematical Society, honorary degrees from the University of Waterloo and Queen's University, and the 1989 Steele Prize for cumulative influence from the American Mathematical Society. The Steele Prize citation says, in part, ". . . he has made striking changes in mathematics and has inspired generations of younger mathematicians." Kaplansky also received the University of Chicago Quantrell Prize for excellence in undergraduate teaching.

22

Finite Fields

This theory [of finite fields] is of considerable interest in its own right and it provides a particularly beautiful example of how the general theory of the preceding chapters fits together to provide a rather detailed description of all finite fields.

Richard A. Dean, *Elements of Abstract Algebra*

CLASSIFICATION OF FINITE FIELDS

In this, our final chapter on field theory, we take up one of the most beautiful and important areas of abstract algebra—finite fields. Finite fields were first introduced by Galois in 1830 in his proof of the unsolvability of the general quintic equation. When Cayley invented matrices a few decades later, it was natural to investigate groups of matrices over finite fields. To this day, matrix groups over finite fields are among the most important classes of groups. In the past forty years, there have been important applications of finite fields in computer science, coding theory, information theory, and cryptography. But, besides the many uses of finite fields in pure and applied mathematics, there is yet another good reason for studying them. They are just plain fun!

The most striking fact about finite fields is the restricted nature of their order and structure. We have already seen that every finite field has prime-power order (exercise 41 in Chapter 13). A converse of sorts is also true.

Theorem 22.1   *Classification of Finite Fields*
> *For each prime p and each positive integer n there is, up to isomorphism, a unique finite field of order p^n.*

> ***Proof.***   Consider the splitting field E of $f(x) = x^{p^n} - x$ over Z_p. We will show that $|E| = p^n$. Since $f(x)$ splits in E, we know that $f(x)$ has exactly p^n zeros in E, counting multiplicity. Moreover, by Theorem 20.5, every zero of $f(x)$ has multiplicity 1. Thus, $f(x)$ has p^n distinct zeros in E. On the other hand, the set of zeros of $f(x)$ in E is closed under addition, subtraction, multiplication, and division by nonzero elements (see exercise 27), so that the set of zeros $f(x)$ is itself a field extension of Z_p in which $f(x)$ splits. Thus, the set of zeros of $f(x)$ is E and, therefore, $|E| = p^n$.

> To show that there is a unique field for each prime-power, suppose that K is any field of order p^n. Then K has a subfield isomorphic to Z_p, and, because the nonzero elements of K form a multiplicative group of order $p^n - 1$, every element of K is a zero of $f(x) = x^{p^n} - x$ (see exercise 18). So, K must be a splitting field for $f(x)$ over Z_p. By the corollary to Theorem 20.4, there is only one such field up to isomorphism.   ∎

The existence portion of Theorem 22.1 appeared in the works of Galois and Gauss in the first third of the 19th century. Rigorous proofs were given by Dedekind in 1857, and by Jordan in 1870 in his classic book on group theory. The uniqueness half of the theorem was proved by E. H. Moore in an 1893 paper concerning finite groups. The mathematics historian E. T. Bell once said that this paper by Moore marked the beginning of abstract algebra in America.

Because there is only one field for each prime-power p^n, we may unambiguously denote it by $GF(p^n)$, in honor of Galois, and call it the *Galois field of order p^n*.

STRUCTURE OF FINITE FIELDS

The next theorem tells us the additive and multiplicative structure of a field of order p^n.

Theorem 22.2   *Structure of Finite Fields*
> *As a group under addition, $GF(p^n)$ is isomorphic to*

$$\underbrace{Z_p \oplus Z_p \oplus \cdots \oplus Z_p}_{n\ factors};$$

> *as a group under multiplication, the set of nonzero elements of $GF(p^n)$ is isomorphic to Z_{p^n-1} (and is, therefore, cyclic).*

Proof. Since GF(p^n) has characteristic p, we know that $px = 0$ for all x in GF(p^n). Thus, every nonzero element of GF(p^n) has additive order p. Clearly, then, under addition, GF(p^n) is isomorphic to a direct product of n copies of Z_p.

To see that the multiplicative group GF(p^n)$^\#$ of nonzero elements of GF(p^n) is cyclic, we first note, by exercise 10 in Chapter 11, that it is isomorphic to a direct product of the form $Z_{n_1} \oplus Z_{n_2} \oplus \cdots \oplus Z_{n_k}$, where each n_{i+1} divides n_i. So, for any element $a = (a_1, a_2, \ldots, a_k)$ in this product, we have

$$a^{n_1} = (n_1 a_1, n_1 a_2, \ldots, n_1 a_k) = (0, 0, \ldots, 0).$$

(Remember, the operation is componentwise addition.) Thus, the polynomial $x^{n_1} - 1$ has $p^n - 1$ zeros in GF(p^n). Since the number of zeroes of a polynomial over a field cannot exceed the degree of the polynomial (Corollary 3 of Theorem 16.2), we know that $p^n - 1 \le n_1$. On the other hand, since GF(p^n)$^\#$ has a subgroup isomorphic to Z_{n_1}, we also have $n_1 \le p^n - 1$. It follows then that GF(p^n)$^\#$ is isomorphic to Z_{p^n-1}. ∎

Since $Z_p \oplus Z_p \oplus \cdots \oplus Z_p$ is a vector space over Z_p with $(1, 0, \ldots, 0), (0, 1, 0, \ldots, 0), \ldots, (0, 0, \ldots, 1)$ as a basis, we have the following useful and aesthetically appealing formula.

Corollary 1 $[GF(p^n){:}GF(p)] = n$

Corollary 2   GF(p^n) *Contains an Element of Degree n*
Let a be a generator of the group of nonzero elements of GF(p^n) *under multiplication. Then a is algebraic over* GF(p) *of degree n.*

Proof.   Observe that $[GF(p)(a){:}GF(p)] = [GF(p^n){:}GF(p)] = n$. ∎

Example 1   Let's examine the field GF(16) in detail. Since $x^4 + x + 1$ is irreducible over Z_2, we know that

$$GF(16) \approx \{ax^3 + bx^2 + cx + d + \langle x^4 + x + 1\rangle \mid a, b, c, d \in Z_2\}.$$

Thus, we may think of GF(16) as the set

$$F = \{ax^3 + bx^2 + cx + d \mid a, b, c, d \in Z_2\},$$

where addition is done as in $Z_2[x]$, but multiplication is done modulo $x^4 + x + 1$. For example,

$$(x^3 + x^2 + x + 1)(x^3 + x) = x^3 + x^2,$$

since the remainder upon dividing

$$(x^3 + x^2 + x + 1)(x^3 + x) = x^6 + x^5 + x^2 + x$$

by $x^4 + x + 1$ in Z_2 is $x^3 + x^2$. An easier way to perform the same calculation is to observe that in this context $x^4 + x + 1$ *is* 0, so

$$x^4 = -x - 1 = x + 1.$$
$$x^5 = x^2 + x,$$

and

$$x^6 = x^3 + x^2.$$

Thus,

$$x^6 + x^5 + x^2 + x = (x^3 + x^2) + (x^2 + x) + x^2 + x = x^3 + x^2.$$

Another way to simplify the multiplication process is to make use of the fact that the nonzero elements of GF(16) form a cyclic group of order 15. To take advantage of this, we must first find a generator of this group. Since any element F^* must have multiplicative order that divides 15, all we need do is find an element α in F^* so that $\alpha^3 \neq 1$ and $\alpha^5 \neq 1$. Obviously, x has these properties. So, we may think of GF(16) as the set $\{0, 1, x, x^2, \ldots, x^{14}\}$ where $x^{15} = 1$. This makes multiplication in F trivial, but, unfortunately, it makes addition more difficult. For example, $x^{10} \cdot x^7 = x^{17} = x^2$, but what is $x^{10} + x^7$? So, we face a dilemma. If we write the elements of F^* in the additive form $ax^3 + bx^2 + cx + d$, then addition is easy and multiplication is hard. On the

Table 22.1   Conversion Table for Addition and Multiplication in GF(16)

Multiplicative Form to Additive Form		Additive Form to Multiplicative Form	
1	1	1	1
x	x	x	x
x^2	x^2	$x + 1$	x^4
x^3	x^3	x^2	x^2
x^4	$x + 1$	$x^2 + x$	x^5
x^5	$x^2 + x$	$x^2 + 1$	x^8
x^6	$x^3 + x^2$	$x^2 + x + 1$	x^{10}
x^7	$x^3 + x + 1$	x^3	x^3
x^8	$x^2 + 1$	$x^3 + x^2$	x^6
x^9	$x^3 + x$	$x^3 + x$	x^9
x^{10}	$x^2 + x + 1$	$x^3 + 1$	x^{14}
x^{11}	$x^3 + x^2 + x$	$x^3 + x^2 + x$	x^{11}
x^{12}	$x^3 + x^2 + x + 1$	$x^3 + x^2 + 1$	x^{13}
x^{13}	$x^3 + x^2 + 1$	$x^3 + x + 1$	x^7
x^{14}	$x^3 + 1$	$x^3 + x^2 + x + 1$	x^{12}

other hand, if we write the elements of F^* in the multiplicative form x^i, then multiplication is easy and addition is hard. Can we have the best of both? Yes, we can. All we need to do is use the relation $x^4 = x + 1$ to make a two-way conversion table, as in Table 22.1.

So, we see from Table 22.1 that

$$x^{10} + x^7 = (x^2 + x + 1) + (x^3 + x + 1)$$
$$= x^3 + x^2 = x^6$$

and

$$(x^3 + x^2 + 1)(x^3 + x^2 + x + 1) = x^{13} \cdot x^{12}$$
$$= x^{25} = x^{10} = x^2 + x + 1. \quad \square$$

Don't be misled by the preceding example into believing that the element x is always a generator for the cyclic multiplicative group of nonzero elements. It is not. (See exercise 10.) Although any two irreducible polynomials of the same degree over $Z_p[x]$ yield isomorphic fields, some are better than others for computation purposes.

SUBFIELDS OF A FINITE FIELD

Theorem 22.1 gave us a complete description of all finite fields. The following theorem gives us a complete description of all the subfields of a finite field. Notice the close analogy between this theorem and Theorem 4.3, which describes the subgroups of a finite cyclic group.

Theorem 22.3 Subfields of a Finite Field
For each divisor m of n, $GF(p^n)$ has a unique subfield of order p^m. Moreover, these are the only subfields of $GF(p^n)$.

Proof. To show the existence portion of the theorem, suppose that m divides n. Then, since

$$p^n - 1 = (p^m - 1)(p^{n-m} + p^{n-2m} + \cdots + p^m + 1),$$

we see that $p^m - 1$ divides $p^n - 1$. This implies that $x^{p^m-1} - 1$ divides $x^{p^n-1} - 1$ in $Z_p[x]$. Thus, every zero of $x(x^{p^m-1} - 1)$ is also a zero of $x(x^{p^n-1} - 1)$. But the proof of Theorem 22.1 shows that the set of zeros of $x(x^{p^m-1} - 1)$ is $GF(p^m)$ and the set of zeros of $x(x^{p^n-1} - 1)$ in $GF(p^n)$ is $GF(p^n)$. Hence, $GF(p^m)$ is a subfield of $GF(p^n)$ whenever m divides n. The uniqueness portion of the theorem follows from the observation that if $GF(p^n)$ had two distinct subfields of order p^m, then the polynomial $x^{p^m} - x$ would have more than p^m zeros in $GF(p^n)$. This contradicts Corollary 3 of Theorem 16.2.

Finally, suppose that F is a subfield of $GF(p^n)$. Then F is isomorphic to $GF(p^m)$ for some m and, by Theorem 21.5,

$$
\begin{aligned}
n &= [GF(p^n){:}GF(p)] \\
&= [GF(p^n){:}GF(p^m)][GF(p^m){:}GF(p)] \\
&= [GF(p^n){:}GF(p^m)]m.
\end{aligned}
$$

Thus, m divides n. ∎

Theorems 22.2 and 22.3, together with Theorem 4.3, make the task of finding the subfields of a field a simple exercise in arithmetic.

Example 2 Let F be the field of order 16 given in Example 1. Then there are exactly three subfields of F, and their orders are 2, 4, and 16. Obviously, the subfield of order 2 is $\{0, 1\}$ and the subfield of order 16 is F itself. To find the subfield of order 4, we merely observe that the three nonzero elements of this subfield must be the cyclic subgroup of $F^* = \langle x \rangle$ of order 3. So the subfield of order 4 is

$$\{0, 1, x^5, x^{10}\} = \{0, 1, x^2 + x, x^2 + x + 1\}. \qquad \square$$

Example 3 If F is a field of order $3^6 = 729$ and α is a generator of F^*, then the subfields of F are

1. $GF(3) = \{0\} \cup \langle \alpha^{364} \rangle = \{0, 1, 2\}$
2. $GF(9) = \{0\} \cup \langle \alpha^{91} \rangle$
3. $GF(27) = \{0\} \cup \langle \alpha^{28} \rangle$
4. $GF(729) = \{0\} \cup \langle \alpha \rangle$ $\qquad \square$

The subfield lattice of $GF(2^{24})$ is illustrated in Figure 22.1.

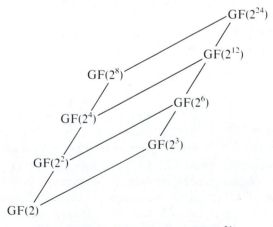

Figure 22.1 Subfield lattice of $GF(2^{24})$.

EXERCISES

We are an intelligent species and the use of our intelligence quite properly gives us pleasure. In this respect the brain is like a muscle. When it is used we feel very good. Understanding is joyous.

Carl Sagan, *Broca's Brain*

1. Find $[GF(729):GF(9)]$ and $[GF(64):GF(8)]$.
2. If m divides n, show that $[GF(p^n):GF(p^m)] = n/m$.
3. Let K be a finite extension field of a finite field F. Show that there is an element a in K such that $K = F(a)$.
4. Let F be as in Example 1. Use the generator $z = x^2 + 1$ for F^* and construct a table that converts polynomials in F^* to powers of z, and vice versa.
5. Let $f(x)$ be a cubic irreducible over Z_2. Prove that the splitting field of $f(x)$ over Z_2 has order 8.
6. Prove that the rings $Z_3[x]/\langle x^2 + x + 2\rangle$ and $Z_3[x]/\langle x^2 + 2x + 2\rangle$ are isomorphic.
7. Prove that the maximum degree of any irreducible factor of $x^8 - x$ over Z_2 is 3.
8. Prove that the maximum degree of any irreducible factor of $x^{p^n} - x$ over Z_p is n.
9. Show that x is a generator of the cyclic group $(Z_3[x]/\langle x^3 + 2x + 1\rangle)^*$.
10. Show that x is not a generator of the cyclic group $(Z_3[x]/\langle x^3 + 2x + 2\rangle)^*$. Find one.
11. Without actually calculating $|x|$, explain why x is a generator of the cyclic group $(Z_2[x]/\langle x^5 + x^3 + 1\rangle)^*$.
12. Suppose that α and β belong to $GF(81)^*$, with $|\alpha| = 5$ and $|\beta| = 16$. Show that $\alpha\beta$ is a generator of $GF(81)^*$.
13. Construct a field of order 27 and carry out the analysis done in Example 1, including the conversion table.
14. Show that any finite subgroup of the multiplicative group of a field is cyclic.
15. Suppose that m and n are positive integers and m divides n. If F is any field, show that $x^m - 1$ divides $x^n - 1$ in $F[x]$.
16. If $g(x)$ is irreducible over $GF(p)$ and $g(x)$ divides $x^{p^n} - x$, prove that deg $g(x)$ divides n.
17. Draw the subfield lattice of $GF(3^{18})$ and of $GF(2^{30})$.
18. Use a purely group-theoretical argument to show that if F is a field of order p^n, then every element of F^* is a zero of $x^{p^n} - x$. (This exercise is referred to in the proof of Theorem 22.1.)

19. How does the subfield lattice of $GF(2^{30})$ compare with the subfield lattice of $GF(3^{30})$?

20. If $p(x)$ is an irreducible polynomial in $Z_p[x]$ with no multiple zeros, show that $p(x)$ divides $x^{p^n} - x$ for some n.

21. Suppose that p is a prime and $p \neq 2$. Let a be a nonsquare in $GF(p)$—that is, a does not have the form b^2 for any b in $GF(p)$. Show that a is a nonsquare in $GF(p^n)$ if n is odd and that a is a square in $GF(p^n)$ if n is even.

22. Show that the *Frobenius mapping* $\phi : GF(p^n) \to GF(p^n)$, given by $a \to a^p$, is an automorphism of order n (that is, ϕ^n is the identity mapping).

23. Show that every element of $GF(p^n)$ can be written in the form a^p for some unique a in $GF(p^n)$.

24. Suppose that F is a field of order 1024 and $F^\# = \langle \alpha \rangle$. List the elements of each subfield of F.

25. Suppose that F is a field of order 125 and $F^\# = \langle \alpha \rangle$. Show that $\alpha^{62} = -1$.

26. Show that no finite field is algebraically closed.

27. Let E be the splitting field of $f(x) = x^{p^n} - x$ over Z_p. Show that the set of zeros of $f(x)$ in E is closed under addition, subtraction, multiplication, and division (by nonzero elements). (This exercise is referred to in the proof of Theorem 22.1.)

28. Suppose that L and K are subfields of $GF(p^n)$. If L has p^s elements and K has p^t elements, how many elements does $L \cap K$ have?

PROGRAMMING EXERCISE

He who labors diligently need never despair; for all things are accomplished by diligence and labor.

Menander of Athens

1. The number of monic irreducible polynomials of degree d over $GF(p)$ is denoted by $I_p(d)$. Use the equation $p^n = \Sigma d I_p(d)$, where the summation is over all divisors d of n, to compute $I_p(d)$. [For example, $3^1 = I_3(1)$; $3^2 = I_3(1) + 2I_3(2)$; $3^3 = I_3(1) + 3I_3(3)$. Thus, $I_3(1) = 3$, $I_3(2) = 3$, and $I_3(3) = 8$.] Make reasonable assumptions on p and d. Run your program for $p = 2, 3, 5, 7$ and $d = 1, 2, 3, 4, 5$.

Suggested Readings

Shalom Feigelstock, "Mersenne Primes and Group Theory," *Mathematics Magazine* 49 (1976): 198–199.

> A group G is called a *hereditary field group* if G and each of its subgroups is the multiplicative group of nonzero elements of a finite field. This paper gives a theorem that describes all hereditary field groups.

Judy L. Smith and J. A. Gallian, "Factoring Finite Factor Rings,"*Mathematics Magazine* 58 (1985): 93–95.

> This paper gives an algorithm for finding the group of units of the ring $F[x]/\langle g(x)^m \rangle$.

L. E. Dickson

One of the books [written by L. E. Dickson] is his major, three-volume *History of the Theory of Numbers* which would be a life's work by itself for a more ordinary man.

A. A. Albert, *Bulletin of the American Mathematical Society*

LEONARD EUGENE DICKSON was born in Independence, Iowa, on January 22, 1874. Dickson was the valedictorian of the 1893 class at the University of Texas. In 1894, he went to the University of Chicago and studied under E. H. Moore. Two years later he received a Ph.D., the first to be awarded in mathematics at Chicago. After spending a few years at the University of California and the University of Texas, he was appointed to the faculty at Chicago and remained there until his retirement in 1939.

Dickson was one of the most prolific mathematicians of this century, writing 267 research papers and 18 books. His three-volume *History of the Theory of Numbers* took nine years to write. His principal interests were matrix groups, finite fields, algebra, and number theory. He published the first extensive exposition of the theory of finite fields.

Although Dickson supervised the Ph.D. dissertations of 64 students, he was not a good teacher in the conventional sense. Rather, he inspired his students to emulate him as a research mathematician.

Dickson had a disdainful attitude toward applicable mathematics; he would often say, "Thank God that number theory is unsullied by any applications." He also had a sense of humor. Dickson would often mention his honeymoon: "It was a great success," he said, "except that I only got two research papers written."

Dickson received many honors in his career. He was the first to be awarded the prize from the American Association for the Advancement of Science for the most notable contribution to the advancement of science, and the first to receive the Cole Prize in algebra from the American Mathematical Society; he was president of the American Mathematical Society, and was given honorary degrees by Harvard and Princeton. The University of Chicago has research instructorships named after him. Dickson died on January 17, 1954.

23
Geometric Constructions

Failure properly to understand the theoretical character of the question of geometrical construction and stubbornness in refusing to take cognizance of well-established scientific facts are responsible for the persistence of an unending line of angle-trisectors and circle-squarers. Those among them who are able to understand elementary mathematics might profit by studying this chapter.

Richard Courant and Herbert Robbins, *What Is Mathematics?*

HISTORICAL DISCUSSION OF GEOMETRIC CONSTRUCTIONS

The ancient Greeks were fond of geometric constructions. They were especially interested in constructions that could be achieved using only a straightedge without markings and a compass. They knew, for example, that any angle can be bisected, and they knew how to construct an equilateral triangle, a square, a regular pentagon, and a regular hexagon. But they did not know how to trisect every angle or how to construct a regular seven-sided polygon (heptagon). Another problem that they attempted was the duplication of the cube—that is, given any cube, they tried to construct a new cube having twice the volume of the given one by means of a straightedge and compass. Legend has it that the ancient Athenians were told by the oracle at Delos that a plague would end if they constructed a new altar to Apollo in the shape of a cube with double the volume of the old altar, which was also a cube. Besides "doubling the cube," the Greeks also attempted to "square the circle"—to construct a square with area equal to that of a given circle. They knew how to solve all these problems using other means, such as a compass and a straightedge with two marks, or a straightedge and a spiral, but they could not achieve any of the constructions with a compass and straightedge alone. These problems vexed mathematicians for over 2000 years.

The resolution of these perplexities was made possible when they were transferred from questions of geometry to questions of algebra in the 19th century. With the introduction of coordinate geometry, it was obvious that the existence of straightedge constructions corresponds to solving linear equations, whereas the use of a compass corresponds to solving quadratic equations. Thus, a straightedge and compass construction is possible if and only if linear and quadratic equations can be solved by a series of additions, subtractions, multiplications, divisions, and extractions of square roots.

The first of the famous problems of antiquity to be solved was that of the construction of regular polygons. It was known since Euclid that regular polygons with a number of sides of the form 2^k, $2^k \cdot 3$, $2^k \cdot 5$, and $2^k \cdot 3 \cdot 5$ could be constructed, and it was believed that no others were possible. In 1796, while still a teenager, Gauss proved that the 17-sided regular polygon is constructible. In 1801 Gauss asserted that a regular polygon of n sides is constructible if and only if n has the form $2^k p_1 p_2 \cdots p_t$, where the p's are distinct primes of the form $2^{2^s} + 1$ and $k \geq 0$.

Thus, regular polygons with 3, 4, 5, 6, 8, 10, 12, 15, 16, 17, and 20 sides are possible to construct, whereas those with 7, 9, 11, 13, 14, 18, and 19 sides are not. How these constructions can be effected is another matter. One person spent 10 years trying to determine a way to construct the 65,537-sided polygon.

Gauss's result of the constructibility of regular n-gons eliminated another of the famous unsolved problems because the ability to trisect a 60° angle enables one to construct a regular 9-gon. Thus, there is no method for trisecting a 60° angle with a straightedge and compass. In 1837 Wantzel proved that it was not possible to double the cube. The last of the four problems, the squaring of the circle, resisted attempts until 1882, when Ferdinand Lindemann proved that π is transcendental since, as we will show, all constructible numbers are algebraic.

CONSTRUCTIBLE NUMBERS

With the field theory we now have, it is an easy matter to solve the following problem. Given an unmarked straightedge, a compass, and a unit length, what other lengths can be constructed? To begin, we call a real number α *constructible* if, by means of a straightedge, a compass, and a line segment of length 1, we can construct a line segment of length $|\alpha|$ in a finite number of steps. It follows from plane geometry that if α and β ($\beta \neq 0$) are constructible numbers, then so are $\alpha + \beta$, $\alpha - \beta$, $\alpha \cdot \beta$, and α/β. (See the exercises for hints.) Thus, the set of constructible numbers contains Q and is a subfield of the real numbers. What we desire is an algebraic characterization of this field. To derive such a characterization, let F be any subfield of the reals. Call the subset $F \times F$ of the real plane the *plane of F*, any line joining two points in the plane of F a *line in F*, and any circle whose center is in the plane of F and whose radius is in F a *circle in F*. Then a line in F has an equation of the form

$$ax + by + c = 0, \qquad \text{where } a, b, c \in F,$$

and a circle in F has an equation of the form

$$x^2 + y^2 + ax + by + c = 0, \qquad \text{where } a, b, c \in F.$$

In particular, note that to find the point of intersection of a pair of lines in F or the points of intersection of a line in F and a circle in F, one need only solve a linear or quadratic equation in F. We now come to the crucial question. Starting with points in the plane of some field F, which points in the real plane can be obtained with a straightedge and compass? Well, there are only three ways to construct points, starting with points in the plane of F.

   1. Intersect two lines in F.
   2. Intersect a circle in F and a line in F.
   3. Intersect two circles in F.

In case 1, we do not obtain any new points, because two lines in F intersect in a point in the plane of F. In case 2, the point of intersection is either the solution to a linear equation in F or a quadratic equation in F. So, the point lies in the plane of F or in the plane of $F(\sqrt{\alpha})$, where $\alpha \in F$ and α is positive. Case 3 can be reduced to case 2 by choosing one of the circles and the line joining the two points of intersection of the circles. [We leave it as an exercise to prove that this line is in F or in $F(\sqrt{\alpha})$, where $\alpha \in F$ and α is positive.]

   It follows then that the only points in the real plane, which can be constructed from the plane of a field F, are those whose coordinates lie in fields of the form $F(\sqrt{\alpha})$, where $\alpha \in F$ and α is positive. Of course, we can start over with $F_1 = F(\sqrt{\alpha})$ and construct points whose coordinates lie in fields of the form $F_2 = F_1(\sqrt{\beta})$, where $\beta \in F_1$ and β is positive. Continuing in this fashion, we see that a real number c is constructible if and only if there is a series of fields $Q = F_1 \subseteq F_2 \subseteq \cdots \subseteq F_n \subseteq \mathbf{R}$ such that $F_{i+1} = F_i(\sqrt{\alpha_i})$, where $\alpha_i \in F_i$ and $c \in F_n$. Since $[F_{i+1}:F_i] = 1$ or 2, we see by Theorem 21.5 that if c is constructible, then $[Q(c):Q] = 2^k$ for some nonnegative integer k.

   We now dispatch the problems that plagued the Greeks. Consider doubling the cube of volume 1. The enlarged cube would have an

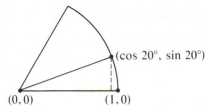

(cos 20°, sin 20°)

(0,0) (1,0)

Figure 23.1

edge of length $\sqrt[3]{2}$. But $[Q(\sqrt[3]{2}):Q] = 3$, so such a cube cannot be constructed.

Next consider the possibility of trisecting a 60° angle. If it were possible to trisect an angle of 60°, then cos 20° would be constructible. (See Figure 23.1.) In particular, $[Q(\cos 20°):Q] = 2^k$ for some k. Now, using the trigonometric identity $\cos 3\theta = 4 \cos^3 \theta - 3 \cos \theta$, with $\theta = 20°$, we see that $1/2 = 4 \cos^3 20° - 3 \cos 20°$, so that cos 20° is a root of $8x^3 - 6x - 1$. But, since $8x^3 - 6x - 1$ is irreducible over Q, we must also have $[Q(\cos 20°):Q] = 3$. This contradiction shows that trisecting a 60° angle is impossible.

The remaining problems are relegated to the exercises.

ANGLE-TRISECTORS AND CIRCLE-SQUARERS

Down through the centuries, hundreds of people have claimed to have achieved one or more of the impossible constructions. In 1775, the Paris Academy, so overwhelmed with these claims, passed a resolution no longer to examine them or to examine machines purported to exhibit perpetual motion. Although it has been more than 100 years since the last of the constructions has been shown to be impossible, there continues to be a steady parade of people who claim to have done one or more of them. Most of these people have heard that this is impossible but refused to believe it. One person insisted he could trisect any angle with a straightedge alone [2, p. 158]. Another found his trisection in 1973 after 12,000 hours of work [2, p. 80]. One got his from God [2, p. 73]. In 1971, a person with a Ph.D. in mathematics asserted that he had a valid trisection method [2, p. 127]. Many people have claimed the hat trick: trisecting the angle, doubling the cube, and squaring the circle. Two men who did this in 1961 succeeded in having their accomplishment noted in the *Congressional Record* [2, p. 110]. Occasionally, newspapers and magazines have run stories about "doing the impossible," often giving the impression that the construction may be valid. Many angle-trisectors and circle-squarers have had their work published at their own expense and distributed to colleges and universities. One had his printed in four languages! There are two delightful books written by mathematicians about their encounters with these people. The books are full of wit, charm, and humor ([1] and [2]).

EXERCISES

> Only prove to me that it is impossible, and I will set about it this very evening.
>
> **Spoken by a member of the audience after De Morgan**
> **gave a lecture on the impossibility of squaring the circle.**

1. If a and b are constructible numbers, give a geometric proof that $a + b$ and $a - b$ are constructible.

2. If a and b are constructible, give a geometric proof that ab is constructible. (*Hint:* Consider the following figure. Notice that all segments in the figure can be made with a straightedge and compass.)

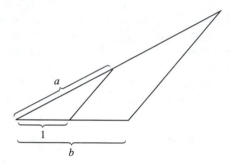

3. If a and b ($b \neq 0$) are constructible numbers, give a geometric proof that a/b is constructible. (*Hint:* Consider the following figure.)

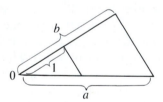

4. Prove that if c is a constructible number, then so is $\sqrt{|c|}$. (*Hint:* Consider the following semicircle with diameter $1 + |c|$.)

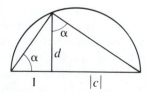

5. Prove that $\sin \theta$ is constructible if and only if $\cos \theta$ is constructible.

6. Prove that an angle θ is constructible if and only if $\sin \theta$ is constructible.

7. Prove that cos 2θ is constructible if and only if cos θ is constructible.
8. Prove that 30° is a constructible angle.
9. Prove that a 45° angle can be trisected with a straightedge and compass.
10. Prove that a 40° angle is not constructible.
11. Show that the point of intersection of two lines in the plane of a field F lie in the plane of F.
12. Show that the points of intersection of a circle in the plane of a field F and a line in the plane of F are points in the plane of F or in the plane of $F(\sqrt{\alpha})$, where $\alpha \in F$ and α is positive. Give an example of a circle and a line in the plane of Q whose points of intersection are not in the plane of Q.
13. Prove that $8x^3 - 6x - 1$ is irreducible over Q.
14. Many of the proposed methods of trisection are equivalent to the assertion that $\sin(\theta/3) = \sin \theta/(2 + \cos \theta)$ is an identity. Show that it is not.
15. Many of the proposed trisections are equivalent to the assertion that $\tan(\theta/3) = 2 \sin(\theta/2)/(2 + \cos(\theta/2))$ is an identity. Show that it is not.
16. Use the fact that $8 \cos^3(2\pi/7) + 4 \cos^2(2\pi/7) - 4 \cos(2\pi/7) - 1 = 0$ to prove that a regular seven-sided polygon is not constructible with a straightedge and a compass.
17. Show that a regular 9-gon cannot be constructed with a straightedge and a compass.
18. (Squaring the Circle) Show that it is impossible to construct with a straightedge and a compass, a square whose area equals that of a circle of radius 1. You may use the fact that π is transcendental over Q.
19. Show that if a regular n-gon is constructible then so is a regular $2n$-gon.
20. Use the fact that $4 \cos^2(2\pi/5) + 2 \cos(2\pi/5) - 1 = 0$ to prove that a regular pentagon is constructible.
21. Can the cube be "tripled"?
22. Can the cube be "quadrupled"?
23. Can the circle be "cubed"?
24. If a, b, and c are constructible, show that the real roots of $ax^2 + bx + c$ are constructible.

REFERENCES

1. Augustus De Morgan, *A Budget of Paradoxes,* 2nd ed., Salem, N.H.: Ayer, 1915.
2. Underwood Dudley, *A Budget of Trisections,* New York: Springer-Verlag, 1987.

Suggested Reading

Underwood Dudley, *A Budget of Trisections,* New York: Springer-Verlag, 1987.
This highly entertaining book includes detailed information about the person-alities of trisectors and their constructions. There is also a chapter on methods for trisecting an angle with tools other than a straightedge and a compass. Some of these will give you a chuckle: one uses a tomahawk, another a watch. According to Dudley, "No one has yet shown how to accomplish the trisection with a digital watch." No doubt there are people working on this right now.

SUPPLEMENTARY EXERCISES FOR CHAPTERS 19–23

Difficulties strengthen the mind, as labor does the body.

Seneca

1. Show that $x^{50} - 1$ has no multiple zeros in any extension of Z_3.
2. Suppose that $p(x)$ is a quadratic polynomial with rational coefficients and is irreducible over Q. Show that $p(x)$ has two zeros in $Q[x]/\langle p(x)\rangle$.
3. Let F be a finite field of order q and let a be a nonzero element in F. If n divides $q - 1$, prove that the equation $x^n = a$ has either no solutions in F or n distinct solutions in F.
4. Without using the Primitive Element Theorem, prove that if $[K{:}F]$ is prime, then K has a primitive element.
5. Let a be a complex zero of $x^2 + x + 1$. Express $(5a^2 + 2)/a$ in the form $c + ba$, where c and b are rational.
6. Describe the elements of the extension $Q(\sqrt[4]{2})$ over the field $Q(\sqrt{2})$.
7. If $[F(a){:}F] = 5$, find $[F(a^3){:}F]$.
8. If $p(x) \in F[x]$ and $\deg p(x) = n$, show that the splitting field for $p(x)$ over F has degree at most $n!$.
9. Let a be a nonzero algebraic element over F of degree n. Show that a^{-1} is also algebraic over F of degree n.
10. Prove that $\pi^2 - 1$ is algebraic over $Q(\pi^3)$.
11. If ab is algebraic over F, prove that a is algebraic over $F(b)$.
12. Let E be an algebraic extension of a field F. If R is a ring and $E \supseteq R \supseteq F$, show that R must be a field.
13. If a is transcendental over F, show that every element of $F(a)$ that is not in F is transcendental over F.
14. What is the order of the splitting field of $x^5 + x^4 + 1 = (x^2 + x + 1)(x^3 + x + 1)$ over Z_2?
15. Show that a finite extension of a finite field is a simple extension.

PART 5

SPECIAL
TOPICS

24

Sylow Theorems

Generally these three results are implied by the expression "Sylow's Theorem." All of them are of fundamental importance. In fact, if the theorems of group theory were arranged in order of their importance Sylow's Theorem might reasonably occupy the second place—coming next to Lagrange's Theorem in such an arrangement.

G. A. Miller, *Theory and Application of Finite Groups*

CONJUGACY CLASSES

In this chapter, we derive several important arithmetic relationships between a group and certain of its subgroups. Recall from Chapter 8 that Lagrange's Theorem was proved by showing that cosets of a subgroup partition the group. Another fruitful method of partitioning the elements of a group is by way of conjugacy classes.

DEFINITION Conjugacy Class of a

Let a and b be elements of a group G. We say that a and b are *conjugate* in G (and call b a *conjugate* of a) if $xax^{-1} = b$ for some x in G. The *conjugacy class of a* is the set $\text{cl}(a) = \{xax^{-1} \mid x \in G\}$.

We leave it to the reader (exercise 1) to prove that conjugacy is an equivalence relation on G, and that the conjugacy class of a is the equivalence class of a under conjugacy. Thus, we may partition any group

349

into disjoint conjugacy classes. Let's look at one example. In D_4 we have

$$\text{cl}(H) = \{R_0 H R_0^{-1}, R_{90} H R_{90}^{-1}, R_{180} H R_{180}^{-1}, R_{270} H R_{270}^{-1},$$
$$HHH^{-1}, VHV^{-1}, DHD^{-1}, D'HD'^{-1}\} = \{H, V\}.$$

Similarly, one may verify that

$$\text{cl}(R_0) = \{R_0\},$$
$$\text{cl}(R_{90}) = \{R_{90}, R_{270}\} = \text{cl}(R_{270}),$$
$$\text{cl}(R_{180}) = \{R_{180}\},$$
$$\text{cl}(V) = \{V, H\} = \text{cl}(H),$$
$$\text{cl}(D) = \{D, D'\} = \text{cl}(D').$$

Theorem 24.1 gives an arithmetic relationship between the size of the conjugacy class of a and the size of the centralizer of a.

Theorem 24.1   *The Number of Conjugates of a*
Let G be a group and let a be an element of G. Then, $|\text{cl}(a)| = |G:C(a)|$.

Proof.   Consider the function T that sends the coset $xC(a)$ to the conjugate xax^{-1} of a. A routine calculation shows that T is well defined, is one-to-one, and maps the set of left cosets onto the conjugacy class of a (exercise). Thus, the number of conjugates of a is the index of the centralizer of a. ∎

THE CLASS EQUATION

Since the conjugacy classes partition a group, the following important counting principle is a corollary to Theorem 24.1.

Corollary   *The Class Equation*
For any finite group G,

$$|G| = \Sigma |G:C(a)|,$$

where the sum runs over one element a from each conjugacy class of G.

In finite group theory, counting principles such as this corollary are powerful tools.* Theorem 24.2 is the single most important fact about finite groups of prime-power order (a group of order p^n, where p is a prime, is called a *p-group*).

Theorem 24.2   *p-Groups Have Nontrivial Centers*
Let G be a finite group whose order is a power of a prime p. Then Z(G) has more than one element.

———
*"Never underestimate a theorem that counts something." John Fraleigh, *A First Course in Abstract Algebra.*

Proof.   First observe that cl(a) = {a} if and only if $a \in Z(G)$. Thus, by culling out these elements, we may write the class equation in the form

$$|G| = |Z(G)| + \Sigma|G:C(a)|,$$

where the sum runs over representatives of all conjugacy classes with more than one element (this set may be empty). But $|G:C(a)| = |G|/|C(a)|$, so each term in $\Sigma|G:C(a)|$ has the form p^k with $k \geq 1$. Hence,

$$|G| - \Sigma|G:C(a)| = |Z(G)|,$$

where each term on the left is divisible by p. It follows, then, that p also divides $|Z(G)|$ and, hence, $|Z(G)| \neq 1$.

Corollary   *Groups of Order p^2 Are Abelian*
If $|G| = p^2$, where p is prime, then G is Abelian.

Proof.   By Theorem 24.2 and Lagrange's Theorem, $|Z(G)| = p$ or p^2. If $|Z(G)| = p^2$, then $G = Z(G)$ and G is Abelian. If $|Z(G)| = p$, then $|G/Z(G)| = p$, so that $G/Z(G)$ is cyclic. But, then, by Theorem 9.3, G is Abelian.   ∎

THE PROBABILITY THAT TWO ELEMENTS COMMUTE

Before proceeding to the main goal of this chapter, we pause for an interesting application of Theorem 24.1 and the class equation. (Our discussion is based on [1] and [2].) Suppose we select two elements at random (with replacement) from a finite group. What is the probability that these two elements commute? Well, suppose that G is a finite group of order n. Then the probability Pr(G) that two elements selected at random from G commute is $|K|/n^2$, where $K = \{(x, y) \in G \oplus G \mid xy = yx\}$. Now notice that for each $x \in G$ we have $(x, y) \in K$ if and only if $y \in C(x)$. Thus,

$$|K| = \sum_{x \in G} |C(x)|.$$

Also, it follows from Theorem 24.1 that if x and y are in the same conjugacy class, then $|C(x)| = |C(y)|$. If, for example, cl(a) = {$a_1, a_2, \ldots, a_t$}, then

$$|C(a_1)| + |C(a_2)| + \cdots + |C(a_t)| = t|C(a)|$$
$$= |G:C(a)| \, |C(a)| = |G| = n.$$

So, by choosing one representative from each conjugacy class, say, $x_1, x_2, \ldots, x_m$, we have

$$|K| = \sum_{i \in G} |C(x)| = \sum_{i=1}^{m} |G:C(x_i)| \, |C(x_i)| = m \cdot n.$$

Thus, the answer to our question is m/n, where m is the number of conjugacy classes in G and n is the number of elements of G.

Obviously, when G is non-Abelian, $\Pr(G)$ is less than 1. But how much less than 1? Clearly, the more conjugacy classes there are, the larger $\Pr(G)$ is. Consequently, $\Pr(G)$ is large when the sizes of the conjugacy classes are small. Noting that $|\text{cl}(a)| = 1$ if and only if $a \in Z(G)$, we obtain the maximum number of conjugacy classes when $|Z(G)|$ is as large as possible and all other conjugacy classes have exactly two elements in each. Since G is non-Abelian, it follows from Theorem 9.3 that $|G/Z(G)| \geq 4$ and, therefore, $|Z(G| \leq |G|/4$. Thus, in the extreme case, we would have $|Z(G)| = |G|/4$, and the remaining $3/4|G|$ elements would be distributed in conjugacy classes with two elements each. So, in a non-Abelian group, the number of conjugacy classes is no more than $|G|/4 + 1/2 \cdot 3/4|G|$, and $\Pr(G)$ is less than or equal to 5/8. The dihedral group D_4 is an example of a group that has probability equal to 5/8.

$$\frac{1}{4}\frac{|G|}{} + \frac{1}{8}|G|$$

THE SYLOW THEOREMS

Now to Sylow's Theorems. Recall that the converse of Lagrange's Theorem is false; that is, if G is a group of order m and n divides m, G need *not* have a subgroup of order n. Our next theorem is a partial converse of Lagrange's Theorem. It, as well as Theorem 24.2, was first proved by the Norwegian mathematician Ludwig Sylow (1832–1918). Sylow's Theorem and Lagrange's Theorem are the two most important results in finite group theory. The first gives a sufficient condition for the existence of subgroups, and the second gives a necessary condition.

Theorem 24.3 *Existence of Subgroups of Prime-Power Order (Sylow's First Theorem, 1872)*
Let G be a finite group and let p be a prime. If p^k divides $|G|$, then G has at least one subgroup of order p^k.

Proof. We proceed by induction of $|G|$. If $|G| = 1$, Theorem 24.3 is trivially true. Now assume that the statement is true for all groups of order less than $|G|$. If G has a proper subgroup H such that p^k divides $|H|$, then, by our inductive assumption, H has a subgroup of order p^k and we are done. Thus, we may henceforth assume that p^k does not divide the order of any proper subgroup of G. Next, consider the class equation for G in the form

$$|G| = |Z(G| + \Sigma|G{:}C(a)|,$$

where we sum over a representative of each conjugacy class $\text{cl}(a)$, where $a \notin Z(G)$. Since p^k divides $|G| = |G{:}C(a)| \, |C(a)|$ and p^k does not divide $|C(a)|$, we know that p must divide $|G{:}C(a)|$ for all $a \notin Z(G)$.

It then follows from the class equation that p divides $|Z(G)|$. The Fundamental Theorem of Finite Abelian Groups (or Theorem 9.5) then guarantees that $Z(G)$ contains an element of order p, say, x. Since x is in the center of G, $\langle x \rangle$ is a normal subgroup of G, and we may form the factor group $G/\langle x \rangle$. Now observe that p^{k-1} divides $|G/\langle x \rangle|$. Thus, by the induction hypothesis, $G/\langle x \rangle$ has a subgroup of order p^{k-1} and, by exercise 43 in Chapter 10, this subgroup has the form $H/\langle x \rangle$, where H is a subgroup of G. Finally, note that $|H/\langle x \rangle| = p^{k-1}$ and $|\langle x \rangle| = p$ imply that $|H| = p^k$, and this completes the proof. ∎

Let's be sure we understand exactly what Sylow's First Theorem means. Say we have a group G of order $2^3 \cdot 3^2 \cdot 5^4 \cdot 7$. Then Sylow's First Theorem says that G must have at least one subgroup of each of the following orders: 2, 4, 8, 3, 9, 5, 25, 125, 625, and 7. On the other hand, Sylow's First Theorem tells us nothing about the possible existence of subgroups of orders 6, 10, 15, 30, or any other divisor of $|G|$ that has two or more distinct prime factors. Because certain subgroups guaranteed by Sylow's First Theorem play a central role in the theory of finite groups, they are given a special name.

DEFINITION Sylow p-Subgroup
Let G be a finite group and let p be a prime divisor $|G|$. If p^k divides $|G|$ and p^{k+1} does not divide $|G|$, then any subgroup of G of order p^k is called a *Sylow p-subgroup of G.*

So, returning to our group G of order $2^3 \cdot 3^2 \cdot 5^4 \cdot 7$, we call any subgroup of order 8 a Sylow 2-subgroup of G, any subgroup of order 625 a Sylow 5-subgroup of G, and so on. Notice that a Sylow p-subgroup of G is a subgroup whose order is the largest power of p consistent with Lagrange's Theorem.

Since any subgroup of order p is cyclic, we have the following corollary, first proved by Cauchy in 1845. His proof ran nine pages!

Corollary *Cauchy's Theorem*
Let G be a finite group and p a prime that divides the order of G. Then G has an element of order p.

Sylow's First Theorem is so fundamental to finite group theory that many different proofs of it have been published over the years [our proof is essentially the one given by Georg Frobenius (1849–1917) in 1895]. Likewise, there are scores of generalizations of Sylow's Theorem.

Observe that the corollary of the Fundamental Theorem of Finite Abelian Groups and Sylow's First Theorem show that the converse of Lagrange's Theorem is true for all finite Abelian groups and all finite groups of prime-power order.

There are two more Sylow theorems that are extremely valuable tools in finite group theory. But first we introduce a new term.

DEFINITION   Conjugate Subgroups

Let H and K be subgroups of a group G. We say that H and K are *conjugate* in G if there is an element g in G such that $H = gKg^{-1}$.

Recall from Chapter 8 that if G is a finite group of permutations on a set S and $i \in S$, then $\text{orb}_G(i) = \{\phi(i) \mid \phi \in G\}$ and $|\text{orb}_G(i)|$ divides $|G|$.

Theorem 24.4   *Sylow's Second Theorem*

If H is a subgroup of a finite group G and $|H|$ is a power of a prime p, then H is contained in some Sylow p-subgroup of G.

Proof.   Let K be a Sylow p-subgroup of G and let $C = \{K = K_1, K_2, \ldots, K_n\}$ be the set of all conjugates of K in G. Since conjugation is an automorphism, each element of C is a Sylow p-subgroup of G. Let S_C denote the group of all permutations of C. For each $g \in G$, define $\phi_g : C \to C$ by $\phi_g(K_i) = gK_ig^{-1}$. It is an easy exercise to show that each $\phi_g \in S_C$.

Now define a mapping $T : G \to S_C$ by $T(g) = \phi_g$. Since $\phi_{gh}(K_i) = (gh)K_i(gh)^{-1} = g(hK_ih^{-1})g^{-1} = g\phi_h(K_i)g^{-1} = \phi_g(\phi_h(K_i)) = (\phi_g\phi_h)(K_i)$, we have $\phi_{gh} = \phi_g\phi_h$ and, therefore, T is a homomorphism from G to S_C.

Next consider $T(H)$, the image of H under T. Since $|H|$ is a power of p, so is $|T(H)|$ (see property 9 of Theorem 10.1). Thus, by the Orbit-Stabilizer Theorem (Theorem 8.2), for each i, $|\text{orb}_{T(H)}(K_i)|$ divides $|T(H)|$ so that $|\text{orb}_{T(H)}(K_i)|$ is a power of p. Now we ask: under what condition does $|\text{orb}_{T(H)}(K_i)| = 1$? Well, $|\text{orb}_{T(H)}(K_i)| = 1$ means that $\phi_g(K_i) = gK_ig^{-1} = K_i$ for all $g \in H$; that is, $|\text{orb}_{T(H)}(K_i)| = 1$ if and only if $H \leq N(K_i)$. But the only elements of $N(K_i)$ that have orders that are powers of p are those of K_i (see exercise 9). Thus, $|\text{orb}_{T(H)}(K_i)| = 1$ if and only if $H \leq K_i$.

So, to complete the proof, all we need to do is show that for some i, $|\text{orb}_{T(H)}(K_i)| = 1$. Analogous to Theorem 24.1, we have $|C| = |G:N(K)|$ (see exercise 38). And since $|G:K| = |G:N(K)| \, |N(K):K|$ is not divisible by p, neither is $|C|$. Because the orbits partition C, $|C|$ is the sum of powers of p. If no orbit has size 1, then p divides each summand and, therefore, p divides $|C|$, which is a contradiction. Thus, there is an orbit of size 1, and the proof is complete. ∎

Theorem 24.5   *Sylow's Third Theorem*

The number of Sylow p-subgroups of G is equal to 1 modulo p and divides $|G|$. Furthermore, any two Sylow p-subgroups of G are conjugate.

Proof.   Let K be any Sylow p-subgroup of G and let $C = \{K = K_1, K_2, \ldots, K_n\}$ be the set of all conjugates of K in G. We first prove that $n = 1 \bmod p$.

Let S_C and T be as in the proof of Theorem 24.4. This time we consider $T(K)$, the image of K under T. As before, we have $|\text{orb}_{T(K)}(K_i)|$ is a power of p for each i and $|\text{orb}_{T(K)}(K_i)| = 1$ if and only if $K \le K_i$. Thus, $|\text{orb}_{T(K)}(K_1)| = 1$ and $|\text{orb}_{T(K)}(K_i)|$ is a power of p greater than 1 for all $i \ne 1$. Since the orbits partition C, it follows that $n = |C| = 1 \bmod p$.

Next we show that every Sylow p-subgroup of G belongs to C. To do this, suppose that H is a Sylow p-subgroup of G that is not in C. Let S_C and T be as in the proof of Theorem 24.4 and this time consider $T(H)$. As in the previous paragraph, $|C|$ is the sum of the orbits' sizes under the action of $T(H)$. However, no orbit has size 1, since H is not in C. Thus, $|C|$ is a sum of terms each divisible by p so that $n = |C| = 0 \bmod p$. This contradiction proves that H belongs to C, and that n is the number of Sylow p-subgroups of G.

Finally, that n divides $|G|$ follows directly from the fact that $n = |G:N(K)|$ (see exercise 20). ∎

It is convenient to let n_p denote the number of Sylow p-subgroups of a group. Observe that the first portion of Sylow's Third Theorem is a counting principle.* As an important consequence of Sylow's Third Theorem, we have the following corollary.

Corollary   *A Unique Sylow p-Subgroup Is Normal*
A Sylow p-subgroup of a finite group G is a normal subgroup of G if and only if it is the only Sylow p-subgroup of G.

We illustrate Sylow's Third Theorem with two examples.

Example 1   Consider the Sylow 2-subgroups of S_3. They are $\{(1), (12)\}$, $\{(1), (23)\}$, and $\{(1), (13)\}$. According to Sylow's Third Theorem, we should be able to obtain the latter two of these from the first by conjugation. Indeed,

$$(13)\{(1), (12)\}(13)^{-1} = \{(1), (23)\}$$
$$(23)\{(1), (12)\}(23)^{-1} = \{(1), (13)\}.$$ ☐

Example 2   Consider the Sylow 3-subgroups of A_4. They are $\{\alpha_1, \alpha_5, \alpha_9\}$, $\{\alpha_1, \alpha_6, \alpha_{11}\}$, $\{\alpha_1, \alpha_7, \alpha_{12}\}$, and $\{\alpha_1, \alpha_8, \alpha_{10}\}$. (See the table on page 94.) Then,

$$\alpha_2\{\alpha_1, \alpha_5, \alpha_9\}\alpha_2^{-1} = \{\alpha_1, \alpha_7, \alpha_{12}\},$$
$$\alpha_3\{\alpha_1, \alpha_5, \alpha_9\}\alpha_3^{-1} = \{\alpha_1, \alpha_8, \alpha_{10}\},$$
$$\alpha_4\{\alpha_1, \alpha_5, \alpha_9\}\alpha_4^{-1} = \{\alpha_1, \alpha_6, \alpha_{11}\}.$$

Thus, the number of Sylow 3-subgroups is 1 modulo 3 and the four Sylow 3-subgroups are conjugate. ☐

---

*"Whenever you can, count." Sir Francis Galton (1822–1911), *The World of Mathematics*.

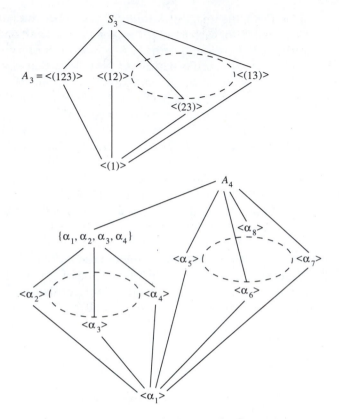

Figure 24.1   Lattice of subgroups for S_3 and A_4.

Figure 24.1 shows the subgroup lattice for S_3 and A_4. We have connected the Sylow *p*-groups with a dashed circle to indicate that they belong to one orbit under conjugation. Notice that the three subgroups of order 2 in A_4 are contained in a Sylow 2-group as required by Sylow's Second Theorem. As it happens, these three subgroups also belong to one orbit under conjugation, but this is not a consequence of Sylow's Third Theorem.

In contrast to the two preceding examples, observe that the dihedral group of order 12 has seven subgroups of order 2, but conjugating $\{R_0, R_{180}\}$ does not yield any of the other six. (Why?)

APPLICATIONS OF SYLOW THEOREMS

A few numerical examples will make the Sylow theorems come to life. Say G is a group of order 40. What do the Sylow theorems tell us about G? A great deal! Since 1 is the only divisor of 40 that is congruent to 1 modulo 5, we know that G has exactly one subgroup of order 5 and that it is normal. Similarly, G has either one or five subgroups of order 8. If there is only one subgroup of order 8, it is normal. If there are five sub-groups of order 8, none is normal and all five can be obtained by starting with any particular one, say, H, by computing xHx^{-1} for various x's. Finally, if we let K denote the normal subgroup of order 5 and H any subgroup of order 8, then $G = HK$. (See exercise 7, Supplementary Exercises for Chapters 5–7.) If H happens to be normal, we can say even more: $G = H \times K$.

What about a group G of order 30? It must have either one or six subgroups of order 5 and one or 10 subgroups of order 3. However, G cannot have both six subgroups of order 5 *and* 10 subgroups of order 3 (for then G would have more than 30 elements). Thus, G has one subgroup of order 3 and one of order 5, and at least one of these is normal in G. It follows, then, that the product of a subgroup of order 3 and one of order 5 is a group of order 15 that is both cyclic (exercise 24) and normal (exercise 7 in chapter 9) in G. [This, in turn, implies that *both* the subgroup of order 3 and the subgroup of order 5 are normal in G (exercise 50 in Chapter 9).] So, if we let y be a generator of the cyclic subgroup of order 15 and let x be an element of order 2 (the existence of which is guaranteed by Cauchy's Theorem), we see that

$$G = \{x^i y^j \mid 0 \le i \le 1, 0 \le j \le 14\}.$$

Note that in these two examples we were able to deduce all of this information from knowing only the order of the group—so many conclusions from one assumption! This is the beauty of finite group theory.

In Chapter 8, we classified all groups of order 6. As a further illustration of the power of the Sylow theorems, we now determine all groups of order $2p$ where p is prime.

Theorem 24.6   *Classification of Groups of Order 2p*
Let $|G| = 2p$, where p is an odd prime. Then G is isomorphic to Z_{2p} or D_p.

Proof.   By Cauchy's Theorem, G has subgroups of orders 2 and p. So, G has an element a of order 2 and an element b of order p. Note that G is generated by a and b. Since $\langle b \rangle$ has index 2, it is normal and therefore $aba^{-1} = b^k$ for some positive integer k less than p. Thus, $b^{k^2} = (b^k)^k = (aba^{-1})^k = ab^ka^{-1} = a(aba^{-1})a^{-1} = a^2ba^{-2}$. Because $|a| = 2$, we must have $b = b^{k^2}$. So $b^{k^2-1} = e$, which implies that p divides $k^2 - 1 = (k-1)(k+1)$. Since $1 \le k < p$, this implies that $k - 1 = 0$ or $k + 1 = p$. In the first case, we have $k = 1$ and $ab = ba$. Thus, $|ab| = 2p$ and G is isomorphic to Z_{2p}. In the second case, the $2p$ elements of G can be written in the form $a^i b^j$, where $0 \le i \le 1$ and $0 \le j \le p - 1$. Furthermore.

$$a^i b^{j_1} a b^{j_2} = a^{i+1} b^{j_2 - j_1},$$

and we leave it as an exercise to show that this guarantees that G is isomorphic to D_p.   ∎

The argument used to prove Theorem 24.6 can be extended to show that there are at most two groups of order pq for any primes p and q. (This was first proved by E. Netto in 1882—a proof is given in [4, p. 204].) One special case of this result of particular interest is the following.

Theorem 24.7   *Cyclic Groups of Order pq*
If G is a group of order pq, where p and q are primes, $p < q$, and p does not divide $q - 1$, then G is cyclic. In particular, G is isomorphic to Z_{pq}.

Proof.   Let H be a Sylow p-subgroup of G and let K be a Sylow q-subgroup of G. Sylow's Third Theorem states that the number of Sylow p-subgroups of G is of the form $1 + kp$ and divides pq. So $1 + kp = 1, p, q$, or pq. From this and the fact that $p \nmid q - 1$, it follows that $k = 0$ and, therefore, H is the only Sylow p-subgroup of G.

Similarly, there is only one Sylow q-subgroup of G. Thus, by the corollary to Theorem 24.5, H and K are normal subgroups of G. Let $H = \langle x \rangle$ and $K = \langle y \rangle$. To show that G is cyclic, it suffices to show that x and y commute, for then $|xy| = |x|\,|y| = pq$. But observe that, since H and K are normal, we have

$$xyx^{-1}y^{-1} = (xyx^{-1})y^{-1} \in Ky^{-1} = K$$

and

$$xyx^{-1}y^{-1} = x(yx^{-1}y^{-1}) \in xH = H.$$

Thus, $xyx^{-1}y^{-1} \in K \cap H = \{e\}$, and hence $xy = yx$.   ∎

Theorems 24.6 and 24.7 demonstrate the power of the Sylow theorems in classifying the finite groups that have small numbers of prime factors. Similar results exist for groups of order p^2q, p^2q^2, p^3, and p^4, where p and q are prime.

For your amusement, Figure 24.2 gives a list of the number of non-isomorphic groups with orders at most 100. Note in particular the large number of groups of order 64. Also observe that, generally speaking, it is not the size of the group that gives rise to a large number of groups of that size but the number of prime factors involved. In all, there are 1047 nonisomorphic groups with 100 or fewer elements. Contrast this with the fact reported in 1989 that there are 2328 groups of order 128 and 56,092 groups of order 256 [3].

As a final application of the Sylow theorems, you might enjoy seeing a determination of the groups of orders 99, 66, and 255. In fact, our arguments serve as a good review of much of our work in group theory.

Example 3   Determination of the Groups of Order 99
Suppose that G is a group of order 99. Let H be a Sylow 3-subgroup of G and let K be a Sylow 11-subgroup of G. Since 1 is the only positive divisor of 99 that is equal to 1 mod 11, we know from Sylow's Third

Order	1	2	3	4	5	6	7	8	9	10	11	12	13	14	15	16	17	18	19	20
Number	1	1	1	2	1	2	1	5	2	2	1	5	1	2	1	14	1	5	1	5

Order	21	22	23	24	25	26	27	28	29	30	31	32	33	34	35	36	37	38	39	40
Number	2	2	1	15	2	2	5	4	1	4	1	51	1	2	1	14	1	2	2	14

Order	41	42	43	44	45	46	47	48	49	50	51	52	53	54	55	56	57	58	59	60
Number	1	6	1	4	2	2	1	52	2	5	1	5	1	15	2	13	2	2	1	13

Order	61	62	63	64	65	66	67	68	69	70	71	72	73	74	75	76	77	78	79	80
Number	1	2	4	267	1	4	1	5	1	4	1	50	1	2	3	4	1	6	1	52

Order	81	82	83	84	85	86	87	88	89	90	91	92	93	94	95	96	97	98	99	100
Number	15	2	1	15	1	2	1	12	1	10	1	4	2	2	1	230	1	5	2	16

Figure 24.2   The number of groups of a given order ≤ 100.

Theorem and its corollary that K is normal in G. Similarly, H is normal in G. It follows that elements from H and K commute and, therefore, $G = H \times K$. Since both H and K are Abelian, G is also Abelian. Thus, G is isomorphic to Z_{99} or $Z_3 \oplus Z_{33}$.   ∎

Example 4   Determination of the Groups of Order 66
Suppose that G is a group of order 66. Let H be a Sylow 3-subgroup of G and let K be a Sylow 11-subgroup of G. Since 1 is the only positive divisor of 66 that is equal to 1 mod 11, we know that K is normal in G. Thus, HK is a subgroup of G of order 33 (exercise 46 of Chapter 9 and exercise 7, Supplementary Exercises for Chapters 5–7). Since any group of order 33 is cyclic (Theorem 24.7), we may write $HK = \langle x \rangle$. Next, let $y \in G$ and $|y| = 2$. Since $\langle x \rangle$ has index 2 in G, we know it is normal. So $yxy^{-1} = x^i$ for some i from 1 to 32. Then, $yx = x^i y$ and, since every member of G is of the form $x^s y^t$, the structure of G is completely determined by the value of i. We claim that there are only four possibilities for i. To prove this, observe that $|x^i| = |x|$ (exercise 6, Supplementary Exercises for Chapters 1–4). Thus, i and 33 are relatively prime. But also, since y has order 2,

$$x = y^{-1}(yxy^{-1})y = y^{-1}x^i y = yx^i y^{-1} = (yxy^{-1})^i = (x^i)^i = x^{i^2}.$$

So $x^{i^2-1} = e$ and, therefore, 33 divides $i^2 - 1$. From this it follows that 11 divides $i \pm 1$ and, therefore, $i = 0 \pm 1$, $i = 11 \pm 1$, $i = 22 \pm 1$, or $i = 33 \pm 1$. Putting this together with the other information we have about i, we see that $i = 1, 10, 23,$ or 32. This proves that there are at most four groups of order 66.
To prove that there are exactly four, we simply observe that Z_{66}, D_{33}, $D_{11} \oplus Z_3$, and $D_3 \oplus Z_{11}$ each has order 66 and that no two are

isomorphic. For example, $D_{11} \oplus Z_3$ has 11 elements of order 2, whereas $D_3 \oplus Z_{11}$ has only three elements of order 2. ☐

Example 5 The Only Group of Order 255 Is Z_{255}

Let G be a group of order $255 = 3 \cdot 5 \cdot 17$, and let H be a Sylow 17-subgroup of G. By Sylow's Third Theorem, H is the only Sylow 17-subgroup of G, so $N(H) = G$. By Example 14 in Chapter 10, $|N(H)/C(H)|$ divides $|\text{Aut}(H)| = |\text{Aut}(Z_{17})|$. By Theorem 6.3, $|\text{Aut}(Z_{17})| = |U(17)| = 16$. Since $|N(H)/C(H)|$ must divide 255 and 16, we have $|N(H)/C(H)| = 1$. Thus, $C(H) = G$. This means that every element of G commutes with every element of H and, therefore, $H \subseteq Z(G)$. Thus, 17 divides $|Z(G)|$, which in turn divides 255. So $|Z(G)| = 17, 51, 85,$ or 255 and $|G/Z(G)| = 15, 5, 3,$ or 1. But the only groups of orders 15, 5, 3, or 1 are the cyclic ones, so we know that $G/Z(G)$ is cyclic. Now the "G/Z theorem" (Theorem 9.3) shows that G is Abelian, and the Fundamental Theorem of Finite Abelian Groups tells us that G is cyclic. ☐

EXERCISES

If I rest, I rust.

Martin Luther

1. Show that conjugacy is an equivalence relation on a group.
2. Calculate all conjugacy classes for the quaternions (see exercise 4, Supplementary Exercises for Chapters 1–4).
3. Show that $\text{cl}(a) = \{a\}$ if and only if $a \in Z(G)$.
4. Describe the conjugacy classes of an Abelian group.
5. Exhibit a Sylow 2-subgroup of S_4. Describe an isomorphism from this group to D_4.
6. If $|G| = 36$ and is non-Abelian, prove that there is more than one Sylow 2-subgroup or more than one Sylow 3-subgroup.
7. Suppose that G is a group of order 48. Show that the intersection of any two distinct Sylow 2-subgroups of G has order 8.
8. Find all the Sylow 3-subgroups of A_4.
9. Let K be a Sylow p-subgroup of a finite group G. Prove that if $x \in N(K)$ and the order of x is a power of p, then $x \in K$. (This exercise is referred to in this chapter.)
10. Let H be a Sylow p-subgroup of G. Prove that H is the only Sylow p-subgroup of G contained in $N(H)$.
11. Suppose that G is a group of order 168. If G has more than one Sylow 7-subgroup, exactly how many does it have?

12. Show that every group of order 56 has a proper nontrivial normal subgroup.

13. What is the smallest composite (that is, nonprime and greater than 1) integer n such that there is a unique group of order n?

14. Let G be a noncyclic group of order 21. How many Sylow 3-subgroups does G have?

15. Prove that a noncyclic group of order 21 must have 14 elements of order 3.

16. How many Sylow 5-subgroups of S_5 are there? Exhibit two.

17. How many Sylow 3-subgroups of S_5 are there? Exhibit five.

18. Prove that a group of order 175 is Abelian.

19. Generalize the argument given in Example 3 to obtain a theorem about groups of order p^2q, where p and q are distinct primes.

20. Let H be a subgroup of a group G. Prove that the number of conjugates of H in G is $|G{:}N(H)|$. (*Hint:* Mimic the proof of Theorem 24.1.)

21. What is the smallest possible odd integer that can be the order of a non-Abelian group?

22. Prove that a group of order 375 has a subgroup of order 15.

23. Prove that a group of order 105 contains a subgroup of order 35.

24. Without using Theorem 24.7, prove that a group of order 15 is cyclic. (This exercise is referred to in the discussion about groups of order 30.)

25. Prove that a group of order 595 has a normal Sylow 17-subgroup.

26. Let G be a group of order 60. Show that G has exactly four elements of order 5 or exactly 24 elements of order 5. Which of these cases holds for A_5?

27. Show that the center of a group of order 60 cannot have order 4.

28. Suppose that G is a group of order 60 and G has a normal subgroup N of order 2. Show that
    a. G has normal subgroups of orders 6, 10, and 30.
    b. G has subgroups of orders 12 and 20.
    c. G has a cyclic subgroup of order 30.

29. Let G be a group of order 60. If the Sylow 3-subgroup is normal, show that the Sylow 5-subgroup is normal.

30. Show that if G is a group of order 168 that has a normal subgroup of order 4, then G has a normal subgroup of order 28.

31. Suppose that p is prime and $|G| = p^n$. Show that G has normal subgroups of order p^k for all k between 1 and n (inclusively).

32. Suppose that p is a prime and $|G| = p^n$. If H is a proper subgroup of G, prove that $N(H) > H$. (This exercise is referred to in Chapter 25.)

33. Suppose that G is a group of order p^n, where p is prime, and G has exactly one subgroup for each divisor of p^n. Show that G is cyclic.

34. Suppose that G is a finite group and that all its Sylow subgroups are normal. Show that G is a direct product of its Sylow subgroups.

35. Let G be a finite group and let H be a normal Sylow p-subgroup of G. Show that $\alpha(H) = H$ for all automorphisms α of G.

36. If H is a normal subgroup of a finite group G and $|H| = p^k$ for some prime p, show that H is contained in every Sylow p-subgroup of G.

37. Let H and K denote a Sylow 3-subgroup and a Sylow 5-subgroup of a group, respectively. Suppose that $|H| = 3$ and $|K| = 5$. If 3 divides $|N(K)|$, show that 5 divides $|N(H)|$.

38. Let G be a group of order p^2q^2, where p and q are distinct primes, $q \nmid p^2 - 1$, and $p \nmid q^2 - 1$. Prove that G is Abelian. List three pairs of primes that satisfy these conditions.

39. Let H be a normal subgroup of a group G. Show that H is the union of the conjugacy classes of the elements of H. Is this true when H is not normal in G?

40. Let p be a prime. If the order of every element of a finite group G is a power of p, prove that G is a power of p. (Such a group is called a p-group.)

41. Prove that all Sylow p-subgroups of a finite group are isomorphic.

42. What is the probability that a randomly selected element from D_4 commutes with V?

43. Let G be a finite group and let $a \in G$. Express the probability that a randomly selected element from G commutes with a in terms of orders of subgroups of G.

44. Prove that if x and y are in the same conjugacy class of a group, then $|C(x)| = |C(y)|$. (This exercise is referred to in the discussion on the probability that two elements from a group commute.)

45. Find $\Pr(D_4)$, $\Pr(S_3)$, and $\Pr(A_4)$.

46. Prove that $\Pr(G \oplus H) = \Pr(G) \cdot \Pr(H)$.

47. Let R be a finite noncommutative ring. Show that the probability that two randomly chosen elements from **R** commute is at most $\frac{5}{8}$. [*Hint:* Mimic the group case and use the fact that the additive group $R/C(R)$ is not cyclic.]

REFERENCES

1. W. H. Gustafson, "What Is the Probability that Two Group Elements Commute?" *The American Mathematical Monthly* 80 (1973): 1031–1034.
2. Desmond MacHale, "How Commutative Can a Non-Commutative Group Be?" *The Mathematical Gazette* 58 (1974): 199–202.
3. E. A. O'Brien, "The Groups of Order Dividing 256," *Bulletin of the Australian Mathematical Society* 39 (1989): 159–160.
4. H. Paley and P. Weichsel, *A First Course in Abstract Algebra,* New York: Holt, Rinehart & Winston, 1966.

SUGGESTED READINGS

J. A. Gallian and D. Moulton, "When Is Z_n the Only Group of Order n?" *Elemente der Mathematik* 48 (1993): 118–120.

> It is shown that Z_n is the only group of order n if and only if n and $\phi(n)$ are relatively prime.

W. H. Gustafson, "What Is the Probability that Two Group Elements Commute?" *The American Mathematical Monthly* 80 (1973): 1031–1034.

> This paper is concerned with the problem posed in the title. It is shown that for all finite non-Abelian groups and certain infinite non-Abelian groups, the probability that two elements from a group commute is at most 5/8. The paper concludes with several exercises.

Dieter Jungnickel, "On the Uniqueness of the Cyclic Group of Order n," *The American Mathematical Monthly* 99 (1992): 545–546.

> In this note, it is shown that Z_n is the only group of order n if and only if n and $\phi(n)$ are relatively prime.

Desmond MacHale, "Commutativity in Finite Rings," *The American Mathematical Monthly* 83 (1976): 30–32.

> In this easy-to-read paper, it is shown that the probability that two elements from a finite noncommutative ring commute is at most 5/8. Several properties of $\Pr(G)$ when G is a finite group are stated. For example, if $H \leq G$, then $\Pr(G) \leq \Pr(H)$. Also, there is no group G such that $7/16 < \Pr(G) < 1/2$.

Ludvig Sylow

Sylow's Theorem is 100 years old. In the course of a century this remarkable theorem has been the basis for the construction of numerous theories.

L. A. Shemetkov

LUDVIG SYLOW (pronounced "SEE-loe") was born on December 12, 1832, in Christiania (now Oslo), Norway. While a student at Christiania University, Sylow won a gold medal for competitive problem-solving. In 1855, he became a high school teacher, and despite the long hours required by his teaching duties, Sylow found time to study the papers of Abel. During the school year 1862–1863, Sylow received a temporary appointment at Christiania University and gave lectures on Galois's theory and permutation groups. Among his students that year was the great mathematician Sophus Lie (pronounced "Lee"), after whom Lie algebras and Lie groups are named. From 1873 to 1881, Sylow, with some help from Lie, prepared a new edition of Abel's works. In 1902, Sylow and Elling Holst published Abel's correspondence.

Sylow's great discovery, Sylow's Theorem, came in 1872. Upon learning of Sylow's result, C. Jordan called it "one of the essential points in the theory of permutations." The result took on greater importance when the theory of abstract groups flowered in the late 19th century and early 20th century.

In 1869, Sylow was offered a professorship at Christiania University, but turned it down. Upon Sylow's retirement at age 65 from high school teaching, Lie mounted a successful campaign to establish a chair for Sylow at Christiania University. Sylow held this position until his death on September 7, 1918.

25

Finite Simple Groups

It is a widely held opinion that the problem of classifying finite simple groups is close to a complete solution. This will certainly be one of the great achievements of mathematics of this century.

Nathan Jacobson

Historical Background

We now come to the El Dorado of finite group theory—the simple groups. Simple group theory is a vast and difficult subject; we call it the El Dorado of group theory because of the enormous effort put forth by hundreds of mathematicians during recent years to discover and classify all finite simple groups. Let's begin our discussion with the definition of a simple group and some historical background.

Definition   Simple Group
A group is *simple* if its only normal subgroups are the identity subgroup and the group itself.

The notion of a simple group was introduced by Galois about 160 years ago. The simplicity of A_5, the group of even permutations on five symbols, played a crucial role in his proof that there is not a solution by radicals of the general fifth-degree polynomial (that is, there is no "quintic formula"). But what makes simple groups important in the theory of groups? They are important because they play a role in group theory somewhat analogous to that of primes in number theory or the elements in chemistry; that is, they serve as the building blocks for all groups. These building blocks may be determined in the following way. Given a finite group G, choose a normal subgroup G_1 of $G = G_0$ of largest order. Then the factor group G_0/G_1 is simple, and we next choose a normal subgroup G_2 of G_1 of largest order. Then G_1/G_2 is also simple, and we continue in this fashion until we arrive at $G_n = \{e\}$. The simple groups $G_0/G_1, G_1/G_2, \ldots, G_{n-1}/G_n$ are called the *composition factors* of G. More than 100 years ago, Jordan and Hölder proved that these factors are independent of the choices of the normal subgroups made in the process described. In a certain sense, a group can be reconstructed from its composition factors, and many of the properties of a group are determined by the nature of its composition factors. This and the fact that many questions about finite groups can be reduced (by induction) to questions about simple groups make clear the importance of determining all finite simple groups.

Just which groups are the simple ones? The Abelian simple groups are precisely Z_n where $n = 1$ or n is prime. This follows directly from the corollary in Chapter 11. Unfortunately, it is not at all easy to describe the non-Abelian simple groups. The best we can do here is to give a few examples and mention a few words about their discovery. It was Galois who first observed that A_n is simple for all $n \geq 5$. The next discoveries were made by Jordan in 1870, when he found four infinite families of simple matrix groups over the field Z_p, where p is prime. Between the years 1892 and 1905, the American mathematician Leonard Dickson (see Chapter 22 for a biography) generalized Jordan's results to arbitrary finite fields and discovered several new infinite families of simple groups. About the same time, it was shown by G. A. Miller and F. N. Cole that a family of five groups first described by E. Mathieu in 1861 were in fact simple groups. Since these five groups were constructed by ad hoc methods that did not yield infinitely many possibilities such as A_n or the matrix groups over finite fields, they were called "sporadic."

The next important discoveries came in the 1950s. In that decade, many new infinite families of simple groups were found and the initial steps down the long and winding road that led to the complete classification of all finite simple groups were taken. The first step was Richard Brauer's observation that the centralizer of an element of order 2 was an important tool for studying simple groups. A few years later, John Thompson, in his Ph.D. thesis, introduced the crucial idea of studying the normalizers of various subgroups of prime-power order.

In the early 1960s came the momentous Feit-Thompson Theorem, which says that a non-Abelian simple group must have even order. This property was first conjectured around 1900 by one of the pioneers of modern group-theoretical methods, the Englishman William Burnside (see Chapter 29 for a biography). The proof of the Feit-Thompson Theorem filled an entire issue of a journal, 255 pages in all (see Figure 25.1, [2]). This result provided the impetus to classify the finite simple groups—that is, a program to discover all finite simple groups and *prove* that there are no more to be found. Throughout the 1960s, the methods introduced in the Feit-Thompson proof were generalized and improved with great success by several mathematicians. Moreover, between 1966 and 1975, 19 new sporadic simple groups were discovered. Despite many spectacular achievements, research in simple group theory in the 1960s was haphazard, and the decade ended with many people believing that the classification would never be completed. (The pessimists feared that the sporadic simple groups would foil all attempts. The anonymously written "song" in Figure 25.1 captures the spirit of the times.) Others, more optimistic, were predicting that it would be accomplished in the 1990s.

The 1970s began with Thompson receiving the Fields Medal for his fundamental contributions to simple group theory. This honor is the highest recognition a mathematician can receive (more information about the Fields Medal is given near the end of this chapter). Within a few years, three major events took place that ultimately led to the classification. First, Thompson published what is regarded as the single most important paper in simple group theory—the N-group paper. Here, Thompson introduced many fundamental techniques and supplied a model for the classification of a broad family of simple groups. Second was Daniel Gorenstein's elaborate outline for the classification, delivered in a series of lectures at the University of Chicago in 1972. Here a program for the overall proof was laid out. The army of researchers now had a battle plan and a commander-in-chief. But this army still needed more and better weapons. Thus, came the third critical development: the involvement of Michael Aschbacher. In a dazzling series of papers, Aschbacher combined his own insight with the methods of Thompson, which had been generalized throughout the 1960s, and a geometric approach pioneered by Bernd Fisher to achieve one brilliant result after another in rapid succession. In fact, so much progress was made by Aschbacher and others that, by 1976, it was clear to nearly everyone involved that enough techniques had been developed to complete the classification. Only details remained.

The 1980s were ushered in with Aschbacher following in the footsteps of Feit and Thompson by winning the American Mathematical Society's Cole Prize in algebra (see the last section of this chapter). A week later, Robert L. Griess made the spectacular announcement that he had constructed the "Monster."* The Monster is the largest of the sporadic simple groups. In fact, it has vastly more elements than there

---

* The name was coined by the English mathematician John H. Conway.

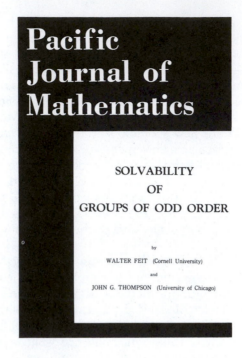

Pacific Journal of Mathematics

SOLVABILITY

OF

GROUPS OF ODD ORDER

by

WALTER FEIT (Cornell University)

and

JOHN G. THOMPSON (University of Chicago)

Oh, what are the orders of all simple groups?
I speak of the honest ones, not of the loops.
It seems that old Burnside their orders has
    guessed
Except for the cyclic ones, even the rest.

CHORUS: Finding all groups that are simple
          is no simple task.

Groups made up with permutes will produce
    some more:
For A_n is simple, if n exceeds 4.
Then, there was Sir Matthew who came into
    view
Exhibiting groups of an order quite new.

Still others have come on to study this thing.
Of Artin and Chevalley now we shall sing.
With matrices finite they made quite a list
The question is: Could there be others they've
    missed?

Suzuki and Ree then maintained it's the case
That these methods had not reached the end of
    the chase.
They wrote down some matrices, just four by
    four.
That made up a simple group. Why not make
    more?

And then came the opus of Thompson and
    Feit
Which shed on the problem remarkable light.
A group, when the order won't factor by two
Is cyclic or solvable. That's what is true.

Suzuki and Ree had caused eyebrows to raise,
But the theoreticians they just couldn't faze.
Their groups were not new: if you added a
    twist,
You could get them from old ones with a flick
    of the wrist.

Figure 25.1

Still, some hardy souls felt a thorn in their side.
For the five groups of Mathieu all reason defied;
Not A_n, not twisted, and not Chevalley,
They called them sporadic and filed them away.

Are Mathieu groups creatures of heaven or hell?
Zvonimir Janko determined to tell.
He found out that nobody wanted to know:
The masters had missed 1 7 5 5 6 0.

The floodgates were opened! New groups were the rage!
(And twelve or more sprouted, to greet the new age.)
By Janko and Conway and Fischer and Held
McLaughlin, Suzuki, and Higman, and Sims.

No doubt you noted the last lines don't rhyme.
Well, that is, quite simply, a sign of the time.
There's chaos, not order, among simple groups;
And maybe we'd better go back to the loops.

Figure 25.1 (continued)

are atoms on the earth! Its order is

$$808,017,424,794,512,875,886,459,904,961,710,757,005,754,368,000,000,000$$

(hence, the name). This is approximately 8×10^{53}. The Monster is a group of rotations in 196,883 dimensions. Thus, each element can be expressed as a $196,883 \times 196,883$ matrix.

The "Twenty-Five Years' War" came to an end in January 1981, when Gorenstein, speaking for "The Team," announced at an American Mathematical Society meeting that all finite simple groups had been classified—that is, that there was now a complete list of all the finite simple groups. The proof that this list is complete runs over 10,000 journal pages. Some of the mathematicians involved in this effort are now searching for ways to simplify the proof. Ten years after the classification had been announced, Scott Radtke, an undergraduate student, completed the simple group song (see Figure 25.2).

NONSIMPLICITY TESTS

In view of the fact that simple groups are the building blocks for all groups, it is surprising how scarce the non-Abelian simple groups are. For example, A_5 is the only one whose order is less than 168; there are only five non-Abelian simple groups of order less than 1000 and only 56 of order less than 1,000,000. In this section, we give a few theorems that are useful in proving that a particular integer is not the order of a non-Abelian simple group. Our first such result is an easy arithmetic test that comes from combining Sylow's Third Theorem and the fact that groups of prime-power order have nontrivial centers.

Completion of Song on the Simple Group War
by Scott C. Radtke
Central Michigan University
Modern Algebra Class Spring 1991

Wait! Don't give up, else all was for naught.
To order this chaos, a leader is sought.
A man with a vision for an overall plan.
Gorenstein's got this outlined, let's make him the man.

With more help from Thompson's fundamental technique,
The army of researchers, better weapons they seek.
They need some new insight, another approach.
Like a team in the field, they needed a coach.

Then Fischer came forward and geometrically preached,
Add insight from Aschbacher and the problem was breached.
Things now happenend quickly, the beast has been tamed.
Classification's now imminent, only details remained.

Some details were dramatic, like the Monster of Griess.
A sporadic group that'll make you look twice.
In fact it's so big that it's given wide berth.
Its order is greater than the atoms of Earth.

When the battles were over, the War came to an end.
Commander-in-Chief Gorenstein, an announcement did send.
We've found all finite simple groups, we sure earned our wages.
The proof that it's true is some 10,000 pages.

Final Chorus: Finding all groups that are simple
                    is finished at last.

Figure 25.2

Theorem 25.1   *Sylow Test for Nonsimplicity*

Let n be a positive integer that is not prime, and let p be a prime divisor of n. If 1 is the only divisor of n that is congruent to 1 modulo p, then there does not exist a simple group of order n.

Proof.   If n is a prime-power, then a group of order n has a nontrivial center and, therefore, is not simple. If n is not a prime-power, then every Sylow subgroup is proper and, by Sylow's Third Theorem, we know that the number of Sylow p-subgroups of a group of order n is congruent to 1 modulo p and divides n. Since 1 is the only such number, the Sylow p-subgroup is unique and, therefore, by the corollary to Sylow's Third Theorem, it is normal.   ∎

How good is this test? Well, if we were to program a computer to apply this criterion to all the nonprime integers between 1 and 200 and eliminate any that satisfy it, only the following would be left as possible orders of finite simple groups: 12, 24, 30, 36, 48, 56, 60, 72, 80, 90, 96, 105, 108, 112, 120, 132, 144, 150, 160, 168, 180, and 192. (In fact, computer experiments have revealed that for large intervals, say, 500 or more, this test eliminates over 90% of the nonprime integers as possible orders of simple groups. See [3] for more on this.)

Our next test rules out 30, 90, and 150.

Theorem 25.2 $2 \cdot$ *Odd Test*

An integer of the form $2 \cdot n$, where n is an odd number greater than 1, is not the order of a simple group.

Proof.   Let G be a group of order $2n$, where n is odd and greater than 1. Recall from the proof of Cayley's Theorem (Example 8 in Chapter 6) that the mapping $g \rightarrow T_g$ is an isomorphism from G to a permutation group on the elements of G (where $T_g(x) = gx$ for all x in G). Since $|G| = 2n$, Cauchy's Theorem guarantees that there is an element g in G of order 2. Then, when the permutation T_g is written in disjoint cycle form, each cycle must have length 1 or 2; otherwise, $|g| \neq 2$. But T_g can contain no 1-cycles, because the 1-cycle (x) would mean $x = T_g(x) = gx$, so $g = e$. Thus, in cycle form, T_g consists of exactly n transpositions, where n is odd. Therefore, T_g is an odd permutation. This means that the set of even permutations in the image of G is a normal subgroup of index 2. (See exercise 19 in Chapter 5 and exercise 7 in Chapter 9.) Hence, G is not simple.   ∎

The next theorem is a broad generalization of Cayley's Theorem. We will make heavy use of its two corollaries.

Theorem 25.3   *Generalized Cayley Theorem*

Let G be a group and H a subgroup of G. Let S be the group of all permutations of the left cosets of H in G. Then there is a homomorphism from G into S whose kernel lies in H and contains every normal subgroup of G that is contained in H.

Proof.   For each $g \in G$, define a permutation T_g of the left cosets of H by $T_g(xH) = gxH$. As in the proof of Cayley's Theorem, it is easy to verify that the mapping of $\alpha\colon g \rightarrow T_g$ is a homomorphism from G into S.

Now, if $g \in \text{Ker } \alpha$, then T_g is the identity map, so $H = T_g(H) = gH$, and, therefore, g belongs to H. Thus, Ker $\alpha \subseteq H$. On the other hand, if K is normal in G and $K \subseteq H$, then for any $k \in K$ and any x in G, there is an element k' in K so that $kx = xk'$. Thus,

$$T_k(xH) = kxH = xk'H = xH$$

and, therefore, T_k is the identity permutation. This means that $k \in$ Ker α. We have proved, then, that every normal subgroup of G contained in H is also contained in Ker α. ∎

As a consequence of Theorem 25.3, we obtain the following very powerful arithmetic test for nonsimplicity.

Corollary 1   Index Theorem

If G is a finite group and H is a proper subgroup of G such that $|G|$ does not divide $|G{:}H|!$, then H contains a nontrivial normal subgroup of G. In particular, G is not simple.

Proof.   Let α be the homomorphism given in Theorem 25.3. Then Ker α is a normal subgroup of G contained in H, and $G/\mathrm{Ker}\ \alpha$ is isomorphic to a subgroup of S. Thus, $|G/\mathrm{Ker}\ \alpha|$ divides $|S| = |G{:}H|!$. Since $|G|$ does not divide $|G{:}H|!$, the order of Ker α must be greater than 1. ∎

Corollary 2   Embedding Theorem

If a finite non-Abelian simple group G has a subgroup of index n, then G is isomorphic to a subgroup of A_n.

Proof.   Let H be the subgroup of index n and let S_n be the group of all permutations of the n left cosets of H in G. By the Generalized Cayley Theorem, there is a nontrivial homomorphism from G into S_n. Since G is simple and the kernel of a homomorphism is a normal subgroup of G, we see that the mapping from G into S_n is one-to-one, so that G is isomorphic to some subgroup of S_n. Recall from exercise 19 in Chapter 5 that any subgroup of S_n consists of even permutations only or half even and half odd. If G were isomorphic to a subgroup of the latter type, the even permutations would be a normal subgroup of index 2 (see exercise 7 in Chapter 9), which would contradict the fact that G is simple. Thus, G is isomorphic to a subgroup of A_n. ∎

Using the Index Theorem with the largest Sylow subgroup for H reduces our list of possible orders of non-Abelian simple groups still further. For example, let G be any group of order $80 = 16 \cdot 5$. We may choose H to be a subgroup of order 16. Since 80 is not a divisor of 5!, there is no simple group of order 80. The same argument applies to 12, 24, 36, 48, 72, 96, 108, 160, and 192, thereby leaving only 56, 60, 105, 112, 120, 132, 144, 168, and 180 as possible orders of non-Abelian simple groups up to 200. Let's consider these. Quite often we may use a counting argument to eliminate an integer. Consider 56. By Sylow's Theorem, we know that a simple group of order $56 = 8 \cdot 7$ would contain eight Sylow 7-subgroups and seven Sylow 2-subgroups. Now, any two Sylow p-subgroups that have order p must intersect in only the identity. So the union of the eight Sylow 7-subgroups yields 48 elements of order 7, and the union of any two Sylow 2-subgroups gives at least $8 + 8 - 4 = 12$ new elements. But there are only 56 elements in all. This contradiction shows that there is not a simple group of order 56. An analogous argument also eliminates the integers 105 and 132.

So, our list of possible orders of non-Abelian simple groups up to 200 is down to 60, 112, 120, 144, 168, and 180. Of these, 60 and 168 do correspond to simple groups. The others can be eliminated with a bit of razzle dazzle.

The easiest case to handle is $112 = 2^4 \cdot 7$. Suppose there were a simple group G of order 112. A Sylow 2-subgroup of G must have index 7. So, by the Embedding Theorem, G is isomorphic to a subgroup of A_7. But 112 does not divide $|A_7|$, which is a contradiction.

Next consider the possibility of a simple group G of order $144 = 9 \cdot 16$. By the Sylow theorems, we know that $n_3 = 4$ or 16 and $n_2 \geq 3$. [Recall that n_p denotes the number of Sylow p-subgroups of G and $n_p = |G:N(H)|$, where H is any Sylow p-subgroup of G.] The Index Theorem rules out the case where $n_3 = 4$, so we know that there are 16 Sylow 3-subgroups. Now, if every pair of Sylow 3-subgroups had only the identity in common, a straightforward counting argument would produce more than 144 elements. So, let H and H' be a pair of Sylow 3-subgroups whose intersection has order 3. Then $H \cap H'$ is a subgroup of both H and H' and, by the corollary to Theorem 24.2 (or by exercise 32 in Chapter 24), we see that $N(H \cap H')$ must contain both H and H' and, therefore, the set HH'. (HH' need not be a subgroup.) Thus,

$$|N(H \cap H')| \geq |HH'| = \frac{|H| \, |H'|}{|H \cap H'|} = \frac{9 \cdot 9}{3} = 27.$$

Now, we have three arithmetic conditions on $k = |N(H \cap H')|$. We know that 9 divides k; k divides 144; and $k \geq 27$. Clearly, then, $k \geq 36$, and so $|G:N(H \cap H')| \leq 4$. The Index Theorem now gives us the desired contradiction.

Finally, suppose that G is a non-Abelian simple group of order $180 = 2^2 \cdot 3^2 \cdot 5$. Then $n_5 = 6$ or 36 and $n_3 = 10$. First, assume that $n_5 = 36$. Then G has $36 \cdot 4 = 144$ elements of order 5. Now, if each pair of the Sylow 3-subgroups intersects in only the identity, then there are 80 more elements in the group, which is a contradiction. So, we may assume that there are two Sylow 3-subgroups L_3 and L_3' whose intersection has order 3. Then, as was the case for integer 144, we have

$$|N(L_3 \cap L_3')| \geq |L_3 L_3'| = \frac{9 \cdot 9}{3} = 27.$$

Thus,

$$|N(L_3 \cap L_3')| = 9 \cdot k,$$

where $k \geq 3$ and k divides 20. Clearly, then,

$$|N(L_3 \cap L_3')| \geq 36$$

and

$$|G:N(L_3 \cap L_3')| \leq 5.$$

The Index Theorem now gives us another contradiction. Hence, we may assume that $n_5 = 6$. In this case, we let H be the normalizer of a Sylow 5-subgroup of G. By Sylow's Third Theorem, we have $6 = |G:H|$, so that $|H| = 30$. In Chapter 24, we proved that every group of order 30 has an element of order 15. On the other hand, since $n_5 = 6$, G has a subgroup of index 6 and the Embedding Theorem tells us that G is isomorphic to a subgroup of A_6. But A_6 has no element of order 15. (See exercise 6 in Chapter 5.)

Unfortunately, the argument for 120 is fairly long and complicated. However, no new techniques are required to do it. We leave this as an exercise. Some hints are given in the answer section.

THE SIMPLICITY OF A_5

Once 120 has been disposed of, we will have shown that the only integers between 1 and 200 that can be the orders of simple groups are 60 and 168. For completeness, we will now prove that A_5, which has order 60, is a simple group. A similar argument can be used to show that the factor group $SL(2, 7)/Z(SL(2, 7))$ is a simple group of order 168. [This group is denoted by $PSL(2, 7)$.]

If A_5 had a nontrivial proper normal subgroup H, then $|H| = 2, 3, 4, 5, 6, 10, 12, 15, 20,$ or 30. By exercise 38 in Chapter 5, A_5 has 24 elements of order 5, 20 elements of order 3, and no elements of order 15. Now, if $|H| = 3, 6, 12,$ or 15, then $|A_5/H|$ is relatively prime to 3, and by exercise 45 in Chapter 9, H would have to contain all 20 elements of order 3. If $|H| = 5, 10,$ or 20, then $|A_5/H|$ is relatively prime to 5 and, therefore, H would have to contain the 24 elements of order 5. If $|H| = 30$, then $|A_5/H|$ is relatively prime to both 3 and 5, and so H would have to contain all the elements of orders 3 and 5. Finally, if $|H| = 2$ or 4, then $|A_5/H| = 30$ or 15. But we know from our results in Chapter 24 that any group of order 30 or 15 has an element of order 15. However, since A_5 contains no such element, neither does A_5/H. This proves that A_5 is simple.

The simplicity of A_5 was known to Galois in 1830, although the first formal proof was done by Jordan in 1870. A few years later, Felix Klein showed that the group of rotations of a regular icosahedron is simple and, therefore, isomorphic to A_5 (see exercise 28). Since then it has frequently been called the *icosahedral group*. Klein was the first to prove that there is a simple group of order 168.

The problem of determining which integers in a certain interval are possible orders for finite simple groups goes back to 1892, when Hölder went up to 200. His arguments for the integers 144 and 180 alone used up 10 pages. By 1975, this problem had been pushed to well beyond 1,000,000. See [4] for a detailed account of this endeavor. Of course, now that all finite simple groups have been classified, this problem is merely a historical curiosity.

THE FIELDS MEDAL

The highest award for mathematical achievement is the Fields Medal. Two to four such awards are bestowed at the opening session of the International Congress of Mathematicians, held once every four years. Although the Fields Medal is considered by most mathematicians as the equivalent of the Nobel Prize, there are great differences between these awards. Besides the huge disparity in publicity and monetary value associated with the two honors, the Fields Medal is restricted to those under 40 years of age.* This tradition stems from John Charles Fields's stipulation, in his will establishing the medal, that the awards should be "an encouragement for further achievement."

More details about the Fields Medal and a list of the recipients can be found in [1] and [5]. The latter article also includes photographs of the medal.

THE COLE PRIZE

Approximately every five years since 1928, the American Mathematical Society awards one or two Cole Prizes for research in algebra and one or two Cole Prizes for research in algebraic number theory. The prize was founded in honor of Frank Nelson Cole on the occasion of his retirement as secretary of the American Mathematical Society. In view of the fact that Cole was one of the first people interested in simple groups, it is interesting to note that no fewer than six recipients of the prize—Dickson, Chevalley, Brauer, Feit, Thompson, and Aschbacher—have made fundamental contributions to simple groups at some time in their careers.

---

* "Take the sum of human achievement in action, in science, in art, in literature—subtract the work of the men above forty, and while we should miss great treasures, even priceless treasures, we would practically be where we are to-day. . . . The effective, moving, vitalizing work of the world is done between the ages of twenty-five and forty." Sir William Osler (1849–1919), *Life of Sir William Osler*, vol. I, chap. 24 (The Fixed Period).

EXERCISES

If you don't learn from your mistakes, there's no sense making them.

Herbert V. Prochnow

1. Prove that there is no simple group of order $210 = 2 \cdot 3 \cdot 5 \cdot 7$.
2. Prove that there is no simple group of order $280 = 2^3 \cdot 5 \cdot 7$.
3. Prove that there is no simple group of order $216 = 2^3 \cdot 3^3$.
4. Prove that there is no simple group of order $300 = 2^2 \cdot 3 \cdot 5^2$.
5. Prove that there is no simple group of order $525 = 3 \cdot 5^2 \cdot 7$.
6. Prove that there is no simple group of order $540 = 2^2 \cdot 3^3 \cdot 5$.
7. Prove that there is no simple group of order $528 = 2^4 \cdot 3 \cdot 11$.
8. Prove that there is no simple group of order $315 = 3^2 \cdot 5 \cdot 7$.
9. Prove that there is no simple group of order $396 = 2^2 \cdot 3^2 \cdot 11$.
10. Prove that there is no simple group of order n where $201 \le n \le 235$.
11. Without using the Generalized Cayley Theorem or its corollaries, prove that there is no simple group of order 112.
12. Without using the "2 · odd" test, prove that there is no simple group of order 210.
13. You may have noticed that all the "hard integers" are even. Choose three odd integers between 200 and 1000. Show that none of these is the order of a simple group unless it is prime.
14. Show that there is no simple group of order pqr, where p, q, and r are primes (p, q, and r need not be distinct).
15. Show that A_5 cannot contain a subgroup of order 30, 20, or 15.
16. Show that S_5 cannot contain a subgroup of order 40 or 30. (This exercise is referred to in Chapter 32.)
17. Prove that a simple group of order 60 has a subgroup of order 6 and a subgroup of order 10.
18. Prove that if G is a finite group and H is a proper normal subgroup of largest order, then G/H is simple.
19. Suppose that H is a subgroup of a finite group G and that $|H|$ and $(|G{:}H| - 1)!$ are relatively prime. Prove that H is normal in G. What does this tell you about a subgroup of index 2 in a finite group?
20. Suppose that p is the smallest prime that divides $|G|$. Show that any subgroup of index p in G is normal in G.
21. Prove that there is no simple group of order $120 = 2^3 \cdot 3 \cdot 5$.
22. Show that the group of rotations of a regular dodecahedron is simple.
23. Show that the group of rotations of a regular icosahedron is simple.
24. Prove that the only nontrivial proper normal subgroup of S_5 is A_5. (This exercise is referred to in Chapter 32.)

25. Show that $PSL(2, 7) = SL(2, 7)/Z(SL(2, 7))$, which has order 168, is a simple group. (This exercise is referred to in this chapter.)

26. Show that the permutations (12) and (12345) generate S_5.

27. Suppose that a subgroup H of S_5 contains a 5-cycle and a 2-cycle. Show that $H = S_5$. (This exercise is referred to in Chapter 32.)

28. Show that (up to isomorphism) A_5 is the only simple group of order 60.

29. Suppose that G is a finite simple group and contains subgroups H and K such that $|G:H|$ and $|G:K|$ are prime. Show that $|H| = |K|$.

PROGRAMMING EXERCISES

One machine can do the work of fifty ordinary men. No machine can do the work of one extraordinary man.

Elbert Hubbard, *Roycraft Dictionary and Book of Epigrams*

1. Program Theorem 25.1. Use a counter M to keep track of how many integers the theorem eliminates in any given interval. Run your program for the following intervals: 1–100; 501–600; 5001–5100; 10,001–10,100. How does M seem to behave as the sizes of the integers grow?

2. Program the Index Theorem. Use a counter M to keep track of how many integers it eliminates in any given interval. Run your program for the same intervals as in exercise 1. How does M seem to behave as the sizes of the integers grow?

REFERENCES

1. H. Edwards, "A Short History of the Fields Medal," *The Mathematical Intelligencer* 1 (1978): 127–129.

2. W. Feit and J. G. Thompson, "Solvability of Groups of Odd Order," *Pacific Journal of Mathematics* 13 (1963): 775–1029.

3. J. A. Gallian, "Computers in Group Theory," *Mathematics Magazine* 49 (1976): 69–73.

4. J. A. Gallian, "The Search for Finite Simple Groups," *Mathematics Magazine* 49 (1976): 163–179.

5. H. S. Tropp, "The Origins and History of the Fields Medal," *Historia Mathematica* 3 (1976): 167–181.

SUGGESTED READINGS

B. Artmann, "A Simple Proof for the Simplicity of A_5," *The American Mathematical Monthly* 95 (1988): 344–349.

In this note, a geometric proof of the simplicity of A_5 is given. It is based on the fact that A_5 is isomorphic to the group of rotations of a dodecahedron.

G. Cornell, N. Pele, and M. Wage, "Simple Groups of Orders Less Than 1000," *Journal of Undergraduate Research* 5 (1973): 77–86.

> In this charming article, three undergraduate students use slightly more theory than was given in this chapter to show that the only integers less than 1000 that could be orders of simple groups are 60, 168, 320, 504, 660, and 720. All but the last one are orders of simple groups. The proof that there is no simple group of order 720 is omitted because it is significantly beyond most undergraduates.

K. David, "Using Commutators to Prove A_5 Is Simple," *The American Mathematical Monthly* 94 (1987): 775–776.

> This note gives an elementary proof that A_5 is simple using commutators.

J. A. Gallian, "The Search for Finite Simple Groups," *Mathematics Magazine* 49 (1976): 163–179.

> A historical account of the search for finite simple groups is given.

Anthony Gardiner, "Groups of Monsters," *New Scientist* April 5 (1979): 34.

> In this article, the author briefly discusses the construction of the sporadic simple groups. He mentions that Charles Sims constructed the "Baby Monster" (order 4,154,781,481,226,426,191,177,580,544,000,000) using a computer, and that Sims has a technique for constructing the "Monster" that would occupy the entire Rutgers University Computer Complex for a year! (Incidentally, Griess's construction was done entirely by hand.)

Martin Gardner, "The Capture of the Monster: A Mathematical Group with a Ridiculous Number of Elements," *Scientific American* 242 (6) (1980): 20–32.

> This article gives an elementary introduction to groups and a discussion of simple groups, including the "Monster."

Daniel Gorenstein, "The Enormous Theorem," *Scientific American* 253 (6) (1985): 104–115.

> You won't find an article on a complex subject better written for the layperson than this one. Gorenstein, the driving force behind the classification, uses concrete examples, analogies, and nontechnical terms to make the difficult subject matter of simple groups accessible.

A. L. Hammond, "Sporadic Groups: Exceptions, or Part of a Pattern?" *Science* 181 (1973): 146–148.

> This article gives a brief discussion of sporadic simple groups and their connection with error-correcting codes such as those used to transmit data from a spacecraft to earth.

Richard Silvestri, "Simple Groups of Finite Order," *Archive for the History of Exact Sciences* 20 (1979): 313–356.

> This article contains a plethora of historical information about the work on simple groups in the 19th century.

Michael Aschbacher

Fresh out of graduate school, he [Aschbacher] had just entered the field, and from that moment he became the driving force behind my program. In rapid succession he proved one astonishing theorem after another. Although there were many other major contributors to this final assault, Aschbacher alone was responsible for shrinking my projected 30-year timetable to a mere 10 years.

Daniel Gorenstein, *Scientific American*

MICHAEL ASCHBACHER was born on April 8, 1944, in Little Rock, Arkansas. Shortly after his birth, his family moved to Illinois, where his father was a professor of accounting and his mother was a high school English teacher. When he was nine years old, his family moved to East Lansing, Michigan; six years later, they moved to Los Angeles.

After high school, Aschbacher enrolled at the California Institute of Technology. In addition to his schoolwork, he passed the first four actuary exams and was employed for a few years as an actuary, full-time in the summers and part-time during the academic year. Two of the Cal Tech mathematicians who influenced him were Marshall Hall and Donald Knuth. In his senior year, Aschbacher took abstract algebra but showed little interest in the course. Accordingly, he received a grade of C.

In 1966, Aschbacher went to the University of Wisconsin for a Ph.D. degree. He completed his dissertation in 1969, and, after spending one year as an assistant professor at the University of Illinois, he returned to Caltech and quickly moved up to the rank of professor.

Aschbacher's dissertation work in the area of combinatorial geometries had led him to consider certain group-theoretical questions. Gradually, he turned his attention more and more to purely group-theoretical problems, particularly those bearing on the classification of finite simple groups. The 1980 Cole Prize Selection Committee said of one of his papers "[It] *lifted the subject to a new plateau and brought the classification within reach.*" In 1990, Aschbacher was elected to the National Academy of Sciences.

Daniel Gorenstein

Gorenstein was one of the most influential mathematicians of the last few decades.

Michael Aschbacher, *Notices of the American Mathematical Society* 39 (1992): 1190

DANIEL GORENSTEIN was born in Boston on January 1, 1923. He became interested in mathematics at the age of 12, when he taught himself calculus. After graduating from the Boston Latin School, he entered Harvard University. His senior thesis was done, under the direction of Saunders MacLane, on finite groups. Upon graduating in 1943, Gorenstein was offered an instructorship at Harvard to teach mathematics to army personnel. After the war ended, he began graduate work at Harvard. He received his Ph.D. degree in 1951, working in algebraic geometry under Oscar Zariski. It was in his dissertation that he introduced the class of rings that is now named after him. In 1951, Gorenstein took a position at Clark University in Worcester, Massachusetts, where he stayed until moving to Northeastern University in 1964. From 1969 until his death he was at Rutgers University.

In 1957, Gorenstein switched from algebraic geometry to finite groups, learning the basic material from I. N. Herstein while collaborating with him over the next few years. A milestone in Gorenstein's development as a group theorist came in 1960–1961, when he was invited to participate in a "Group Theory Year" at the University of Chicago. It was there that Gorenstein, assimilating the revolutionary techniques then being developed by John Thompson, began his fundamental work that contributed to the classification of finite simple groups.

Through his pioneering research papers, his dynamic lectures, his numerous personal contacts, and his influential book on finite groups, Gorenstein became the leader in the 25-year effort, by hundreds of mathematicians, that led to the classification of the finite simple groups.

Among the honors received by Gorenstein are the Steele Prize from the American Mathematical Society and election to membership in the National Academy of Sciences and the American Academy of Arts and Sciences.

In 1988, Gorenstein was named director of the newly created DIMACS, a national science and technology center in discrete mathematics and theoretical computer science funded by the National Science Foundation. DIMACS is a consortium of Princeton and Rutgers Universities, AT&T Bell Laboratories, and Bellcore. Gorenstein died on August 26, 1992.

John Thompson

There seemed to be no limit to his power.

Daniel Gorenstein

JOHN G. THOMPSON was born on October 13, 1932, in Ottawa, Kansas. In 1951, he entered Yale University as a divinity student, but switched to mathematics in his sophomore year. In 1955, he began graduate school at the University of Chicago and obtained his Ph.D. degree four years later. After one year on the faculty at Harvard, Thompson returned to Chicago. He remained there until 1968, when he moved to Cambridge University in England. In 1993, Thompson accepted an appointment at the University of Florida.

Thompson's brilliance was evident early. In his dissertation, he verified a 50-year-old conjecture about finite groups possessing a certain kind of automorphism. (An article about his achievement appeared in the *New York Times*!) The novel methods Thompson used in his dissertation foreshadowed the revolutionary ideas he would later introduce in the Feit-Thompson paper and the classification of minimal simple groups (i.e., simple groups that contain no proper non-Abelian simple subgroups). The assimilation and extension of Thompson's methods by others throughout the 1960s and 1970s ultimately led to the classification of finite simple groups.

In the late 1970s, Thompson made significant contributions to coding theory, the theory of finite projective planes, and the theory of modular functions. His recent work on Galois groups is considered the most important in the field in the last half of this century.

John Thompson and Walter Feit received the Cole Prize in algebra from the American Mathematical Society in 1965, and Thompson won the Fields Medal in 1970. He has received honorary doctorates from Yale University, the University of Illinois, and Oxford University. He was elected to the National Academy of Sciences in 1967 and to the Royal Society of London in 1979. In 1982, he received the Senior Berwick Prize and in 1987 the Sylvester Prize. In 1992, he was presented a $50,000 Wolf Foundation Prize by the President of Israel. The prize committee cited his profound contributions to all aspects of finite group theory and connections with other branches of mathematics. Also in 1992, the Académie des Sciences de Paris awarded Thompson its "Henri Poincaré" golden medal. Thompson is only the third recipient of the medal in its 30-year history.

26

Generators and Relations

One cannot escape the feeling that these mathematical formulae have an independent existence and an intelligence of their own, that they are wiser than we are, wiser even than their discoverers, that we get more out of them than we originally put into them.

Heinrich Hertz

MOTIVATION

In this chapter, we present a convenient way to define a group with certain prescribed properties. Simply put, we begin with a set of elements that we want to generate the group, and a set of equations (called *relations*) that specify the conditions that these generators are to satisfy. Among all such possible groups, we will select one that is as large as possible. This will uniquely determine the group up to isomorphism.

To provide motivation for the theory involved, we begin with a concrete example. Consider D_4, the group of symmetries of a square. Recall that $R = R_{90}$ and H, a reflection across a horizontal axis, generate the group. Observe that R and H are related in the following ways:

$$R^4 = H^2 = (RH)^2 = R_0 \quad \text{(the identity).} \quad (1)$$

Other relations between R and H, such as $HR = R^3H$ and $RHR = H$, also exist, but they can be derived from those given in (1). For example,

$(RH)^2 = R_0$ yields $HR = R^{-1}H^{-1}$, and $R^4 = H^2 = R_0$ gives $R^{-1} = R^3$ and $H^{-1} = H$. So, $HR = R^3H$. In fact, every relation between R and H can be derived from those given in Equation (1).

Thus, D_4 is a group that is generated by a pair of elements a and b subject to the relations $a^4 = b^2 = (ab)^2 = e$ and such that all other relations between a and b can be derived from these. This last stipulation is necessary because the subgroup $\{R_0, R_{180}, H, V\}$ of D_4 is generated by two elements satisfying the relations in Equation (1) with $a = R_{180}$ and $b = H$. However, the "extra" relation $a^2 = e$ satisfied by this subgroup cannot be derived from the original ones (since $R_{90}^2 \neq R_0$). It is natural to ask whether this description of D_4 applies to some other group as well. The answer is no. Any other group generated by two elements α and β satisfying only the relations $\alpha^4 = \beta^2 = (\alpha\beta)^2 = e$, and those that can be derived from these, is isomorphic to D_4.

Similarly, one can show that the group $Z_4 \oplus Z_2$ is generated by two elements a and b such that $a^4 = b^2 = e$ and $ab = ba$, and any other relation between a and b can be derived from these. The purpose of this chapter is to show that this procedure can be reversed; that is, we can begin with any set of generators and relations among the generators and construct a group that is uniquely described by these generators and relations, subject to the stipulation that all other relations among the generators can be derived from the original ones.

DEFINITIONS AND NOTATION

We begin with some definitions and notation. For any set $S = \{a, b, c, \dots\}$ of distinct symbols, we create a new set $S^{-1} = \{a^{-1}, b^{-1}, c^{-1}, \dots\}$ by replacing each x in S by x^{-1}. Define the set $W(S)$ to be the collection of all formal finite strings of the form $x_1x_2 \cdots x_k$, where each $x_i \in S \cup S^{-1}$. The elements of $W(S)$ are called *words from S*. We also permit the string with no elements to be in $W(S)$. This word is called the *empty word* and is denoted by e.

We may define a binary operation on the set $W(S)$ by juxtaposition; that is, if $x_1x_2 \cdots x_k$ and $y_1y_2 \cdots y_t$ belong to $W(S)$, then so does $x_1x_2 \cdots x_ky_1y_2 \cdots y_t$. Observe that this operation is associative and the empty word is the identity. Also, notice that a word such as aa^{-1} is not the identity, because we are treating the elements of $W(S)$ as formal symbols with no implied meaning.

At this stage we have everything we need to make a group out of $W(S)$ except inverses. Here a difficulty arises since it seems reasonable that the inverse of the word ab, say, should be $b^{-1}a^{-1}$. But $abb^{-1}a^{-1}$ is not the empty word! You may recall that we faced a similar obstacle long ago when we carried out the construction of the field of quotients of an integral domain. There we had formal symbols of the form a/b and we wanted the inverse of a/b to be b/a. But their product, ab/ba, was a formal symbol not the same as the formal symbol 1/1, the identity. So, we proceed here as we did there—by way of equivalence classes.

DEFINITION   Equivalence Classes of Words
For any pair of elements u and v of $W(S)$, we say that u is related to v if v can be obtained from u by a finite sequence of insertions or deletions of words of the form xx^{-1} or $x^{-1}x$, where $x \in S$.

We leave it as an exercise to show that this relation is an equivalence relation on $W(S)$.

Example 1   Let $S = \{a, b, c\}$. Then $acc^{-1}b$ is equivalent to ab; $aab^{-1}bbaccc^{-1}$ is equivalent to $aabac$; the word $a^{-1}aabb^{-1}a^{-1}$ is equivalent to the empty word; and the word $ca^{-1}b$ is equivalent to $cc^{-1}caa^{-1}a^{-1}bbca^{-1}ac^{-1}b^{-1}$. Note, however, that $cac^{-1}b$ is not equivalent to ab.   ◻

FREE GROUP

Theorem 26.1   *Equivalence Classes Form a Group*
Let S be a set of distinct symbols. For any word u in W(S), let $\bar{u}$ denote the set of all words in W(S) equivalent to u (that is, $\bar{u}$ is the equivalence class containing u). Then the set of all equivalence classes of elements of W(S) is a group under the operation $\bar{u} \cdot \bar{v} = \overline{uv}$.

Proof.   This proof is left as an exercise.   ∎

The group defined in Theorem 26.1 is called a *free group on S*. Theorem 26.2 shows why free groups are important.

Theorem 26.2   *The Universal Mapping Property*
Every group is a homomorphic image of a free group.

Proof.   Let G be a group and let S be a set of generators for G. (Such a set exists, because we may take S to be G itself.) Now let F be the free group on S. Unfortunately, since any word in $W(S)$ is also an element of G, we have created a notational problem for ourselves. So, to distinguish between these two cases, we will denote the word $x_1x_2 \cdots x_n$ in $W(S)$ by $(x_1x_2 \cdots x_n)_F$ and the product $x_1x_2 \cdots x_n$ in G by $(x_1x_2 \cdots x_n)_G$. As before, $\overline{x_1x_2 \cdots x_n}$ denotes the equivalence class in F containing the word $x_1x_2 \cdots x_n$ in $W(S)$. Notice that $(x_1x_2 \cdots x_n)_F$ and $(x_1x_2 \cdots x_n)_G$ may be entirely different elements, since the operations on $W(S)$ and G are different.

Now consider the mapping from F into G given by

$$\phi((\overline{x_1x_2 \cdots x_n})_F) = (x_1x_2 \cdots x_n)_G.$$

Clearly, ϕ is well defined, for inserting or deleting expressions of the form xx^{-1} or $x^{-1}x$ in elements of $W(S)$ corresponds to inserting or deleting the identity in G. To check that ϕ is operation-preserving, observe that

$$\phi((\overline{x_1 x_2 \cdots x_n})_F \, (\overline{y_1 y_2 \cdots y_m})_F) = \phi((\overline{x_1 x_2 \cdots x_n y_1 y_2 \cdots y_m})_F)$$
$$= (x_1 x_2 \cdots x_n y_1 y_2 \cdots y_m)_G$$
$$= (x_1 x_2 \cdots x_n)_G (y_1 y_2 \cdots y_m)_G.$$

[All we are doing is taking a product in F and viewing it as a product in G. For example, if G is the cyclic group of order 4 generated by a, then

$$\phi((\overline{aaaaa})_F) = (aaaaa)_G = a.]$$

Finally, ϕ is onto G because S generates G.   ∎

The following corollary is an immediate consequence of Theorem 26.2 and the First Isomorphism Theorem for Groups.

Corollary   *Universal Factor Group Property*
Every group is isomorphic to a factor group of a free group.

GENERATORS AND RELATIONS

We have now laid the foundation for defining a group by way of generators and relations. Before doing so, we will illustrate the basic idea with an example.

Example 2   Let F be the free group on the set $\{a, b\}$ and let N be the smallest normal subgroup of F containing the set $\{a^4, b^2, (ab)^2\}$. We will show that F/N is isomorphic to D_4. We begin by observing that the mapping ϕ from F onto D_4, which takes a to R_{90} and b to H (horizontal reflection), is a homomorphism whose kernel contains N. Thus, $F/\mathrm{Ker}\, \phi$ is isomorphic to D_4. On the other hand, we claim that the set

$$K = \{N, aN, a^2N, a^3N, bN, abN, a^2bN, a^3bN\}$$

of left cosets of N is F/N itself. To see this, it suffices to show that K is closed under multiplication on the left by a and b. Clearly, every member of F/N can be generated by starting with N and successively multiplying on the left by various combinations of a's and b's. So, once we have generated K, the closure property implies that no further elements of F/N can be generated. It is trivial that K is closed under left multiplication by a. For b, we will do only one of the eight cases. The others can be done in a similar fashion. Consider $b(aN)$. From $(ab)^2N = N$ and $a^4N = N$, we deduce $babN = a^{-1}N = a^3N$. From the normality of N, we obtain $babN = baNb$. So, $baNb = a^3N$ and, therefore, $baNb^2 = a^3Nb$. Finally, since $b^2N = N$ and N is normal, this last relation yields

$b(aN) = a^3bN$. Upon completion of the other cases, we know that F/N has at most eight elements. At the same time, we know that $F/\text{Ker } \phi$ has exactly eight elements. Since $F/\text{Ker } \phi$ is a factor group of F/N [indeed, $F/\text{Ker } \phi \approx (F/N)/(\text{Ker } \phi/N)$], it follows that F/N also has eight elements and $F/N = F/\text{Ker } \phi \approx D_4$. ❑

DEFINITION   Generators and Relations
Let G be a group generated by some set $A = \{a_1, a_2, \ldots, a_n\}$ and let F be the free group on A. Let $W = \{w_1, w_2, \ldots, w_t\}$ be a subset of F and let N be the smallest normal subgroup of F containing W. We say that G is *given by the generators* $a_1, a_2, \ldots, a_n$ *and the relations* $w_1 = w_2 = \cdots = w_t = e$ if there is an isomorphism from F/N onto G that carries a_iN to a_i.

The notation for this situation is

$$G = \langle a_1, a_2, \ldots, a_n \mid w_1 = w_2 = \cdots = w_t = e \rangle.$$

As a matter of convenience, we have restricted the number of generators and relations in our definition to be finite. This restriction is not necessary, however. Also, it is often more convenient to write a relation in implicit form. For example, the relation $a^{-1}b^{-3}ab = e$ is often written as $ab = b^3a$. In practice, one does not bother writing down the normal subgroup N that contains the relations. Instead, one just thinks of anything in N as the identity, as our notation suggests. Rather than saying that G is given by

$$\langle a_1, a_2, \ldots, a_n \mid w_1 = w_2 = \cdots = w_t = e \rangle,$$

many authors prefer to say that G has the *presentation*

$$\langle a_1, a_2, \ldots, a_n \mid w_1 = w_2 = \cdots = w_t = e \rangle.$$

Notice that a free group is "free" of relations; that is, the equivalence class containing the empty word is the only relation. We mention in passing the fact that a subgroup of a free group is also a free group. (See [3, p. 242] for a proof.) Free groups are of fundamental importance in a branch of algebra known as combinatorial group theory.

Example 3   The discussion in Example 2 can now be summed up by writing

$$D_4 = \langle a, b \mid a^4 = b^2 = (ab)^2 = e \rangle.$$ ❑

Example 4   The group of integers is the free group on one letter; that is, $Z \approx \langle a \rangle$. (This is the only nontrivial Abelian group that is free.) ❑

The next theorem formalizes the argument used in Example 2 to prove that the group defined there had eight elements.

Theorem 26.3   *(Dyck, 1882)*
   Let

$$G = \langle a_1, a_2, \ldots, a_n \mid w_1 = w_2 = \cdots = w_t = e \rangle$$

and let

$$\overline{G} = \langle a_1, a_2, \ldots, a_n \mid w_1 = w_2 = \cdots = w_t$$
$$= w_{t+1} = \cdots = w_{t+k} = e \rangle.$$

Then $\overline{G}$ *is a homomorphic image of G.*

Proof.   See Exercise 5.   ∎

Corollary   *Largest Group Satisfying Defining Relations*
   If K is a group satisfying the defining relations of a finite group G and $|K| \geq |G|$, *then K is isomorphic to G.*

Proof.   See Exercise 5.   ∎

Example 5   Quaternions
   Consider the group $G = \langle a, b \mid a^2 = b^2 = (ab)^2 \rangle$. What does G look like? Formally, of course, G is isomorphic to F/N, where F is free on $\{a, b\}$ and N is the smallest normal subgroup of F containing $b^{-2}a^2$ and $(ab)^{-2}a^2$. But as we have already said, we need not use the N. We just think of the elements of G as words in a and b where $a^2 = b^2 = (ab)^2$. Now, let $H = \langle b \rangle$ and $S = \{H, aH\}$. Then, just as in Example 2, it follows that S is closed under multiplication by a and b from the left. So, as in Example 2, we have $G = H \cup aH$. Thus, we can determine the elements of G once we know exactly how many elements there are in H. (Here again, the three relations come in.) To do this, first observe that $b^2 = (ab)^2 = abab$ implies $b = aba$. Then $a^2 = b^2 = (aba)(aba) = aba^2ba = ab^4a$ and, therefore, $b^4 = e$. Hence, H has at most four elements and therefore G has at most eight—namely, $e, b, b^2, b^3, a, ab, ab^2,$ and ab^3. It is conceivable, however, that not all of these eight are distinct. For example, $Z_2 \oplus Z_2$ satisfies the defining relations and has only four elements. Perhaps it is the largest group satisfying the relations. How can we show that the eight elements listed above are distinct? Well, consider the group $\overline{G}$ generated by the matrices

$$A = \begin{bmatrix} 0 & 1 \\ -1 & 0 \end{bmatrix} \quad \text{and} \quad B = \begin{bmatrix} 0 & i \\ i & 0 \end{bmatrix},$$

where $i = \sqrt{-1}$. Direct calculations show that in $\overline{G}$ the elements $e, B,$ $B^2, B^3, A, AB, AB^2,$ and AB^3 are distinct and $\overline{G}$ satisfies the relations $A^2 = B^2 = (AB)^2$. So, it follows from the corollary to Dyck's Theorem that G has order 8.   □

The next example illustrates why, in Examples 2 and 5, it is necessary to show that the eight elements listed for the group are distinct.

Example 6   Let

$$G = \langle a, b \mid a^3 = b^9 = e, a^{-1}ba = b^{-1} \rangle.$$

Once again, we let $H = \langle b \rangle$ and observe that $G = H \cup aH \cup a^2H$. Thus,

$$G = \{a^i b^j \mid 0 \le i \le 2, 0 \le j \le 8\}$$

and, therefore, G has at most 27 elements. But this time we will not be able to find some concrete group of order 27 satisfying the same relations that G does, for notice that $b^{-1} = a^{-1}ba$ implies

$$b = (a^{-1}ba)^{-1} = a^{-1}b^{-1}a.$$

Hence,

$$b = ebe = a^{-3}ba^3 = a^{-2}(a^{-1}ba)a^2 = a^{-2}b^{-1}a^2$$
$$= a^{-1}(a^{-1}b^{-1}a)a = a^{-1}ba = b^{-1}.$$

So, the original three relations imply the additional relation $b^2 = e$. But $b^2 = e = b^9$ further implies $b = e$. It follows, then, that G has only three distinct elements—namely, e, a, and a^2.   ◻

We hope Example 6 convinces you of the fact that, once a list of the elements of the group given by a set of generators and relations has been obtained, one must further verify that this list has no duplications. Typically, this is accomplished by exhibiting a specific group that satisfies the given set of generators and relations and that has the same size as the list. Obviously, experience plays a role here.

CLASSIFICATION OF GROUPS OF ORDER UP TO 15

The next theorem illustrates the utility of the ideas presented in this chapter.

Theorem 26.4   *Classification of Groups of Order 8 (Cayley, 1859)*
Up to isomorphism, there are only five groups of order 8: Z_8, $Z_4 \oplus Z_2$, $Z_2 \oplus Z_2 \oplus Z_2$, D_4, and the quaternions.

Proof.   The Fundamental Theorem of Finite Abelian Groups takes care of the Abelian cases. Now, let G be a non-Abelian group of order 8. Also, let $G_1 = \langle a, b \mid a^4 = b^2 = (ab)^2 = e \rangle$ and let $G_2 = \langle a, b \mid a^2 = b^2 = (ab)^2 \rangle$. We know from the preceding examples that G_1 is isomorphic to D_4 and G_2 is isomorphic to the quaternions. Thus, it suffices to show that G must satisfy the defining relations for G_1 or G_2. It follows from exercise 38 in Chapter 2 and Lagrange's Theorem that

G has an element of order 4; call it a. Then, if b is any element of G not in $\langle a \rangle$, we know that

$$G = \langle a \rangle \cup \langle a \rangle b = \{e, a, a^2, a^3, b, ab, a^2b, a^3b\}.$$

Consider the element b^2 of G. Which of the eight elements of G can it be? Not b, ab, a^2b, or a^3b, by cancellation. Not a, for b^2 commutes with b and a does not. Not a^3, for the same reason. Thus, $b^2 = e$ or $b^2 = a^2$. Suppose $b^2 = e$. Since $\langle a \rangle$ is a normal subgroup of G, we know that $bab^{-1} \in \langle a \rangle$. From this and the fact that $|bab^{-1}| = |a|$, we then conclude that $bab^{-1} = a$ or $bab^{-1} = a^{-1}$. The first relation would mean that G is Abelian, so we know that $bab^{-1} = a^{-1}$. But then, since $b^2 = e$, we have $(ab)^2 = e$ and, therefore, G satisfies the defining relations for G_1.

Finally, if $b^2 = a^2$ holds instead of $b^2 = e$, we can show by an argument like that already given that $(ab)^2 = a^2$, and, therefore, G satisfies the defining relations for G_2. ∎

The classification of the groups of order 8, together with our results on groups of order p^2, $2p$, and pq from Chapter 24, allow us to classify the groups of order up to 15 with the exception of those of order 12. We already know four groups of order 12—namely, Z_{12}, $Z_6 \oplus Z_2$, D_6, and A_4. An argument along the lines of Theorem 26.4 can be given to show that there is only one more group of order 12. This group, called the *dicyclic group of order 12* and denoted by Q_6, has presentation $\langle a, b \mid a^6 = e, a^3 = b^2, b^{-1}ab = a^{-1} \rangle$. Table 26.1 lists the groups of order at most 15. We use Q_4 to denote the quaternions (see Example 5 in this chapter).

Table 26.1 Classification of Groups of Order up to 15

Order	Abelian Groups	Non-Abelian Groups
1	Z_1	
2	Z_2	
3	Z_3	
4	Z_4, $Z_2 \oplus Z_2$	
5	Z_5	
6	Z_6	D_3
7	Z_7	
8	Z_8, $Z_4 \oplus Z_2$, $Z_2 \oplus Z_2 \oplus Z_2$	D_4, Q_4
9	Z_9, $Z_3 \oplus Z_3$	
10	Z_{10}	D_5
11	Z_{11}	
12	Z_{12}, $Z_6 \oplus Z_2$	D_6, A_4, Q_6
13	Z_{13}	
14	Z_{14}	D_7
15	Z_{15}	

CHARACTERIZATION OF DIHEDRAL GROUPS

As another nice application of generators and relations, we will now give a characterization of the dihedral groups that has been known for over 100 years. For $n \geq 3$, we have used D_n to denote the group of symmetries of a regular n-gon. Imitating Example 2, one can show that $D_n \approx \langle a, b \mid a^n = b^2 = (ab)^2 = e \rangle$ (see exercise 8). By analogy, these generators and relations serve to define D_1 and D_2 also. (These are also called dihedral groups.) Finally, we define the infinite dihedral group D_∞ as $\langle a, b \mid a^2 = b^2 = e \rangle$. The elements of D_∞ can be listed as $e, a, b, ab, ba,$ $(ab)a, (ba)b, (ab)^2, (ba)^2, (ab)^2 a, (ba)^2 b, (ab)^3, (ba)^3, \ldots .$

Theorem 26.5   *Characterization of Dihedral Groups*
Any group generated by a pair of elements of order 2 is dihedral.

Proof.   Let G be a group generated by a pair of elements of order 2, say, a and b. We consider the order of ab. If $|ab| = \infty$, then G is infinite and satisfies the relations of D_∞. We will show that G is isomorphic to D_∞. By Dyck's Theorem, G is isomorphic to some factor group of D_∞, say, D_∞/H. Now, suppose $x \in H$ and $x \neq e$. Since every element of D_∞ has one of the forms $(ab)^i, (ba)^i, (ab)^i a,$ or $(ba)^i b$, by symmetry, we may assume that $x = (ab)^i$ or $x = (ab)^i a$. If $x = (ab)^i$, then

$$H = (ab)^i H = (abH)^i,$$

so that $(abH)^{-1} = (abH)^{i-1}$. But

$$(abH)^{-1} H = (abH)^{-1} = b^{-1} a^{-1} H = baH$$

and it follows that

$$aHabHaH = baH = (abH)^{-1}.$$

Thus,

$$D_\infty/H = \langle aH, bH \rangle = \langle aH, abH \rangle$$

(see exercise 7) and D_∞/H satisfies the defining relations for D_i (use exercise 8 with $x = aH$ and $y = abH$). In particular, G is finite—an impossibility.
  If $x = (ab)^i a$, then

$$H = (ab)^i aH = (ab)^i HaH$$

and, therefore,

$$(abH)^i = (ab)^i H = (aH)^{-1} = a^{-1} H = aH.$$

It follows that

$$\langle aH, bH \rangle = \langle aH, abH \rangle \subseteq \langle abH \rangle.$$

However,

$$(abH)^{2i} = (aH)^2 = a^2 H = H,$$

so that D_∞/H is again finite. This contradiction forces $H = \{e\}$ and G to be isomorphic to D_∞.

Finally, suppose that $|ab| = n$. Since $G = \langle a, b \rangle = \langle a, ab \rangle$, we can show that G is isomorphic to D_n by proving that $b(ab)b = (ab)^{-1}$, which is the same as $ba = (ab)^{-1}$ (see exercise 8). But $(ab)^{-1} = b^{-1}a^{-1} = ba$, since a and b have order 2. ■

REALIZING THE DIHEDRAL GROUPS WITH MIRRORS

A geometric realization of D_∞ can be obtained by placing two mirrors in a parallel position, as shown in Figure 26.1.* If we let a and b denote reflections in mirrors A and B, respectively, then ab, viewed as the composition of a and b, represents a translation through twice the distance between the two mirrors to the left, and ba is the translation through the same distance to the right.

The finite dihedral groups can also be realized with a pair of mirrors. For example, if we place a pair of mirrors at a 45° angle, we obtain the group D_4. Notice that in Figure 26.2, the effect of reflecting an object in mirror A, then mirror B, is a rotation of twice the angle between the two mirrors (that is, 90°).

In Figure 26.3, we see a portion of the pattern produced by reflections in a pair of mirrors set at a 1° angle. The corresponding group is D_{180}. In general, reflections in a pair of mirrors set at the angle $180°/n$ correspond to the group D_n. As n becomes larger and larger, the mirrors approach a parallel position. In the limiting case, we have the group D_∞.

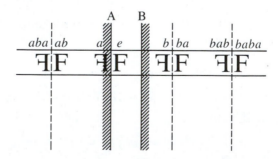

Figure 26.1   The group D_∞—reflections in parallel mirrors.

*Perhaps the most spectacular illustration of this is the photograph in *Time* magazine of Ann-Margaret in a dancing costume between a pair of parallel mirrors. The result is infinitely many images of Ann-Margaret! *Time,* September 18 (1978): 96.

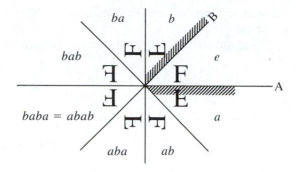

Figure 26.2   The group D_4—reflections in mirrors at a 45° angle.

The ideas discussed in this section are relevant to the design of kaleidoscopes.* These are mechanical devices that utilize mirrors to produce pleasing images. The reader may find information about building various kaleidoscopes in [2] and [4, pp. 68–69, 171–172, 200–201].

We conclude this chapter by commenting on the advantages and disadvantages of using generators and relations to define groups. The principal advantage is that in many situations—particularly in knot theory, algebraic topology, and geometry—groups defined by way of generators and relations arise in a natural way. Within group theory itself, it is often convenient to construct examples and counterexamples with generators and relations. Among the disadvantages in defining a group by generators and relations is the fact that it is often difficult to decide whether or not the group is finite, or even whether or not a particular element is the identity. Furthermore, the same group can be defined with entirely different sets of generators and relations, and, given two groups defined by generators and relations, it is often extremely difficult to decide whether or not these two groups are isomorphic. Nowadays,

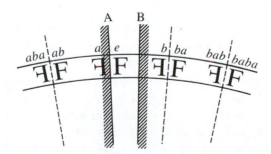

Figure 26.3   The group D_{180}—reflections in mirrors at a 1° angle.

---

*The word *kaleidoscope* is derived from three Greek words meaning "beautiful," "form," and "to see." The term was coined by Sir David Brewster, who wrote a treatise on the kaleidoscope's theory and history.

these questions are frequently tackled with the aid of a powerful computer.

Generators and relations for many well-known groups can be found in [1].

EXERCISES

It don't come easy. **Title of a song by Ringo Starr, May 1971**

1. Let n be an even integer. Prove that $D_n/Z(D_n)$ is isomorphic to $D_{n/2}$.
2. Let S be a set of distinct symbols. Show that the relation defined on $W(S)$ in this chapter is an equivalence relation.
3. Show that $\langle a, b \mid a^5 = b^2 = e, ba = a^2b \rangle$ is isomorphic to Z_2.
4. Verify that the set K in Example 2 is closed under multiplication on the left by b.
5. Prove Theorem 26.3 and its corollary.
6. Let G be the group $\{\pm 1, \pm i, \pm j, \pm k\}$ with multiplication defined as in exercise 44 in Chapter 9. Show that G is isomorphic to $\langle a, b \mid a^2 = b^2 = (ab)^2 \rangle$. (Hence, the name "quaternions.")
7. In any group, show that $\langle a, b \rangle = \langle a, ab \rangle$. (This exercise is referred to in the proof of Theorem 26.5.)
8. Prove that $G = \langle x, y \mid x^2 = y^n = e, xyx = y^{-1} \rangle$ is isomorphic to D_n. (This exercise is referred to in the proof of Theorem 26.5.)
9. Let

$$M = \frac{1}{3} \begin{bmatrix} 0 & -2 & 1 \\ 2 & 0 & -2 \\ -1 & 2 & 0 \end{bmatrix} \quad \text{and} \quad N = \frac{1}{3} \begin{bmatrix} 1 & -2 & 0 \\ -2 & 0 & 2 \\ 0 & 2 & -1 \end{bmatrix}$$

   Show that the group generated by M and N is isomorphic to D_4.
10. What is the minimum number of generators needed for $Z_2 \oplus Z_2 \oplus Z_2$? Find a set of generators and relations for this group.
11. Suppose that $x^2 = y^2 = e$ and $yz = zxy$. Show that $xy = yx$.
12. Let $G = \langle a, b \mid a^2 = b^4 = e, ab = b^3a \rangle$.
   a. Express $a^3b^2abab^3$ in the form b^ia^j.
   b. Express b^3abab^3a in the form b^ia^j.
13. Let $G = \langle a, b \mid a^2 = b^2 = (ab)^2 \rangle$.
   a. Express $a^3b^2abab^3$ in the form b^ia^j.
   b. Express b^3abab^3a in the form b^ia^j.
14. Let G be the group defined by the following table. Show that G is isomorphic to D_n.
15. Let $G = \langle x, y \mid x^8 = y^2 = e, yxyx^3 = e \rangle$. Show that $|G| \leq 16$. Find the center of G. Assuming G does have 16 elements, find the order of xy.

	1	2	3	4	5	6	$\cdots$	2n
1	1	2	3	4	5	6	$\cdots$	2n
2	2	1	2n	2n − 1	2n − 2	2n − 3	$\cdots$	3
3	3	4	5	6	7	8	$\cdots$	2
4	4	3	2	1	2n	2n − 1	$\cdots$	5
5	5	6	7	8	9	10	$\cdots$	4
6	6	5	4	3	2	1	$\cdots$	7
⋮	⋮	⋮	⋮	⋮	⋮	⋮	⋮	⋮
2n	2n	2n − 1	2n − 2	2n − 3	2n − 4	2n − 5	$\cdots$	1

16. Classify all groups of order ≤ 11.

17. Let G be defined by some set of generators and relations. Show that every factor group of G satisfies the generators and relations defining G.

18. Let $G = \langle s, t \mid sts = tst \rangle$. Show that the permutations (23) and (13) satisfy the defining relations of G. Explain why this proves that G is non-Abelian.

19. Let G be generated by a and b and suppose that $\langle b \rangle$ is normal in G. Show that every element of G can be written in the form $a^i b^j$.

20. Let $G = \langle x, y \mid x^{2n} = e, x^n = y^2, y^{-1}xy = x^{-1} \rangle$. Show that $Z(G) = \{e, x^n\}$. Assuming that $|G| = 4n$, show that $G/Z(G)$ is isomorphic to D_n. (The group G is called the *dicyclic* group of order $4n$.)

21. Let $G = \langle a, b \mid a^6 = b^3 = e, b^{-1}ab = a^3 \rangle$. How many elements does G have? To what familiar group is G isomorphic?

22. Let $G = \langle x, y \mid x^4 = y^4 = e, xyxy^{-1} = e \rangle$. Show that $|G| \leq 16$. Find the center of G. Assuming that $|G| = 16$, show that $G/\langle y^2 \rangle$ is isomorphic to D_4.

23. Determine the orders of the elements of D_∞.

24. The group defined by the following table is isomorphic to one of the groups listed in Table 26.1. Which one is it?

1	2	3	4	5	6	7	8	9	10	11	12
2	3	4	5	6	1	8	9	10	11	12	7
3	4	5	6	1	2	9	10	11	12	7	8
4	5	6	1	2	3	10	11	12	7	8	9
5	6	1	2	3	4	11	12	7	8	9	10
6	1	2	3	4	5	12	7	8	9	10	11
7	12	11	10	9	8	4	3	2	1	6	5
8	7	12	11	10	9	5	4	3	2	1	6
9	8	7	12	11	10	6	5	4	3	2	1
10	9	8	7	12	11	1	6	5	4	3	2
11	10	9	8	7	12	2	1	6	5	4	3
12	11	10	9	8	7	3	2	1	6	5	4

25. Let $G = \langle a, b, c, d \mid ab = c, bc = d, cd = a, da = b \rangle$. Determine $|G|$.

26. Let $G = \left\{ \begin{bmatrix} 1 & a & b \\ 0 & 1 & c \\ 0 & 0 & 1 \end{bmatrix} \;\middle|\; a, b, c \in Z_2 \right\}$. Prove that G is isomorphic

to D_4.

REFERENCES

1. H. S. M. Coxeter and W. O. J. Moser, *Generators and Relations for Discrete Groups,* 4th ed., Berlin: Springer-Verlag, 1980.
2. J. Kennedy and D. Thomas, *Kaleidoscope Math,* Palo Alto: Creative Publications, 1978.
3. Joseph J. Rotman, *The Theory of Groups: An Introduction,* Boston: Allyn and Bacon, 1965.
4. A. V. Shubnikov and V. A. Koptsik, *Symmetry in Science,* New York: Plenum Press, 1974.

SUGGESTED READINGS

Alexander H. Fran, Jr., and David Singmaster, *Handbook of Cubik Math,* Hillside, New Jersey: Enslow, 1982.

> This book is replete with the group-theoretical aspects of the Magic Cube. It uses permutation group theory and generators and relations to discuss the solutions to the cube and related results. The book has numerous challenging exercises stated in group-theoretical terms.

I. Kleiner, "The Evolution of Group Theory: A Brief Survey," *Mathematics Magazine* 59 (1986): 195–215.

> This award-winning paper outlines the origins of the main concepts in group theory.

Lee Neuwirth, "The Theory of Knots," *Scientific American* 240 (1979): 110–124.

> This article shows how a unique group can be associated with a knotted string. Mathematically, a knot is just a one-dimensional curve situated in three-

The cloverleaf knot.

dimensional space. The theory of knots—a branch of topology—seeks to classify and analyze the different ways of tracing such a curve. Around the turn of this century, Henri Poincaré observed that important geometric characteristics of knots could be described in terms of group generators and relations—the so-called "knot group." Among others, Neuwirth describes the construction of the knot group for the cloverleaf knot pictured. One set of generators and relations for this group is $\langle x, y, z \mid xy = yz, zx = yz \rangle$.

B. L. van der Waerden, "Hamilton's Discovery of Quaternions," *Mathematics Magazine* 49 (1976): 227–234.

This award-winning paper uses Hamilton's papers and letters to describe how he came to discover the quaternions. (van der Waerden is the author of a classic text on modern algebra published in 1930. It is one of the most influential mathematics books written in this century.)

Marshall Hall, Jr.

Professor Hall was a
mathematician in the broadest
sense of the word but with a
predilection for group theory,
geometry and combinatorics.

Hans Zassenhaus, *Notices of the
American Mathematical Society*

MARSHALL HALL, JR. was born on September 17, 1910, in St. Louis, Missouri.
He demonstrated interest in mathematics at the age of 11 when he constructed a
seven-place table of logarithms for the positive integers up to 1000. He completed
a B.A. degree in 1932 at Yale. After spending a year at Cambridge University, where
he worked with Philip Hall, Harold Davenport, and G. H. Hardy, he returned to
Yale for his Ph.D. degree. After a year at the Institute for Advanced Study, he
returned to Yale as an instructor until the outbreak of World War II, when he
joined Naval Intelligence. During the war, Hall obtained significant results deciphering both the Japanese codes and the German Enigma messages. These successes helped to turn the tide of the war. In 1945, Hall returned to Yale for a year
before going to Ohio State University. Hall spent the years from 1959 to 1981 at
Cal Tech and the years from 1981 until his death on July 4, 1990, at Emory
University.

Hall's highly regarded books on group theory and combinatorial theory are
classics. His mathematical legacy includes over 120 research papers on group theory, coding theory, and design theory. His 1943 paper on projective planes ranks
among the most cited papers in mathematics. Several fundamental concepts as well

as a sporadic simple group are identified with Hall's name. One of Hall's most celebrated results is his solution to the "Burnside Problem" for exponent 6—that is, a finitely generated group in which the order of every element divides 6 must be finite. Hall influenced both John Thompson and Michael Aschbacher, finite group theory's two greatest contributors. It was Hall who suggested Thompson's Ph.D. dissertation problem. Hall's Ph.D. students at Cal Tech included Donald Knuth and Robert McEliece.

Among the many honors accorded to Hall were two Guggenheim Fellowships, Yale University's Wilbur Cross Medal, membership in the American Academy of Arts and Sciences, designation as a Fellow of the American Association for the Advancement of Science, and honorary degrees from Emory University and Ohio State University.

27

Symmetry Groups

It [group theory] provides a sensitive instrument for
investigating symmetry, one of the most pervasive and
elemental phenomena of the real world.

M. I. Kargapolov and Ju. I. Marzljakov, *Fundamentals of the*
Theory of Groups

ISOMETRIES

In the early chapters of this book, we briefly discussed symmetry groups. In this chapter and the next, we examine this fundamentally important concept in some detail. It is convenient to begin such a discussion with the definition of an isometry (from the Greek *isometros,* meaning "equal measure") in $\mathbf{R}^n$.

DEFINITION Isometry

An *isometry* of n-dimensional space $\mathbf{R}^n$ is a function from $\mathbf{R}^n$ onto $\mathbf{R}^n$ that preserves distance.

In other words, a function T from $\mathbf{R}^n$ onto $\mathbf{R}^n$ is an isometry if, for every pair of points p and q in $\mathbf{R}^n$, the distance from $T(p)$ to $T(q)$ is the same as the distance from p to q. With this definition, we may now make precise the definition of the symmetry group of an n-dimensional figure.

DEFINITION Symmetry Group of a Figure in **R**^{n}

Let F be a set of points in **R**^{n}. The *symmetry group of F* in **R**^{n} is the set of all isometries of **R**^{n} that carry F onto itself under the operation function composition.

It is important to realize that the symmetry group of an object depends not only on the object, but also on the space in which we view it. For example, the symmetry group of a line segment in **R**^{1} has order 2; the symmetry group of a line segment considered as a set of points in **R**^{2} has order 4; and the symmetry group of a line segment viewed as a set of points in **R**^{3} has infinite order (see exercise 9).

Although we have formulated our definitions for all finite dimensions, our chief interest will be the 2-dimensional case. It has been known since 1831 that every isometry of **R**^{2} is one of four types: rotation, reflection, translation, and glide-reflection (see [1, p. 46]). Rotation about a point in a plane needs no explanation. A *reflection across a line L* is that transformation that leaves every point of L fixed and takes every point Q, not on L, to the point Q' so that L is the perpendicular bisector of the line segment from Q to Q' (see Figure 27.1). The line L is called the *axis of reflection*. In an x, y coordinate plane, the transformation $(x, y) \rightarrow (x, -y)$ is a reflection across the x-axis, whereas $(x, y) \rightarrow (y, x)$ is a reflection across the line $y = x$. Some authors call an axis of reflective symmetry L a *mirror* because L acts like a two-sided mirror; that is, the image of a point Q in a mirror placed on the line L is, in fact, the image of Q under the reflection across the line L. Reflections are called *opposite* isometries because they reverse orientation. For example, the reflected image of a clockwise spiral is a counterclockwise spiral. Similarly, the reflected image of a right hand is a left hand. (See Figure 27.1.)

A *translation* is simply a function that carries all points the same distance in the same direction. For example, if p and q are points in a plane and T is a translation, then the two vectors joining p to $T(p)$ and q to $T(q)$ have the same length and direction. A *glide-reflection* is the product of a translation and a reflection across the line containing the translation vector. In Figure 27.2, the vector gives the direction and length of the translation, and is contained in the axis of reflection. A glide-reflection is also an opposite isometry. Successive footprints in wet sand are related by a glide-reflection.

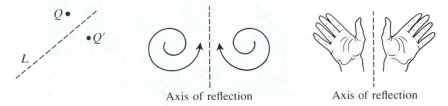

Axis of reflection   Axis of reflection

Figure 27.1 Reflected images.

$p \, \bullet$ $\bullet \, T(p)$

Figure 27.2   Glide-reflection.

CLASSIFICATION OF FINITE PLANE SYMMETRY GROUPS

Our first goal in this chapter is to classify all finite plane symmetry groups. As we have seen in earlier chapters, the dihedral group D_n is the plane symmetry group of a regular n-gon. (For convenience, call D_2 the plane symmetry group of a nonsquare rectangle and D_1 the plane symmetry group of the letter "V." In particular, $D_2 \approx Z_2 \oplus Z_2$ and $D_1 \approx Z_2$.) The cyclic groups Z_n are easily seen to be plane symmetry groups also. Figure 27.3 is an illustration of an organism whose plane symmetry group consists of four rotations and is isomorphic to Z_4. The surprising fact is that the cyclic groups and dihedral groups are the only finite plane symmetry groups. The famous mathematician Hermann Weyl attributes the following theorem to Leonardo da Vinci (1452–1519).

Theorem 27.1   *Finite Symmetry Groups in the Plane*
The only finite plane symmetry groups are Z_n and D_n.

Proof.   Let G be a finite plane symmetry group of some figure. We first observe that G cannot contain a translation or a glide-reflection, because in either case G would be infinite. Next, we show that there is some point in the plane that is left fixed by every member of G. To do this, let us

Figure 27.3   *Aurelia insulinda.* An organism whose plane symmetry group is Z_4.

suppose that the plane is coordinatized, and let $S = \{(x_1, y_1), (x_2, y_2), \ldots, (x_m, y_m)\}$ be the orbit of $(0, 0)$ under G (that is, $S = \{\phi((0, 0)) \mid \phi \in G\}$). Then,

$$(\bar{x}, \bar{y}) = \left(\frac{1}{m} \sum_{i=1}^{m} x_i, \frac{1}{m} \sum_{i=1}^{m} y_i \right)$$

is the centroid of the system of points in S. Since every member of G preserves distances, it follows that, for any ρ in G, the point $\rho((\bar{x}, \bar{y}))$ is the centroid of $\rho(S)$. But, because $\{\rho\phi \mid \phi \in G\} = G$, we have $\rho(S) = \{\rho(\phi((0, 0))) \mid \phi \in G\} = \{(\rho\phi)((0, 0)) \mid \phi \in G\} = \{\alpha((0, 0)) \mid \alpha \in G\} = S$, and therefore $\rho((\bar{x}, \bar{y})) = (\bar{x}, \bar{y})$. So, $(\bar{x}, \bar{y})$ is left fixed by every element of G. In particular, for any reflection in G, the axis of reflection must contain $(\bar{x}, \bar{y})$, and, for any rotation in G, the center of rotation is $(\bar{x}, \bar{y})$.

For convenience, let us denote a rotation about $(\bar{x}, \bar{y})$ of σ degrees by R_σ. Now, among all rotations in G, let β be the smallest positive angle of rotation. (Such an angle exists, since G is finite and R_{360} belongs to G.) We claim that every rotation in G is some power of R_β. To see this, suppose that R_σ is in G. We may assume $0° < \sigma \leq 360°$. Then, $\beta \leq \sigma$ and there is some integer t such that $t\beta \leq \sigma < (t + 1)\beta$. But, then, $R_{\sigma-t\beta} = R_\sigma \circ (R_\beta)^{-t}$ is in G and $0 \leq \sigma - t\beta < \beta$. Since β represents the smallest positive angle of rotation among the elements of G, we must have $\sigma - t\beta = 0$, and therefore, $R_\sigma = (R_\beta)^t$. This verifies the claim.

For convenience, let us say that $|R_\beta| = n$. Now, if G has no reflections, we have proved that $G = \langle R_\beta \rangle \approx Z_n$. If G has at least one reflection, say, α, then

$$\alpha, \alpha R_\beta, \alpha(R_\beta)^2, \ldots, \alpha(R_\beta)^{n-1}$$

are also reflections. Furthermore, this is the entire set of reflections of G. For if γ is any reflection in G, then $\alpha\gamma$ is a rotation, and so $\alpha\gamma = (R_\beta)^k$ for some k. Thus, $\gamma = \alpha^{-1}(R_\beta)^k = \alpha(R_\beta)^k$. So

$$G = \{R_0, R_\beta, (R_\beta)^2, \ldots, (R_\beta)^{n-1}, \alpha, \alpha R_\beta, \alpha(R_\beta)^2, \ldots, \alpha(R_\beta)^{n-1}\},$$

and G is generated by the pair of reflections α and αR_β. Hence, by our characterization of the dihedral groups (Theorem 26.5), G is the dihedral group D_n. ■

CLASSIFICATION OF FINITE GROUPS OF ROTATIONS IN $\mathbf{R}^3$

One might think that the set of all possible finite symmetry groups in three dimensions is much more diverse. Surprisingly, this is not the case. For example, moving to three dimensions only introduces three new groups of rotations. This observation was first made by the physicist and mineralogist Auguste Bravais in 1849, in his study of possible structures of crystals.

Theorem 27.2   *Finite Groups of Rotations in* $\mathbf{R}^3$.
     Up to isomorphism, the finite groups of rotations in $\mathbf{R}^3$ *are* Z_n, D_n, A_4, S_4, *and* A_5.

Theorem 27.2, together with the Orbit-Stabilizer Theorem (Theorem 8.2), make easy work of determining the group of rotations of an object in $\mathbf{R}^3$.

Example 1   We determine the group G of rotations of the solid in Figure 27.4, which is comprised of six congruent squares and eight congruent equilateral triangles. We begin by singling out any one of the squares. Obviously, there are four rotations that fix this square, and the designated square can be rotated to the location of any of the other five. So, by the Orbit-Stabilizer Theorem (Theorem 8.2), the rotation group has order $4 \cdot 6 = 24$. By Theorem 27.2, G is one of Z_{24}, D_{12}, and S_4. But each of the first two groups has exactly two elements of order 4, whereas G has more than two. So, G is isomorphic to S_4. $\square$

The group of rotations of a tetrahedron (the *tetrahedral group*) is isomorphic to A_4; the group of rotations of a cube or an octahedron (the *octahedral group*) is isomorphic to S_4; the group of rotations of a dodecahedron or an icosahedron (the *icosahedral group*) is isomorphic to A_5. (Coxeter [1, pp. 271–273] specifies which portions of the polyhedra are being permuted in each case.) These five solids are illustrated in Figure 27.5.

That these five solids are the only possible regular solids (that is, all faces are congruent and all solid angles at the vertices are equal) was one of the great discoveries of the ancient Greeks. Euclid discussed them at length in his book *Elements*. Plato theorized that all matter was made

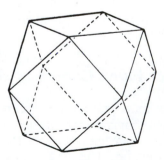

Figure 27.4

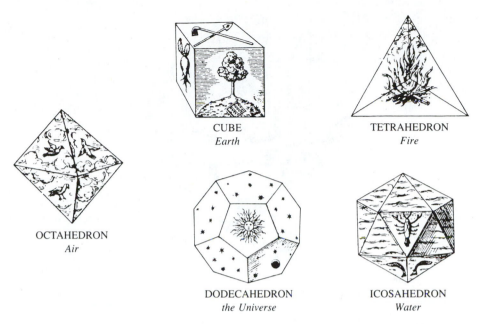

CUBE
Earth

TETRAHEDRON
Fire

OCTAHEDRON
Air

DODECAHEDRON
the Universe

ICOSAHEDRON
Water

Figure 27.5 The five regular solids as depicted by Johannes Kepler in *Harmonices Mundi, Book II* (1619).

up of minute particles of earth, air, fire, and water. He believed that earth particles were cubes, air particles were octahedra, fire particles were tetrahedra, and water particles were icosahedra. Pythagoras and his followers mystically associated the dodecahedron with the cosmos. They believed that understanding of the dodecahedron was too danger-ous for ordinary people and tried to restrict this knowledge to their own cult. Over 2000 years later, the astronomer Johannes Kepler tried, in vain, to model the planetary system after the way in which the five reg-ular solids can be inscribed in one another. It was his conjecture that the spheres of the planets were circumscribed and inscribed by these solids. (See Figure 27.6.)

A complete list of the finite symmetry groups in $\mathbf{R}^3$ (not just the groups of rotations) is given in [1, p. 413]. We mention in passing that, with the exception of the tetrahedron, the symmetry groups of the Pla-tonic solids are just direct products of the rotation subgroup and the nor-mal subgroup consisting of the identity and inversion in the center of the solid. [An *inversion* in a point Q is the isometry that fixes Q and carries every other point P to the point P' such that Q bisects the line segment from P to P'. For example, in $\mathbf{R}^3$, the transformation $(x, y, z) \rightarrow (-x, -y, -z)$ is an inversion in the origin.]

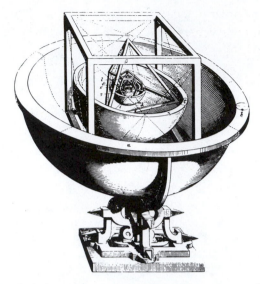

Figure 27.6   Kepler's cosmic mystery pictures the spheres of the six planets nested in the five perfect solids of Pythagoras and Plato. The outermost solid is the cube.

EXERCISES

Perhaps the most valuable result of all education is the ability to make yourself do the thing you have to do, when it ought to be done, whether you like it or not.

Thomas Henry Huxley, *Technical Education*

1. Show that a function from $\mathbf{R}^n$ to $\mathbf{R}^n$ that preserves distance (see page 402) is a one-to-one function.
2. Show that the translations in $\mathbf{R}^n$ form a group.
3. Exhibit a plane figure whose plane symmetry group is Z_5.
4. Show that the group of rotations in $\mathbf{R}^3$ of a 3-prism (that is, a prism with equilateral ends, as in the following figure) is isomorphic to D_3.

5. What is the order of the (entire) symmetry group in $\mathbf{R}^3$ of a 3-prism?
6. What is the order of the symmetry group in $\mathbf{R}^3$ of a 4-prism (a box with square ends that is not a cube)?

7. What is the order of the symmetry group in $\mathbf{R}^3$ of an n-prism?
8. Show that the symmetry group in $\mathbf{R}^3$ of a box of dimensions $2'' \times 3'' \times 4''$ is isomorphic to $Z_2 \oplus Z_2 \oplus Z_2$.
9. Describe the symmetry group of a line segment viewed as
   a. a subset of $\mathbf{R}^1$,
   b. a subset of $\mathbf{R}^2$,
   c. a subset of $\mathbf{R}^3$.
   (This exercise is referred to in this chapter.)
10. Determine the group of rotations for each of the solids shown in exercise 50 in Chapter 8.
11. Exactly how many elements of order 4 does the group in Example 1 have?
12. Why is inversion not listed as one of the four kinds of isometries in $\mathbf{R}^2$?
13. In $\mathbf{R}^2$, inversion through a point is the same as a rotation. Explain why inversion through a point in $\mathbf{R}^3$ cannot be duplicated by a rotation in $\mathbf{R}^3$.
14. Reflection in a line L in $\mathbf{R}^3$ is the isometry that takes each point Q to the point Q' with the property that L is a perpendicular bisector of the line segment joining Q and Q'. Describe a rotation that has this same effect.
15. In $\mathbf{R}^2$, a rotation fixes a point; in $\mathbf{R}^3$, a rotation fixes a line. In $\mathbf{R}^4$, what does a rotation fix? Generalize these observations to $\mathbf{R}^n$.
16. Show that an isometry of a plane preserves angles.
17. Show that an isometry of a plane is completely determined by the image of three noncolinear points.
18. Suppose that an isometry of a plane leaves three noncolinear points fixed. Which isometry is it?
19. Suppose that an isometry of a plane fixes exactly one point. What type of isometry must it be?
20. Suppose that A and B are rotations of 180° about the points a and b, respectively. What is A followed by B? How is the composite motion related to the points a and b?

REFERENCE

1. H. S. M. Coxeter, *Introduction to Geometry,* 2nd ed., New York: Wiley, 1969.

SUGGESTED READINGS*

Lorraine Foster, "On the Symmetry Group of the Dodecahedron," *Mathematics Magazine* 63 (1990): 106–107.

It is shown that the group of rotations of a dodecahedron and the group of rotations of an icosahedron are both A_5.

———

*See also the Suggested Readings for Chapter 1.

J. Rosen, *Symmetry Discovered*, Cambridge: Cambridge University Press, 1975.

> This excellent book was written for first- or second-year college students. It includes sections on group theory, spatial symmetry, temporal symmetry, color symmetry, and chapters on symmetry in nature and the uses of symmetry in science.

Marjorie Senechal, "Finding the Finite Groups of Symmetries of the Sphere," *The American Mathematical Monthly* 97 (1990): 329–335.

> The author shows how to find all finite subgroups of the group of symmetries of the sphere.

Doris Schattschneider, "The Taxicab Group," *The American Mathematical Monthly* 91 (1984): 423–428.

> The *taxicab metric* is the function d_t, defined on $\mathbf{R}^2$ by $d_t(a, b) = |a_1 - b_1| + |a_2 - b_2|$ for any pair of points $a = (a_1, a_2)$ and $b = (b_1, b_2)$. A *taxicab isometry* is a mapping ϕ from $\mathbf{R}^2$ to $\mathbf{R}^2$ that preserves the taxicab metric [that is, $d_t(a, b) = d_t(\phi(a), \phi(b))$. In this note it is shown that the group of taxicab isometries is the semidirect product of the dihedral group of order 8 and the group of all translations of the plane. (G is the *semidirect product* of A and B if $G = AB$, B is normal in G, and $A \cap B = \{e\}$.)

Andrew Watson, "The Mathematics of Symmetry," *New Scientist,* October (1990): 45–50.

> This article discusses how chemists use group theory to understand molecular structure and how physicists use it to study the fundamental forces and particles.

SUGGESTED FILMS

Dihedral Kaleidoscopes with H. S. M. Coxeter, International Film Bureau, $13\frac{1}{2}$ minutes, in color.

> See Chapter 2 for a description of this film.

Symmetries of the Cube, H. S. M. Coxeter and W. O. J. Moser, International Film Bureau, $13\frac{1}{2}$ minutes, in color.

> In this beautiful film, the symmetry properties of the cube and the octahedron are explored. At one point in the film, a cube is cut along its nine planes of symmetry to yield 48 congruent tetrahedra. One of the tetrahedra is then placed in an octahedral kaleidoscope made from three mirrors. The image produced is the entire cube, thereby proving that the symmetry group of a cube has order 48 and is generated by three reflections. The film also shows how the cube and the octahedron are related and why they have the same symmetry group. The use of stop-action photogrpahy and ultraviolet light to illuminate the models yields particularly dramatic effects.

28

Frieze Groups and
Crystallographic Groups

Symmetry, considered as a law of regular composition of
structural objects, is similar to harmony. More precisely,
symmetry is one of its components, while the other component
is dissymmetry. In our opinion the whole esthetics of scientific
and artistic creativity lies in the ability to feel this where others
fail to perceive it.

A. V. Shubnikov and V. A. Koptsik,
Symmetry in Science and Art

THE FRIEZE GROUPS

In this chapter, we discuss an interesting collection of infinite symmetry
groups that arise from periodic designs in a plane. There are two types
of such groups. The *discrete frieze groups* are the plane symmetry
groups of patterns whose subgroup of translations is isomorphic to Z.
These kinds of designs are the ones used for decorative strips [5] and for
patterns on jewelry, as illustrated in Figure 28.1. In mathematics, famil-
iar examples include the graphs of $y = \sin x$, $y = \tan x$, $y = |\sin x|$,
and $|y| = \sin x$.

In previous chapters, it was our custom to view two isomorphic
groups as the same group since we could not distinguish between them

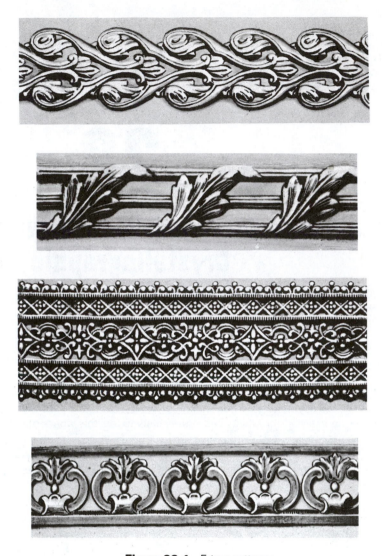

Figure 28.1   Frieze patterns.

algebraically. In the case of the frieze groups, we will soon see that, although some of them are isomorphic as groups (that is, algebraically the same), geometrically they are quite different. To emphasize this difference, we will treat them separately. In each of the following cases, the given pattern extends infinitely far in both directions.

Pattern I (Figure 28.2) consists of translations only. Letting x denote a translation to the right of one unit (that is, the distance between two consecutive R's), we may write the symmetry group of pattern I as

$$F_1 = \{x^n \mid n \in Z\}.$$

R   R   R   R   ,

Figure 28.2 Pattern I.

The group for pattern II (Figure 28.3), like that of pattern I, is infinitely cyclic. Letting x denote a glide-reflection, we may write the symmetry group of pattern II as

$$F_2 = \{x^n \mid n \in Z\}.$$

Notice that the translation subgroup is just $\langle x^2 \rangle$.

The symmetry group for pattern III (Figure 28.4) is generated by a translation x and a reflection y across the dotted vertical line. (There are infinitely many axes of reflective symmetry, including those midway between consecutive pairs of opposite-facing R's. Any one will do.) The entire group (the operation is function composition) is

$$F_3 = \{x^n y^m \mid n \in Z, m = 0 \text{ or } 1\}.$$

Note that the pair of elements xy and y have order 2, they generate F_3, and their product $(xy)y = x$ has infinite order. Thus, by Theorem 26.5, F_3 is the infinite dihedral group. A geometric fact about pattern III worth mentioning is that the distance between consecutive pairs of vertical reflection axes is half the length of the smallest translation vector.

In pattern IV (Figure 28.5), the symmetry group is generated by a translation x and a rotation y of 180° about a point p midway between consecutive R's (such a rotation is often called a *half-turn*). This group, like F_3, is also infinitely dihedral. (Another rotation point lies between a top and bottom R. As in pattern III, the distance between consecutive points of rotational symmetry is half the length of the smallest translation vector.)

$$F_4 = \{x^n y^m \mid n \in Z, m = 0 \text{ or } 1\}.$$

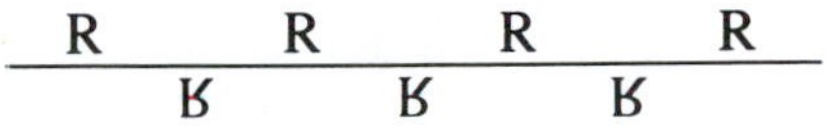

R   R   R   R

Я   Я   Я

Figure 28.3 Pattern II.

ЯR   ЯR   Я|R   ЯR   ЯR

Figure 28.4 Pattern III.

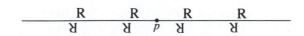

R   R   R   R

Я   Я *p* Я   Я

Figure 28.5 Pattern IV.

The group for pattern V (Figure 28.6) is yet another infinite dihedral group generated by a glide-reflection x and a rotation y of 180° about the point p. Notice that pattern V has vertical reflection symmetry xy. The rotation points are midway between the vertical reflection axes.

$$F_5 = \{x^n y^m \mid n \in Z, m = 0 \text{ or } 1\}.$$

The symmetry group for pattern VI (Figure 28.7) is generated by a translation x and a horizontal reflection y. The elements are

$$F_6 = \{x^n y^m \mid n \in Z, m = 0 \text{ or } 1\}.$$

Note that, since x and y commute, F_6 is not infinite dihedral. In fact, F_6 is isomorphic to $Z \oplus Z_2$. Pattern VI is left invariant under a glide-reflection also, but in this case the glide-reflection is called *trivial* since it is the product of x and y. (Conversely, a glide-reflection is *nontrivial* if its translation component and reflection component are not elements of the symmetry group.)

The symmetry group of pattern VII (Figure 28.8) is generated by a translation x, a horizontal reflection y, and a vertical reflection z. It is isomorphic to the direct product of the infinite dihedral group and Z_2. The product of y and z is a 180° rotation.

$$F_7 = \{x^n y^m z^k \mid n \in Z, m = 0 \text{ or } 1, k = 0 \text{ or } 1\}.$$

The preceding discussion is summarized in Figure 28.9. Figure 28.10 provides an identification algorithm for the frieze patterns.

In describing the seven frieze groups, we have not explicitly said how multiplication is done algebraically. However, each group element corresponds to some isometry, so multiplication is the same as function

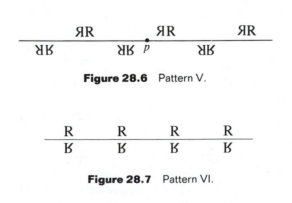

Figure 28.6   Pattern V.

Figure 28.7   Pattern VI.

Figure 28.8   Pattern VII.

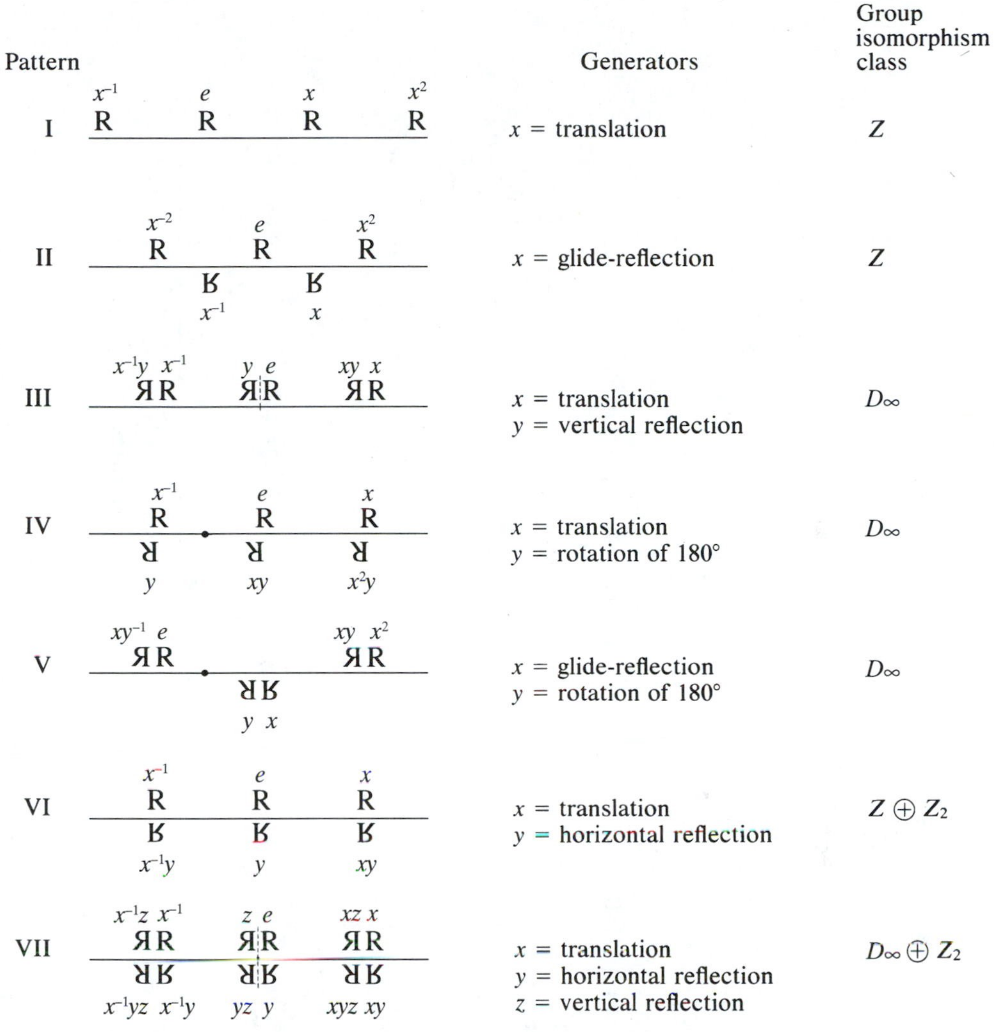

Pattern		Generators	Group isomorphism class

I

$$\begin{array}{cccc} x^{-1} & e & x & x^2 \\ R & R & R & R \end{array}$$

$x = $ translation Z

II

$$\begin{array}{ccc} x^{-2} & e & x^2 \\ R & R & R \end{array}$$
$$\begin{array}{cc} Я & Я \\ x^{-1} & x \end{array}$$

$x = $ glide-reflection Z

III

$$\begin{array}{ccc} x^{-1}y\ x^{-1} & y\ e & xy\ x \\ Я R & Я R & Я R \end{array}$$

$x = $ translation
$y = $ vertical reflection D_∞

IV

$$\begin{array}{ccc} x^{-1} & e & x \\ R & R & R \\ Я & Я & Я \\ y & xy & x^2y \end{array}$$

$x = $ translation
$y = $ rotation of 180° D_∞

V

$$\begin{array}{cc} xy^{-1}\ e & xy\ x^2 \\ Я R & Я R \end{array}$$
$$\begin{array}{c} Я К \\ y\ x \end{array}$$

$x = $ glide-reflection
$y = $ rotation of 180° D_∞

VI

$$\begin{array}{ccc} x^{-1} & e & x \\ R & R & R \\ К & К & К \\ x^{-1}y & y & xy \end{array}$$

$x = $ translation
$y = $ horizontal reflection $Z \oplus Z_2$

VII

$$\begin{array}{ccc} x^{-1}z\ x^{-1} & z\ e & xz\ x \\ Я R & Я R & Я R \\ Я К & Я К & Я К \\ x^{-1}yz\ x^{-1}y & yz\ y & xyz\ xy \end{array}$$

$x = $ translation
$y = $ horizontal reflection
$z = $ vertical reflection $D_\infty \oplus Z_2$

Figure 28.9   The seven frieze patterns and their groups of symmetries.

composition. Thus, we can always use the geometry to determine the product of any particular string of elements.

For example, we know that every element of F_7 can be written in the form $x^n y^m z^k$. So, just for fun, let's determine the appropriate values for n, m, and k for the element $g = (x^{-1}yz)(xz)$. We may do this simply by looking at the effect that g has on pattern VII. For convenience, we will pick out a particular R in the pattern and trace the action of g one step at a time. To distinguish this R, we enclose it in a shaded box. Also, we draw the axis of the vertical reflection z as a dotted line segment. See Figure 28.11.

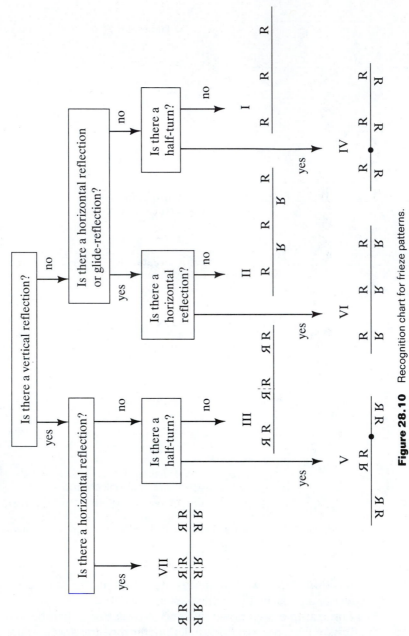

Figure 28.10   Recognition chart for frieze patterns.

Adapted from [11, p. 83].

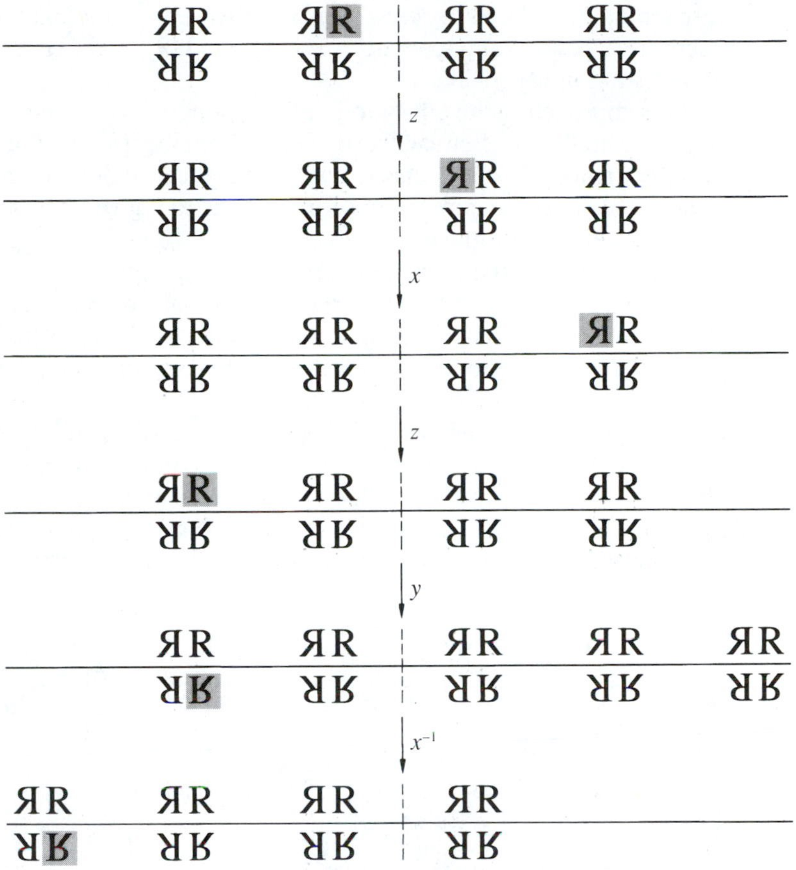

Figure 28.11

Now, comparing the starting position of the shaded R with its final position, we see that $x^{-1}yzxz = x^{-2}y$. Exercise 7 suggests how one may arrive at the same result through purely algebraic manipulation.

THE CRYSTALLOGRAPHIC GROUPS

The seven frieze groups catalog all symmetry groups that leave a design invariant under all multiples of just one translation. However, there are 17 additional kinds of discrete plane symmetry groups that arise from infinitely repeating designs in a plane. These groups are the symmetry groups of plane patterns whose subgroups of translations are isomorphic to $Z \oplus Z$. Consequently, the patterns are invariant under linear combinations of two linearly independent translations. These 17 groups were first studied by 19th-century crystallographers and are often called the *plane crystallographic groups*. Another term occasionally used for these groups is *wallpaper groups*. A group-theoretic proof that there are exactly 17 types of wallpaper patterns is given in [9]. A geometric proof can be found in [2].

Our approach to the crystallographic groups will be geometric. It is adapted from the excellent article by Schattschneider [8] and the monograph by Crowe [2]. Our goal is to enable the reader to determine which of the 17 plane symmetry groups corresponds to a given periodic pattern. We begin with some examples.

The simplest of the 17 crystallographic groups contains translations only. In Figure 28.12, we present an illustration of a representative pattern for this group (imagine the pattern repeated to fill the entire plane). The crystallographic notation for it is *p*1. (This notation is explained in [8].)

The symmetry group of the pattern in Figure 28.13 contains translations and glide-reflections. This group has no (nonzero) rotational or reflective symmetry. The crystallographic notation for it is *pg*.

Figure 28.14 has translational symmetry and threefold rotational symmetry (that is, the figure can be rotated 120° about certain points

Figure 28.12 *Study of Regular Division of the Plane with Fish and Birds*, 1938. Escher graphic with symmetry group *p*1. The arrows are translation vectors.

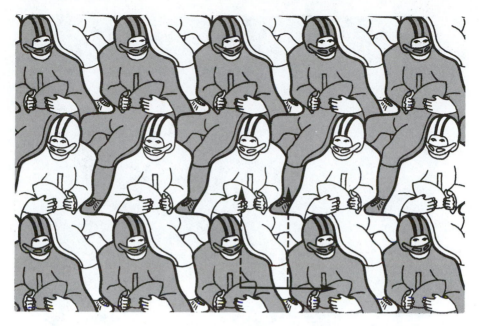

Figure 28.13 Escher-like tessellation by J. L. Teeters, with symmetry group *pg* (disregarding shading). The solid arrow is a translation vector. The dashed arrows are glide-reflection vectors.

and be brought into coincidence with itself). The notation for this group is *p3*.

Representative patterns for all 17 plane crystallographic groups, together with their notation, are given in Figures 28.15 and 28.16. Figure 28.17 illustrates the 17 classes of symmetry patterns generated in a triangle motif.

IDENTIFICATION OF PLANE PERIODIC PATTERNS

To decide which of the 17 classes any particular plane periodic pattern belongs to, we may use the flowchart presented in Figure 28.18. This is done by determining the rotational symmetry, and whether or not the pattern has reflection symmetry or nontrivial glide-reflection symmetry. These three pieces of information will narrow the list of candidates to at most two. The final test, if necessary, is to determine the locations of the centers of rotation.

For example, consider the two patterns in Figure 28.19 generated in a hockey stick motif. Both patterns have a smallest positive rotational symmetry of 120°; both have reflectional and nontrivial glide-reflectional symmetry. Now, according to Figure 28.18, these patterns must be of type *p3m1* or *p31m*. But notice that the pattern on the left has all

Figure 28.14   *Study of Regular Division of the Plane with Human Figures*, 1938.
Escher graphic with symmetry *p*3 (disregarding shading). The
inserted arrows are translation vectors.

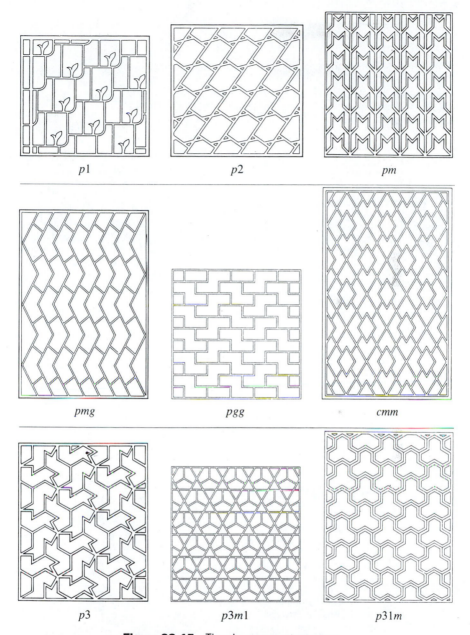

Figure 28.15 The plane symmetry groups.

All designs in Figures 28.15 and 28.16 except *pm*, *p3*, and *pg* are found in [3]. The designs for *p3* and *pg* are based on elements of Chinese lattice designs found in [3]; the design for *pm* is based on a weaving pattern from the Sandwich Islands, found in [5].

pg

cm

pmm

p4

p4m

p4g

p6

p6m

Figure 28.16 The plane symmetry groups.

Figure 28.17   The 17 plane periodic patterns formed in a triangle motif.

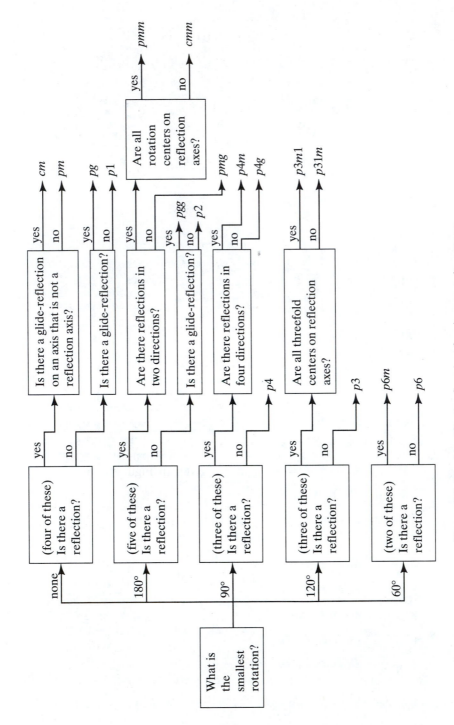

Figure 28.18   Identification flowchart for plane periodic patterns.

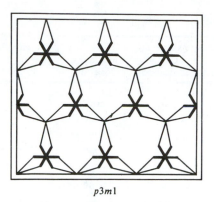
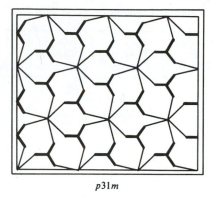

*p3m*1   *p*31*m*

Figure 28.19   Patterns generated in a hockey stick motif.

its threefold centers of rotation on the reflection axis, whereas in the pattern on the right the points where the three blades meet are not on a reflection axis. Thus, the left pattern is *p3m*1, and the right pattern is *p*31*m*.

Table 28.1 (reproduced from [8, p. 443]) can also be used to determine the type of periodic pattern and contains two other features that are often useful. A *lattice of points* of a pattern is a set of images of any particular point acted on by the translation group of the pattern. The possible lattices for periodic patterns in a plane, together with lattice units, are shown in Figure 28.20. A *generating region* (or *fundamental region*) of a periodic pattern is the smallest portion of the lattice unit whose images under the full symmetry group of the pattern cover the plane. Examples of generating regions for the patterns represented in Figures 28.12, 28.13, and 28.14 are given in Figure 28.21. In Figure 28.21, the portion of the lattice unit with vertical bars is the generating region. The only symmetry pattern in which the lattice unit and the generating unit coincide is the *p*1 pattern illustrated in Figure 28.12. Figure 28.18 tells which proportion of the lattice unit constitutes the generating region of each plane-periodic plane pattern.

Notice that Table 28.1 reveals that the only possible *n*-fold rotational symmetries occur when *n* = 1, 2, 3, 4, and 6. This fact is commonly called the *crystallographic restriction*. The first proof of this was given by the Englishman W. Barlow. The information in Table 28.1 can also be used in reverse to create patterns with a specific symmetry group. The patterns in Figure 28.19 were made in this way. Generators and relations for the crystallographic groups are given in [1, p. 136].

In sharp contrast to the situation for finite symmetry groups, the transition from two-dimensional crystallographic groups to three-dimensional crystallographic groups introduces a great many more possibilities since the motif is repeated indefinitely by three independent translations. Indeed, there are 230 three-dimensional crystallographic

Table 28.1   Identification Chart for Plane Periodic Patterns[a]

Type	Lattice	Highest Order of Rotation	Reflections	Nontrivial Glide-Reflections	Generating Region	Helpful Distinguishing Properties
*p*1	parallelogram	1	no	no	1 unit	
*p*2	parallelogram	2	no	no	$\frac{1}{2}$ unit	
pm	rectangular	1	yes	no	$\frac{1}{2}$ unit	
pg	rectangular	1	no	yes	$\frac{1}{2}$ unit	
cm	rhombic	1	yes	yes	$\frac{1}{2}$ unit	
pmm	rectangular	2	yes	no	$\frac{1}{4}$ unit	
pmg	rectangular	2	yes	yes	$\frac{1}{4}$ unit	parallel reflection axes
pgg	rectangular	2	no	yes	$\frac{1}{4}$ unit	
cmm	rhombic	2	yes	yes	$\frac{1}{4}$ unit	perpendicular reflection axes
*p*4	square	4	no	no	$\frac{1}{4}$ unit	
*p*4*m*	square	4	yes	yes	$\frac{1}{8}$ unit	fourfold centers on reflection axes
*p*4*g*	square	4	yes	yes	$\frac{1}{8}$ unit	fourfold centers not on reflection axes
*p*3	hexagonal	3	no	no	$\frac{1}{3}$ unit	
*p*3*m*1	hexagonal	3	yes	yes	$\frac{1}{6}$ unit	all threefold centers on reflection axes
*p*31*m*	hexagonal	3	yes	yes	$\frac{1}{6}$ unit	not all threefold centers on reflection axes
*p*6	hexagonal	6	no	no	$\frac{1}{6}$ unit	
*p*6*m*	hexagonal	6	yes	yes	$\frac{1}{12}$ unit	

[a]A rotation through an angle of 360°/*n* is said to have order *n*. A glide-reflection is nontrivial if its component translation and reflection are not symmetries of the pattern.

groups (often called *space groups*). These were independently determined by Fedorov, Schonflies, and Barlow in the 1890s. For information on these groups, refer to [10] or [12]. David Hilbert, one of the leading mathematicians of this century, focused attention on the crystallographic groups in his famous lecture in 1900 at the International Congress of Mathematicians at Paris. One of 23 problems he posed was whether or not the number of crystallographic groups in *n*-dimensions is always finite. This was answered affirmatively by L. Bieberbach in 1910. We mention in passing that in four dimensions, there are 4783 symmetry groups for infinitely repeating patterns.

As one might expect, the crystallographic groups are fundamentally

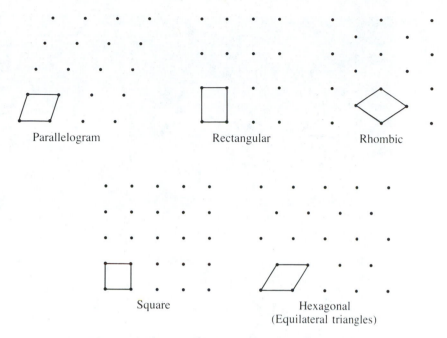

Parallelogram   Rectangular   Rhombic

Square   Hexagonal
                     (Equilateral triangles)

Figure 28.20   Possible lattices for periodic plane patterns.

important in the study of crystals. In fact, a crystal is defined as a rigid body in which the component particles are arranged in a pattern that repeats in three directions (the repetition is caused by the chemical bonding). A grain of salt and a grain of sugar are two examples of common crystals. In crystalline materials, the motif units are atoms, ions, ionic groups, clusters of ions, or molecules.

Perhaps it is fitting to conclude this chapter by recounting two episodes in the history of science in which an understanding of symmetry groups was crucial to a great discovery. In 1912, Max von Laue, a young German physicist, hypothesized that a narrow beam of x-rays directed onto a crystal with a photographic film behind it would be deflected (the technical term is "diffracted") by the unit cell (made up of atoms or ions) and would show up on the film as spots. (See Figure 1.2.) Shortly thereafter, two British scientists, Sir William Henry Bragg and his 22-year-old son William Lawrence Bragg, who was a student, noted that von Laue's diffraction spots together with the known information about crystallographic space groups could be used to calculate the shape of the internal array of atoms. This discovery marked the birth of modern mineralogy. From the first crystal structures deduced by the Braggs to the present, x-ray diffraction has been the means by which the internal structures of crystals are determined. Von Laue was awarded the Nobel Prize in physics in 1914, and the Braggs were jointly awarded the Nobel Prize in physics in 1915.

Figure 28.21   A lattice unit and generating region for the patterns in Figures 28.12, 28.13, and 28.14. Generating regions are shaded with vertical bars.

Our second episode took place in the early 1950s, when a handful of scientists were attempting to learn the structure of the DNA molecule—the basic genetic material. One of these was a graduate student named Francis Crick; another was an x-ray crystallographer, Rosalind Franklin. On one occasion, Crick was shown one of Franklin's research reports and an x-ray diffraction photograph of DNA. At this point, we let Horace Judson [6, pp. 165–166], our source, continue the story.

> Crick saw in Franklin's words and numbers something just as important, indeed eventually just as visualizable. There was drama, too: Crick's insight began with an extraordinary coincidence. Crystallographers distinguish

230 different space groups, of which the face-centered monoclinic cell with its curious properties of symmetry is only one—though in biological substances a fairly common one. The principal experimental subject of Crick's dissertation, however, was the X-ray diffraction of the crystals of a protein that was of exactly the same space group as DNA. So Crick saw at once the symmetry that neither Franklin nor Wilkins had comprehended, that Perutz, for that matter, hadn't noticed, that had escaped the theoretical crystallographer in Wilkins' lab, Alexander Stokes—namely, that the molecule of DNA, rotated a half turn, came back to congruence with itself. The structure was dyadic, one half matching the other half in reverse.

This was a crucial fact. Shortly thereafter, James Watson and Crick built an accurate model of DNA. In 1962, Watson, Crick, and Maurice Wilkins received the Nobel Prize in medicine and physiology for their discovery. The opinion has been expressed that, had Franklin correctly recognized the symmetry of the DNA molecule, she might have been the one to unravel the mystery and receive the Nobel prize [6, p. 172].

EXERCISES

You can see a lot just by looking.
>   **Yogi Berra**

1. Show that the frieze group F_6 is isomorphic to $Z \oplus Z_2$.
2. How many nonisomorphic frieze groups are there?
3. In the frieze group F_7, write x^2yzxz in the form $x^ny^mz^k$.
4. In the frieze group F_7, write $x^{-3}zxyz$ in the form $x^ny^mz^k$.
5. In the frieze group F_7, show that $yz = zy$ and $xy = yx$.
6. In the frieze group F_7, show that $zxz = x^{-1}$.
7. Use the results of exercises 5 and 6 to do exercises 3 and 4 through symbol manipulation only (that is, without referring to the pattern). (This exercise is referred to in this chapter.)
8. Prove that in F_7 the cyclic subgroup generated by x is a normal subgroup.
9. Quote a previous result that tells why the subgroups $\langle x, y \rangle$ and $\langle x, z \rangle$ must be normal in F_7.
10. Look up the word *frieze* in an ordinary dictionary. Explain why the frieze groups are appropriately named.
11. Determine which of the seven frieze groups is the symmetry group of each of the following patterns.

a.

b.

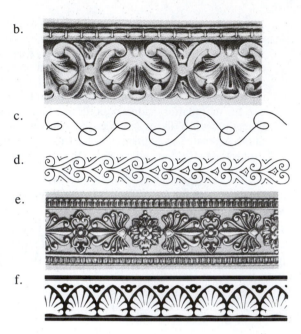

c.

d.

e.

f.

12. Determine the frieze group corresponding to each of the following patterns:
    a. $y = \sin x$,
    b. $y = |\sin x|$,
    c. $|y| = \sin x$,
    d. $y = \tan x$,
    e. $y = \csc x$.

13. Determine the symmetry group of the tessellation of the plane exemplified by the brickwork shown.

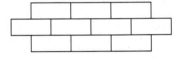

14. Determine the plane symmetry group for each of the patterns in Figure 28.16.

15. Determine which of the 17 crystallographic groups is the symmetry group of each of the following patterns.

(a)

(b)

(c) 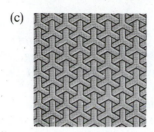 (d)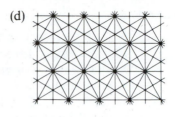

16. Look up Escher's "Flying Fish" print *Depth*, 1955 [7, p. 21]. Describe the isomorphism class of the three-dimensional symmetry group of the pattern depicted in the print.

17. Look up Escher's print *Cubic Space Division*, 1952 [7, p. 117]. Is the symmetry group of the pattern depicted the same as that of the "Flying Fish" print? Notice that the symmetry group of the structure depicted in this print contains subgroups isomorphic to $Z \oplus Z \oplus Z$ and S_4. Explain why the entire group is not the direct product of these two subgroups.

18. Determine the plane symmetry group of the Escher print *Study of Regular Division of the Plane with Birds*, 1955 [7, p. 125].

19. Determine the plane symmetry group of the Escher print *Study of Regular Division of the Plane with Angels and Devils*, 1941 [7, p. 141].

20. Determine the plane symmetry group of the Escher print *Study of Regular Division of the Plane with Human Figures*, 1936 [7, p. 77, No. 84].

21. Determine which of the frieze groups is the symmetry group of each of the following patterns.
    a. · · · D D D D · · ·
    b. · · · V Λ V Λ · · ·
    c. · · · L L L L · · ·
    d. · · · V V V V · · ·
    e. · · · N N N N · · ·
    f. · · · H H H H · · ·
    g. · · · L ⌐ L ⌐ · · ·

22. In the following figure is a point labelled 1. Let α be the translation of the plane that carries the point labelled 1 to the point labelled α, and let β be the translation of the plane that carries the point labelled 1 to the point labelled β. The image of 1 under the composition of α and β is labelled $\alpha\beta$. In the corresponding fashion, label the remaining points in the figure in the form $\alpha^i \beta^j$.

REFERENCES

1. H. S. M. Coxeter and W. O. J. Moser, *Generators and Relations for Discrete Groups,* 4th ed., Berlin: Springer-Verlag, 1980.
2. Donald Crowe, *Symmetry, Rigid Motions, and Patterns,* Arlington: COMAP, 1986.
3. Daniel S. Dye, *A Grammar of Chinese Lattice,* Harvard-Yenching Institute Monograph Series, vol. VI, Cambridge, Mass.: Harvard University Press, 1937. (Reprinted as *Chinese Lattice Designs,* New York: Dover, 1974.)
4. Bruno Ernst, *The Magic Mirror of M. C. Escher,* New York: Random House, 1976. (Paperback edition, New York: Ballantine Books, 1977.)
5. Owen Jones, *The Grammar of Ornament,* New York: Van Nostrand Reinhold, 1972. (Reproduction of the same title, first published in 1856 and reprinted in 1910 and 1928.)
6. Horace Freeland Judson, *The Eighth Day of Creation,* New York: Simon and Schuster, 1979.
7. J. L. Locher, editor, *The World of M. C. Escher.* (Paperback edition, New York: Harry N. Abrams, 1971.)
8. D. Schattschneider, "The Plane Symmetry Groups: Their Recognition and Notation," *The American Mathematical Monthly* 85 (1978): 439–450.
9. R. L. E. Schwarzenberger, "The 17 Plane Symmetry Groups," *Mathematical Gazette* 58 (1974): 123–131.
10. A. V. Shubnikov and V. A. Koptsik, *Symmetry in Science and Art,* New York: Plenum Press, 1974.
11. D. K. Washburn and D. W. Crowe, *Symmetries of Culture: Theory and Practice of Plane Pattern Analysis,* Seattle: University of Washington Press, 1988.
12. H. Weyl, *Symmetry,* Princeton: Princeton University Press, 1952.

SUGGESTED READINGS

Vladimir Dubrovsky, "Ornamental Groups," *Quantum,* November/December (1991): 32–35.
    An introduction to crystallographic groups illustrated with Escher prints.

M. Gardner, "The Eerie Mathematical Art of Maurits C. Escher," *Scientific American,* April (1966): 110–121.
    The author analyzes many of Escher's most famous works.

S. Garfunkel et al., *For All Practical Purposes,* 2d ed., New York: W. H. Freeman, 1991.
    This book has a well-written, richly illustrated chapter on symmetry in art and nature.

W. G. Jackson, "Symmetry in Automobile Tires and the Left-Right Problem," *Journal of Chemical Education,* 69 (1992): 624–626.

> This article uses automobile tires as a tool for introducing and explaining the symmetry terms and concepts important in chemistry.

C. MacGillivray, *Fantasy and Symmetry—The Periodic Drawings of M. C. Escher,* New York: Harry N. Abrams, 1976.

> This is a collection of Escher's periodic drawings together with a mathematical discussion of each one.

B. Rose and R. Stafford, "An Elementary Course in Mathematical Symmetry," *The American Mathematical Monthly* 88 (1981): 59–64.

> This article contains identification algorithms for the frieze groups and the wallpaper groups.

D. Schattschneider, *Visions of Symmetry,* New York: Freeman, 1990.

> A loving, lavish, encyclopedic book on the drawings of M. C. Escher.

H. von Baeyer, "Impossible Crystals," *Discover* 11 (2) (1990): 69–78.

> This article tells how the discovery of nonperiodic tilings of the plane led to the discovery of quasicrystals. The x-ray diffraction patterns of quasicrystals exhibit fivefold symmetry—something that had been thought to be impossible.

SUGGESTED FILMS

Adventures in Perception, Film Productie, 22 minutes, in color, 1989. Distributed by BFA Educational Media, Santa Monica, Calif.

> This excellent film features some 50 of Escher's works, with Escher himself discussing some of them.

Maurits Escher: Painter of Fantasies, A Coronet Film, Chicago, $26\frac{1}{4}$ minutes, in color, 1970.

> This award-winning film features Escher talking informally about his work. On several occasions, he mentions mathematics and at one point he constructs a Möbius strip. The film has pleasant background music and presents a close-up view of numerous prints that can serve as starting points for mathematical discussions.

M. C. Escher

I never got a pass mark in math. The funny thing is I seem to latch on to mathematical theories without realizing what is happening. No indeed, I was a pretty poor pupil at school. And just imagine—mathematicians now use my prints to illustrate their books. Fancy me consorting with all these learned folk, as though I were their long-lost brother. I guess they are quite unaware of the fact that I'm ignorant about the whole thing.

M. C. Escher

M. C. ESCHER was born on June 17, 1898, in the Netherlands. His artistic work prior to 1937 was dominated by the representation of visible reality, such as landscapes and buildings. Gradually, he became less and less interested in the visible world and became increasingly absorbed in an inventive approach to space. He studied the abstract space-filling patterns used in the Moorish mosaics in the Alhambra in Spain. He also studied the mathematician George Polya's paper on the 17 plane crystallographic groups. Instead of the geometric motifs used by the Moors and Polya, Escher preferred to use animals, plants, or people in his space-filling prints.

Escher was fond of incorporating various mathematical ideas into his works. Among these are infinity, Möbius bands, stellations, deformations, reflections, Platonic solids, spirals, and the hyperbolic plane. This latter idea was suggested to Escher by a figure in a paper by the geometer H. S. M. Coxeter.

Although Escher originals are now quite expensive, it was not until 1951 that he derived a significant portion of his income from his prints. Today, Escher is widely known and appreciated as a graphic artist. His graphics have appeared on postage stamps, bank notes, a candy box (in the shape of an icosahedron!), note cards, T-shirts, jigsaw puzzles, record album covers, and covers of dozens of scientific publications. His prints have been used to illustrate ideas in hundreds of scientific works. Despite this popularity among scientists, Escher has never been held in high esteem in traditional art circles. Escher died on March 27, 1973, in Holland.

29

Symmetry and Counting

Let us pause to slake our thirst one last time at symmetry's bubbling spring.

Timothy Ferris, *Coming of Age in the Milky Way*

MOTIVATION

Permutation groups naturally arise in many situations involving symmetrical designs or arrangements. Consider, for example, the task of coloring the six vertices of a regular hexagon so that three are black and three are white. Figure 29.1 shows the 20 possibilities. However, if these designs appeared on one side of hexagonal ceramic tiles, it would be nonsensical to count the designs shown in Figure 29.1(a) as different since all six designs shown there can be obtained from one of them by rotating. (A manufacturer would only make one of the six.) In this case we say that the designs in Figure 29.1(a) are *equivalent* under the group of rotations of the hexagon. Similarly the designs in Figure 29.1(b) are equivalent under the group of rotations, as are the designs in Figure 29.1(c) and (d). And, since no design from any one of those in Figure 29.1(a)–(d) can be obtained from a design from a different figure by rotation, we see that the designs within each figure are equivalent to each other but nonequivalent to any design in another figure. In contrast, the designs in Figure 29.1(b) and (c) are equivalent under the dihedral group D_6, since the designs in Figure 29.1(b) can be reflected to give the designs

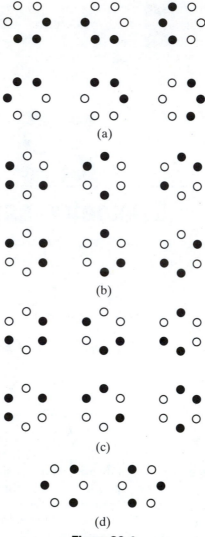

Figure 29.1

in Figure 29.1(c). For example, for purposes of arranging three black beads and three white beads to form a necklace, the designs shown in Figure 29.1(b) and (c) would be considered equivalent.

In general, we say that two designs (arrangements) A and B are *equivalent under a group G* of permutations if there is an element ϕ in G such that $\phi(A) = B$. That is, two designs are equivalent under G if they are in the same orbit of G. It follows, then, that the number of non-equivalent designs under G is simply the number of orbits of G. (The set being permuted is the set of all possible designs or arrangements.)

Notice that the designs in Figure 29.1 divide into four orbits under the group of rotations but only three orbits under the group D_6 since the designs in Figure 29.1(b) and (c) form a single orbit under D_6. Thus, we could obtain all 20 tile designs from just four tiles, but we could obtain all 20 necklaces from just three of them.

BURNSIDE'S THEOREM

Although the problems we have just posed are simple enough to solve by observation, more complicated ones require a more sophisticated approach. Such an approach was provided by the English mathematician William Burnside in his classic book on group theory published in 1911. But first, a definition.

DEFINITION   Elements Fixed by ϕ

For any group G of permutations on a set S and any ϕ in G, we let $\text{fix}(\phi) = \{i \in S \mid \phi(i) = i\}$. This set is called the *elements fixed by ϕ* (or more simply, "fix of ϕ").

Theorem 29.1   *(Burnside)*

If G is a finite group of permutations on a set S, then the number of orbits of G on S is

$$\frac{1}{|G|} \sum_{\phi \in G} |\text{fix}(\phi)|.$$

Proof.   Let n denote the number of pairs (ϕ, i), with $\phi \in G$, $i \in S$, and $\phi(i) = i$. We begin by counting these pairs in two ways. First, for each particular ϕ in G, the number of such pairs is exactly $|\text{fix}(\phi)|$ as i runs over S. So,

$$n = \sum_{\phi \in G} |\text{fix}(\phi)|. \tag{1}$$

Second, for each particular i in S, observe that $|\text{stab}_G(i)|$ is exactly the number of pairs of the form (ϕ, i), as ϕ runs over G. So,

$$n = \sum_{i \in S} |\text{stab}_G(i)|. \tag{2}$$

It follows from exercise 33 in Chapter 8 that if s and t are in the same orbit of G, then $\text{orb}_G(s) = \text{orb}_G(t)$ and $|\text{stab}_G(s)| = |\text{stab}_G(t)|$. So, if we choose $s \in S$, sum over $\text{orb}_G(s)$, and appeal to the Orbit-Stabilizer Theorem (Theorem 8.2), we have

$$\sum_{t \in \text{orb}_G(s)} |\text{stab}_G(t)| = |\text{orb}_G(s)| \, |\text{stab}_G(s)| = |G|. \tag{3}$$

Finally, by summing over all the elements of G, one orbit at a time, it follows from Equations (1), (2), and (3) that

$$\sum_{g \in G} |\text{fix}(\phi)| = \sum_{i \in S} |\text{stab}_G(i)| = |G| \cdot \text{(number of orbits)}$$

and the result follows.   ∎

APPLICATIONS

To illustrate how to apply Burnside's Theorem, let us return to the ceramic tile and necklace problems. In the case of counting hexagonal tiles with three black vertices and three white vertices, the set of objects being permutated is the 20 possible designs, whereas the group of permutations is the group of six rotational symmetries of a hexagon. Obviously, the identity fixes all 20 designs. We see from Figure 29.1 that rotations of 60°, 180°, or 300° fix none of the 20 designs. Finally, Figure 29.2 shows fix(ϕ) for the rotations of 120° and 240°. These data are collected in Table 29.1.

Figure 29.2   Tile designs fixed by 120° rotation and 240° rotation.

So, applying Burnside's Theorem, we obtain that the number of orbits under the group of rotations is

$$\frac{1}{6}(20 + 0 + 2 + 0 + 2 + 0) = 4.$$

Now let's use Burnside's Theorem to count the number of necklace arrangements consisting of three black beads and three white beads. (For the purposes of analysis, we may arrange the beads in the shape of a regular hexagon.) For this problem, two arrangements are equivalent

Table 29.1

Element	Number of Designs Fixed by Element
Identity	20
Rotation of 60°	0
Rotation of 120°	2
Rotation of 180°	0
Rotation of 240°	2
Rotation of 300°	0

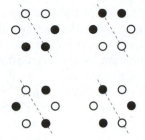

Figure 29.3 Bead arrangements fixed by the reflection across a diagonal.

if they are in the same orbit under D_6. Figure 29.3 shows the arrangements fixed by a reflection across a diagonal. Table 29.2 summarizes the information needed to apply Burnside's Theorem.

So, there are

$$\frac{1}{12}(1 \cdot 20 + 1 \cdot 0 + 2 \cdot 2 + 2 \cdot 0 + 3 \cdot 4 + 3 \cdot 0) = 3$$

nonequivalent ways to string three black beads and three white beads on a necklace.

Now that we have gotten our feet wet on a few easy problems, let's try a more difficult one. Suppose that we have the colors red(R), white(W), and blue(B) that can be used to color the edges of a regular tetrahedron (see Figure 5.1). First, observe that there are $3^6 = 729$ colorings without regard to equivalence. How shall we decide when two colorings of the tetrahedron are nonequivalent? Certainly, if we were to pick up a tetrahedron colored in a certain manner, rotate it, and put it back down, we would think of the tetrahedron as being positioned differently rather than as being colored differently (just as if we picked up a die labeled in the usual way and rolled it we would not say that the die is now differently labeled). So our permutation group for this problem is just the group of 12 rotations of the tetrahedron shown in Figure 5.1.

Table 29.2

Type of Element	Number of Elements of This Type	Number of Arrangements Fixed by Type of Element
Identity	1	20
Rotation of order 2 (180°)	1	0
Rotation of order 3 (120° or 240°)	2	2
Rotation of order 6 (60° or 300°)	2	0
Reflection across diagonal	3	4
Reflection across side bisector	3	0

(The group consists of the identity; eight elements of order 3, each of which fixes one vertex; and three elements of order 2, each of which fixes no vertex.) Every rotation permutes the 729 colorings, and to apply Burnside's theorem we must determine the size of fix(ϕ) for each of the 12 rotations of the group.

Clearly, the identity fixes all 729 colorings. Next, consider the element (234) of order 3 shown in the bottom row, second from the left in Figure 5.1. Suppose that a specific coloring is fixed by this element (that is, the tetrahedron appears to be colored the same before and after this rotation). Since (234) carries edge 12 to edge 13, edge 13 to edge 14, and edge 14 to edge 12, these three edges must agree in color (edge ij is the edge joining vertex i and vertex j). The same argument shows that the three edges 23, 34, and 42 also must agree in color. So, $|\text{fix}(234)| = 3^2$ since there are three choices for each of these two sets of three edges. The nine columns of Table 29.3 show the possible colorings of the two sets of three edges. The analogous analysis applies to the other seven elements of order 3.

Now consider the rotation (12)(34) of order 2. (See the second tetrahedron in the top row in Figure 5.1.) Since edges 12 and 34 are fixed, they may be colored in any way and will appear the same after the rotation (12)(34). This gives $3 \cdot 3$ choices for those two edges. Since edge 13 is carried to edge 24, they must agree in color. Similarly, edges 23 and 14 must agree. This means that we have $3 \cdot 3 \cdot 3 \cdot 3$ ways to color the tetrahedron that will be equivalent under (12)(34). (Table 29.4 gives the complete list of 81 colorings.) So, $|\text{fix}((12)(34))| = 3^4$, and the other two elements of order 2 yield the same results.

Now that we have analyzed the three types of group elements, we can apply Burnside's Theorem. In particular,

$$\frac{1}{12}[1 \cdot 3^6 + 8 \cdot 3^2 + 3 \cdot 3^4] = 87.$$

Surely it would be a difficult task to solve this problem without Burnside's Theorem.

Just as surely, you are wondering who besides mathematicians is

Table 29.3   Nine Colorings Fixed by (234)

Edge	Colorings								
12	R	R	R	W	W	W	B	B	B
13	R	R	R	W	W	W	B	B	B
14	R	R	R	W	W	W	B	B	B
23	R	W	B	W	R	B	B	R	W
34	R	W	B	W	R	B	B	R	W
24	R	W	B	W	R	B	B	R	W

Table 29.4 81 Colorings Fixed by (12)(34) (X and Y can be any of R, W, and B)

Edge	Colorings								
12	X	X	X	X	X	X	X	X	X
34	Y	Y	Y	Y	Y	Y	Y	Y	Y
13	R	R	R	W	W	W	B	B	B
24	R	R	R	W	W	W	B	B	B
23	R	W	B	W	R	B	B	R	W
14	R	W	B	W	R	B	B	R	W

interested in counting problems such as the ones we have discussed. Well, chemists are. Indeed, a benzene molecule can be viewed as six carbon atoms arranged in a hexagon with any of the three radicals NH_2, COOH, or OH attached at each carbon atom. See Figure 29.4 for one example.

So Burnside's Theorem enables a chemist to determine the number of benzene molecules (see exercise 4). Another kind of molecule that chemists consider is visualized as a regular tetrahedron with a carbon atom at the center and any of the four radicals $HOCH_2$ (hydroxymethyl), C_2H_5 (ethyl), Cl (chlorine) or H (hydrogen) at the four vertices. Again, the number of such molecules can be easily counted using Burnside's Theorem.

GROUP ACTION

Our informal approach to counting the number of objects that are considered nonequivalent can be made formal as follows. If G is a group and S is a set of objects, we say that *G acts on S* if there is a

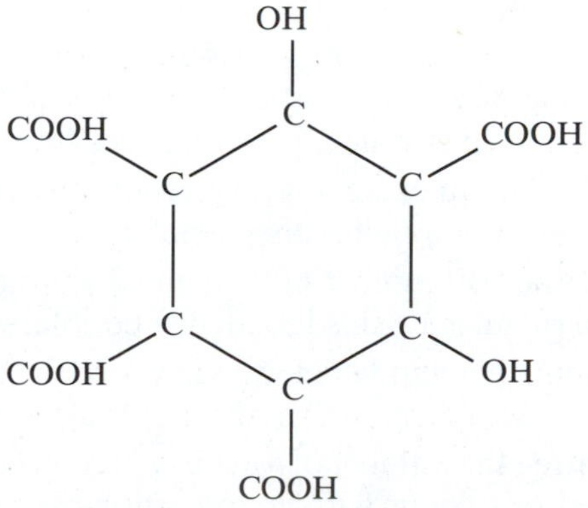

Figure 29.4 A benzene molecule.

homomorphism ϕ from G to sym(S), the group of all permutations on S. (The homomorphism is sometimes called the *group action*.) For convenience, we denote the image of g under ϕ as ϕ_g. Then two objects x and y in S are viewed as equivalent under the action of G if and only if $\phi_g(x) = y$ for some g in G. Notice that when ϕ is one-to-one the elements of G may be regarded as permutations on S. On the other hand, when ϕ is not one-to-one the elements of G may still be regarded as permutations on S but there are distinct elements g and h in G such that ϕ_g and ϕ_h induce the same permutation on S [that is, $\phi_g(x) = \phi_h(x)$ for all x in S]. Thus, a group acting on a set is a natural generalization of the permutation group concept.

EXERCISES

When things go wrong, as they sometimes will,
When the road you're trudging seems all up hill,
. . . When care is pressing you down a bit,
Rest, if you must—but don't you quit.
. . . Often the goal is nearer than
It seems to a faint and faltering man,
Often the struggler has given up
When he might have captured the victor's cup.

Anonymous

1. Determine the number of ways in which the four corners of a square can be colored with two colors. (It is permissible to use a single color on all four corners.)

2. Determine the number of different necklaces that can be made using 13 white beads and three black beads.

3. Determine the number of ways in which the vertices of an equilateral triangle can be colored with five colors so that at least two colors are used.

4. A benzene molecule can be viewed as six carbon atoms arranged in a regular hexagon. At each carbon atom, one of three radicals (NH_2, COOH, or OH) can be attached. How many such compounds are possible?

5. Suppose that in exercise 4 we permit only two choices for the radicals. How many compounds are possible?

6. Determine the number of ways in which the faces of a regular dodecahedron (regular 12-sided solid) can be colored with three colors.

7. Determine the number of ways in which the edges of a square can be colored with six colors so that no color is used on more than one edge.

8. Determine the number of ways in which the edges of a square can be colored with six colors with no restriction placed on the number of times a color can be used.

9. Let G be a finite group and sym(G) the group of all permutations on G. For each g in G, let ϕ_g denote the element of sym(G) defined by $\phi_g(x) = gxg^{-1}$ for all x in G. Show that G acts on itself under the action $g \to \phi_g$. Give an example in which the mapping $g \to \phi_g$ is not one-to-one.

10. Let G be a finite group, H a subgroup of G, and S the set of left cosets of H in G. For each g in G, let ϕ_g denote the element of sym(S) defined by $\phi_g(xH) = gxH$. Show that G acts on S under the action $g \to \phi_g$.

SUGGESTED READING

Norman Biggs, *Discrete Mathematics,* Oxford: Clarendon Press, 1989.
Chapter 20 of this book presents a more detailed treatment of the subject of symmetry and counting.

William Burnside

Burnside, during a life of steadfast devotion to his science, has contributed to many an issue. In one of the most abstract domains of thought, he has systematized and amplified its range so that, there, his work stands as a landmark in the widening expanse of knowledge. Whatever be the estimate of Burnside made by posterity, contemporaries salute him as a Master among the mathematicians of his own generation.

A. R. Forsyth, *Journal of the London Mathematical Society*

WILLIAM BURNSIDE was born on July 2, 1852, in London. In 1871, he entered Cambridge University and was considered the best of his college class. After graduating in 1875, Burnside was appointed lecturer at Cambridge, where he stayed until 1885. He then accepted a position at the Royal Naval College at Greenwich and spent the rest of his career in that post.

Burnside wrote more than 150 research papers in many fields. Most of his early papers were devoted to applied mathematics, principally hydrodynamics. He also published papers on differential geometry, elliptic functions, and probability theory. He is best remembered, however, for his pioneering work in group theory, which appeared in some 50 papers, and his classic book entitled *Theory of Groups*. Because of his emphasis on the abstract approach, many consider Burnside to have been the first pure group theorist.

One mark of greatness in a mathematician is the ability to pose important and challenging problems—problems that open up new areas of research for future generations. Here, Burnside excelled. It was he who first conjectured that a group of odd order is solvable (that is, that the group G has a series of normal subgroups, $G = G_0 \geq G_1 \geq G_2 \geq \cdots \geq G_n = \{e\}$, such that G_i/G_{i+1} is Abelian). This extremely important conjecture was finally proved more than 50 years later by Feit and Thompson in a 255-page paper (see Chapter 25 for more on this). Another of Burnside's conjectures concerns the unique group $B_{m,n}$ with the following three properties: (1) the group has m generators; (2) $x^n = e$ for all x in the group; and (3) every group satisfying properties 1 and 2 is a factor group of $B_{m,n}$. In 1902, Burnside conjectured that $B_{m,n}$ is finite for all m and n. So far it is known that $B_{m,n}$ is finite when $n = 1, 2, 3, 4$, or 6 and that $B_{m,n}$ is infinite when $m \geq 2$ and n is odd and greater than or equal to 655.

Burnside was elected a Fellow of the Royal Society and awarded two Royal medals. He served as president of the Council of the London Mathematical Society and received their De Morgan medal. Burnside died on August 21, 1927.

30

Cayley Digraphs of Groups

The important thing in science is not so much to obtain new
facts as to discover new ways of thinking about them.

Sir William Lawrence Bragg, *Beyond Reductionism*

MOTIVATION

In this chapter, we introduce a graphical representation of a group given
by a set of generators and relations. The idea was originated by Cayley
in 1878. Although this topic is not usually covered in an abstract algebra
book, we include it for four reasons: it provides a method of visualizing
a group (in fact, the word *graph* is Latin for "picture"); it connects two
important branches of modern mathematics—groups and graphs; it
gives a review of some of our old friends—cyclic groups, dihedral
groups, direct products, and generators and relations; and most impor-
tantly, it is fun!

Intuitively, a directed graph (or digraph) is a finite set of points,
called *vertices,* and a set of arrows, called *arcs,* connecting some of the
vertices. Although there is a rich and important general theory of
directed graphs with many applications (see [2]), we are interested only
in those that arise from groups.

THE CAYLEY DIGRAPH OF A GROUP

DEFINITION   Cayley Digraph of a Group

Let G be a finite group and S a set of generators for G. We define a digraph Cay(S:G), called the *Cayley digraph of G with generating set S*, as follows.

1. Each element of G is a vertex of Cay(S:G).
2. For x and y in G, there is an arc from x to y if and only if $xs = y$ for some $s \in S$.

    To tell from the digraph which particular generator connects two vertices, Cayley proposed that each generator be assigned a color, and that the arrow joining x to xs be colored with the color assigned to s. He called the resulting figure the *color graph of the group*. This terminology is still occasionally used. Rather than use colors to distinguish the different generators, we will use solid arrows, dashed arrows, and dotted arrows. In general, if there is an arc from x to y, there need not be an arc from y to x. An arrow emanating from x and pointing to y indicates that there is an arc from x to y.

    Following are numerous examples of Cayley digraphs. Note that there are several ways to draw the digraph of a group given by a particular generating set. However, it is not the appearance of the graph that is relevant but the manner in which the vertices are connected. These connections are uniquely determined by the generating set. Thus, distances between vertices and angles formed by the arcs have no significance. (In the digraph below, a headless arrow joining two vertices x and y indicates that there is an arc from x to y and an arc from y to x. This occurs when the generating set contains both an element and its inverse. For example, a generator of order 2 is its own inverse.)

Example 1 $Z_6 = \langle 1 \rangle$.

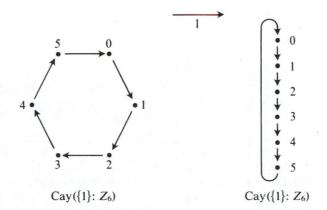

Cay({1}: Z_6)   Cay({1}: Z_6)

Example 2 $Z_3 \oplus Z_2 = \langle (1, 0), (0, 1) \rangle$.

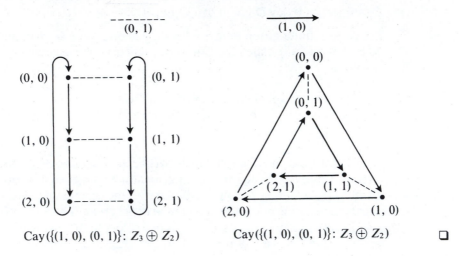

Cay$(\{(1, 0), (0, 1)\}: Z_3 \oplus Z_2)$ Cay$(\{(1, 0), (0, 1)\}: Z_3 \oplus Z_2)$ ❑

Example 3 $D_4 = \langle R_{90}, H \rangle$.

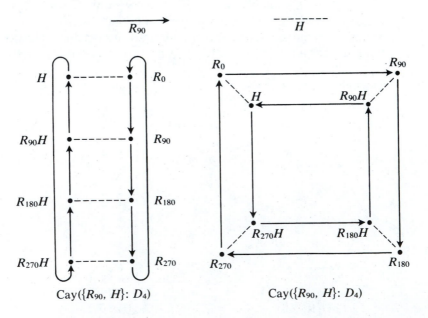

Cay$(\{R_{90}, H\}: D_4)$ Cay$(\{R_{90}, H\}: D_4)$ ❑

Example 4 $S_3 = \langle (12,), (123) \rangle$.

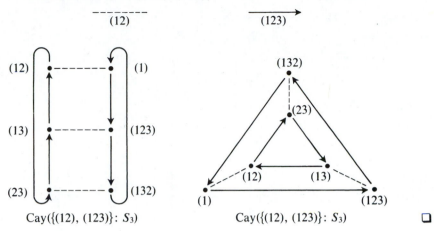

Cay({(12), (123)} : S_3)   Cay({(12), (123)} : S_3)   ❑

Example 5 $S_3 = \langle (12), (13) \rangle$.

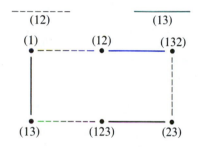

❑

Example 6 $A_4 = \langle (12)(34), (123) \rangle$.

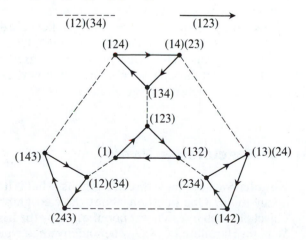

❑

Example 7 $Q_4 = \langle a, b \mid a^4 = e, a^2 = b^2, b^{-1}ab = a^3 \rangle.$

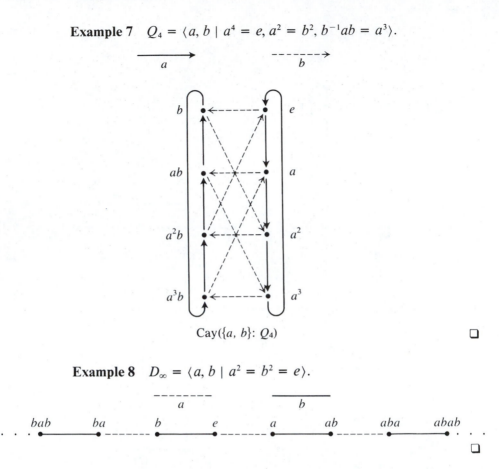

Cay($\{a, b\}$: Q_4)   ❏

Example 8 $D_\infty = \langle a, b \mid a^2 = b^2 = e \rangle.$

bab	*ba*	*b*	*e*	*a*	*ab*	*aba*	*abab*

❏

The Cayley digraph provides a quick and easy way to determine the value of any product of the generators and their inverses. Consider, for example, the product $ab^3ab^{-1}ab^{-1}b^{-3}$ from the group given in Example 7. To reduce this to one of the eight elements used to label the vertices, we need only begin at the vertex e and follow the arcs from each vertex to the next as specified in the given product. Of course, a^{-1} means traverse the a arc in reverse. (Observations such as $b^{-3} = b$ also help.) Tracing the product through, we obtain ab. Similarly, one can verify or discover relations among the generators.

HAMILTONIAN CIRCUITS AND PATHS

Now that we have these directed graphs, what is it that we care to know about them? One question about directed graphs that has been the object of much research was popularized by the Irish mathematician Sir William Hamilton in 1859, when he invented a puzzle called "Around

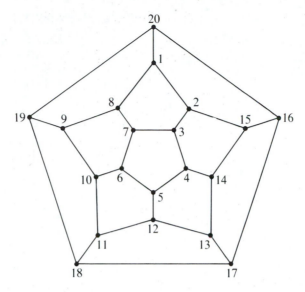

Figure 30.1 Around the world.

the World." His idea was to label the 20 vertices of a regular dodeca-hedron with the names of famous cities. One solves this puzzle by start-ing at any particular city (vertex) and travelling "around the world," moving along the arcs in such a way that each other city is visited exactly once before returning to the original starting point. One solution to this puzzle is given in Figure 30.1, where the vertices are visited in the order indicated.

Obviously, this idea can be applied to any digraph; that is, one starts at some vertex and attempts to traverse the digraph by moving along arcs in such a way that each vertex is visited exactly once before return-ing to the starting vertex. (To go from x to y, there must be an arc from x to y.) Such a sequence of arcs is called a *Hamiltonian circuit* in the diagraph. A sequence of arcs that passes through each vertex exactly once without returning to the starting point is called a *Hamiltonian path*. In the remainder of this chapter, we concern ourselves with the existence of Hamiltonian circuits and paths in Cayley digraphs.

Figures 30.2 and 30.3 show a Hamiltonian path for the digraph given in Example 2 and a Hamiltonian circuit for the diagraph given in Example 7, respectively.

Is there a Hamiltonian circuit in

$$\text{Cay}(\{(1, 0), (0, 1)\}: Z_3 \oplus Z_2)?$$

More generally, let us investigate the existence of Hamiltonian circuits in

$$\text{Cay}(\{(1, 0), (0, 1)\}: Z_m \oplus Z_n),$$

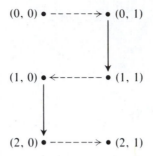

Figure 30.2   Hamiltonian path in Cay({(1, 0), (0, 1)}: $Z_3 \oplus Z_2$) from (0, 0) to (2, 1).

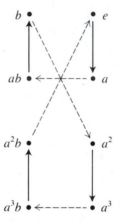

Figure 30.3   Hamiltonian circuit in Cay({a, b}: Q_4).

where m and n are relatively prime and both greater than 1. Visualize the Cayley digraph as a rectangular grid coordinatized with $Z_m \oplus Z_n$, as in Figure 30.4. Suppose there is a Hamiltonian circuit in the digraph and (a, b) is some vertex where the circuit exits horizontally. (Clearly, such a vertex exists.) Then the circuit must exit $(a - 1, b + 1)$ horizontally also, for otherwise the circuit passes through $(a, b + 1)$ twice—see Figure 30.5. Repeating this argument again and again, we see that the circuit exits horizontally from each of the vertices (a, b), $(a - 1, b + 1)$, $(a - 2, b + 2), \ldots$, which is just the coset $(a, b) + \langle(-1, 1)\rangle$. But when m and n are relatively prime, $\langle(-1, 1)\rangle$ is the entire group. Obviously there cannot be a Hamiltonian circuit consisting entirely of horizontal moves. Let us record what we have just proved.

Theorem 30.1   *A Necessary Condition*

Cay({(1, 0), (0, 1)}: $Z_m \oplus Z_n$) does not have a Hamiltonian circuit when m and n are relatively prime and greater than 1.

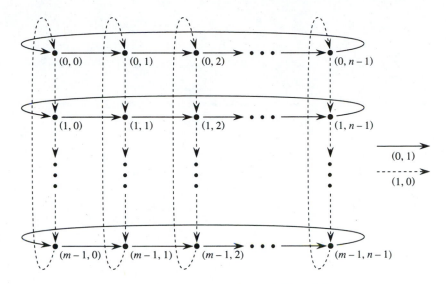

Figure 30.4 Cay({(1, 0), (0, 1)}: $Z_m \oplus Z_n$).

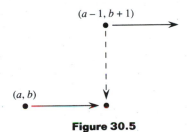

Figure 30.5

What about when m and n are not relatively prime? In general, the answer is somewhat complicated (see [7]), but the following special case is easy to prove.

Theorem 30.2 *A Sufficient Condition*
Cay({(1, 0), (0, 1)}: $Z_m \oplus Z_n$) has a Hamiltonian circuit when n divides m.

Proof. Say, $m = kn$. Then we may think of $Z_m \oplus Z_n$ as k blocks of size $n \times n$. (See Figure 30.6 for an example.) Start at $(0, 0)$ and cover the vertices of the top block as follows. Use the generator $(0, 1)$ to move horizontally across the first row to the end. Then use the generator $(1, 0)$ to move vertically to the point below, and cover the remaining points in the second row by moving horizontally. Keep this process up until arriving at the point $(n - 1, 0)$—the lower left-hand corner of the first block. Next, move vertically to the second block and repeat the process used

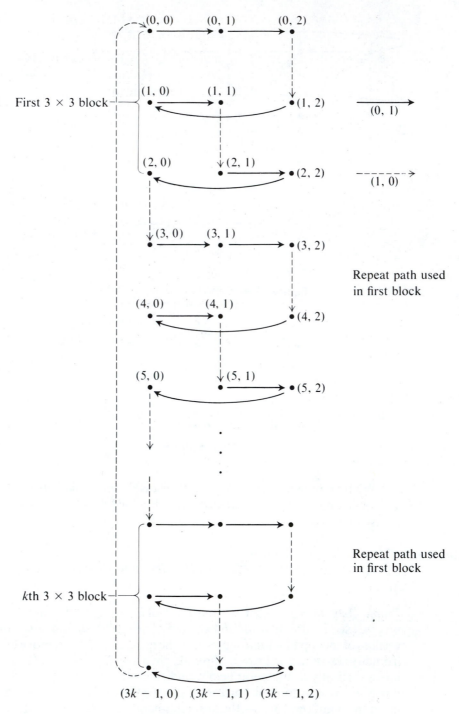

Figure 30.6 Cay({(1, 0), (0, 1)}: $Z_{3k} \oplus Z_3$).

in the first block. Keep this up until the bottom block is covered. Complete the circuit by moving vertically back to (0, 0). ∎

Notice that the circuit given in the proof of Theorem 30.2 is easy to visualize but somewhat cumbersome to describe in words. A much more convenient way to describe a Hamiltonian path or circuit is to specify the starting vertex and the sequence of generators in the order in which they are to be applied. In Example 5, for instance, we may start at (1) and alternate the generators (12) and (13) until we return to (1). In Example 3, we may start at R_0 and successively apply R_{90}, R_{90}, R_{90}, H, R_{90}, R_{90}, R_{90}, H. When k is a positive integer and a, b, . . . , c is a sequence of group elements, we use $k * (a, b, . . . , c)$ to denote the concatenation of k copies of the sequence $(a, b, . . . , c)$. Thus, $2 *$ $(R_{90}, R_{90}, R_{90}, H)$ and $2 * (3 * R_{90}, H)$ both mean R_{90}, R_{90}, R_{90}, H, R_{90}, R_{90}, R_{90}, H. With this notation, we may conveniently denote the Hamiltonian circuit given in Theorem 30.2 as

$$m * ((n - 1) * (0, 1), (1, 0)).$$

We leave it as an exercise to show that if $x_1, x_2, . . . , x_n$ is a sequence of generators determining a Hamiltonian circuit starting at some vertex, then the same sequence determines a Hamiltonian circuit for any starting vertex.

From Theorem 30.1, we know that there are some Cayley digraphs of Abelian groups that do not have any Hamiltonian circuits. But Theorem 30.3 (first proved by Holsztyński and Nathanson [5]) shows that each of these Cayley digraphs does have a Hamiltonian path. There are some Cayley digraphs for *non-Abelian* groups that do not even have Hamiltonian paths, but we will not discuss them here. (See [5] for more information.)

Theorem 30.3 *Abelian Groups Have Hamiltonian Paths*
Let G be a finite Abelian group, and let S be any (nonempty*) generating set for G. Then Cay *(S:G) has a Hamiltonian path.*

Proof. We induct on $|S|$. If $|S| = 1$, say, $S = \{a\}$, then the digraph is just a circle labeled with e, a, a^2, . . . , a^{m-1}, where $|a| = m$. Obviously, there is a Hamiltonian path for this case. Now assume that $|S| > 1$. Choose some $s \in S$. Let $T = S - \{s\}$—that is, T is S with s removed—and set $H = \langle T \rangle$. (Notice that H may be equal to G.)

Because $|T| < |S|$ and H is a finite Abelian group, the induction hypothesis guarantees that there is a Hamiltonian path $(a_1, a_2, . . . , a_k)$ in Cay($T:H$). We will show that

$$(a_1, a_2, . . . , a_k, s, a_1, a_2, . . . , a_k, s, . . . , a_1, a_2, . . . , a_k, s, a_1, a_2, . . . , a_k),$$

---

*If S is the empty set, it is customary to define $\langle S \rangle$ as the identity group. We prefer to ignore this trivial case.

where $a_1, a_2, \ldots, a_k$ occurs $|G|/|H|$ times and s occurs $|G|/|H| - 1$ times, is a Hamiltonian path in Cay(S:G).

Because $S = T \cup \{s\}$ and T generates H, the coset Hs generates the factor group G/H. (Since G is Abelian, this group exists.) Hence, the cosets of H are $H, Hs, Hs^2, \ldots, Hs^n$, where $n = |G|/|H| - 1$. Starting from the identity element of G, the path given by $(a_1, a_2, \ldots, a_k)$ visits each element of H exactly once [because $(a_1, a_2, \ldots, a_k)$ is a Hamiltonian path in Cay(T:H)]. The generator s then moves us to some element of the coset Hs. Starting from there, the path $(a_1, a_2 \ldots, a_k)$ visits each element of Hs exactly once. Then, s moves us to the coset Hs^2, and we visit each element of this coset exactly once. Continuing this process, we successively move to $Hs^3, Hs^4, \ldots, Hs^n$, visiting each vertex in each of these cosets exactly once. Because each vertex of Cay(S:G) is in exactly one coset Hs^i, this implies that we visit each vertex of Cay(S:G) exactly once. Thus we have a Hamiltonian path. ∎

We next look at three generated Cayley digraphs.

Example 9   Let

$$D_3 = \langle r, f \mid r^3 = f^2 = e, rf = fr^2 \rangle.$$

Then a Hamiltonian circuit in

$$\text{Cay}(\{(r, 0), (f, 0), (e, 1)\}: D_3 \oplus Z_6)$$

is given in Figure 30.7.   ☐

Although it is not easy to prove, it is true that

$$\text{Cay}(\{(r, 0), (f, 0), (e, 1)\}: D_n \oplus Z_m)$$

has a Hamiltonian circuit for all n and m. (See [9].) Example 10 shows the circuit for this digraph when m is even.

Example 10   Let

$$D_n = \langle r, f \mid r^n = f^2 = e, rf = fr^{-1} \rangle.$$

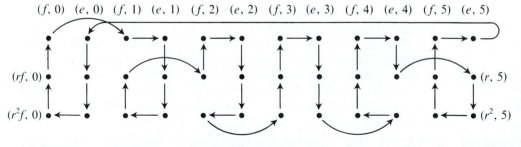

Figure 30.7

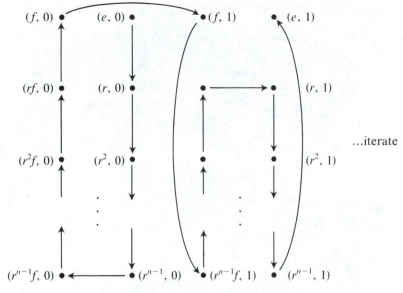

Figure 30.8

Then a Hamiltonian circuit in

$$\text{Cay}(\{(r, 0), (f, 0), (e, 1)\}: D_n \oplus Z_m)$$

with m even is traced in Figure 30.8. The sequence of generators that traces the circuit is

$$m * [(n - 1) * (r, 0), (f, 0), (n - 1) * (r, 0), (e, 1)].$$ ❑

SOME APPLICATIONS

Hamiltonian paths and circuits in Cayley digraphs have arisen in a variety of group theory contexts. A Hamiltonian path in a Cayley digraph of a group is simply an ordered listing of the group elements without repetition. The vertices of the digraph are the group elements, and the arcs of the path are generators of the group. In 1948, R. A. Rankin used these ideas (although not the terminology) to prove that certain bell-ringing exercises could not be done by the traditional methods employed by bell ringers. (See [3, Chap. 22] for the group-theoretical aspects of bell ringing.) In 1981, Hamiltonian paths in Cayley digraphs were used in an algorithm for creating computer graphics of Escher-type repeating patterns in the hyperbolic plane [4]. This program can produce repeating hyperbolic patterns in color from among five infinite classes of symmetry groups. The program has now been improved so that the user may choose from many kinds of color symmetry. Two Escher drawings and their computer-drawn counterparts are given in Figures 30.9–30.12.

Figure 30.9 M. C. Escher's *Circle Limit I* [6]

Figure 30.10   A computer duplication of the pattern of M. C. Escher's *Circle Limit I* [6]. The program used a Hamiltonian path in a Cayley digraph of the underlying symmetry group.

Figure 30.11   M. C. Escher's *Circle Limit IV* [6]

Figure 30.12 A computer drawing inspired by the pattern of M. C. Escher's *Circle Limit IV* [6]. The program used a Hamiltonian path in a Cayley digraph of the underlying symmetry group.

In this chapter, we have shown how one may construct a directed graph from a group. It is also possible to associate a group—called the *automorphism group*—with every directed graph. (See [8, pp. 16–20] or [1, pp. 104–107] for details.) In fact, several of the 26 sporadic simple groups were first constructed in this way.

EXERCISES

It is the function of creative men to perceive the relations between thoughts, or things, or forms of expression that may seem utterly different, and to be able to combine them into some new forms—the power to connect the seemingly unconnected.

William Plomer

1. Find a Hamiltonian circuit in the digraph given in Example 7 different from the one in Figure 30.3.
2. Find a Hamiltonian circuit in

$$\text{Cay}(\{(a, 0), (b, 0), (e, 1)\}: Q_4 \oplus Z_2).$$

3. Find a Hamiltonian circuit in

$$\text{Cay}(\{(a, 0), (b, 0), (e, 1)\}: Q_4 \oplus Z_m),$$

where m is even.

4. Write the sequence of generators for each of the circuits found in exercises 1, 2, and 3.
5. Use the Cayley digraph in Example 7 to evaluate the product $a^3ba^{-1}ba^3b^{-1}$.
6. Let x and y be two vertices of a Cayley digraph. Explain why two paths from x to y in the digraph yield a group relation.
7. Use the Cayley digraph in Example 7 to verify the relation $aba^{-1}b^{-1}a^{-1}b^{-1} = a^2ba^3$.
8. Identify the following Cayley digraph of a familiar group.

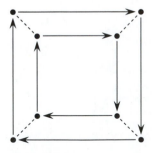

9. Let $D_4 = \langle r, f \mid r^4 = e = f^2, rf = fr^{-1} \rangle$. Verify that

$$6 * [3 * (r, 0), (f, 0), 3 * (r, 0), (e, 1)]$$

 is a Hamiltonian circuit in

$$\text{Cay}(\{(r, 0), (f, 0), (e, 1)\}: D_4 \oplus Z_6).$$

10. Draw a picture of $\text{Cay}(\{2, 5\}: Z_8)$.

11. If $s_1, s_2, \ldots, s_n$ is a sequence of generators that determines a Hamiltonian circuit beginning at some vertex, explain why the same sequence determines a Hamiltonian circuit beginning at any point.

12. Show that the Cayley digraph given in Example 7 has a Hamiltonian path from e to a.

13. Show that there is no Hamiltonian path in

$$\text{Cay}(\{(1, 0), (0, 1)\}: Z_3 \oplus Z_2)$$

 from $(0, 0)$ to $(2, 0)$.

14. Draw $\text{Cay}(\{2, 3\}: Z_6)$. Is there a Hamiltonian circuit in this digraph?

15. **a.** Let G be a group of order n generated by a set S. Show that a sequence $s_1, s_2, \ldots, s_{n-1}$ from S is a Hamiltonian path in $\text{Cay}(S:G)$ if and only if, for all i and j with $1 \leq i \leq j < n$, we have $s_i s_{i+1} \cdots s_j \neq e$.
    b. Show that the sequence in part a is a Hamiltonian circuit if and only if $s_1 s_2 \cdots s_n = e$, and that whenever $1 \leq i \leq j < n$, we have $s_i s_{i+1} \cdots s_j \neq e$.

16. Let $D_4 = \langle a, b \mid a^2 = b^2 = (ab)^4 = e \rangle$. Draw $\text{Cay}(\{a, b\}: D_4)$. Why is it reasonable to say that this digraph is undirected?

17. Let D_n be as in Example 10. Show that $2 * [(n-1) * r, f]$ is a Hamiltonian circuit in $\text{Cay}(\{r, f\}: D_n)$.

18. Let $Q_8 = \langle a, b \mid a^8 = e, a^4 = b^2, b^{-1}ab = a^{-1} \rangle$. Find a Hamiltonian circuit in $\text{Cay}(\{a, b\}: Q_8)$.

19. Let Q_8 be as in exercise 18. Find a Hamiltonian circuit in

$$\text{Cay}(\{(a, 0), (b, 0), (e, 1)\}: Q_8 \oplus Z_5).$$

20. Prove that the Cayley digraph given in Example 6 does not have a Hamiltonian circuit. Does it have a Hamiltonian path?

21. Find a Hamiltonian circuit in

$$(\text{Cay}(\{(R_{90}, 0), (H, 0), (R_0, 1)\}: D_4 \oplus Z_3).$$

 Does this circuit generalize to the case $D_{n+1} \oplus Z_n$ for all $n \geq 3$?

22. Let Q_8 be as in exercise 18. Find a Hamiltonian circuit in

$$\text{Cay}(\{(a, 0), (b, 0), (e, 1)\}: Q_8 \oplus Z_m) \text{ for all even } m.$$

23. Find a Hamiltonian circuit in

$$\text{Cay}(\{(a, 0), (b, 0), (e, 1)\}: Q_4 \oplus Z_3).$$

24. Find a Hamiltonian circuit in

$$\text{Cay}(\{(a, 0), (b, 0), (e, 1)\}: Q_4 \oplus Z_m) \text{ for all odd } m \geq 3.$$

25. Write the sequence of generators that describes the Hamiltonian circuit in Example 9.

26. Let D_n be as in Example 10. Find a Hamiltonian circuit in

$$\text{Cay}(\{(r, 0), (f, 0), (e, 1)\}: D_4 \oplus Z_5).$$

Does your circuit generalize to the case $D_n \oplus Z_{n+1}$ for all $n \geq 4$?

27. Prove that $\text{Cay}(\{(0, 1), (1, 1)\}: Z_m \oplus Z_n)$ has a Hamiltonian circuit for all m and n greater than 1.

28. Suppose that a Hamiltonian circuit exists for $\text{Cay}(\{(1, 0), (0, 1)\}: Z_m \oplus Z_n)$ and this circuit exits from vertex (a, b) vertically. Show that the circuit exits from every member of the coset $(a, b) + \langle (1, -1) \rangle$ vertically.

29. Let $D_2 = \langle r, f \mid r^2 = f^2 = e, rf = fr^{-1} \rangle$. Find a Hamiltonian circuit in $\text{Cay}(\{(a, 0), (b, 0), (e, 1)\}: D_2 \oplus Z_3)$.

30. Let Q_8 be as in exercise 18. Find a Hamiltonian circuit in $\text{Cay}(\{(a, 0), (b, 0), (e, 1)\}: Q_8 \oplus Z_3)$.

31. A finite group is called *Hamiltonian* if all of its subgroups are normal. (One non-Abelian example is Q_4.) Show that Theorem 30.3 can be generalized to include all Hamiltonian groups.

32. (Factor Group Lemma) Let S be a generating set for a group G, let N be a cyclic normal subgroup of G, and let

$$\overline{S} = \{sN \mid s \in S\}.$$

If $(a_1N, \ldots, a_rN)$ is a Hamiltonian circuit in $\text{Cay}(\overline{S}:G/N)$ and the product $a_1 \cdots a_r$ generates N, prove that

$$|N| * (a_1, \ldots, a_r)$$

is a Hamiltonian circuit in $\text{Cay}(S:G)$.

REFERENCES

1. Sabra S. Anderson, *Graph Theory and Finite Combinatorics,* Chicago: Markham, 1970.
2. Claude Berge, *Graphs and Hypergraphs,* Amsterdam: North-Holland, 1973.
3. F. J. Budden, *The Fascination of Groups,* Cambridge: Cambridge University Press, 1972.
4. Douglas Dunham, John Lindgren, and David Witte, "Creating Repeating Hyperbolic Patterns." *Computer Graphics* 15 (1981): 215–223.
5. W. Holsztyński and R. F. E. Strube, "Paths and Circuits in Finite Groups," *Discrete Mathematics* 22 (1978): 263–272.
6. J. L. Locher, ed., *The World of M. C. Escher,* New York: Harry N. Abrams. 1971.

7. W. T. Trotter, Jr., and P. Erdös, "When the Cartesian Product of Directed Cycles Is Hamiltonian," *Journal of Graph Theory* 2 (1978): 137–142.
8. Arthur T. White, *Graphs, Groups and Surfaces,* New York: Elsevier Science, 1984.
9. David Witte, Gail Letzter, and Joseph A. Gallian, "On Hamiltonian Circuits in Cartesian Products of Cayley Digraphs," *Discrete Mathematics* 43 (1983): 297–307.

SUGGESTED READINGS

Frank Budden, "Cayley Graphs for Some Well-Known Groups," *The Mathematical Gazette* 69 (1985): 271–278.
> This article contains the Cayley graphs of A_4, Q_4, and S_4 using a variety of generators and relations.

E. L. Burrows and M. J. Clark, "Pictures of Point Groups," *Journal of Chemical Education* 51 (1974): 87–90.
> Chemistry students may be interested in reading this article. It gives a comprehensive collection of the Cayley digraphs of the important point groups.

Douglas Dunham, John Lindgren, and David Witte, "Creating Repeating Hyperbolic Patterns," *Computer Graphics* 15 (1981): 215–223.
> In this beautifully illustrated paper, a process for creating repeating patterns of the hyperbolic plane is described. The paper is a blend of group theory, geometry, and art.

Joseph A. Gallian, "Circuits in Directed Grids," *The Mathematical Intelligencer* 13 (1991) 40–43.
> This article surveys research done on variations of the themes discussed in this chapter.

Joseph A. Gallian and David Witte, "Hamiltonian Checkerboards," *Mathematics Magazine* 57 (1984): 291–294.
> This paper gives some additional examples of Hamiltonian circuits in Cayley digraphs.

Paul Hoffman, "The Man Who Loves Only Numbers," *The Atlantic Monthly* 260 (1987): 60–74.
> A charming portrait of Paul Erdös, the most prolific and most eccentric mathematician in the world.

Henry Levinson, "Cayley Diagrams," in *Mathematical Vistas: Papers from the Mathematics Section,* New York Academy of Sciences, J. Malkevitch and D. McCarthy, eds., 1990: 62–68.
> This richly illustrated article presents Cayley digraphs of many of the groups that appear in this text.

A. T. White, "Ringing the Cosets," *The American Mathematical Monthly* 94 (1987): 721–746.
This article analyzes the practice of bell ringing by way of Cayley digraphs.

David Witte and Joseph A. Gallian, "A Survey: Hamiltonian Cycles in Cayley Graphs," *Discrete Mathematics* 51 (1984): 293–304.
This paper surveys the results, techniques, applications, and open problems in the field.

William Rowan Hamilton

After Isaac Newton, the greatest mathematician of the English-speaking peoples is William Rowan Hamilton.

Sir Edmund Whittaker,
Scientific American

$$j^2 = j^2 = k^2 = -1$$
$$ij = k \quad jk = i \quad ki = j$$
$$ji = -k \quad kj = -i \quad ik = -j$$

EUROPA

ÉIRE 29

Quaternions discovery by **Hamilton** 1843

This stamp featuring the quaternions was issued in 1983.

WILLIAM ROWAN HAMILTON was born on August 3, 1805, in Dublin, Ireland. Although Hamilton did not attend school before entering college, he was widely recognized as a child prodigy. At three, he was skilled at reading and arithmetic. At five, he read and translated Latin, Greek, and Hebrew; at 14, he had mastered 14 languages, including Arabic, Sanskrit, Hindustani, Malay, and Bengali.

Hamilton's undergraduate career at Trinity College in Dublin was brilliant. At the age of 21, he wrote a paper entitled *A Theory of Systems of Rays,* introducing techniques that have become indispensable in physics and have created the field of mathematical optics. This work was so extraordinary that a year later, when the chair of professor of astronomy at Trinity—the holder of which is given the title of Royal Astronomer of Ireland—became vacant, Hamilton was unanimously elected to the position, over many distinguished candidates, in spite of the fact that he was still an undergraduate and had not applied for the appointment!

In 1833, Hamilton provided the first modern treatment of complex numbers. In 1843, he made what he considered his greatest discovery—the algebra of quaternions. The quaternions represent a natural generalization of the complex numbers with three numbers $i, j,$ and k whose squares are -1. With these, rotations in three and four dimensions can be algebraically treated. Of greater significance, however, is the fact that the quaternions are noncommutative under multiplication. This was the first ring to be discovered in which the commutative property does not hold. After 10 years of fruitless thought, the essential idea for the quaternions suddenly came to him. Many years later, in a letter to his son, Hamilton described the circumstances of his discovery:

> But on the 16th day of the same month [October 1843]—which happened to be a Monday and a Council day of the Royal Irish Academy—I was walking to attend and preside, and your mother was walking with me, along the Royal Canal, to which she had perhaps been driven; and although she talked with me now and then, yet an under-current of thought was going on in my mind, which gave at last a result, whereof it is not too much to say that I felt at once the importance. An electric circuit seemed to close; and a spark flashed forth, the herald (as I foresaw immediately) of many long years to come of definitely directed thought and work, by myself if spared, and at all events on the part of others, if I should ever be allowed to live long enough distinctly to communicate the discovery. I pulled out on the spot a pocket-book, which still exists, and made an entry there and then. Nor could I resist the impulse—unphilosophical as it may have been—to cut with a knife on a stone of Brougham Bridge, as we passed it, the fundamental formula with the symbols i, j, k;
>
> $$i^2 = j^2 = k^2 = ijk = -1,$$
>
> which contains the solution of the Problem, but of course as an inscription, has long since mouldered away.

Today Hamilton's name is attached to several concepts, such as the Hamiltonian function, which represents the total energy in a physical system; the Hamilton-Jacobi differential equations; and the Cayley-Hamilton Theorem from linear algebra. He also coined the terms *vector, scalar,* and *tensor.*

In his later years, Hamilton was plagued by alcoholism. He died on September 2, 1865, at the age of 60.

Paul Erdös

Paul Erdös is a socially helpless Hungarian who has thought about more mathematical problems than anyone else in history.

The Atlantic Monthly

PAUL ERDÖS (pronounced AIR-dish) is one of the best known and most highly respected mathematicians of this century. Unlike most of his contemporaries who concentrate on theory building, Erdös focuses on problem solving and problem posing. The problems and methods of solution of Erdös—like those of Euler, whose solutions to special problems pointed the way to much of the mathematical theory we have today—have helped pioneer new theories, such as combinatorial and probabilistic number theory, combinatorial geometry, probabilistic and transfinite combinatorics, and graph theory.

Erdös was born on March 26, 1913, in Hungary. Both of his parents were high school mathematics teachers. They provided his early training and encouraged the development of his mathematical talent. Erdös was a mathematical prodigy. At the age of four, he told his mother that if you take 250 away from 100, you have 150 below 0. His first research paper, published when he was 18, gave a new proof that, for every positive integer n, there must exist a prime number between n and $2n$. Erdös, a Jew, left Hungary in 1934 at the age of 21 because of the rapid rise of anti-Semitism in Europe. Ever since then, he has been traveling. Erdös has no family, no property, no fixed address. He travels from place to place, never staying more than a month, giving lectures for small honoraria and staying with mathematicians. All that he owns he carries with him in a medium-sized suitcase, frequently visiting as many as 15 places in a month. He discusses his record for travel as follows. "There was a Saturday meeting in Winnipeg on number theory and computing. On Saturday evening we had a dinner in a Hungarian restaurant, a farewell dinner for the speaker. Then on Sunday morning I flew to Toronto. I was met at the airport, and we went to Waterloo to a picnic. In the evening I was taken back to Toronto, and I flew to London where I lectured at 11 o'clock at Imperial College." His motto is "Another roof, another proof." Although in his eighties, he puts in 19-hour days doing mathematics.

One might say that Erdös has lived a sheltered life. He buttered his first piece of bread at age 21. He has never cooked anything or even boiled water. He has never driven a car.

One of Erdös's customs is to offer cash prizes for the solutions to unsolved problems. These awards range from $5 to $10,000, depending on how difficult he judges them to be.

Erdös usually dispenses with common greetings and small talk. Typically, he will greet someone with, "Consider the following problem." He sends off over 1000 letters a year, all to fellow mathematicians and exclusively about mathematics.

Erdös has written over 1000 research papers. It is believed that the previous record was held by Cayley, at 927. He has coauthored papers with over 250 people. These people are said to have Erdös number 1. People who do not have Erdös number 1, but who have written a paper with someone who does, are said to have Erdös number 2, and so on inductively. Of course, these numbers keep changing. The highest known Erdös number in 1987 was 7.

Erdös's papers include a broad range of topics, but the majority have been on number theory, combinatorics, and graph theory. In 1951, he received the American Mathematical Society's Cole Prize for number theory. In 1984, he received the prestigious $50,000 Wolf Prize. He kept $750, established a scholarship fund, and gave the rest away. Erdös has also established a prize given biannually by the Israel Mathematical Union to the best young Israeli mathematician using "Fields Medal rules."

Among his friends Erdös is admired for his great generosity, his extraordinary concern for human rights, and his keen interest in helping promising young mathematicians.

31

Introduction to Algebraic Coding Theory

Damn it, if the machine can detect an error, why can't it locate the position of the error and correct it?

Richard W. Hamming

MOTIVATION

One of the most interesting and important applications of finite fields has been the development of algebraic coding theory. This theory, originated in the late 1940s, was created in response to practical communication problems. (Algebraic coding has nothing to do with secret codes.) Algebraic codes are now used in compact disc players, fax machines, modems, and bar code scanners, and are essential to computer maintenance.

To motivate this theory, imagine that we wish to transmit one of two possible signals to a spacecraft approaching Mars. If video pictures reveal the conditions of the proposed landing site and they look unfavorable, we will command the craft to orbit the planet; otherwise, we will command the craft to land. The signal for orbiting will be a 0, and the signal for landing will be a 1. But it is possible that some sort of interference (called *noise*) could cause an incorrect message to be received.

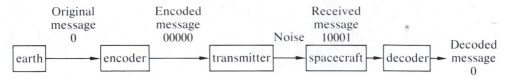

Figure 31.1   Encoding and decoding by fivefold repetition.

To decrease the chance of this happening, redundancy is built into the transmission process. For example, if we wish the craft to orbit Mars, we could send five 0's. The craft's onboard computer is programmed to take any five-digit message received and decode the result by majority rule. So, if 00000 is sent and 10001 is received, the computer decides that 0 was the intended message. Notice that, for the computer to make the wrong decision, at least three errors must occur during transmission. If we assume that errors occur independently, it is less likely that three errors will occur than two or fewer errors. For this reason, this decision process is frequently called the *maximum-likelihood decoding* procedure. Our particular situation is illustrated in Figure 31.1. The general coding procedure is illustrated in Figure 31.2.

In practice, the means of transmission are telephone, telegraph, radiowave, microwave, or even a magnetic tape or disc. The noise might be human error, cross talk, lightning, thermal noise, impulse noise, or deterioration of a tape or disc. Throughout this chapter, we assume that errors in transmission occur independently. Different methods are needed when this is not the case.

Now, let's consider a more complicated situation. This time, assume that we wish to send a sequence of 0's and 1's of length 500. Further, suppose that the probability that an error will be made in the transmission of any particular digit is .01. If we send this message directly without any redundancy, the probability that it will be received error-free is $(.99)^{500}$, or approximately .0066.

On the other hand, if we adopt a threefold repetition scheme by sending each digit three times and decoding each block of three digits received by majority rule, we can do much better. For example, the sequence 1011 is encoded as 111000111111. If the received message is 011000001110, the decoded message is 1001. Now, what is the probability that our 500-digit message will be error-free? Well, if a 1, say, is sent, it will be decoded as a 0 only if the block received is 001, 010, 100,

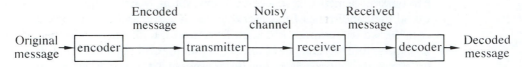

Figure 31.2   General encoding-decoding.

or 000. The probability that this will occur is

$$(.01)(.01)(.99) + (.01)(.99)(.01) + (.99)(.01)(.01) + (.01)(.01)(.01)$$
$$= (.01)^2[3(.99) + .01]$$
$$= .000298 < .0003.$$

Thus, the probability that any particular digit in the sequence will be decoded correctly is greater than .9997, and it follows that the probability of the entire 500-digit message being decoded correctly is greater than $(.9997)^{500}$, or approximately .86—a dramatic improvement over .0066.

This example illustrates the three basic features of a code. One has a set of messages, a method of encoding these messages, and a method of decoding the received messages. The encoding procedure builds some redundancy into the original messages; the decoding procedure corrects or detects certain prescribed errors. Repetition codes have the advantage of simplicity of encoding and decoding, but they are too inefficient. In a fivefold repetition code, 80% of all transmitted information is redundant. The goal of coding theory is to devise message encoding and decoding methods that are reliable, efficient, and reasonably easy to implement.

Before plunging into the formal theory, it is instructive to look at a sophisticated example.

Example 1 The Hamming (7, 4) Code
This time, our message set consists of all possible 4-tuples of 0's and 1's (that is, we wish to send a sequence of 0's and 1's of length 4). Encoding will be done by viewing these messages as 1×4 matrices with entries from Z_2 and multiplying each of the 16 messages on the right by the matrix

$$G = \begin{bmatrix} 1 & 0 & 0 & 0 & 1 & 1 & 0 \\ 0 & 1 & 0 & 0 & 1 & 0 & 1 \\ 0 & 0 & 1 & 0 & 1 & 1 & 1 \\ 0 & 0 & 0 & 1 & 0 & 1 & 1 \end{bmatrix}.$$

(All arithmetic is done modulo 2.) The resulting 7-tuples are called *code words*. (See Table 31.1.)

Notice that the first four digits of each code word constitute just the original message corresponding to the code word. The last three digits of the code word constitute the redundancy features. For this code, we use the *nearest-neighbor* decoding method (which, in the case that the errors occur independently, is the same as the maximum-likelihood decoding procedure). For any received word v, we assume that the word sent is the code word v', which differs from v in the fewest number of positions. If the choice of v' is not unique, we can decide not to decode or arbitrarily choose one of the code words closest to v. (The first option is usually selected when retransmission is practical.) Once we have

Table 31.1

Message	Encoder G	Code Word	*wt*
0000	→	0000000	0
0001	→	0001011	3
0010	→	0010111	4
0100	→	0100101	3
1000	→	1000110	3
1100	→	1100011	4
1010	→	1010001	3
1001	→	1001101	4
0110	→	0110010	3
0101	→	0101110	4
0011	→	0011100	3
1110	→	1110100	4
1101	→	1101000	3
1011	→	1011010	4
0111	→	0111001	4
1111	→	1111111	7

decoded the received word, we can obtain the message by deleting the last three digits of v'. For instance, suppose that 1000 were the intended message. It would be encoded and transmitted as $u = 1000110$. If the received word were $v = 1100110$ (an error in the second position), it would still be decoded as u, since v and u differ in only one position, whereas v and any other code word would differ in at least three positions. Similarly, the intended message 1111 would be encoded as 1111111. If, instead of this, the word 0111111 were received, our decoding procedure would still give us the intended message 1111. ❑

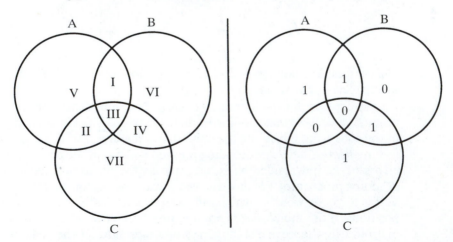

Figure 31.3 Venn diagram of the message 1001 and the encoded message 1001101.

The code in Example 1 is one of an infinite class of important codes discovered by Richard Hamming in 1948. The Hamming codes are the most widely used codes.

The Hamming (7, 4) encoding scheme can be conveniently illustrated with the use of a Venn diagram, as shown in Figure 31.3. Begin by placing the four message digits in the four overlapping regions I, II, III, and IV, with the digit in position 1 in region I, the digit in position 2 in region II, and so on. For regions V, VI, and VII, assign 0 or 1 so that the total number of 1's in each circle is even.

Consider the Venn diagram of the received word 0001101:

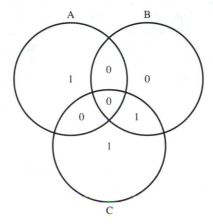

How may we detect and correct an error? Well, observe that each of the circles *A* and *B* has an odd number of 1's. This tells us that something is wrong. At the same time, we note that circle *C* has an even number of 1's. Thus, the portion of the diagram in both *A* and *B* but not in *C* is the source of the error. See Figure 31.4.

Quite often, codes are used to detect errors rather than correct

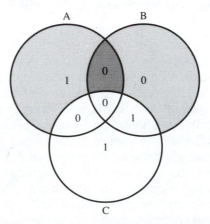

Figure 31.4 Circles *A* and *B* but not *C* have wrong parity.

them. This is especially appropriate when it is easy to retransmit a message. If a received word is not a code word, we have detected an error. For example, computers are designed to use a parity check for numbers. Inside the computer, each number is represented by a string of 0's and 1's. If there is an even number of 1's in this representation, a 0 is attached to the string; if there is an odd number of 1's in the representation, a 1 is attached to the string. Thus, each number stored in the computer memory has an even number of 1's. Now, when the computer reads a number from memory, it performs a parity check. If the read number has an odd number of 1's, the computer will know that an error has been made, and it will reread the number. Note that an even number of errors will not be detected by a parity check.

The methods of error detection introduced in Chapters 0 and 2 are based on the same principle. An extra character is appended to a string of numbers so that a particular condition is satisfied. If we find that such a string does not satisfy that condition, we know that an error has occurred.

LINEAR CODES

We now formalize some of the ideas introduced in the preceding discussion.

DEFINITION   Linear Code
An (n, k) *linear code* over a finite field F is a k-dimensional subspace V of the vector space

$$F^n = \underbrace{F \oplus F \oplus \cdots \oplus F}_{n \text{ copies}}$$

over F. The members of V are called the *code words*. The ratio k/n is called the *information rate* of the code. When F is Z_2, the code is called *binary*.

One should think of an (n, k) linear code over F as a set of n-tuples from F, where each n-tuple is comprised of two parts: the message part, consisting of k digits; and the redundancy part, consisting of the remaining $n - k$ digits. Note that an (n, k) linear code over a finite field F of order q has q^k code words, since every member of the code is uniquely expressible as a linear combination of the k basis vectors with coefficients from F. The set of q^k code words is closed under addition and scalar multiplication by members of F. Also, since errors in transmission may occur in any of the n positions, there are q^n possible vectors that can be received. Where there is no possibility of confusion, it is customary to denote an n-tuple $(a_1, a_2, \ldots, a_n)$ more simply as $a_2 a_2 \cdots a_n$, as we did in Example 1.

Example 2   The set

{0000000, 0010111, 0101011, 1001101,

1100110, 1011010, 0111100, 1110001}

is a (7, 3) binary code. This code has information rate 3/7.   ❑

Example 3   The set {0000, 0101, 1010, 1111} is a (4, 2) binary code. ❑

Although binary codes are by far the most important ones, other codes are occasionally used.

Example 4   The set

{0000, 0121, 0212, 1022, 1110, 1201, 2011, 2102, 2220}

is a (4, 2) linear code over Z_3. A linear code over Z_3 is called a *ternary code.*   ❑

To facilitate our discussion of the error-correcting and error-detecting capability of a code, we introduce the following terminology.

DEFINITION   Hamming Distance, Hamming Weight
The *Hamming distance* between two vectors of a vector space is the number of components in which they differ. The *Hamming weight* of a vector is the number of nonzero components of the vector. The *Hamming weight* of a linear code is the minimum weight of any nonzero vector in the code.

We will use $d(u, v)$ to denote the Hamming distance between the vectors u and v, and wt(u) for the Hamming weight of the vector u.

Example 5   Let $s = 0010111$, $t = 0101011$, $u = 1001101$, and $v = 1101101$. Then, $d(s, t) = 4$, $d(s, u) = 4$, $d(s, v) = 5$, $d(u, v) = 1$; and wt(s) = 4, wt(t) = 4, wt(u) = 4, wt(v) = 5.   ❑

The Hamming distance and Hamming weight have the following important properties.

Theorem 31.1   *Properties of Hamming Distance and Hamming Weight*
For any vectors u, v, and w, $d(u, v) \leq d(u, w) + d(w, v)$ and $d(u, v) =$ wt($u - v$).

Proof.   See exercise 4.   ∎

With the preceding definitions and Theorem 31.1, we can now explain why the codes given in Examples 1, 2, and 4 will correct any single error, but why the code in Example 3 will not.

Theorem 31.2   *Correcting Capability of a Linear Code*

Mutually
exclusive

If the Hamming weight of a linear code is at least $2t + 1$, then the code can correct any t or fewer errors. Furthermore, the same code can detect any $2t$ or fewer errors.

Proof. We will use nearest-neighbor decoding; that is, for any received vector v, we will assume that the corresponding code word sent is a code word v' such that the Hamming distance $d(v, v')$ is a minimum. (If there is more than one such v', we do not decode.) Now, suppose that a transmitted code word u is received as the vector v and that at most t errors have been made in transmission. Then, by the definition of distance between u and v, we have $d(u, v) \leq t$. If w is any code word other than u, then $w - u$ is a nonzero code word. Thus, by assumption,

p.476

$$2t + 1 \leq \text{wt}(w - u) = d(w, u) \leq d(w, v) + d(v, u) \leq d(w, v) + t,$$

and it follows that $t + 1 \leq d(w, v)$. So, the code word closest to the received vector v is u, and, therefore, v is correctly decoded as u.

To show that the code can detect $2t$ errors, we suppose that a transmitted code word u is received as the vector v and that at least one error, but no more than $2t$ errors, was made in transmission. Because only code words are transmitted, an error will be detected whenever a received word is not a code word. But v cannot be a code word, since $d(v, u) \leq 2t$ whereas we know that the minimum distance between distinct code words is at least $2t + 1$. ∎

Theorem 31.2 is often misinterpreted to mean that a linear code with Hamming weight $2t + 1$ can correct any t errors *and* detect any $2t$ or fewer errors simultaneously. This is not the case. The user must choose one or the other role for the code. Consider, for example, the Hamming $(7, 4)$ code given in Table 31.1. By inspection, the Hamming weight of the code is $3 = 2 \cdot 1 + 1$, so we may elect either to correct any single error or to detect any one or two errors. To understand why we can't do both, consider the received word 0001010. The intended message could have been 0000000, in which case two errors were made (likewise for the intended messages 1011010 and 0101110), or the intended message could have been 0001011, in which case one error was made. But there is no way for us to know which of these possibilities occurred. If our choice were error correction, we would assume—perhaps mistakenly—that 0001011 was the intended message. If our choice were error detection, we simply would not decode. (Typically, one would request retransmission.)

Inclusive

On the other hand, if we write the Hamming weight of a linear code in the form $2t + s + 1$, we can correct any t errors *and* detect any $t + s$ or fewer errors. Thus, for a code with Hamming weight 5, our options include any one of the following:

1. Detect any four errors ($t = 0$, $s = 4$).
2. Correct any one error and detect any two or three errors ($t = 1$, $s = 2$).
3. Correct any two errors ($t = 2$, $s = 0$).

Example 6 Since the Hamming weight of the linear code given in Example 2 is 4, it will correct any single error and detect any two errors ($t = 1$, $s = 1$) or detect any three errors ($t = 0$, $s = 3$). ❑

It is natural to wonder how the matrix G used to produce the Hamming code in Example 1 was chosen. Better yet, in general, how can one find a matrix G that carries a subspace V of F^k to a subspace of F^n in such a way that for any k-tuple v in V, the vector vG will agree with v in the first k components and build in some redundancy in the last $n - k$ components? Such a matrix is a $k \times n$ matrix of the form

$$\begin{bmatrix} 1 & 0 & \cdots & 0 & a_{11} & \cdots & a_{1n-k} \\ 0 & 1 & \cdots & 0 & \cdot & & \cdot \\ \cdot & \cdot & & \cdot & \cdot & & \cdot \\ \cdot & \cdot & & \cdot & \cdot & & \cdot \\ 0 & 0 & \cdots & 1 & a_{k1} & \cdots & a_{kn-k} \end{bmatrix},$$

where the a_{ij}'s belong to F and the rows are linearly independent. A matrix of this form is called the *standard generator matrix* (or *standard encoding matrix*) for the resulting code.

Any $k \times n$ matrix whose rows are linearly independent will transform the subspace of F^k to a k-dimensional subspace of F^n that could be used to build redundancy, but using the standard generator matrix has the advantage that the original message constitutes the first k components of the transformed vectors. An (n, k) linear code in which the k information digits occur at the beginning of each code word is called a *systematic code*. Schematically, we have

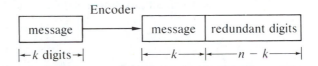

Notice that, by definition, a standard generator matrix produces a systematic code.

Example 7 From the set of messages

$$\{000, 001, 010, 100, 110, 101, 011, 111\},$$

Table 31.2

Message	Encoder G	Code Word
000	$\rightarrow$	000000
001	$\rightarrow$	001111
010	$\rightarrow$	010101
100	$\rightarrow$	100110
110	$\rightarrow$	110011
101	$\rightarrow$	101001
011	$\rightarrow$	011010
111	$\rightarrow$	111100

we may construct a (6, 3) linear code over Z_2 with the standard generator matrix

$$G = \begin{bmatrix} 1 & 0 & 0 & 1 & 1 & 0 \\ 0 & 1 & 0 & 1 & 0 & 1 \\ 0 & 0 & 1 & 1 & 1 & 1 \end{bmatrix}.$$

The resulting code words are given in Table 31.2.

Since the minimum weight of any nonzero code word is 3, this code will correct any single error or detect any double error.   ❏

Example 8   Here we take a set of messages as

$$\{00, 01, 02, 10, 11, 12, 20, 21, 22\},$$

and we construct a (4, 2) linear code over Z_3 with the standard generator matrix

$$G = \begin{bmatrix} 1 & 0 & 2 & 1 \\ 0 & 1 & 2 & 2 \end{bmatrix}.$$

The resulting code words are given in Table 31.3.

This code will correct any single error or detect any double error.   ❏

PARITY-CHECK MATRIX DECODING

Now that we can conveniently encode messages with a standard generator matrix, we need a convenient method for decoding the received messages. Unfortunately, this is not as easy to do; however, in the case where at most one error per code word has occurred, there is a fairly simple method for decoding. (When more than one error occurs in a code word, our decoding method fails.)

Table 31.3

Message	Encoder G	Code Word
00	→	0000
01	→	0122
02	→	0211
10	→	1021
11	→	1110
12	→	1202
20	→	2012
21	→	2101
22	→	2220

To describe this method, suppose that V is a systematic linear code over the field F given by the standard generator matrix $G = [I_k \mid A]$, where I_k represents the $k \times k$ identity matrix and A is the $k \times (n - k)$ matrix obtained from G by deleting the first k columns of G. Then, the $n \times (n - k)$ matrix

$$H = \left[\frac{-A}{I_{n-k}} \right],$$

where $-A$ is the negative of A and I_{n-k} is the $(n - k) \times (n - k)$ identity matrix, is called the *parity-check matrix* for V. (In the literature, the transpose of H is called the parity-check matrix, but H is much more convenient for our purposes.) The decoding procedure is:

1. For any received word w, compute wH.
2. If wH is the zero vector, assume that no error was made.
3. If there is exactly one instance of a nonzero element $s \in F$ and a row i of H so that wH is s times row i, assume that the sent word was $w - 0 \cdots s \cdots 0$, where s occurs in the ith component. If there is more than one such instance, do not decode.
3'. When the code is binary, category 3 reduces to the following. If wH is the ith row of H for exactly one i, assume that an error was made in the ith component of w. If wH is more than one row of H, do not decode.
4. If wH does not fit into either category 2 or category 3, we know that at least two errors occurred in transmission and we do not decode.

Example 9 Consider the Hamming (7, 4) code given in Example 1. The generator matrix is

$$G = \begin{bmatrix} 1 & 0 & 0 & 0 & 1 & 1 & 0 \\ 0 & 1 & 0 & 0 & 1 & 0 & 1 \\ 0 & 0 & 1 & 0 & 1 & 1 & 1 \\ 0 & 0 & 0 & 1 & 0 & 1 & 1 \end{bmatrix},$$

and the corresponding parity-check matrix is

$$H = \begin{bmatrix} 1 & 1 & 0 \\ 1 & 0 & 1 \\ 1 & 1 & 1 \\ 0 & 1 & 1 \\ 1 & 0 & 0 \\ 0 & 1 & 0 \\ 0 & 0 & 1 \end{bmatrix}.$$

Now, if the received vector is $v = 0000110$, we find $vH = 110$. Since this is the first row of H and no other row, we assume that an error has been made in the first position of v. Thus, the transmitted code word is assumed to be 1000110, and the corresponding message is assumed to be 1000. Similarly, if $w = 1011111$ is the received word, then $wH = 101$ and we assume that an error has been made in the second position. So, we assume that 1111111 was sent and that 1111 was the intended message. If the encoded message 1001100 is received as $z = 1001010$ (with errors in the fifth and sixth positions), we find $zH = 111$. Since this matches the third row of H, we decode z as 1011010 and incorrectly assume that the message 1011 was intended. On the other hand, nearest-neighbor decoding would yield the same result.   □

Notice that when only one error was made in transmission, the parity-check decoding procedure gave us the originally intended message. We will soon see under what conditions this is true, but first we need an important fact relating a code given by a generator matrix and its parity-check matrix.

Lemma   *Orthogonality Relation*

Let C be an (n, k) linear code over F with generator matrix G and parity-check matrix H. Then, for any vector v in F^n, we have $vH = 0$ (the zero vector) if and only if v belongs to C.

Proof.   First note that, since H has rank $n - k$, we may think of H as a linear transformation from F^n onto F^{n-k}. Therefore, it follows from the dimension theorem for linear transformations that $n = n - k + \dim \text{Ker } H$, so that Ker H has dimension k. (Alternatively, one can use a group theory argument to show that $|\text{Ker } H| = |F|^k$.) Then, since the dimension of C is also k, it suffices to show that $C \subseteq \text{Ker } H$. To do this, let $G = [I_k \mid A]$, so that $H = \begin{bmatrix} -A \\ \hline I_{n-k} \end{bmatrix}$. Then,

$$GH = [I_k \mid A] \begin{bmatrix} -A \\ \hline I_{n-k} \end{bmatrix} = -A + A = [0] \qquad \text{(the zero matrix)}.$$

Now, by definition, any vector v in C has the form mG, where m is a message vector. Thus, $vH = (mG)H = m[0] = 0$ (the zero vector). ∎

Because of the way H was defined, the parity-check matrix method correctly decodes any received word in which no error has been made. But it will do more.

Theorem 31.3 *Parity-Check Matrix Decoding*
Parity-check matrix decoding will correct any single error if and only if the rows of the parity-check matrix are nonzero and no one row is a scalar multiple of any other.

Proof. For simplicity's sake, we prove only the binary case. In this special situation, the condition on the rows is that they are nonzero and distinct. So, let H be the parity-check matrix, and let's assume that this condition holds for the rows. Suppose that the transmitted code word w was received with only one error, and that this error occurred in the ith position. Denoting the vector that has a 1 in the ith position and 0's elsewhere by e_i, we may write the received word as $w + e_i$. Now, using the Orthogonality Lemma, we obtain

$$(w + e_i)H = wH + e_iH = 0 + e_iH = e_iH.$$

But this last vector is precisely the ith row of H. Thus, if there was exactly one error in transmission, we can use the rows of the parity-check matrix to identify the location of the error, provided that these rows are distinct. (If two rows, say, the ith and jth, are the same, we know that the error occurred in either the ith position or the jth position, but we do not know in which.)

Conversely, suppose that the parity-check matrix method correctly decodes all received words in which at most one error has been made in transmission. If the ith row of the parity-check matrix H were the zero vector and if the code word $u = 0 \cdots 0$ were received as e_i, we would find $e_iH = 0 \cdots 0$, and we would erroneously assume that the vector e_i was sent. Thus, no row of H is the zero vector. Now, suppose that the ith row of H and the jth row of H are equal and $i \neq j$. Then, if some code word w is transmitted and the received word is $w + e_i$ (that is, a single error in the ith position), we find

$$(w + e_i)H = wH + e_iH = i\text{th row of } H = j\text{th row of } H.$$

Thus, our decoding procedure tells us not to decode. This contradicts our assumption that the method correctly decodes all received words in which at most one error has been made. ∎

COSET DECODING

There is another convenient decoding method that utilizes the fact that an (n, k) linear code C over a finite field F is a subgroup of the additive group of $V = F^n$. This method was devised by David Slepian in 1956 and is called *coset decoding* (or *standard decoding*). To use this method, we proceed by constructing a table, called a *standard array*. The first row of the table is the set C of code words beginning in column 1 with the identity $0 \cdots 0$. To form additional rows of the table, choose an element v of V not listed in the table thus far. Among all the elements of the coset $v + C$, choose one of minimum weight, say, v'. Complete the next row of the table by placing under the column headed by the code word c the vector $v' + c$. Continue this process until all the vectors in V have been listed in the table. [Note that an (n, k) linear code over Z_p will have $|V:C| = p^{n-k}$ rows.] The words in the first column are called the *coset leaders*. The decoding procedure is simply to decode any received word w as the code word at the head of the column containing w.

Example 10   Consider the $(6, 3)$ binary linear code

$$C = \{000000, 100110, 010101, 001011, 110011, 101101, 011110, 111000\}.$$

The first row of the standard array is just the elements of C. Obviously, 100000 is not in C and has minimum weight among the elements of $100000 + C$, so it can be used to lead the second row. Table 31.4 is the completed table.

If the word 101001 is received, it is decoded as 101101, since 101001 lies in the column headed by 101101. Similarly, the received word 011001 is decoded as 111000. ◻

Recall that the first method of decoding that we introduced was the nearest-neighbor method; that is, any received word w is decoded as the code word c such that $d(w, c)$ is a minimum, provided that there is only one code word c so that $d(w, c)$ is a minimum. The next result shows that in this situation coset decoding is the same as nearest-neighbor decoding.

Theorem 31.4   *Coset Decoding Is Nearest-Neighbor Decoding*
In coset decoding, a received word w is decoded as a code word c such that $d(w, c)$ is a minimum.

Table 31.4   A Standard Array for a $(6, 3)$ Linear Code

Coset Leaders	Words						
000000	100110	010101	001011	110011	101101	011110	111000
100000	000110	110101	101011	010011	001101	111110	011000
010000	110110	000101	011011	100011	111101	001110	101000
001000	101110	011101	000011	111011	100101	010110	110000
000100	100010	010001	001111	110111	101001	011010	111100
000010	100100	010111	001001	110001	101111	011100	111010
000001	100111	010100	001010	110010	101100	011111	111001
100001	000111	110100	101010	010010	001100	111111	011001

Proof. Let C be a linear code and let w be any received word. Suppose that v is the coset leader for the coset $w + C$. Then, $w + C = v + C$, so $w = v + c$ for some c in C. Thus, using coset decoding, w is decoded as c. Now, if c' is any code word, then $w - c' \in w + C = v + C$, so that $\text{wt}(w - c') \geq \text{wt}(v)$, since the coset leader v was chosen as a vector of minimum weight among the members of $v + C$. Therefore,

$$d(w, c') = \text{wt}(w - c') \geq \text{wt}(v) = \text{wt}(w - c) = d(w, c).$$

So, using coset decoding, w is decoded as a code word c such that $d(w, c)$ is a minimum. ∎

When we know a parity-check matrix for a linear code, coset decoding can be considerably simplified.

DEFINITION   Syndrome

If an (n, k) linear code over F has parity-check matrix H, then for any vector u in F^n, the vector uH is called the *syndrome** of u.

The importance of syndromes stems from the following property.

Theorem 31.5   *Same Coset—Same Syndrome*
Let C be an (n, k) linear code over F with a parity-check matrix H. Then, two vectors of F^n are in the same coset of C if and only if they have the same syndrome.

Proof. Two vectors u and v are in the same coset of C if and only if $u - v$ is in C. So, by the Orthogonality Lemma, u and v are in the same coset if and only if $0 = (u - v)H = uH - vH$. ∎

We may now use syndromes for decoding any received word w:

1. Calculate wH, the syndrome of w.
2. Find the coset leader v so that $wH = vH$.
3. Assume that the vector sent was $w - v$.

With this method, we can decode any received word with a table that has only two rows—one row of coset leaders and another with the corresponding syndromes.

Example 11   Consider the code given in Example 10. The parity-check matrix for this code is

$$H = \begin{bmatrix} 1 & 1 & 0 \\ 1 & 0 & 1 \\ 0 & 1 & 1 \\ 1 & 0 & 0 \\ 0 & 1 & 0 \\ 0 & 0 & 1 \end{bmatrix}.$$

*This term was coined by D. Hagelbarger in 1959.

The list of coset leaders and corresponding syndromes is

Coset leader	000000	100000	010000	001000	000100	000010	000001	100001
Syndromes	000	110	101	011	100	010	001	111

So, to decode the received word $v = 101001$, we compute $vH = 100$. Since the coset leader 000100 has 100 as its syndrome, we assume that $v - 000100 = 101101$ was sent. If the received word is $w = 011001$, we compute $wH = 111$ and assume that $w - 100001 = 111000$ was sent. Notice that these answers are in agreement with those obtained by using the standard-array method of Example 10. ◻

The term "syndrome" is a descriptive one. In medicine, it is used to designate a collection of symptoms that typify a disorder. In coset decoding, the syndrome typifies an error pattern.

In this chapter, we have presented algebraic coding theory in its simplest form. A more sophisticated treatment would make substantially greater use of group theory, ring theory, and especially finite-field theory. For example, Gorenstein (see Chapter 25 for a biography) and Zierler, in 1961, made use of the fact that the multiplicative subgroup of a finite field is cyclic. They associated each digit of certain codes with a field element in such a way that an algebraic equation could be derived with its zeros determining the locations of the errors.

In some instances, two error-correcting codes are employed. The European Space Agency space probe Giotto, which came within 370 miles of the nucleus of Halley's Comet in 1986, had two error-correcting codes built into its electronics. One code checked for independently occurring errors, and another—a so-called Reed-Solomon code—checked for bursts of errors. Giotto achieved an error-detection rate of 0.999999. Reed-Solomon codes are also used on compact discs. They can correct thousands of consecutive errors.

HISTORICAL NOTE: REED-SOLOMON CODES

We conclude this chapter with an adapted version of an article by Barry A. Cipra about the Reed-Solomon codes [1]. It was the first in a series of articles called "Mathematics that Counts" in *SIAM News,* the news journal of the Society for Industrial and Applied Mathematics. The articles highlight developments in mathematics that have led to products and processes of substantial benefit to industry and the public.

The Ubiquitous Reed-Solomon Codes

Irving Reed and Gustave Solomon monitor the encounter of Voyager II with Neptune at the Jet Propulsion Laboratory in 1989.

In this "Age of Information," no one need be reminded of the importance not only of speed but also of accuracy in the storage, retrieval, and transmission of data. Machines *do* make errors, and their non-man-made mistakes can turn otherwise flawless programming into worthless, even dangerous, trash. Just as architects design buildings that will remain standing even through an earthquake, their computer counterparts have come up with sophisticated techniques capable of counteracting digital disasters.

The idea for the current error-correcting techniques for everything from computer hard disk drives to CD players was first introduced in 1960 by Irving Reed and Gustave Solomon, then staff members at MIT's Lincoln Laboratory. Their paper titled "Polynomial Codes over Certain Finite Fields" [2] was published that year in the *Journal of the Society for Industrial and Applied Mathematics.*

Reed-Solomon codes (plus a lot of engineering wizardry, of course) made possible the stunning pictures of the outer planets sent back by the space probe Voyager II. They make it possible to scratch a compact disc and still enjoy the music. And in the not-too-distant future, they will enable the profit mongers of cable television to squeeze more than 500 channels into their systems, making a vast wasteland vaster yet.

"When you talk about CD players and digital audio tape and now digital television, and various other digital imaging systems that are coming—all of those need Reed-Solomon [codes] as an integral part of the system," says Robert McEliece, a coding theorist in the electrical engineering department at Caltech.

Why? Because digital information, virtually by definition, consists of strings of "bits"—0's and 1's—and a physical device, no matter how capably manufactured, may occasionally confuse the two. Voyager II, for example, was transmitting data at incredibly low power—barely a whisper—over tens of millions of miles. Disk drives pack data so densely that a read/write head can (almost) be excused if it can't tell where one bit stops and the next 1 (or 0) begins. Careful engineering can reduce the error rate to what may sound like a negligible level—the industry standard for hard disk drives is 1 in 10 billion—but given the

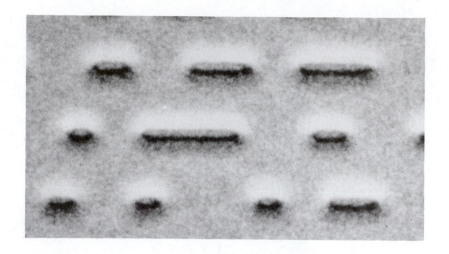

Microscope photograph (magnification 2500) showing the molded pits on the bottom of a compact disc. When the laser beam that focuses on the revolving disc strikes a pit, its light scatters and the level of light reflected back changes. This change (along with that resulting from a transition from pit to land) is interpreted as a binary 1. The varying path lengths between transitions in the level of reflected light are interpreted as corresponding numbers of 0's. Errors in this pattern of binary digits are corrected by a Reed-Solomon code.

volume of information processing done these days, that "negligible" level is an invitation to daily disaster. Error-correcting codes are a kind of safety net—mathematical insurance against the vagaries of an imperfect material world.

In 1960, the theory of error-correcting codes was only about a decade old. The basic theory of reliable digital communication had been set forth by Claude Shannon in the late 1940s. At the same time, Richard Hamming introduced an elegant approach to single-error correction and double-error detection. Through the 1950s, a number of researchers began experimenting with a variety of error-correcting codes. But with their SIAM journal paper, McEliece says, Reed and Solomon "hit the jackpot."

The payoff was a coding system based on groups of bits—such as bytes—rather than individual 0's and 1's. That feature makes Reed-Solomon codes particularly good at dealing with "bursts" of errors: six consecutive bit errors, for example, can affect at most two bytes. Thus, even a double-error-correction version of a Reed-Solomon code can provide a comfortable safety factor.

(Current implementations of Reed-Solomon codes in CD technology are able to cope with error bursts as long as 4000 consecutive bits.)

Mathematically, Reed-Solomon codes are based on the arithmetic of finite fields. Indeed, the 1960 paper begins by defining a code as "a mapping from a vector space of dimension m over a finite field K into a vector space of higher dimension over the same field." Starting from a "message" $(a_0, a_1, \ldots, a_{m-1})$, where each a_k is an element of the field K, a Reed-Solomon code produces $(P(0), P(g), P(g^2), \ldots, P(g^{N-1}))$, where N is the number of elements in K, g is a generator of the (cyclic) group of nonzero elements in K, and $P(x)$ is the polynomial $a_0 + a_1 x + \cdots + a_{m-1} x^{m-1}$. If N is greater than m, then the values of P overdetermine the polynomial, and the properties of finite fields guarantee that the coefficients of P—i.e., the original message—can be recovered from any m of the values.

Conceptually, the Reed-Solomon code specifies a polynomial by "plotting" a large number of points. And just as the eye can recognize and correct for a couple of "bad" points in what is otherwise clearly a smooth parabola, the Reed-Solomon code can spot incorrect values of P and still recover the original message. A modicum of combinatorial reasoning (and a bit of linear algebra) establishes that this approach can cope with up to s errors, as long as m, the message length, is strictly less than $N - 2s$.

In today's byte-sized world, for example, it might make sense to let K be the field of degree 8 over Z_2, so that each element of K corresponds to a single byte (in computerese, there are four bits to a nibble and two nibbles to a byte). In that case, $N = 2^8 = 256$, and hence messages up to 251 bytes long can be recovered even if two errors occur in transmitting the values $P(0)$, $P(g)$, ..., $P(g^{255})$. That's a lot better than the 1255 bytes required by the say-everything-five-times approach.

Despite their advantages, Reed-Solomon codes did not go into use immediately—they had to wait for the hardware technology to catch up. "In 1960, there was no such thing as fast digital electronics"—at least not by today's standards, says McEliece. The Reed-Solomon paper "suggested some nice ways to process data, but nobody knew if it was practical or not, and in 1960 it probably wasn't practical."

But technology did catch up, and numerous researchers began to work on implementing the codes. One of the key individuals was Elwyn Berlekamp, a professor of electrical engineering at the University of California at Berkeley, who invented an efficient algorithm for decoding the Reed-Solomon code. Berlekamp's algorithm was used by Voyager II and is the basis for decoding in CD players. Many other bells and whistles (some of fundamental theoretic significance) have also been added. Compact discs, for example, use a version of a Reed-Solomon code.

Reed, now a professor of electrical engineering at the University of Southern California, is still working on problems in coding theory. Solomon, recently retired from the Hughes Aircraft Company, consults for the Jet Propulsion Laboratory. Reed was among the first to recognize the significance of abstract algebra as the basis for error-correcting codes.

"In hindsight it seems obvious," he told *SIAM News*. However, he added, "coding theory was not a subject when we published that paper." The two authors knew they had a nice result; they didn't know what impact the paper would have.

Three decades later, the impact is clear. The vast array of applications, both current and pending, has settled the question of the practicality and significance of Reed-Solomon codes. "It's clear they're practical, because everybody's using them now," says Berlekamp. Billions of dollars in modern technology depend on ideas that stem from Reed and Solomon's original work. In short, says McEliece, "it's been an extraordinarily influential paper."

EXERCISES

The New Testament offers the basis for modern computer coding theory, in the form of an affirmation of the binary number system.

> "But let your communication be yea, yea; nay, nay: for whatsoever is more than these cometh of evil."

Anonymous

1. Find the Hamming weight of each code in Example 1.
2. Find the Hamming distance between the following pairs of vectors: {1101, 0111}, {0220, 1122}, {11101, 00111}.
3. Referring to Example 1, use the nearest-neighbor method to decode the received words 0000110 and 1110100.

4. For any vector space V and any u, v, w in V, prove that the Hamming distance has the following properties:
   a. $d(u, v) = \text{wt}(u - v)$
   b. $d(u, v) = d(v, u)$ *(symmetry)*
   c. $d(u, v) = 0$ if and only if $u = v$
   d. $d(u, v) \leq d(u, w) + d(w, v)$ *(triangle inequality)*
   e. $d(u, v) = d(u + w, v + w)$ *(translation invariance)*

5. Determine the (6, 3) binary linear code with generator matrix

$$G = \begin{bmatrix} 1 & 0 & 0 & 0 & 1 & 1 \\ 0 & 1 & 0 & 1 & 0 & 1 \\ 0 & 0 & 1 & 1 & 1 & 0 \end{bmatrix}.$$

6. Show that for binary vectors, $\text{wt}(u + v) \geq \text{wt}(u) - \text{wt}(v)$ and equality occurs if and only if the ith component of u is 1 whenever the ith component of v is 1.

7. If the minimum weight of any nonzero code word is 2, what can we say about the error-correcting capability of the code?

8. Suppose that C is a linear code with Hamming weight 3 and that C' is one with Hamming weight 4. What can C' do that C can't?

9. Let C be a binary linear code. Show that the code words of even weight form a subcode of C. (A *subcode* of a code is a subset of the code that is itself a code.)

10. Let

$$C = \{0000000, 1110100, 0111010, 0011101, 1001110,$$
$$0100111, 1010011, 1101001\}.$$

What is the error-correcting capability of C? What is the error-detecting capability of C?

11. Suppose that the parity-check matrix of a binary linear code is

$$H = \begin{bmatrix} 1 & 0 \\ 0 & 1 \\ 1 & 1 \\ 1 & 0 \\ 0 & 1 \end{bmatrix}.$$

Can the code correct any single error?

12. Use the generator matrix

$$G = \begin{bmatrix} 1 & 0 & 1 & 1 \\ 0 & 1 & 2 & 1 \end{bmatrix}$$

to construct a (4, 2) ternary linear code. What is the parity-check matrix for this code? What is the error-correcting capability of this code? What is the error-detecting capability of this code? Use parity-check decoding to decode the received word 1201.

13. Find all code words of the (7, 4) binary linear code whose generator matrix is

$$G = \begin{bmatrix} 1 & 0 & 0 & 0 & 1 & 1 & 1 \\ 0 & 1 & 0 & 0 & 1 & 0 & 1 \\ 0 & 0 & 1 & 0 & 1 & 1 & 0 \\ 0 & 0 & 0 & 1 & 0 & 1 & 1 \end{bmatrix}.$$

Find the parity-check matrix of this code. Will this code correct any single error?

14. Show that in a binary linear code, either all the code words end with 0, or exactly half end with 0. What about the other components?

15. Suppose that a code word v is received as the vector u. Show that coset decoding will decode u as the code word v if and only if $u - v$ is a coset leader.

16. Consider the binary linear code

$$C = \{00000, 10011, 01010, 11001, 00101, 10110, 01111, 11100\}.$$

Construct the standard array for C. Use nearest-neighbor decoding to decode 11101 and 01100. If the received word 11101 has exactly one error, can we determine the intended code word? If the received word 01100 has exactly one error, can we determine the intended code word?

17. Construct a (6, 3) binary linear code with generator matrix

$$G = \begin{bmatrix} 1 & 0 & 0 & 1 & 1 & 0 \\ 0 & 1 & 0 & 0 & 1 & 1 \\ 0 & 0 & 1 & 1 & 0 & 1 \end{bmatrix}.$$

Decode each of the received words

$$001001, 011000, 000110, 100001$$

by the following methods:
a. nearest-neighbor method,
b. parity-check matrix method,
c. coset decoding using the standard array,
d. coset decoding using the syndrome method.

18. Suppose that the minimum weight of any nonzero code word in a linear code is 6. How many errors can the code correct? How many errors can the code detect?

19. Using the code and the parity-check matrix given in Example 9, show that parity-check matrix decoding cannot detect any multiple errors (that is, two or more errors).

20. Suppose that the last row of the standard array for a binary linear code is

$$10000 \quad 00011 \quad 11010 \quad 01001 \quad 10101 \quad 00110 \quad 11111 \quad 01100.$$

Complete the array.

21. How many code words are there in a (6, 4) linear ternary code? How many possible received words are there for this code?

22. If the parity-check matrix for a binary linear code is

$$H = \begin{bmatrix} 1 & 1 & 0 \\ 0 & 1 & 1 \\ 1 & 0 & 1 \\ 1 & 0 & 0 \\ 0 & 1 & 0 \\ 0 & 0 & 1 \end{bmatrix},$$

will the code correct any single error? Why?

23. Suppose that the parity-check matrix for a ternary code is

$$\begin{bmatrix} 2 & 1 \\ 2 & 2 \\ 1 & 2 \\ 1 & 0 \\ 0 & 1 \end{bmatrix}.$$

Can the code correct all single errors? Give a reason for your answer.

24. Prove that for nearest-neighbor decoding the converse of Theorem 31.2 is true.

25. Can a (6, 3) binary linear code be double-error-correcting using the nearest-neighbor method? Do not assume that the code is systematic.

26. Prove that there is no 2×5 standard generator matrix G that will produce a (5, 2) linear code over Z_3 capable of detecting all possible triple errors.

27. Why can't the nearest-neighbor method with a (4, 2) binary linear code correct all single errors?

28. Suppose that one row of a standard array for a binary code is

000100   110000   011110   111101   101010   001001   100111   010011.

Determine the row that contains 100001.

29. Use the field $F = Z_2[x]/\langle x^2 + x + 1 \rangle$ to construct a (5, 2) linear code that will correct any single error.

30. Find the standard generator matrix for a (4, 2) linear code over Z_3 that encodes 20 as 2012 and 11 as 1100. Determine the entire code and the parity-check matrix for the code.

31. Assume that C is an (n, k) binary linear code and, for each position $i = 1, 2, \ldots, n$, the code C has at least one vector with a 1 in the ith position. Show that the average weight of a code word is $n/2$.

32. Let C be an (n, k) linear code over F such that the minimum weight of any nonzero code word is $2t + 1$. Show that not every vector of weight $t + 1$ in F^n can occur as a coset leader.

33. Let C be an (n, k) binary linear code over F. If $v \in F^n$ but $v \notin C$, show that $C \cup (v + C)$ is a linear code.

34. Let C be a binary linear code. Show that either every member of C has even weight or exactly half the members of C have even weight. (Compare with exercise 19 in Chapter 5.)

35. Let C be an (n, k) linear code. For each i with $1 \leq i \leq n$, let $C_i = \{v \in C \mid$ the ith component of v is 0$\}$. Show that C_i is a subcode of C.

REFERENCES

1. Barry A. Cipra, "The Ubiquitous Reed-Solomon Codes," *SIAM News* 26 (January 1993), 1, 11.

2. Irving S. Reed and Gustave Solomon, "Polynomial Codes over Certain Finite Fields," *Journal of the Society for Industrial and Applied Mathematics* 8 (June 1960), 300–304.

SUGGESTED READINGS

Norman Levinson, "Coding Theory: A Counterexample to G. H. Hardy's Conception of Applied Mathematics," *The American Mathematical Monthly* 77 (1970): 249–258.

The eminent mathematician G. H. Hardy insisted that "real" mathematics was almost wholly useless. In this article, the author argues that coding theory refutes Hardy's notion. Levinson uses the finite field of order 16 to construct a linear code that can correct any three errors.

R. J. McEliece, "The Reliability of Computer Memories," *Scientific American* 252 (1) (1985): 88–95.

This well-written article discusses why and how error-correcting codes are employed in computer memories.

W. W. Peterson, "Error-Correcting Codes," *Scientific American* 206 (2) (1962): 96–108.

This article gives a lucid discussion of the Hamming (15, 11) binary code.

T. M. Thompson, *From Error-Correcting Codes Through Sphere Packing to Simple Groups*, Washington, D.C.: The Mathematical Association of America, 1983.

Chapter 1 of this award-winning book gives a fascinating historical account of the origins of error-correcting codes.

Richard W. Hamming

For introduction of error-correcting codes, pioneering work in operating systems and programming languages, and the advancement of numerical computation.

Citation for the Piore Award, 1979

RICHARD W. HAMMING was born in Chicago, Illinois, on February 11, 1915. He graduated from the University of Chicago with a B.S. degree in mathematics. In 1939, he received an M.A. degree in mathematics from the University of Nebraska, and in 1942, a Ph.D. in mathematics from the University of Illinois.

During the latter part of World War II, Hamming was at Los Alamos, where he was involved in computing atomic-bomb designs. In 1946, he joined Bell Telephone Laboratories, where he worked in mathematics, computing, engineering, and science.

Richard Hamming was one of the first users of early electronic computers. His patch wiring for the IBM CPC became widely used. His work on the IBM 650 in 1956 led to the development of a programming language that was the precursor of modern, high-level languages. It first demonstrated many of the format conversions for numbers, overflow, and fault conventions that are used in today's high-level languages.

In 1950, Hamming published his famous paper on error-detecting and error-correcting codes. This work started a branch of information theory. The Hamming codes are used in many modern computers. Hamming's work in the field of numerical analysis has also been of fundamental importance. The Hamming window for smoothing data prior to Fourier analysis is widely used today.

Hamming retired from Bell Laboratories in 1976 and took up teaching at the Naval Postgraduate School. Thus far, he has written more than 75 research papers and a half dozen books, and he has received numerous prestigious awards, including the Turing Prize from the Association for Computing Machinery, the Piore Award from the Institute of Electrical and Electronics Engineers (IEEE), and the Oender Award from the University of Pennsylvania. IEEE also named its Hamming Medal, which carries a $10,000 prize, after Hamming and made him the first recipient.

32

An Introduction to Galois Theory

Galois theory is a showpiece of mathematical unification,
bringing together several different branches of the subject and
creating a powerful machine for the study of problems of
considerable historical and mathematical importance.

Ian Stewart, *Galois Theory*

FUNDAMENTAL THEOREM OF GALOIS THEORY

The Fundamental Theorem of Galois Theory is one of the most elegant
theorems in mathematics. Look at Figures 32.1 and 32.2. Figure 32.1
pictures the lattice of subgroups of the group of automorphisms of
$Q(\sqrt[4]{2}, i)$. The integer along an upward lattice line from H_1 to H_2 is the
index of H_1 in H_2. Figure 32.2 shows the lattice in subfields of
$Q(\sqrt[4]{2}, i)$. The integer along an upward line from K_1 to K_2 is the degree
of K_2 over K_1. Notice that the lattice in Figure 32.2 is the lattice of Figure
32.1 turned upside down. This is only one of many relationships
between these two lattices. Under suitable conditions, the Fundamental
Theorem of Galois Theory relates, in a multitude of ways, the lattice of
subfields of an algebraic extension E of a field F to the subgroup structure of the group of automorphisms of E that send each element of F to

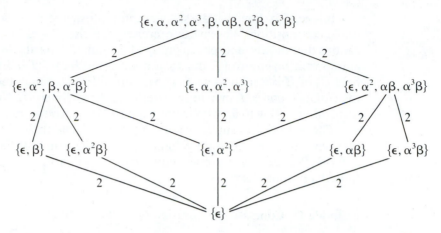

Figure 32.1 Lattice of subgroups of the group of field automorphisms of $Q(\sqrt[4]{2}, i)$, where $\alpha : i \to i$ and $\sqrt[4]{2} \to -i\sqrt[4]{2}$, $\beta : i \to -i$ and $\sqrt[4]{2} \to \sqrt[4]{2}$.

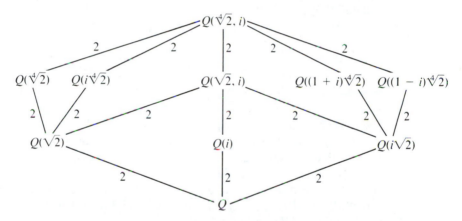

Figure 32.2 Lattice of subfields of $Q(\sqrt[4]{2}, i)$.

itself. Historically, this relationship was discovered in the process of attempting to solve a polynomial equation $f(x) = 0$ by radicals.

Before we can give a precise statement of the Fundamental Theorem of Galois Theory, we need some terminology and notation.

DEFINITIONS Automorphism, Group Fixing F, Fixed Field of H
Let E be an extension field of the field F. An *automorphism of E* is a ring isomorphism from E onto E. The *automorphism group of E fixing F*, $G(E/F)$, is the set of all automorphisms of E that take every element of F to itself. If H is a subgroup of $G(E/F)$, the set

$$E_H = \{x \in E \mid \phi(x) = x \text{ for all } \phi \in H\}$$

is called the *fixed field of H*.

It is easy to show that the set of automorphisms of E forms a group under composition. We leave as exercises (exercises 2 and 3) the verifications that the automorphism group of E fixing F is a subgroup of the automorphism group of E and, for any subgroup H of $G(E/F)$, the fixed field E_H of H is a subfield of E. The group $G(E/F)$ is sometimes called the *Galois group of E over F*. Be careful not to misinterpret $G(E/F)$ as something having to do with factor rings or factor groups. It does not.

The following examples will help you assimilate these definitions. In each example, we simply indicate how the automorphisms are defined. We leave as exercises the verifications that the mappings are indeed automorphisms.

Example 1   Consider the extension $Q(\sqrt{2})$ of Q. Since

$$Q(\sqrt{2}) = \{a + b\sqrt{2} \mid a, b \in Q\}$$

and any automorphism of a field containing Q must act as the identity on Q (exercise 1), an automorphism ϕ of $Q(\sqrt{2})$ is completely determined by $\phi(\sqrt{2})$. Thus,

$$2 = \phi(2) = \phi(\sqrt{2}\sqrt{2}) = (\phi(\sqrt{2}))^2$$

and, therefore, $\phi(\sqrt{2}) = \pm\sqrt{2}$. This proves that the group $G(Q(\sqrt{2})/Q)$ has two elements, the identity mapping and the mapping that sends $a + b\sqrt{2}$ to $a - b\sqrt{2}$.   ❑

Example 2   Consider the extension $Q(\sqrt[3]{2})$ of Q. An automorphism ϕ of $Q(\sqrt[3]{2})$ is completely determined by $\phi(\sqrt[3]{2})$. By an argument analogous to that in Example 1, we see that $\phi(\sqrt[3]{2})$ must be a cube root of 2. Since $Q(\sqrt[3]{2})$ is a subset of the real numbers and $\sqrt[3]{2}$ is the only real cube root of 2, we must have $\phi(\sqrt[3]{2}) = \sqrt[3]{2}$. Thus, ϕ is the identity automorphism and $G(Q(\sqrt[3]{2})/Q)$ has only one element. Obviously, the fixed field of $G(Q\sqrt[3]{2}/Q)$ is $Q(\sqrt[3]{2})$.   ❑

Example 3   Consider the extension $Q(\sqrt[4]{2}, i)$ of $Q(i)$. Any automorphism ϕ of $Q(\sqrt[4]{2}, i)$ fixing $Q(i)$ is completely determined by $\phi(\sqrt[4]{2})$. Since

$$2 = \phi(2) = \phi((\sqrt[4]{2})^4) = (\phi(\sqrt[4]{2}))^4,$$

we see that $\phi(\sqrt[4]{2})$ must be a fourth root of 2. Thus, there are at most four possible automorphisms of $Q(\sqrt[4]{2}, i)$ fixing $Q(i)$. If we define ϕ so that $\phi(i) = i$ and $\phi(\sqrt[4]{2}) = i\sqrt[4]{2}$, then $\phi \in G(Q(\sqrt[4]{2}, i)/Q(i))$ and ϕ has order 4. Thus, $G(Q(\sqrt[4]{2}, i)/Q(i))$ is a cyclic group of order 4. The fixed field of $\{\varepsilon, \phi^2\}$ (where ε is the identity automorphism) is $Q(\sqrt{2}, i)$. The lattice of subgroups of $G(Q(\sqrt[4]{2}, i)/Q(i))$ and the lattice of subfields of $Q(\sqrt[4]{2}, i)$ containing $Q(i)$ are shown in Figure 32.3. As in Figures 32.1 and 32.2, the integers along the lines in the group lattice represent the index of a subgroup in the group above it, and the integers along the

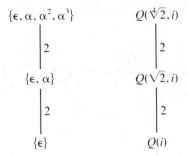

Figure 32.3 Lattice of subgroups of $G(Q(\sqrt[4]{2}, i)/Q(i))$ and lattice of subfields of $Q(\sqrt[4]{2}, i)$ containing $Q(i)$.

lines of the field lattice represent the degree of the extension of a field over the field below it. ❏

Example 4 Consider the extension $Q(\sqrt{3}, \sqrt{5})$ of Q. Since

$$Q(\sqrt{3}, \sqrt{5}) = \{a + b\sqrt{3} + c\sqrt{5} + d\sqrt{3}\sqrt{5} \mid a, b, c, d \in Q\},$$

any automorphism ϕ of $Q(\sqrt{3}, \sqrt{5})$ is completely determined by the two values $\phi(\sqrt{3})$ and $\phi(\sqrt{5})$. This time there are four automorphisms:

ε	α	β	$\alpha\beta$
$\sqrt{3} \to \sqrt{3}$	$\sqrt{3} \to -\sqrt{3}$	$\sqrt{3} \to \sqrt{3}$	$\sqrt{3} \to -\sqrt{3}$
$\sqrt{5} \to \sqrt{5}$	$\sqrt{5} \to \sqrt{5}$	$\sqrt{5} \to -\sqrt{5}$	$\sqrt{5} \to -\sqrt{5}$

Obviously, $G(Q(\sqrt{3}, \sqrt{5})/Q)$ is isomorphic to $Z_2 \oplus Z_2$. The fixed field of $\{\varepsilon, \alpha\}$ is $Q(\sqrt{5})$, the fixed field of $\{\varepsilon, \beta\}$ is $Q(\sqrt{3})$, and the fixed field of $\{\varepsilon, \alpha\beta\}$ is $Q(\sqrt{3}\sqrt{5})$. The lattice of subgroups of $G(Q(\sqrt{3}, \sqrt{5})/Q)$ and the lattice of subfields of $Q(\sqrt{3}, \sqrt{5})$ are shown in Figure 32.4. ❏

Example 5 is a bit more complicated than our previous examples. In particular, the automorphism group is non-Abelian.

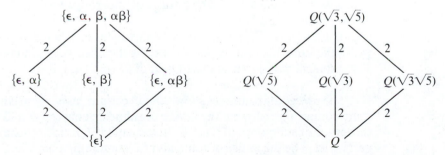

Figure 32.4 Lattice of subgroups of $G(Q(\sqrt{3}, \sqrt{5})/Q)$ and lattice of subfields of $Q(\sqrt{3}, \sqrt{5})$.

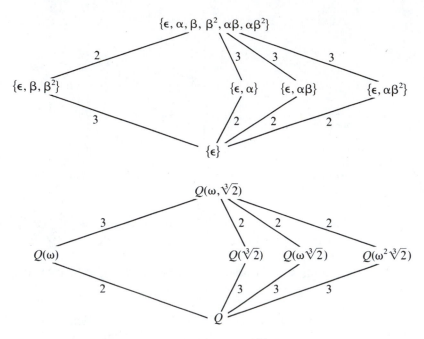

Figure 32.5   Lattice of subgroups of $G(Q(\omega, \sqrt[3]{2})/Q)$ and lattice of subfields of $Q(\omega, \sqrt[3]{2})$, where $\omega = -1/2 + i\sqrt{3}/2$.

Example 5   Direct calculations show that $\omega = -1/2 + i\sqrt{3}/2$ satisfies the equations $\omega^3 = 1$ and $\omega^2 + \omega + 1 = 0$. Now, consider the extension $Q(\omega, \sqrt[3]{2})$ of Q. We may describe the automorphisms of $Q(\omega, \sqrt[3]{2})$ by specifying how they act on ω and $\sqrt[3]{2}$. There are six in all:

ε	α	β	β^2	$\alpha\beta$	$\alpha\beta^2$
$\omega \to \omega$	$\omega \to \omega^2$	$\omega \to \omega$	$\omega \to \omega$	$\omega \to \omega^2$	$\omega \to \omega^2$
$\sqrt[3]{2} \to \sqrt[3]{2}$	$\sqrt[3]{2} \to \sqrt[3]{2}$	$\sqrt[3]{2} \to \omega\sqrt[3]{2}$	$\sqrt[3]{2} \to \omega^2\sqrt[3]{2}$	$\sqrt[3]{2} \to \omega^2\sqrt[3]{2}$	$\sqrt[3]{2} \to \omega\sqrt[3]{2}$

Since $\alpha\beta \neq \beta\alpha$, we know that $G(Q(\omega, \sqrt[3]{2})/Q)$ is isomorphic to S_3. (See Example 4 in Chapter 8.) The lattices of subgroups and subfields are shown in Figure 32.5.

The lattices in Figure 32.5 have been arranged so that the field occupying the same position as some group is the fixed field of that group. For instance, $Q(\omega\sqrt[3]{2})$ is the fixed field of $\{\varepsilon, \alpha\beta\}$.   □

The preceding examples show that, in certain cases, there is an intimate connection between the lattice of subfields between E and F and the lattice of subgroups of $G(E/F)$. In general, if E is an extension of F, and we let $\mathcal{F}$ be the lattice of subfields of E containing F and let $\mathcal{G}$ be the lattice of subgroups of $G(E/F)$, then for each K in $\mathcal{F}$, the group $G(E/K)$ is in $\mathcal{G}$ and, for each H in $\mathcal{G}$, the field E_H is in $\mathcal{F}$. Thus, we may define a mapping $g: \mathcal{F} \to \mathcal{G}$ by $g(K) = G(E/K)$ and a mapping $f: \mathcal{G} \to \mathcal{F}$ by $f(H) = E_H$. It is easy to show that if K and L belong to $\mathcal{F}$ and $K \subseteq L$, then $g(K) \supseteq g(L)$. Similarly, if G and H belong to $\mathcal{G}$ and $G \subseteq H$, then $f(G) \supseteq f(H)$. Thus, f and g are inclusion-reversing mappings between $\mathcal{F}$ and $\mathcal{G}$. We leave it as an exercise to show that for any K in $\mathcal{F}$, we have $(fg)(K) \supseteq K$ and, for any G in $\mathcal{G}$, we have $(gf)(G) \supseteq G$. When E is an arbitrary extension of F, these inclusions may be strict. (See Example 2.) However, when E is a suitably chosen extension of F, the Fundamental Theorem of Galois Theory, Theorem 32.1, says that f and g are inverses of each other so that the inclusions are equalities. In particular, f and g are inclusion-reversing isomorphisms between the lattices $\mathcal{F}$ and $\mathcal{G}$. A stronger result than that given in Theorem 32.1 is true, but our theorem illustrates the fundamental principles involved. The student is referred to [1, p. 455] for additional details and proofs.

Theorem 32.1   *Fundamental Theorem of Galois Theory*
Let F be a field of characteristic 0 or a finite field. If E is the splitting field over F for some polynomial in F[x], then the mapping from the set of subfields of E containing F to the set of subgroups of G(E/F) given by $K \to G(E/K)$ is a one-to-one correspondence. Furthermore, for any subfield K of E containing F,

1. $[E:K] = |G(E/K)|$ *and* $[K:F] = |G(E/F)|/|G(E/K)|$. *[The index of G(E/K) in G(E/F) equals the degree of K over F.]*
2. *If K is the splitting field of some polynomial in F[x], then G(E/K) is a normal subgroup of G(E/F) and G(K/F) is isomorphic to G(E/F)/G(E/K).*
3. $K = E_{G(E/K)}$. *[The fixed field of G(E/K) is K.]*
4. *If H is a subgroup of G(E/F), then $H = G(E/E_H)$. (The automorphism group of E fixing E_H is H.)*

Generally speaking, it is much easier to determine a lattice of subgroups than a lattice of subfields. For example, it is usually quite difficult to determine, directly, how many subfields a given field has, and it is often difficult to decide whether or not two field extensions are the same. The corresponding questions about groups are much more tractable. Hence, the Fundamental Theorem of Galois Theory can be a great labor-saving device. Here is an illustration.

Example 6   Let $\omega = \cos(360°/7) + i \sin(360°/7)$, so that $\omega^7 = 1$, and consider the field $Q(\omega)$. How many subfields does it have and what are they? First, observe that $Q(\omega)$ is the splitting field of $x^7 - 1$ over Q, so that we may apply the Fundamental Theorem of Galois Theory. A simple calculation shows that the automorphism ϕ that sends ω to ω^3 has order 6. Thus,

$$[Q(\omega):G] = |G(Q(\omega)/Q)| \geq 6.$$

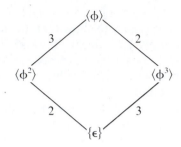

Figure 32.6 Lattice of subgroups of $G(Q(\omega)/Q)$, where $\omega = \cos(360°/7) + i\sin(360°/7)$.

Also, since

$$x^7 - 1 = (x - 1)(x^6 + x^5 + x^4 + x^3 + x^2 + x + 1)$$

and ω is a zero of $x^7 - 1$, we see that

$$|G(Q(\omega)/Q)| = [Q(\omega):Q] \leq 6.$$

Thus, $G(Q(\omega)/Q)$ is a cyclic group of order 6. So, the lattice of subgroups of $G(Q(\omega)/Q)$ is trivial to compute. See Figure 32.6.

This means that $Q(\omega)$ contains exactly two proper extensions of Q, one of degree 3 and one of degree 2. Noting that $\omega + \omega^6$ is fixed by ϕ^3, it follows that $Q \subsetneq Q(\omega + \omega^6) \subseteq Q(\omega)_{\langle\phi^3\rangle}$. Since $[Q(\omega)_{\langle\phi^3\rangle}:Q] = 3$ and $[Q(\omega + \omega^6):Q]$ divides $[Q(\omega)_{\langle\phi^3\rangle}:Q]$, we see that $Q(\omega + \omega^6) = Q(\omega)_{\langle\phi^3\rangle}$. A similar argument shows that $Q(\omega^3 + \omega^5 + \omega^6)$ is the fixed field of ϕ^2. Thus, we have found all subfields of $Q(\omega)$. ☐

SOLVABILITY OF POLYNOMIALS BY RADICALS

For Galois, the elegant correspondence between groups and fields given by Theorem 32.1 was only a means to an end. The goal Galois sought was the solution to a problem that had stymied mathematicians for centuries. Methods for solving linear and quadratic equations were known thousands of years ago (the quadratic formula). In the 16th century, Italian mathematicians developed formulas for solving any third- or fourth-degree equation. Their formulas involved only the operations of addition, subtraction, multiplication, division, and extraction of roots (radicals). For example, the equation

$$x^3 + bx + c = 0$$

has the three solutions

$$A + B,$$
$$-(A + B)/2 + (A - B)\sqrt{-3}/2,$$
$$-(A + B)/2 - (A - B)\sqrt{-3}/2,$$

where

$$A = \sqrt[3]{\frac{-c}{2} + \sqrt{\frac{b^3}{27} + \frac{c^2}{4}}} \quad \text{and}$$

$$B = \sqrt[3]{\frac{-c}{2} - \sqrt{\frac{b^3}{27} + \frac{c^2}{4}}}.$$

The formulas for the general cubic $x^3 + ax^2 + bx + c = 0$ and the general quartic (fourth-degree polynomial) are even more complicated, but nevertheless can be given in terms of radicals of rational expressions of the coefficients. (See [2, 392–394].)

Both Abel and Galois proved that there is no general solution of a fifth-degree equation by radicals. In particular, there is no "quintic formula." Before discussing Galois's method, which provided a group-theoretic criterion for the solution of an equation by radicals and led to the modern-day Galois Theory, we need a few definitions.

DEFINITION   Solvable by Radicals
Let F be a field, and let $f(x) \in F[x]$. We say that $f(x)$ is *solvable by radicals over* F if $f(x)$ splits in some extension $F(a_1, a_2, \ldots, a_n)$ of F and there exist positive integers $k_1, \ldots, k_n$ such that $a_1^{k_1} \in F$ and $a_i^{k_i} \in F(a_1, \ldots, a_{i-1})$ for $i = 2, \ldots, n$.

So, a polynomial in $F[x]$ is solvable by radicals if we can obtain all of its zeros by adjoining nth roots (for various n) to F. In other words, each zero of the polynomial can be written as an expression (usually a messy one) involving elements of F combined by the operations of addition, subtraction, multiplication, division, and extraction of roots.

Thus, the problem of solving a polynomial equation for its zeros can be transformed into a problem about field extensions. At the same time, we can use the Fundamental Theorem of Galois Theory to transform a problem about field extensions into a problem about groups. This is exactly how Galois showed that there are fifth-degree polynomials that cannot be solved by radicals, and this is exactly how we will do it. Before giving an example of such a polynomial, we need some additional group theory.

DEFINITION   Solvable Group
We say that a group G is *solvable* if G has a series of subgroups

$$\{e\} = H_0 \subset H_1 \subset H_2 \subset \cdots \subset H_k = G,$$

where, for each $0 \le i < k$, H_i is normal in H_{i+1} and H_{i+1}/H_i is Abelian.

Obviously, Abelian groups are solvable. So are the dihedral groups and any group whose order has the form p^n, where p is a prime. The monumental Feit-Thompson Theorem (see Chapter 25) says that every group of odd order is solvable. In a certain sense, solvable groups are almost Abelian. On the other hand, it follows directly from the definitions that any non-Abelian simple group is not solvable. In particular, A_5 is not solvable. It follows from exercise 24 in Chapter 25 that S_5 is not solvable.

Theorem 32.2   *Factor Group of a Solvable Group Is Solvable*
A factor group of a solvable group is solvable.

Proof.   Suppose that G has a series of subgroups

$$\{e\} = H_0 \subset H_1 \subset H_2 \subset \cdots \subset H_k = G,$$

where, for each $0 \leq i < k$, H_i is normal in H_{i+1} and H_{i+1}/H_i is Abelian. If N is any normal subgroup of G, then

$$\{e\} = H_0N/N \subset H_1N/N \subset H_2N/N \subset \cdots \subset H_kN/N = G/N$$

is the requisite series of subgroups that guarantees that G/N is solvable. (See exercise 26.)   ∎

We now are able to make the critical connection between solvability of polynomials by radicals and solvable groups.

Theorem 32.3   *Solvable by Radicals Implies a Solvable Group*
Let F be a field of characteristic 0 and let $f(x) \in F[x]$. Let E be the splitting field for $f(x)$ over F. If $f(x)$ is solvable by radicals over F, then the Galois group $G(E/F)$ is solvable.

Proof.   The idea behind this proof is to construct a sequence of field extensions

$$F = L_0 \subset L_1 \subset L_2 \subset \cdots \subset L_t = L$$

such that L contains E and the group $G(L/F)$ is solvable. We then obtain the desired conclusion from Theorem 32.2 by observing that $G(E/F)$ is a factor group of $G(L/F)$.

To do this, suppose that $f(x)$ splits in $F(a_1, a_2, \ldots, a_r)$, where $a_1^{n_1} \in F$ and $a_i^{n_i} \in F(a_1, \ldots, a_{i-1})$ for $i = 2, \ldots, r$. The first thing we want to do is to construct an extension of F that contains all of the roots of unity that we will need. To do this, let n be the largest n_i for $i = 1, \ldots, r$. For each $j = 3, 4, \ldots, n$, let α_j be a primitive jth root of unity. (See Example 2 in Chapter 16.) We construct a sequence of field extensions of F as follows.

$$F_0 = F, \quad F_1 = F_0(\alpha_3), \quad F_2 = F_1(\alpha_4), \ldots, \quad F_{n-2} = F_{n-3}(\alpha_n).$$

Then F_{i+1} is a splitting field of $x^{i+3} - 1$ over F_i, and $G(F_{i+1}/F_i)$ is Abelian. (See exercise 15.) Let K_0 denote F_{n-2}, and observe that K_0 contains all jth roots of unity for $j \leq n$. We next construct another sequence of field extensions, as follows. Let

$$K_1 = K_0(a_1), \quad K_2 = K_1(a_2), \ldots, \quad K_r = K_{r-1}(a_r).$$

For each $i = 1, \ldots, r$, let $b_i = a_i^{n_i}$. Then, since K_i contains all of the n_ith roots of unity, it follows that K_i is the splitting field of $x^{n_i} - b_i$ over K_{i-1}. We further claim that $G(K_i/K_{i-1})$ is Abelian. To see this, observe that any automorphism in $G(K_i/K_{i-1})$ is completely determined by its action on a_i. Also, since a_i is a zero of $x^{n_i} = b_i$, we know that any element of $G(K_i/K_{i-1})$ sends a_i to another zero of $x^{n_i} - b_i$. Since the zeros of $x^{n_i} - b_i$ are $a_i, \alpha_{n_i} a_i, \alpha_{n_i}^2 a_i, \ldots, \alpha_{n_i}^{n_i-1} a_i$, any element of $G(K_i/K_{i-1})$ sends a_i to $\alpha_{n_i}^j a_i$ for some j. Let ϕ and σ be two elements of $G(K_i/K_{i-1})$. Then $\phi(a_i) = \alpha_{n_i}^j a_i$ and $\sigma(a_i) = \alpha_{n_i}^k a_i$ for some j and k. Thus,

$$(\sigma\phi)(a_i) = \sigma(\alpha_{n_i}^j a_i) = \alpha_{n_i}^j \alpha_{n_i}^k a_i = \alpha_{n_i}^{j+k} a_i,$$

whereas

$$(\phi\sigma)(a_i) = \phi(\alpha_{n_i}^k a_i) = \alpha_{n_i}^k \alpha_{n_i}^j a_i = \alpha_{n_i}^{j+k} a_i,$$

so that $\phi\sigma$ and $\sigma\phi$ agree on a_i and K_{i-1}. This shows that $\phi\sigma = \sigma\phi$, and, therefore, $G(K_i/K_{i-1})$ is Abelian.

At this point, we have now constructed a sequence of field extensions $F = L_0 \subset L_1 \subset \cdots \subset L_t = L$ such that L contains E and $G(L_{i+1}/L_i)$ is Abelian. Since the Fundamental Theorem of Galois Theory tells us that $G(L_{i+1}/L_i)$ is isomorphic to $G(L/L_i)/G(L/L_{i+1})$, the series of subgroups of $G(L/F)$,

$$\{e\} = G(L/L_t) \subset G(L/L_{t-1}) \subset \cdots \subset G(L/L_0) = G(L/F),$$

demonstrates that $G(L/F)$ is solvable. Finally, because $G(E/F)$ is isomorphic to $G(L/F)/G(L/E)$, Theorem 32.2 guarantees that $G(E/F)$ is solvable. ■

It is worth remarking that the converse of Theorem 32.3 is true also; that is, if E is the splitting field of a polynomial $f(x)$ over a field F of characteristic 0 and $G(E/F)$ is solvable, then $f(x)$ is solvable by radicals over F.

One of the major unsolved problems in algebra, first posed by Emmy Noether, is determining which finite groups can occur as Galois groups over Q. Many people suspect that the answer is "all of them." It is known that every solvable group is a Galois group. John Thompson has recently proved that certain kinds of simple groups, including the Monster, are Galois groups. The suggested reading by Ian Stewart provides more information on this topic.

INSOLVABILITY OF A QUINTIC

We will finish our introduction to Galois theory by explicitly exhibiting a polynomial that has integer coefficients and that is not solvable by radicals over Q.

Consider $g(x) = 3x^5 - 15x + 5$. By Eisenstein's Criterion (Theorem 17.4), $g(x)$ is irreducible over Q. Since $g(x)$ is continuous and $g(-2) = -61$ and $g(-1) = 17$, we know that $g(x)$ has a real zero between -2 and -1. A similar analysis shows that $g(x)$ also has real zeros between 0 and 1 and between 1 and 2.

Each of these real zeros has multiplicity 1, as can be verified by long division or by appealing to Theorem 20.6. Furthermore, $g(x)$ has no more than three real zeros, because Rolle's Theorem from calculus guarantees that between each pair of real zeros of $g(x)$ there must be a zero of $g'(x) = 15x^4 - 15$. So, for $g(x)$ to have four real zeros, $g'(x)$ would have to have three real zeros, and it does not. Thus, the other two zeros of $g(x)$ are nonreal complex numbers, say, $a + bi$ and $a - bi$. (See exercise 50 in Chapter 15.)

Now, let's denote the five zeros of $g(x)$ by a_1, a_2, a_3, a_4, a_5. Since any automorphism of $K = Q(a_1, a_2, a_3, a_4, a_5)$ is completely determined by its action on the a's and must permute the a's, we know that $|G(K/Q)|$ is isomorphic to a subgroup of S_5, the symmetric group on five symbols. Since a_1 is a zero of an irreducible polynomial of degree 5 over Q, we know that $[Q(a_1):Q] = 5$, and, therefore, 5 divides $[K:Q]$. Thus, the Fundamental Theorem of Galois Theory tells us that 5 also divides $G(K/Q)$. So, by Cauchy's Theorem (corollary to Theorem 24.3), we may conclude that $G(K/Q)$ has an element of order 5. Since the only elements in S_5 of order 5 are the 5-cycles, we know that $G(K/Q)$ contains a 5-cycle. The mapping from $\mathbf{C}$ to $\mathbf{C}$, sending $a + bi$ to $a - bi$, is also an element of $G(K/Q)$. Since this mapping fixes the three real zeros and interchanges the two complex zeros of $g(x)$, we know that $G(K/Q)$ contains a 2-cycle. But, the only subgroup of S_5 that contains both a 5-cycle and a 2-cycle is S_5. (See exercise 27 in Chapter 25.) So, $G(K/Q)$ is isomorphic to S_5. Finally, since S_5 is not solvable (see exercise 21), we have succeeded in exhibiting a fifth-degree polynomial that is not solvable by radicals.

Exercises

Seeing much, suffering much, and studying much are the three pillars of learning.

Benjamin Disraeli

1. Let E be an extension field of Q. Show that any automorphism of E acts as the identity on Q. (This exercise is referred to in this chapter.)
2. Let E be a field extension of the field F. Show that the automorphism group of E fixing F is indeed a group. (This exercise is referred to in this chapter.)
3. Let E be a field extension of a field F and let H be a subgroup of $G(E/F)$. Show that the fixed field of H is indeed a field. (This exercise is referred to in this chapter.)

4. Referring to Example 6, show that the automorphism ϕ has order 6. Show that $\omega + \omega^6$ is fixed by ϕ^3 and $\omega^3 + \omega^5 + \omega^6$ is fixed by ϕ^2.

5. Let $f(x) \in F[x]$ and let the zeros of $f(x)$ be $a_1, a_2, \ldots, a_n$. If $K = F(a_1, a_2, \ldots, a_n)$, show that $G(K/F)$ is isomorphic to a group of permutations of the a_i's. [When K is the splitting field of $f(x)$ over F, the group $G(K/F)$ is called the *Galois group of* $f(x)$.]

6. Show that the Galois group of a polynomial of degree n has order dividing $n!$.

7. Let E be the splitting field of $x^4 + 1$ over Q. Find $G(E/Q)$. Find all subfields of E. Find the automorphisms of E that have fixed fields $Q(\sqrt{2})$, $Q(\sqrt{-2})$, and $Q(i)$. Is there an automorphism of E whose fixed field is Q?

8. Determine the group of field automorphisms of GF(4).

9. Let $E = Q(\sqrt{2}, \sqrt{5})$. What is the order of the group $G(E/Q)$? What is the order of $G(Q(\sqrt{10})/Q)$?

10. Given that the automorphism group of $Q(\sqrt{2}, \sqrt{5}, \sqrt{7})$ is isomorphic to $Z_2 \oplus Z_2 \oplus Z_2$, determine the number of subfields of $Q(\sqrt{2}, \sqrt{5}, \sqrt{7})$ that have degree 4 over Q.

11. Suppose that F is a field of characteristic 0 and E is the splitting field for some polynomial over F. If $G(E/F)$ is isomorphic to A_4, show that there is no subfield K of E such that $[K:F] = 2$.

12. Show that the Galois group of $x^3 - 3$ over Q is isomorphic to S_3.

13. Suppose that K is the splitting field of some polynomial over a field F of characteristic 0. If $[K:F] = p^2q$, where p and q are distinct primes, show that K has subfields $L_1, L_2,$ and L_3 such that $[L_1:F] = pq$, $[L_2:F] = p^2$, and $[L_3:F] = q$.

14. Suppose that E is the splitting field of some polynomial over a field F of characteristic 0. If $G(E/F)$ is isomorphic to D_6, draw the subfield lattice for the fields between E and F.

15. Let F be a field of characteristic 0. If K is the splitting field of $x^n - 1$ over F, prove that $G(K/F)$ is Abelian. (This exercise is referred to in the proof of Theorem 32.3.)

16. Suppose that E is the splitting field of some polynomial over a field F of characteristic 0. If $[E:F]$ is finite, show that there is only a finite number of fields between E and F.

17. Suppose that E is the splitting field of some polynomial over a field F of characteristic 0. If $G(E/F)$ is an Abelian group of order 10, draw the subfield lattice for the fields between E and F.

18. Let ω be a nonreal complex number such that $\omega^5 = 1$. If ϕ is the automorphism of $Q(\omega)$ that carries ω to ω^4, find the fixed field of $\langle \phi \rangle$.

19. Determine the isomorphism class of the group $G(\text{GF}(64)/\text{GF}(2))$.

20. Determine the isomorphism class of the group $G(\text{GF}(729)/\text{GF}(9))$.

21. Show that S_5 is not solvable.

22. Show that the dihedral groups are solvable.

23. Show that a group of order p^n, where p is prime, is solvable.

24. Show that S_n is solvable when $n \leq 4$.

25. Show that a subgroup of a solvable group is solvable.

26. Complete the proof of Theorem 32.2 by showing that the given series of groups satisfies the definition for solvability.

REFERENCES

1. J. B. Fraleigh, *A First Course in Abstract Algebra,* 4th ed., Reading, Mass.: Addison-Wesley, 1989.

2. Samuel M. Selby, *Standard Mathematical Tables,* Cleveland: The Chemical Rubber Company, 1965.

SUGGESTED READINGS

Lisl Gaal, *Classical Galois Theory with Examples,* Chicago: Markham, 1971.
This book has a large number of examples pertaining to Galois theory worked out in great detail.

D. G. Mead, "The Missing Fields," *The American Mathematical Monthly* 94 (1987): 12–13.
This article uses Galois theory to show that, for any positive integer n, there is an extension K of Q with $[K:Q] = n$, and yet there is no field properly between K and Q.

Tony Rothman, "The Short Life of Évariste Galois," *Scientific American,* April (1982): 136–149.
This article gives an elementary discussion of Galois's proof that the general fifth-degree equation cannot be solved by radicals. The article also goes into detail about Galois's controversial life and death. In this regard, Rothman refutes several accounts given by other Galois biographers.

Ian Stewart, "The Duelist and the Monster," *Nature* 317 (1985): 12–13.
This nontechnical article discusses recent work of John Thompson pertaining to the question of "which groups can occur as Galois groups."

Philip Hall

He was pre-eminent as a group theorist and made many fundamental discoveries; the conspicuous growth of interest in group theory in this century owes much to him.

J. E. Roseblade

PHILIP HALL was born on April 11, 1904, in London. Abandoned by his father shortly after his birth, Hall was raised by his mother, a dressmaker. He demonstrated academic prowess early by winning a scholarship to Christ's Hospital, where he had several outstanding mathematics teachers. At Christ's Hospital, Hall won a medal for the best English essay, the gold medal in mathematics, and a scholarship to King's College, Cambridge.

Although abstract algebra was a field neglected at King's College, Hall studied Burnside's book *Theory of Groups* and some of Burnside's later papers. After graduating in 1925, he stayed on at King's College for further study and was elected to a fellowship in 1927. That same year, Hall discovered a major "Sylow-like" theorem about solvable groups: if a solvable group has order mn, where m and n are relatively prime, then every subgroup whose order divides m is contained in a group of order m and all subgroups of order m are conjugate. Over the next three decades, Hall developed a general theory of finite solvable groups that had a profound influence on John Thompson's spectacular achievements of the 1960s. In the 1930s, Hall also developed a general theory of groups of prime-power order that has become a foundation of modern finite group theory. In addition to his fundamental contributions to finite groups, Hall wrote many seminal papers on infinite groups.

509

Among the concepts that have Hall's name attached to them are Hall subgroup, Hall algebra, Hall-Littlewood polynomials, Hall divisors, the marriage theorem from graph theory, and the Hall commutator collecting process. Beyond his own discoveries, Hall had an enormous influence on algebra through his research students. No fewer than one dozen have become eminent mathematicians in their own right.

Hall had a deep love of poetry, flowers, and country walks. He enjoyed music and art. He died on December 30, 1982.

33

An Introduction to Boolean Algebras

In recent years much interest has arisen in what are known as
"Boolean algebras." The principal reason for this great
interest is that many applications of this discipline have been
found in connection with various systems of automation.

John T. Moore, *Elements of Abstract Algebra*

MOTIVATION

To be worthy of study, an abstract mathematical system should have
several concrete realizations. Groups, for instance, arise in connection
with symmetries, matrices with nonzero determinants, and permutations. Rings encompass the integers, polynomials, and matrices. Fields
generalize familiar systems such as the real numbers, the integers modulo a prime, and polynomial rings modulo a maximal ideal. In this
chapter, we introduce another system, called a Boolean algebra, that has
three major concrete models—the algebra of sets, the algebra of electrical circuits, and the algebra of logic. We will briefly discuss these three
models.

Informally, a Boolean algebra is a set of objects with two binary
operations and one unary operation that satisfy certain conditions.
Examples 1–3 motivate the formal definition of Boolean algebra.

511

Example 1   Algebra of Sets

Let X be a nonempty set and let $P(X)$ denote the set of all subsets of X. For any pair A, B in $P(X)$, we let

$$A \cap B = \{x \mid x \in A \text{ and } x \in B\} \quad \text{(intersection)},$$
$$A \cup B = \{x \in A \text{ or } x \in B\} \quad \text{(union)},$$
$$A' = \{x \in X \mid x \notin A\} \quad \text{(complement)}.$$

The student should check that the operations $\cap$ and $\cup$ are commutative and associative, and that each is distributive over the other; that is,

$$A \cup B = B \cup A \quad \text{and} \quad A \cap B = B \cap A;$$
$$A \cup (B \cup C) = (A \cup B) \cup C \quad \text{and} \quad A \cap (B \cap C) = (A \cap B) \cap C;$$
$$A \cup (B \cap C) = (A \cup B) \cap (A \cup C) \quad \text{and}$$
$$A \cap (B \cup C) = (A \cap B) \cup (A \cap C).$$

Also, note that for any A in $P(X)$ we have

$$A \cup \emptyset = A, \qquad A \cap X = A$$

and

$$A \cup A' = X, \qquad A \cap A' = \emptyset. \qquad \square$$

Example 2   Algebra of Switching Functions

For any positive integer n, a function from the Cartesian product of n copies of $\{0, 1\}$ to $\{0, 1\}$ is called a *switching function on n variables*. (Later in this chapter, we will see how an electrical circuit with n switches defines such a function.) If f and g are switching functions on n variables, then so are the functions defined by

$$(f + g)(x) = \max\{f(x), g(x)\}$$

and

$$(f \cdot g)(x) = \min\{f(x), g(x)\}.$$

Furthermore, for any switching function f, we may define a new switching function f' by

$$f'(x) = \begin{cases} 0 & \text{if } f(x) = 1, \\ 1 & \text{if } f(x) = 0. \end{cases}$$

Again, we may observe that switching function addition and multiplication are commutative, associative, and distributive. Moreover, the constant functions $z = 0$ and $u = 1$ have the properties that

$$z + f = f, \qquad u \cdot f = f$$

and

$$f + f' = u, \qquad f \cdot f' = z.$$

Table 33.1 shows all switching functions on two variables.

Table 33.1   All Switching Functions on Two Variables

	f_0	f_1	f_2	f_3	f_4	f_5	f_6	f_7	f_8	f_9	f_{10}	f_{11}	f_{12}	f_{13}	f_{14}	f_{15}
(0, 0)	0	1	0	1	0	1	0	1	0	1	0	1	0	1	0	1
(1, 0)	0	0	1	1	0	0	1	1	0	0	1	1	0	0	1	1
(0, 1)	0	0	0	0	1	1	1	1	0	0	0	0	1	1	1	1
(1, 1)	0	0	0	0	0	0	0	0	1	1	1	1	1	1	1	1

We can compute the sum of any pair of these functions by adding their corresponding columns. The resulting column is the column of the sum. For instance, $f_3 + f_5 = f_7$ (keep in mind that $1 + 1 = 1$).
Similarly, $f_3 \cdot f_5 = f_1$ and $f_3' = f_{12}$.   ❑

Example 3   Divisors of 105
Consider the set S of all positive divisors of 105. Specifically,

$$S = \{1, 3, 5, 7, 15, 21, 35, 105\}.$$

For any pair a, b in S, define $a \vee b = \operatorname{lcm}(a, b)$ and $a \wedge b = \gcd(a, b)$. Finally, for any a in S, let $a' = 105/a$. Then the operations $\vee$ and $\wedge$ are commutative, associative, and distributive. Furthermore, for any a in S,

$$a \vee 1 = a, \qquad a \wedge 105 = a$$

and

$$a \vee a' = 105, \qquad a \wedge a' = 1. \qquad ❑$$

Examples 1–3 are illustrations of Boolean algebras.

DEFINITION AND PROPERTIES

DEFINITION   Boolean Algebra
A *Boolean algebra* is a set B with two binary operations $\vee$ (read "or") and $\wedge$ (read "and") and a unary operation ' (read "complement") that satisfy the following axioms (for all a, b, c in B).

1. Commutativity

$$a \vee b = b \vee a, \qquad a \wedge b = b \wedge a.$$

2. Associativity

$$a \vee (b \vee c) = (a \vee b) \vee c, \qquad a \wedge (b \wedge c) = (a \wedge b) \wedge c.$$

3. Distributivity

$$a \wedge (b \vee c) = (a \wedge b) \vee (a \wedge c),$$
$$a \vee (b \wedge c) = (a \vee b) \wedge (a \vee c).$$

4. Existence of zero and unity. There are elements 0 and 1 in B such that, for all a in B,

$$a \vee 0 = a \quad \text{and} \quad a \wedge 1 = a.$$

5. Complementation

$$a \vee a' = 1; \quad a \wedge a' = 0.$$

It is worth mentioning that a Boolean algebra is almost a commutative ring with unity. The only ring property that a Boolean algebra lacks is the existence of inverses for one of the operations. On the other hand, a commutative ring with unity is usually not a Boolean algebra because addition does not distribute over multiplication; that is,

$$a + bc \neq (a + b)(a + c).$$

Before stating some basic properties of Boolean algebras, we look at one nonexample.

Example 4   Let $X = \{1, 2, 4, 7, 14, 28\}$—the divisors of 28. For any a, b in X, define

$$a \vee b = \text{lcm}(a, b) \quad \text{and} \quad a \wedge b = \text{gcd}(a, b).$$

The operations $\wedge$ and $\vee$ satisfy axioms 1–4 of the definition of Boolean algebra with 1 as the zero and 28 as the unity. However, axiom 5 cannot be satisfied, no matter how we define a'. For instance, consider the possibilities for $14'$. The only x's that yield $14 \vee x = \text{lcm}(14, x) = 28$ are 4 and 28. So, $14' = 4$ or $14' = 28$. But $14 \wedge 4 = \text{gcd}(14, 4) = 2$ and $14 \wedge 28 = \text{gcd}(14, 28) = 14$. Since neither 2 nor 14 is the zero element, there is no complement for 14.   ❑

Theorem 33.1 collects many elementary properties of a Boolean algebra. The proofs are left as exercises.

Theorem 33.1   *Properties of Boolean Algebras*
Let B be a Boolean algebra and let a, b belong to B. Then

1. $a \wedge 0 = 0$.
2. $a \vee 1 = 1$.
3. *(Absorption laws)*

$$a \wedge (a \vee b) = a, \quad a \vee (a \wedge b) = a.$$

4. *(Idempotent laws)*

$$a \wedge a = a, \quad a \vee a = a.$$

5. *If $a \vee b = 1$ and $a \wedge b = 0$, then $b = a'$.*
6. *(De Morgan's Laws)*

$$(a \wedge b)' = a' \vee b', \quad (a \vee b)' = a' \wedge b'.$$

7. *(Involution)*

$$(a')' = a.$$

8. *The zero and unity elements are unique.*

The perceptive student may have noticed that there is a symmetry in properties 3, 4, and 6 of Theorem 33.1. In particular, in each case, one may obtain the right-hand equation from the left-hand equation by interchanging ∧ for ∨. Similarly, property 2 can be obtained from property 1 by replacing ∧ by ∨ and 0 by 1. Indeed, the axioms for a Boolean algebra share this property. Consequently, the *duality principle* holds for Boolean algebras; that is, for any proposition that is true, there is a corresponding proposition obtained by making these interchanges (a *dual*) that is also true.

Example 5 The dual of $a \wedge b = 0$ is $a \vee b = 1$; the dual of $a \wedge (b \vee c) = (a \wedge b) \vee c$ is $a \vee (b \wedge c) = (a \vee b) \wedge c$. ☐

THE ALGEBRA OF ELECTRICAL CIRCUITS

We next give a brief indication of how Boolean algebra is used in electrical engineering. On-off electrical switches, diodes, magnetic dipoles, and transistors are examples of two-state devices called *switches*. In electrical networks, these two states may be "current flows" (on) and "current does not flow" (off), magnetized and not magnetized, high potential and low potential, or closed (current flows) and open (current does not flow). Abstractly, we represent the "on" state by 1 and the "off" state by 0. We use letters $a, b, c, \ldots$ to denote the states of switches that could be on or off so that these variables can take on the values 0 and 1. (See Figure 33.1.)

Two switches, a and b, connecting two terminals are said to be connected in *series* if current will flow between the two terminals only when both a and b are on. This situation is diagrammed in Figure 33.2 and is denoted by $a \cdot b$ (or just ab).

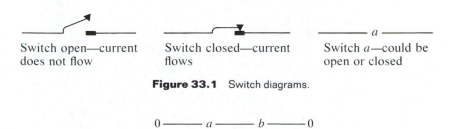

Switch open—current does not flow

Switch closed—current flows

Switch *a*—could be open or closed

Figure 33.1 Switch diagrams.

$$0 \underline{\hspace{1cm}} a \underline{\hspace{1cm}} b \underline{\hspace{1cm}} 0$$
$$a \cdot b$$

Figure 33.2 Switches *a* and *b* connected in series.

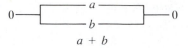

Figure 33.3   Switches *a* and *b* connected in parallel.

Table 33.2   Operation Table for + and ·

a	b	a + b	ab
0 (off)	0 (off)	0 (off)	0 (off)
1 (on)	0 (off)	1 (on)	0 (off)
0 (off)	1 (on)	1 (on)	0 (off)
1 (on)	1 (on)	1 (on)	1 (on)

Switches *a* and *b* connecting two terminals are said to be connected in *parallel* if current will flow between the two terminals when either *a* or *b* is on. This is depicted in Figure 33.3 and is denoted by $a + b$

Following our convention of using 0 for off and 1 for on, the diagrams in Figures 33.2 and 33.3 and the notation defined previously suggest that we may define two binary operations + and · on {0, 1}, as in Table 33.2.

With these operations, we can now combine switches in series and parallel to make more complicated devices called *series-parallel switching circuits.* One such example is given in Figure 33.4.

It is also convenient to define a unary operation on circuits. For any circuit *C* made up of switches *a, b, c, . . . ,* the circuit *C'* is made up of the same switches, but for any choice of states for *a, b, c, . . . , C'* has the opposite state as *C*. So, for example, if $a + b$ is on, $(a + b)'$ is off. We say that two circuits, C_1 and C_2, made up of switches *a, b, c, . . . ,* are equivalent (and write $C_1 = C_2$) if, for any states for *a, b, c, . . . ,* current flows through C_1 if and only if current flows through C_2. Figure 33.5 shows two equivalent circuits.

It should not come as a surprise that the set of all equivalence classes of circuits (made up of a finite number of switches) with the binary operations + and · and the unary operation ′ is a Boolean algebra. Of course, we may determine whether or not current will flow through a circuit composed of switches *a, b, c, . . . ,* by simply replacing each variable rep-

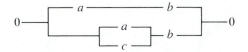

Figure 33.4   The diagram for $ab + (a + c)b$.

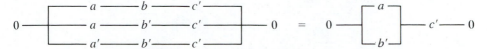

Figure 33.5 $abc' + ab'c' + a'b'c' = (a + b')c'$.

resenting an off switch with 0 and each variable representing an on switch with 1 and using the operations

$$0 + 0 = 0, \qquad 1 + 0 = 1, \qquad 1 + 1 = 1,$$
$$0 \cdot 0 = 0, \qquad 1 \cdot 0 = 0, \qquad 1 \cdot 1 = 1.$$

Table 33.3 gives an example.

In computer design, switches are symbolically represented by so-called "gates," as shown in Figure 33.6. The gate representations for the circuits in Figures 33.4 and 33.5 are shown in Figures 33.7 and 33.8.

Table 33.3 Values for the Circuit $ab + (a + c)b'$

a	b	c	ab	$a + c$	b'	$(a + c)b'$	$ab + (a + c)b'$
0	0	0	0	0	1	0	0
1	0	0	0	1	1	1	1
0	1	0	0	0	0	0	0
0	0	1	0	1	1	1	1
1	1	0	1	1	0	0	1
1	0	1	0	1	1	1	1
0	1	1	0	1	0	0	0
1	1	1	1	1	0	0	1

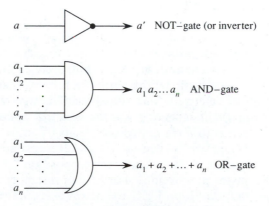

Figure 33.6

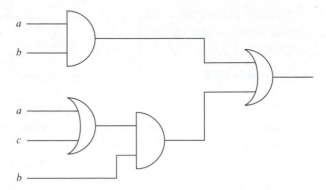

Figure 33.7   Gate diagram for $ab + (a + c)b$.

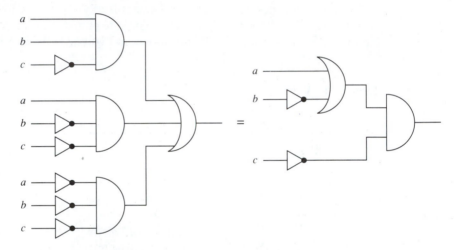

Figure 33.8   Gate diagram for $abc' + ab'c' + a'b'c' = (a + b')c'$

The Algebra of Logic

Just as the set of all equivalence classes of circuits forms a Boolean algebra, so too does the set of all equivalence classes of propositions. For our
purposes, a *proposition* is a declarative statement that we can decide (in
theory, at least) is either true or false. For example, the proposition
"John F. Kennedy was assassinated" is true, whereas the proposition
"Franklin D. Roosevelt was assassinated" is false. Propositions can be
combined to form new propositions by using the logical connectives
"and" and "or." In general, if A and B are propositions, mathematicians
often use $A \wedge B$ to denote the proposition "A and B," and $A \vee B$ to
denote the proposition "A or B." For instance, if we use A to denote the
proposition "John F. Kennedy was assassinated" and B to denote the
proposition "Franklin D. Roosevelt was assassinated," then $A \wedge B$ rep-

resents the proposition "John F. Kennedy was assassinated and Franklin D. Roosevelt was assassinated," which is false. On the other hand, the symbolism $A \vee B$ represents the proposition "John F. Kennedy was assassinated or Franklin D. Roosevelt was assassinated," which is true.

We define $A \wedge B$ to be true if and only if both A and B are true, and $A \vee B$ to be true if and only if either A or B (or both) is true. The latter operation is sometimes called the *inclusive or.* (Another term for it is *and/or*). The *exclusive or* connecting two propositions means that one of the propositions is true and one is false. For any proposition A, there is another proposition, often denoted by A' and called the *negation of A*, that asserts that proposition A is false. For example, if A denotes the proposition "John F. Kennedy was assassinated," then A' is the proposition "It is false that John F. Kennedy was assassinated," or more naturally, "John F. Kennedy was not assassinated." In general, proposition A' is true only when proposition A is false, and vice versa.

If A and B are propositions formed by combining propositions P_1, P_2, P_3, . . . , and P'_1, P'_2, P'_3, . . . , with the connectives $\wedge$ and $\vee$, we say that A is *equivalent* to B (and write $A = B$) if, for every assignment of truth or falsity to the propositions P_1, P_2, P_3, . . . , A is true if and only if B is true. With this definition of equivalence, the set of all equivalence classes of propositions is a Boolean algebra. The zero element is the equivalence class containing $P \wedge P'$, where P is any statement. (Any element in this equivalence class is called a *contradiction.*) The unity element is the equivalence class containing $P \vee P'$, where P is any statement. (Any statement in this equivalence class is called a *tautology.*)

As a consequence of this, we can decide mechanically whether or not a complicated proposition made by combining other propositions is true or false, if we know the truth or falsity of the propositions involved. This can be done easily by assigning the letter T to each proposition that is true and the letter F to each proposition that is false. We then use the operations

$$F \wedge F = F, \qquad F \vee F = F, \qquad F \wedge T = F,$$
$$F \vee T = T, \qquad T \wedge T = T, \qquad T \vee T = T,$$

and the axioms for a Boolean algebra on the set $\{T, F\}$, to reduce the resulting expression to a single T or a single F. If it reduces to T, the proposition under consideration is true; otherwise, it is false. Table 33.4 gives several examples. Such a table is often called a *truth table.*

FINITE BOOLEAN ALGEBRAS

Our last theorem provides an important arithmetic criterion that every finite Boolean algebra must satisfy. Notice that, with this result, we may instantly reject the set in Example 4 as a Boolean algebra, no matter how $\vee$ and $\wedge$ are defined.

Table 33.4   Truth Table for Various Propositions

P	Q	R	$P \vee Q$	$P \wedge Q$	$(P \wedge Q) \vee [(P \vee R) \wedge Q]$	$(P \vee Q') \wedge R'$
F	F	F	F	F	F	T
F	F	T	F	F	F	F
F	T	F	T	F	F	F
F	T	T	T	F	T	F
T	F	F	T	F	F	T
T	F	T	T	F	F	F
T	T	F	T	T	T	T
T	T	T	T	T	T	F

Theorem 33.2 $|B| = 2^n$.

A finite Boolean algebra has 2^n elements for some integer n.

Proof.   Let B be any finite Boolean algebra with more than one element. For any pair of elements a and b in B, we define

$$a + b = (a \vee b) \wedge (a \wedge b)'.$$

(This operation is called the *symmetric difference* of a and b. See exercise 22.) One may tediously verify that B is in fact an Abelian group under $+$. Since $a + a = 0$ for all a, we see that every nonzero element of B has order 2. It now follows from Theorem 9.5 (or the corollary to Theorem 24.3) that the only prime divisor of $|B|$ is 2. Thus, $|B| = 2^n$ for some positive integer n.   ∎

Although it appears that there are many diverse models for Boolean algebras (e.g., algebra of sets, algebra of switching functions, algebra of circuits, algebra of logic), it can be proved (see [1, p. 378]) that every finite Boolean algebra is isomorphic to an algebra of sets. This is analogous to the fact that every group is isomorphic to a group of permutations. (See Cayley's Theorem in Chapter 6.)

EXERCISES

The road to wisdom?—Well it's plain and simple to express:
Err
and err
and err again
but less
and less
and less.

       Piet Hein, "The Road to Wisdom," *Grooks* **(1966)**

1. Prove that the following identities hold in a Boolean algebra.
   a. $a \vee (a' \wedge b) = a \vee b$
   b. $(a \wedge b) \vee (a \wedge b') \vee (a' \wedge b') = a \vee b'$
   c. $(a \vee b) \wedge (a \vee b') \wedge (a' \vee b') = a \wedge b'$

2. Verify that the set and operations given in Example 3 satisfy axiom 4 for a Boolean algebra.

3. Show that the set of divisors of 36 cannot be a Boolean algebra.

4. Prove Theorem 33.1.

5. Referring to Example 2, calculate the following:
   a. $f_2 + f_8, f_2 \cdot f_8$
   b. $f_7, + f_{14}, f_7 \cdot f_{14}$
   c. f_7', f_{14}'

6. Verify that the following two circuits are equivalent.

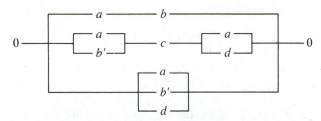

7. Write an algebraic expression for the following circuit.

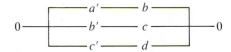

8. Simplify the expression $([a' \wedge b') \vee c] \wedge (a \vee c))'$.

9. Write an expression that represents the following circuit.

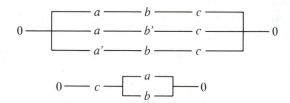

10. Draw the gate representation for the circuit in Exercise 7.

11. Draw the gate representation for the circuit in Exercise 9.

12. In a Boolean algebra containing a and b, prove that $a = b$ if and only if $(a \wedge b') \vee (a' \wedge b) = 0$.

13. Let $S = \{1, 3, 7, 21\}$. For x, y in S, define $x \vee y = \text{lcm}(x, y)$ and $x \wedge y = \gcd(x, y)$. Define $x' = 21/x$. Show that S with these operations satisfies axiom 4 for a Boolean algebra. (In fact, all the axioms are satisfied.)

14. (Poretzky's Law) Let B be a Boolean algebra, and let a belong to B. Prove that $a = 0$ if and only if $b = (a \wedge b') \vee (a' \wedge b)$ for each b in B.

15. Show, by example, that in a Boolean algebra the cancellation law does not hold; that is, $a \vee c = b \vee c$ does not always imply $a = b$.

16. For any a, b, c from a Boolean algebra, prove that
 a. $a \vee c = b \vee c$ and $a \vee c' = b \vee c'$ imply $a = b$.
 b. $a \vee c = b \vee c$ and $a \wedge c = b \wedge c$ imply $a = b$.

17. How would you define an isomorphism for a Boolean algebra?

18. Suppose that ϕ is an isomorphism from a Boolean algebra A onto a Boolean algebra B. Prove that ϕ carries the unity of A to the unity of B. Also prove that $\phi(a') = (\phi(a))'$ for every a in A follows from the other conditions.

19. Draw a circuit diagram that will be closed if and only if exactly one of a, b, and c is closed.

20. Draw a circuit diagram that will be closed if and only if at most one of a, b, and c is closed.

21. Draw a circuit and a gate diagram that represent the expression $(a + b)c + ab'$.

22. Draw a Venn diagram for the symmetric difference of sets A and B. (See the proof of Theorem 33.2 for the definition.)

23. Verify Table 33.4.

REFERENCE

1. Harold S. Stone, *Discrete Mathematical Structures and Their Applications,* Chicago: Science Research Associates, 1973.

SUGGESTED READINGS

F. Hohn, "Some Mathematical Aspects of Switching," *The American Mathematical Monthly* 62 (1955): 75–90.
 In this article, the author discusses the Boolean algebra of switching circuits in greater detail than we have done.

F. Hohn, *Applied Boolean Algebra: An Elementary Introduction,* New York: MacMillan, 1960.
 This brief monograph covers circuits, propositional logic, and the algebra of sets.

Claude E. Shannon

Probably no single work in this century has more profoundly altered man's understanding of communication than C. E. Shannon's "A mathematical theory of communication" first published in 1948. There resulted theorems of great power, elegance, generality and beauty. They have shed much understanding on the elusive true nature of the communication process and have delineated its interest limitations.

David Slepian, *Key Papers in the Development of Information Theory*

CLAUDE E. SHANNON was born on April 30, 1916, in Petoskey, Michigan, and grew up in nearby Gaylord. He received a B.S. degree in electrical engineering from the University of Michigan in 1936. In 1940, he received an M.S. degree in electrical engineering and a Ph.D. degree in mathematics, simultaneously, from the Massachusetts Institute of Technology. In his master's thesis, he showed that Boolean algebra provides an excellent tool for analyzing electrical switching circuits. After spending 1941 at the Institute for Advanced Study at Princeton, Shannon went to Bell Laboratories as a research mathematician. There he made contributions to coding theory, cryptography, computing circuit design, and information theory—a field he founded. In 1957, Shannon became a faculty member at MIT and served as a consultant to Bell Laboratories until 1972.

Shannon's theory defined information. It explained crucial relationships among the elements of a communications system—signal power, bandwidth (the frequency range of an information channel), and noise (static). The computer term *bit* was coined by Shannon.

Shannon has received prestigious awards from the American Institute of Electrical Engineers, the Institute of Electrical and Electronics Engineers, the Research Corporation, the Institute of Radio Engineers, the Franklin Institute, and Rice University. He has received the National Medal of Science as well as honorary degrees from Yale, Princeton, Northwestern, the University of Michigan, the University of Pittsburgh, and Edinburgh University in Scotland. He is a member of the National Academy of the Sciences and the National Academy of Engineering.

SUPPLEMENTARY EXERCISES FOR CHAPTERS 24–33

Nothing worthwhile comes easily. . . . Work, continuous work and hard work, is the only way to accomplish results that last.

Hamilton Holt

1. Let $G = \langle x, y \mid x = (xy)^3, y = (xy)^4 \rangle$. To what familiar group is G isomorphic?

2. Let $G = \langle z \mid z^6 = 1 \rangle$ and $H = \langle x, y \mid x^2 = y^3 = 1, xy = yx \rangle$. Show that G and H are isomorphic.

3. Show that a group of order $315 = 3^3 \cdot 5 \cdot 7$ has a subgroup of order 45.

4. Let G be a group of order p^2q, where p and q are primes, $q < p$, and q does not divide $p^2 - 1$. Show that G is Abelian.

5. Let H denote a Sylow 7-subgroup of a group G and K a Sylow 5-subgroup of G. Assume that $|H| = 49$, $|K| = 5$, and K is a subgroup of $N(H)$. Show that H is a subgroup of $N(K)$.

6. Determine all groups of order 30.

7. Suppose that K is a normal Sylow p-subgroup of H and that H is a normal subgroup of G. Prove that K is a normal subgroup of G. (Compare this with exercise 45 in Chapter 9.)

8. Prove that A_5, the group of even permutations on five objects, cannot have a subgroup of order 15.

9. Let H and K be subgroups of G. Prove that HK is a subgroup of G if $H \leq N(K)$.

10. Suppose that H is a subgroup of a finite group G and that H contains $N(P)$, where P is some Sylow p-subgroup of G. Prove that $N(H) = H$.

11. Prove that a simple group G of order 168 cannot contain an element of order 21.

12. Prove that the only group of order 561 is Z_{561}.

13. Prove that the center of a non-Abelian group of order 105 has order 5.

14. Let n be an odd integer at least 3. Prove that every Sylow subgroup of D_n is cyclic.

15. Let G be the digraph obtained from $\text{Cay}(\{(1, 0), (0, 1)\}: Z_3 \oplus Z_5)$ by deleting the vertex $(0, 0)$. [Also, delete each arc to or from $(0, 0)$.] Prove that G has a Hamiltonian circuit.

16. Prove that the digraph obtained from $\text{Cay}(\{(1, 0), (0, 1)\}: Z_4 \oplus Z_7)$ by deleting the vertex $(0, 0)$ has a Hamiltonian circuit.

17. Let G be a finite group generated by a and b. Let $s_1, s_2, \ldots, s_n$ be the arcs of a Hamiltonian circuit in the digraph $\text{Cay}(\{a, b\}: G)$. We say that the vertex $s_1 s_2 \cdots s_i$ *travels by* a if $s_{i+1} = a$. Show that if a vertex x travels by a, then every vertex in $x\langle ab^{-1} \rangle$ travels by a.

18. Recall that the dot product $u \cdot v$ of two vectors $u = (u_1, u_2, \ldots, u_n)$ and $v = (v_1, v_2, \ldots, v_n)$ from F^n is

$$u_1 v_1 + u_2 v_2 + \cdots + u_n v_n$$

(where the addition and multiplication is that of F). Let C be an (n, k) linear code. Show that

$$C^{\perp} = \{v \in F^n \mid v \cdot u = 0 \text{ for all } u \in C\}$$

is an $(n, n - k)$ linear code. This code is called the *dual* of C.

19. Find the dual of each of the following binary codes:
    a. $\{00, 11\}$,
    b. $\{000, 011, 101, 110\}$,
    c. $\{0000, 1111\}$,
    d. $\{0000, 1100, 0011, 1111\}$.

20. Let C be a binary linear code such that $C \subseteq C^{\perp}$. Show that wt(v) is even for all v in C.

21. Let C be an (n, k) binary linear code. If v is a binary n-tuple, but $v \notin C^{\perp}$, show that $v \cdot u = 0$ for exactly half of the elements u in C.

22. Suppose that C is an (n, k) binary linear code and the vector $11 \cdots 1 \in C^{\perp}$. Show that wt($v$) is even for every v in C.

23. Suppose that C is an (n, k) binary linear code and $C = C^{\perp}$. (Such a code is called *self-dual*.) Prove that n is even. Prove that $11 \cdots 1$ is a code word.

24. If G is a finite solvable group, show that there exist subgroups

$$\{e\} = H_0 \vartriangleleft H_1 \vartriangleleft H_2 \vartriangleleft \cdots \vartriangleleft H_n = G$$

such that H_{i+1}/H_i has prime order.

25. If a group G has a normal subgroup N such that both N and G/N are solvable, show that G is solvable.

26. Show that the polynomial $x^5 - 6x + 3$ over Q is not solvable by radicals.

27. Let a, b, c belong to a Boolean algebra. If $a \vee b \vee c = a \wedge b \wedge c$, prove that $a = b = c$.

The End.

Title of song by John Lennon and Paul McCartney,
***Abbey Road,* side 2, October 1969**

Text Credits

Page vii. Line from "Paperback Writer," words and music by John Lennon and Paul McCartney. © Copyright 1966 by Northern Songs Limited. All rights controlled and administered by MCA Music Publishing, a division of MCA Inc., New York, NY 10019. Under license from ATV Music. Used by permission. All rights reserved.

Page 12. "Brain Boggler" by Maxwell Carver, © 1988 by *Discover Magazine.*

Page 66. Poem from *The Compleat Computer* by Denise L. Van Tassel and Cynthia Van Tassel. Copyright © 1983, 1976, Science Research Associates, Inc. Reprinted by permission of the publisher.

Page 96. Poem "T.T.T" from Grooks, 1966 by Piet Hein. Used with permission of Piet Hein.

Page 185. Line from "The Ballad of John and Yoko," words and music by John Lennon and Paul McCartney. © Copyright 1969 by Northern Songs Limited. All rights controlled and administered by MCA Music Publishing, a division of MCA Inc., New York, NY 10019. Under license from ATV Music. Used by permission. All rights reserved.

Page 254. Poem "Problems" from Grooks, 1966 by Piet Hein. Used with permission of Piet Hein.

Page 279. Newspaper articles in collage copyright © 1993 by The New York Times Company. Reprinted by permission.

Page 370. "Completion of Song on the Simple Group War" by Scott C. Radtke from *American Mathematical Monthly* 99 (December 1992). Reprinted by permission of the author.

Page 416. Adaptation of figure from Dorothy K. Washburn and Donald W. Crowe, *Symmetries of Culture: Theory and Practice of Plane Analysis,* 1988, University of Washington Press.

Pages 487–489. Article adapted with permission from Barry A. Cipra, "The Ubiquitous Reed-Solomon Codes," *SIAM News,* 26 (January 1993), pp. 1, 11. Copyright 1993 by the Society for Industrial and Applied Mathematics. All rights reserved.

Page 520. Poem "The Road to Wisdom" from Grooks, 1966 by Piet Hein. Used with permission of Piet Hein.

Page A7. Line from "All You Need Is Love," words and music by John Lennon and Paul McCartney. © Copyright 1967 by Northern Songs Limited. All rights controlled and administered by MCA Music Publishing, a division of MCA Inc., New York, NY 10019. Under license from ATV Music. Used by permission. All rights reserved.

Page A9. Line from "Let It Be," words and music by John Lennon and Paul McCartney. © Copyright 1970 by Northern Songs Limited. All rights controlled and administered by MCA Music Publishing, a division of MCA Inc., New York, NY 10019. Under license from ATV Music. Used by permission. All rights reserved.

Photo Credits

Page 80 Stock Montage, Inc.

Page 101 Stock Montage, Inc.

Page 115 The Granger Collection

Page 149 Culver Pictures

Page 168 The Granger Collection

Page 182 The Granger Collection

Page 210 Courtesy of St. John's University Public Information Office

Page 232 The Granger Collection

Page 272 The Granger Collection

Page 289 Stock Montage, Inc.

Page 301 Michael Artin

Page 379 American Mathematical Society

Page 381 American Mathematical Society

Page 400 Courtesy of Jonathan Hall

Page 404 Plenum Publishing Corp.

Page 407 (top) Worth Publications

Page 412 Worth Publications

Page 418 © 1988 M. C. Escher heirs/Cordon Art, Baarn-Holland. Collection Haags, Gemeentemuseum, The Hague

Page 419 Creative Publications

Page 420 © 1988 M. C. Escher heirs/Cordon Art, Baarn-Holland. Collection Haags, Gemeentemuseum, The Hague

Pages 421 and 422 The American Mathematical Monthly

Page 423 Plenum Publishing Corp.

Page 434 The Granger Collection

Page 444 The Masters and Fellows of Pembroke College, Cambridge, England

Page 458 © 1988 M. C. Escher heirs/Cordon Art, Baarn-Holland. Collection Haags, Gemeentemuseum, The Hague

Page 459 Douglas Dunham

Selected Answers

When I make a mistake it's a beaut!   **Fiorello LaGuardia**

Many of the proofs given below are merely sketches. In these cases, the student should supply the complete proof.

CHAPTER 0

To make headway, improve your head.   **B. C. Forbes**

1. $\{1, 3, 5, 7\}; \{1, 5, 7, 11\}; \{1, 3, 7, 9, 11, 13, 17, 19\}; \{1, 2, 3, 4, 6, 7, 8, 9, 11, 12, 13, 14, 16, 17, 18, 19, 21, 22, 23, 24\}$

3. $1; 0; 4; 5$

5. 1942, June 18; 1953, December 13.

7. By using 0 as an exponent if necessary, we may write $a = p_1^{m_1} \cdots p_k^{m_k}$ and $b = p_1^{n_1} \cdots p_k^{n_k}$ where the p's are distinct primes and the m's and n's are nonnegative. Then $\mathrm{lcm}(a, b) = p_1^{s_1} \cdots p_k^{s_k}$ where $s_i = \max(m_i, n_i)$ and $\gcd(a, b) = p_1^{t_1} \cdots p_k^{t_k}$ where $t_i = \min(m_i, n_i)$. Then $\mathrm{lcm}(a, b) \cdot \gcd(a, b) = p_1^{m_1+n_1} \cdots p_k^{m_k+n_k} = ab$.

9. Any common divisor of a and b would also divide $at + bs = 1$.

11. $7(5n + 3) - 5(7n + 4) = 1$.

13. Use the "GCD is a linear combination" theorem.

15. Since st divides $a - b$, both s and t divide $a - b$. The converse is true when $\gcd(s, t) = 1$.

17. $\gcd(34, 126) = 2; 2 = 26 \cdot 34 - 7 \cdot 126$

19. Use proof by contradiction.

21. Let S be a set with $n + 1$ elements and pick some a in S. By induction, S has 2^n subsets that do not contain a. But there is a one-to-one correspondence between the subsets of S that do not contain a and those that do. So there are $2 \cdot 2^n = 2^{n+1}$ subsets in all.

23. Say $p_1 p_2 \ldots p_r = q_1 q_2 \ldots q_s$ where the p's and q's are positive primes. By the Generalized Euclid's Lemma, p_1 divides some q_i, say q_1 (we may relabel the q's if necessary).

Then $p_1 = q_1$ and $p_2 \ldots p_r = q_2 \ldots q_s$. Repeating this argument at each step we obtain $p_2 = q_2, \ldots, p_r = q_r$ and $r = s$.

25. By the Second Principle of Mathematical Induction, $f_n = f_{n-1} + f_{n-2} < 2^{n-1} + 2^{n-2} = 2^{n-2}(2+1) < 2^n$.

27. The statement is true for any divisor of $8^4 - 4 = 4092$.

29. Observe that the number with the decimal representation $a_9 a_8 \ldots a_1 a_0$ is $a_9 \cdot 10^9 + a_8 \cdot 10^8 + \cdots + a_1 \cdot 10 + a_0$. Then use exercise 12 and the fact that $a_i 10^i \bmod 9 = a_i$ to deduce that the check digit is $(a_9 + a_8 + \cdots + a_1 + a_0) \bmod 9$.

31. For the case that the check digit is not involved see the answer to exercise 29. If a transposition involving the check digit $c = a_1 + a_2 + \cdots + a_{10}$ goes undetected then $a_{10} = a_1 + a_2 + \cdots + a_9 + c$. Substitution yields $2(a_1 + \cdots + a_9) = 0 \bmod 9$. Therefore, $10(a_1 + \cdots + a_9) = a_1 + \cdots + a_9 = 0 \bmod 9$. It follows that $c = a_{10}$. In this case the transposition does not yield an error.

33. Say the number is $a_8 a_7 \ldots a_1 a_0 = a_8 \cdot 10^8 + a_7 \cdot 10^7 + \cdots + a_1 \cdot 10 + a_0$. Then the error is undetected if and only if $a_i 10^i - a_i' 10^i = 0 \bmod 7$. Multiplying both sides by 5^i and noting that $50 = 1 \bmod 7$ we obtain $a_i - a_i' = 0 \bmod 7$.

35. One need only verify the equation for $n = 0, 1, 2, 3, 4, 5$. Alternatively, observe that $n^3 - n = n(n-1)(n+1)$.

37. Observe that $1^2 = 3^2 = 5^2 = 7^2 = 1 \bmod 8$. Alternatively, note that $(2k+1)^2 - 1 = 4k^2 + 4k = 4k(k+1)$.

39. $a^2 + b^2 = a^2 + b^2$ gives $(a, b)R(a, b)$; since $a^2 + b^2 = c^2 + d^2$ implies $c^2 + d^2 = a^2 + b^2$, R is symmetric; if $a^2 + b^2 = c^2 + d^2$ and $c^2 + d^2 = e^2 + f^2$ then $a^2 + b^2 = c^2 + f^2$ and R is transitive. The equivalence classes are circles in the plane centered at $(0, 0)$.

41. No. $(1, 0) \in R$ and $(0, -1) \in R$, but $(1, -1) \notin R$.

43. Consider the set of integers with aRb if $|a - b| \leq 1$.

45. Let $S = \{1, 2, 3\}$ and $R = \{(1, 1), (2, 2), (1, 2), (2, 1)\}$.

CHAPTER 1

> It requires a very unusual mind to make an analysis of the obvious. **Alfred North Whitehead**

1. Three rotations: 0°, 120°, 240°, and three reflections across lines from vertices to midpoints of opposite sides.

3. no

5. D_n has n rotations of the form $k(360°/n)$ where $k = 0, \ldots, n-1$. D_n has n reflections. When n is odd the axes of reflection are the lines from the vertices to the midpoints of the opposite sides. When n is even, half of the axes of reflection are obtained by joining opposite vertices; the other half, by joining midpoints of opposite sides.

7. A rotation followed by a rotation either fixes every point (and so is the identity) or fixes only the center of rotation. However, a reflection fixes a line.

9. Observe that $1 \cdot 1 = 1; 1(-1) = -1; (-1)1 = -1; (-1)(-1) = 1$. These relationships also hold when 1 is replaced by "rotation" and -1 is replaced by "reflection."

11. $HD = DV$

13. R_0, R_{180}, H, V

15. See answer for exercise 13.

17. In each case the group is D_6.

19. cyclic

21. D_{11}

23. D_{28}

CHAPTER 2

It's easy! **John Lennon and Paul McCartney,** *All You Need Is Love,* **single**

1. Does not contain the identity; closure fails.

3. Suppose $\begin{bmatrix} a & b \\ c & d \end{bmatrix}$ is the inverse. Then, $\begin{bmatrix} 2 & 2 \\ 1 & 1 \end{bmatrix}\begin{bmatrix} a & b \\ c & d \end{bmatrix} = \begin{bmatrix} 1 & 0 \\ 0 & 1 \end{bmatrix}$ so that $2a + 2c = 1$ and $a + c = 0$.

5. Apply the definition.

7. (i) $2a + 3b$; (ii) $-2a + 2(-b + c)$; (iii) $-3(a + 2b) + 2c = 0$

9.

	0	1	2	3
0	0	1	2	3
1	1	2	3	0
2	2	3	0	1
3	3	0	1	2

11. Under modulo 4, 2 does not have an inverse. Under modulo 5, each element has an inverse.

13. $\begin{bmatrix} 9 & 9 \\ 10 & 8 \end{bmatrix}$

15. Use exercise 5 to verify closure. It also follows from exercise 5 that $(\det A^{-1}) = (\det A)^{-1}$. Thus $\det A^{-1} = \pm 1$ whenever $\det A = \pm 1$.

17. 29

19. $(ab)^n$ need not equal $a^n b^n$ in a non-Abelian group.

21. Use exercise 20.

23. The identity is 25.

25. $\{1, 3, 5, 9, 13, 15, 19, 23, 25, 27, 39, 45\}$

27.

	1	5	7	11
1	1	5	7	11
5	5	1	11	7
7	7	11	1	5
11	11	7	5	1

29. Use exercise 28.

31. aca^{-1}

33. If $x^3 = e$ and $x \neq e$, then $(x^{-1})^3 = e$ and $x \neq x^{-1}$. So, nonidentity solutions come in pairs. If $x^2 \neq e$, then $x^{-1} \neq x$ and $(x^{-1})^2 \neq e$. So solutions to $x^2 \neq e$ come in pairs.

35. $(\phi_g \phi_h)(x) = \phi_g(hxh^{-1}) = ghxh^{-1}g^{-1} = ghx(gh)^{-1} = \phi_{gh}(x)$

37. Use associativity. For example, $4 * 5 = 4 * (4 * 1) = (4 * 4) * 1 = 0 * 1 = 1$.

39. a. r^3
     b. r
     c. $r^5 f$

41. If n is not prime, the set is not closed under multiplication modulo n. If n is prime the set is closed and for every r in the set there are integers s and t such that $1 = rs + nt = rs \bmod n$.

43. 4

47. a. 3; b. dot product is not 0; the correct number cannot be determined; yes. c. no; the dot product is 0. d. Use Theorem 2.4.

49. 2; Say a_i' is substituted for a_i ($a_i' \neq a_i$). If i is even, argue as in the proof of Theorem 2.4. Otherwise consider cases: $0 \leq a_i, a_i' < 5$; $5 \leq a_i, a_i' < 9$; $0 \leq a_i < 5$ and $5 \leq a_i' < 9$; $0 \leq a_i' < 5$ and $5 \leq a_i < 9$. In every case a_i' and a_i contribute different amounts to the dot product. Theorem 2.4 does not apply because of the r term. All transposition errors except $09 \leftrightarrow 90$ are detected.

51. The check digit would be the same.

CHAPTER 3

The brain is as strong as its weakest think. **Eleanor Doan**

1. $|Z_{12}| = 12$; $|U(10)| = 4$; $|U(12)| = 4$; $|U(20)| = 8$; $|D_4| = 8$
     In Z_{12}, $|0| = 1$; $|1| = |5| = |7| = |11| = 12$; $|2| = |10| = 6$; $|3| = |9| = 4$; $|4| = |8| = 3$; $|6| = 2$.
     In $U(10)$, $|1| = 1$; $|3| = |7| = 4$; $|9| = 2$.
     In $U(20)$, $|1| = 1$; $|3| = |7| = |13| = |17| = 4$; $|9| = |11| = |19| = 2$.
     In D_4, $|R_0| = 1$; $|R_{90}| = |R_{270}| = 4$; $|R_{180}| = |H| = |V| = |D| = |D'| = 2$.
     In each case notice that the order of the element divides the order of the group.

3.   In Q, $|0| = 1$ and all other elements have infinite order. In Q^*, $|1| = 1$, $|-1| = 2$, and all other elements have infinite order.

5.   Each is the inverse of the other.

7. $U(14) = \{1, 3, 5, 9, 11, 13\}$
     $\langle 3 \rangle = \{3, 3^2, 3^3, 3^4, 3^5, 3^6\} = \{3, 9, 13, 11, 5, 1\} = U(14)$;
     $U(14) \neq \langle 11 \rangle$

9.   By brute force show that $k^4 = 1$ for all k.

11. For any integer $n \geq 3$, D_n contains elements a and b of order 2 with $|ab| = n$. In general, there is no relationship among $|a|$, $|b|$, and $|ab|$.

13. Write $k = sm$. If $x = 1 \bmod k$, then $x - 1 = tk = tsm$ so $x = 1 \bmod m$.

15. If $x \in Z(G)$, then $x \in C(a)$ for all a, so $x \in \bigcap_{a \in G} C(a)$. If $x \in \bigcap_{a \in G} C(a)$, then $xa = ax$ for all a in G so $x \in Z(G)$.

17. a. $C(5) = G$; $C(7) = \{1, 3, 5, 7\}$
     b. $Z(G) = \{1, 5\}$
     c. $|2| = 2$; $|3| = 4$. They divide the order of the group.

19. Mimic the proof of Theorem 3.5.

21. No. In D_4, $C(R_{180}) = D_4$.

23. For the first part see Example 4. For the second part, use D_4.

27. 2

29. Note that $\begin{bmatrix} 1 & 1 \\ 0 & 1 \end{bmatrix}^n = \begin{bmatrix} 1 & n \\ 0 & 1 \end{bmatrix}$.

31. For any positive integer n, a rotation of $360°/n$ has order n. A rotation of $\sqrt{2}°$ has infinite order.

33. $\langle R_0 \rangle, \langle R_{90} \rangle, \langle R_{180} \rangle, \langle D \rangle, \langle D' \rangle, \langle H \rangle, \langle V \rangle$. $\langle R_{270} \rangle$ is not on the list since $\langle R_{90} \rangle = \langle R_{270} \rangle$. $\{R_0, R_{180}, D, D'\}$ and $\{R_0, R_{180}, H, V\}$ are not cyclic.

35. Certainly, $(a^{n/k})^k = a^n = e$. If $(a^{n/k})^t = e$ for some positive $t < k$, then $a^{nt/k} = e$ and $nt/k < n$, a contradiction.

37. No. $7 \in H$ but $7 \cdot 7 \notin H$.

41. $|\langle 3 \rangle| = 4$

43. Let $\begin{bmatrix} a & b \\ c & d \end{bmatrix}$ and $\begin{bmatrix} a' & b' \\ c' & d' \end{bmatrix}$ belong to H. It suffices to show that $a - a' + b - b' + c - c' + d - d' = 0$. This follows from $a + b + c + d = 0 = a' + b' + c' + d'$. If 0 is replaced by 1, H is not a subgroup.

45. If 2^a and $2^b \in K$, then $2^a(2^b)^{-1} = 2^{a-b} \in K$ since $a - b \in H$.

47. $\begin{bmatrix} 2 & 0 \\ 0 & 2 \end{bmatrix}^{-1} = \begin{bmatrix} \frac{1}{2} & 0 \\ 0 & \frac{1}{2} \end{bmatrix}$ is not in H.

49. If $a + bi$ and $c + di \in H$, then $(a + bi)(c + di)^{-1} = (ac + bd) + (bc - ad)i$ and $(ac + bd)^2 + (bc - ad)^2 = 1$ so that H is a subgroup. H is the unit circle in the complex plane.

51. a. $\left\{ \begin{bmatrix} a & a \\ 0 & 0 \end{bmatrix} \middle| a \neq 0, a \in R \right\}$

    b. $\left\{ \begin{bmatrix} a & b \\ b & a \end{bmatrix} \middle| a^2 \neq b^2, a, b \in R \right\}$

    c. $\left\{ \begin{bmatrix} a & 0 \\ 0 & a \end{bmatrix} \middle| a \neq 0, a \in R \right\}$

CHAPTER 4

> There will be an answer, let it be.   **John Lennon and Paul McCartney, *Let It Be*, single**

1. For Z_6, generators are 1 and 5; for Z_8, generators are 1, 3, 5, and 7; for Z_{20}, generators are 1, 3, 7, 9, 11, 13, 17, and 19.

3. $\langle 20 \rangle = \{20, 10, 0\}$
   $\langle 10 \rangle = \{10, 20, 0\}$

5. $\langle 3 \rangle = \{3, 9, 7, 1\}$
   $\langle 7 \rangle = \{7, 9, 3, 1\}$

7. $U(8)$ or D_3

9. Six subgroups; generators are the divisors of 20.
   Six subgroups; generators are a^k where k is a divisor of 20.

11. Certainly, $a^{-1} \in \langle a \rangle$. So $\langle a^{-1} \rangle \subseteq \langle a \rangle$. So, by symmetry, $\langle a \rangle \subseteq \langle a^{-1} \rangle$ as well.

13. Let $k = \text{lcm}(m, n) \bmod 24$. Then $\langle a^m \rangle \cap \langle a^n \rangle = \langle a^k \rangle$.

15. $|g|$ divides 12 is equivalent to $g^{12} = e$. So, if $a^{12} = e$ and $b^{12} = e$, then $(ab^{-1})^{12} = a^{12}(b^{12})^{-1} = ee^{-1} = e$. The general result is given in exercise 23 of Chapter 3.

17. $\langle 1 \rangle, \langle 7 \rangle, \langle 11 \rangle, \langle 17 \rangle, \langle 19 \rangle, \langle 29 \rangle$

19. a. $|a|$ divides 12. b. $|a|$ divides m. c. By Theorem 4.3, $|a| = 1, 2, 3, 4, 6, 8, 12,$ or 24. If $|a| = 2$, then $a^8 = (a^2)^4 = e^4 = e$. A similar argument eliminates all other possibilities except 24.

21. Yes, by Theorem 4.3. The subgroups of Z are of the form $\{0, \pm n, \pm 2n, \pm 3n, \ldots\}$ where n is any integer.

23. Since k is a multiple of $\gcd(n, k)$ we have $\langle a^k \rangle \subseteq \langle a^{\gcd(n,k)} \rangle$. By the "GCD is a linear combination" Theorem from Chapter 0 we may write $\gcd(n, k) = ns + kt$. Thus $a^{\gcd(n,k)} = (a^n)^s(a^k)^t = (a^k)^t \in \langle a^k \rangle$. This proves the first equality. The second part of the exercise follows from the first part.

25. Two: a and a^{-1}

27. 1000000, 3000000, 5000000, 7000000
   By Theorem 4.3, $\langle 1000000 \rangle$ is the unique subgroup of order 8 and only those on the list are generators.
29. Let $G = \{a_1, a_2, \ldots, a_k\}$. Now let $|a_i| = n_i$. Consider $n = n_1 n_2 \ldots n_k$.
31. Mimic exercise 30.
33. Mimic exercise 32.
35. Suppose a and b are relatively prime positive integers and $\langle a/b \rangle = Q^+$. Then there is some positive integer n such that $(a/b)^n = 2$. Clearly $n \neq 0$, 1, or -1. If $n > 1$, $a^n = 2b^n$ so that 2 divides a. But then 2 divides b as well. A similar contradiction occurs if $n < -1$.
37.

	7	35	49	77
7	49	77	7	35
35	77	49	35	7
49	7	35	49	77
77	35	7	77	49

   The identity is 49. The group is not cyclic.
39. Let $t = \text{lcm}(m, n)$ and $|ab| = s$. Then $(ab)^t = a^t b^t = e$ and therefore s divides t. Also, $e = (ab)^s = a^s b^s$ so that $a^s = b^{-s}$ and therefore a^s and b^{-s} belong to $\langle a \rangle \cap \langle b \rangle = e$. Thus, m divides s and n divides s, and, therefore, t divides s. This proves that $s = t$. For the second part try D_3.
41. An infinite cyclic group does not have an element of prime order. A finite cyclic group can have only one subgroup for each divisor of its order. A subgroup of order p has exactly $p - 1$ elements of order p. Another element of order p would give another subgroup of order p.
43. $4, 3 \cdot 4, 7 \cdot 4, 9 \cdot 4$
45. 1 of order 1; 33 of order 2; 2 of order 3; 10 of order 11; 20 of order 33
47. 1, 2, 10, 20
49. If $|a| = 2$ and $|b| = 2$ and a and b commute, then $\{e, a, b, ab\}$ is a subgroup. The subgroup is not cyclic.
51. Use exercise 14 of Chapter 3 and Theorem 4.3.
53. 1 and 2
55. In a cyclic group there are at most n solutions to the equation $x^n = e$.
57. First observe that 1 and $n - 1$ are generators. We must find another. If $n = p$ or $2p$, where p is prime, then 3 is a generator. Otherwise, apply Bertrand's Postulate to the largest prime divisor of n.
59. Use exercise 23.
61. Observe that $\langle a \rangle \cap H \subseteq \langle a \rangle$ so that $\langle a \rangle \cap H$ has the form $\langle a^k \rangle$ where k divides n. Since $a^k \in H$, $k = n$.

SUPPLEMENTARY EXERCISES FOR CHAPTERS 1–4

> I have learned throughout my life as a composer chiefly through my mistakes and pursuits of false assumptions, not by my exposure to founts of wisdom and knowledge.
> **Igor Stravinsky, "Contingencies,"** *Themes and Episodes*

1. a. Let xh_1x^{-1} and xh_2x^{-1} belong to xHx^{-1}. Then $(xh_1x^{-1})(xh_2x^{-1}) = xh_1h_2x^{-1} \in xHx^{-1}$ also.

b. Let $\langle h \rangle = H$. Then $\langle xhx^{-1} \rangle = xHx^{-1}$.

c. $(xh_1x^{-1})(xh_2x^{-1}) = xh_1h_2x^{-1} = xh_2h_1x^{-1} = (xh_2x^{-1})(xh_1x^{-1})$

3. Suppose $\text{cl}(a) \cap \text{cl}(b) \neq \emptyset$. Say $xax^{-1} = yby^{-1}$. Then $(y^{-1}x)a(y^{-1}x)^{-1} = b$. Thus, for any ubu^{-1} in $\text{cl}(b)$, we have $ubu^{-1} = (uy^{-1}x)a(uy^{-1}x)^{-1} \in \text{cl}(a)$. This shows that $\text{cl}(b) \subseteq \text{cl}(a)$. By symmetry, $\text{cl}(a) \subseteq \text{cl}(b)$. Because $a = eae^{-1} \in \text{cl}(a)$, the union of the conjugacy classes is G.

5. If both ab and ba have infinite order, we are done. Suppose $|ab| = k$. Then

$$\underbrace{(ab)(ab) \ldots (ab)}_{k \text{ factors}} = e.$$

Thus,

$$b[\underbrace{(ab)(ab) \ldots (ab)}_{k \text{ factors}}]a = bea = ba.$$

So $(ba)^{k+1} = ba$, and $(ba)^k = e$. This proves $|ba| \leq |ab|$. By symmetry, $|ab| \leq |ba|$.

7. By exercise 6, for every x in G, $|xax^{-1}| = |a|$ so that $xax^{-1} = a$ or $xa = ax$.

9. 1 of order 1, 15 of order 2, 8 of order 15, 4 of order 5, 2 of order 3.

11. Let $|G| = 5$. Let $a \neq e$ belong to G. If $|a| = 5$ we are done. If $|a| = 3$, then $\{e, a, a^2\}$ is a subgroup of G. Let b be either of the remaining two elements of G. Then the set $\{e, a, a^2, b, ab, a^2b\}$ consists of six different elements, a contradiction. Thus, $|a| \neq 3$. Similarly, $|a| \neq 4$. We may now assume that every nonidentity element of G has order 2. Pick $a \neq e$ and $b \neq e$ in G with $a \neq b$. Then $\{e, a, b, ab\}$ is a subgroup of G. Let c be the remaining element of G. Then $\{e, a, b, ab, c, ac, bc, abc\}$ is a set of eight distinct elements of G, a contradiction. It now follows that if $a \in G$, and $a \neq e$, then $|a| = 5$.

13. $a^n(b^n)^{-1} = (ab^{-1})^n$, so G^n is a subgroup. For the non-Abelian group try D_3.

15. Suppose $G = H \cup K$. Pick $h \in H$ with $h \notin K$. Pick $k \in K$, but $k \notin H$. Then, $hk \in G$, but $hk \notin H$ and $hk \notin K$. $U(8)$ is the union of the three subgroups.

17. If $|a| = p^k$ and $|b| = p^r$ with $k \leq r$, say, then $|ab^{-1}|$ divides p^r.

19. Note that $ba^2 = ab$ and $a^3 = b^2 = e$ imply $ba = a^2b$. Thus, every member of the group can be written in the form a^ib^j. Therefore, the group is $\{e, a, a^2, b, ab, a^2b\}$. D_3 satisfies these conditions.

21. $xy = yx$ if and only if $xyx^{-1}y^{-1} = e$. But, $(xy)x^{-1}y^{-1} = x^{-1}(xy)y^{-1} = ee = e$.

23. Let $x \in N(gHg^{-1})$. Then $x(gHg^{-1})x^{-1} = gHg^{-1}$. Thus $g^{-1}xgHg^{-1}x^{-1}g = g^{-1}xgH(g^{-1}xg)^{-1} = H$. This means that $g^{-1}xg \in N(H)$. So $x \in gN(H)g^{-1}$. Reverse the argument to show $gN(H)g^{-1} \subseteq N(gHg^{-1})$.

25. Look at D_{11}.

27. Solution from the *Mathematics Magazine:** "Yes. Let a be an arbitrary element of S. The set $\{a^n \mid n = 1, 2, 3, \ldots \}$ is finite, and therefore $a^m = a^n$ for some m, n with $m > n \geq 1$. By cancellation we have $a^{r(a)} = a$, where $r(a) = m - n + 1 > 1$. If x is any element of S, then $aa^{r(a)-1}x = a^{r(a)}x = ax$, and this implies that $a^{r(a)-1}x = x$. Similarly, we see that $xa^{r(a)-1} = x$, and the element $e = a^{r(a)-1}$ is an identity. The identity element is unique, for if e' is another identity, then $e = ee' = e'$. If $r(a) > 2$ then $a^{r(a)-2}$ is an inverse of a, and if $r(a) = 2$ then $a^2 = a = e$ and a is its own inverse. Thus S is a group."

29. $1 \in H$. Let $a, b \in H$. Then $(ab^{-1})^2 = a^2(b^2)^{-1}$, which is the product of two rationals.

31. Use $\det(AB) = \det A \det B$ to prove H is a subgroup. H is not a subgroup when $\det A$ is an integer since $\det A^{-1}$ need not be an integer.

33. Suppose G is not cyclic. Choose $x \neq e$ and $y \notin \langle x \rangle$. Then $G = \langle x \rangle \cup \langle y \rangle$. But then

---

*Mathematics Magazine 63 (April 1990): 136.

$xy \in \langle y \rangle$ so that $\langle x \rangle \subseteq \langle y \rangle$ and therefore $G = \langle y \rangle$, a contradiction. To prove that $|G| = pq$ or p^3, use Theorem 4.3.

35. 8

CHAPTER 5

You cannot have the success without the failures.
H. G. Hasler, *The Observer*

1. a. 2   b. 3   c. 5
3. a. 3   b. 3   c. 6   d. 12
5. 12
7. $|(123)(45678)| = 15$
9. a. even   b. odd   c. even   d. odd   e. even
11. even; odd
13. An even number of two cycles followed by an even number of two cycles gives an even number of two cycles in all. So the finite subgroup test is verified.
15. even
17.

$$\alpha^{-1} = \begin{bmatrix} 1 & 2 & 3 & 4 & 5 & 6 \\ 2 & 1 & 3 & 5 & 4 & 6 \end{bmatrix}$$

$$\beta\alpha = \begin{bmatrix} 1 & 2 & 3 & 4 & 5 & 6 \\ 1 & 6 & 2 & 3 & 4 & 5 \end{bmatrix}$$

$$\alpha\beta = \begin{bmatrix} 1 & 2 & 3 & 4 & 5 & 6 \\ 6 & 2 & 1 & 5 & 3 & 4 \end{bmatrix}$$

19. Suppose H contains at least one odd permutation, say, σ. Let A be the set of even permutations in H, and B be the set of odd permutations in H. Then $\sigma A \subseteq B$ and $|\sigma A| = |A|$, so $|A| \le |B|$. Also, $\sigma B \subseteq A$ and $|\sigma B| = |B|$, so $|B| \le |A|$.
21. No. The identity is even.
23. $C(\alpha_3) = \{\alpha_1, \alpha_2, \alpha_3, \alpha_4\}$, $C(\alpha_{12}) = \{\alpha_1, \alpha_7, \alpha_{12}\}$
25. $(123)(321) = (1), (1478)(8741) = (1)$
27. Let $\alpha, \beta \in \text{stab}(a)$. Then $\alpha\beta(a) = \alpha(\beta(a)) = \alpha(a) = a$. Also, $\alpha(a) = a$ implies $\alpha^{-1}(\alpha(a)) = \alpha^{-1}(a)$ or $a = \alpha^{-1}(a)$.
29. m is a multiple of 6 but not a multiple of 30.
31. $6!/5 = 144$
33. 3, 7, 9
35. Let $\alpha = (123)$ and $\beta = (145)$.
39. Say $\alpha = a_1 a_2 \ldots a_n$ and $\beta = b_1 \ldots b_m$ where the a's and b's are *cycles*. Then $\alpha\beta^{-1} = a_1 a_2 \ldots a_n b_m^{-1} \ldots b_1^{-1}$ is a finite number of cycles.
41. Hint: $(13)(12) = (123)$ and $(12)(34) = (324)(132)$.
43. 4
45. 2; Adapt proof of Theorem 2.4; $09 \leftrightarrow 90$ is undetected; the methods are essentially the same.
47. Use Theorem 5.3 and exercise 11.
49. The product of an element from $Z(A_4)$ of order 2 and an element of order 3 would have order 6. The product of an element from $Z(A_4)$ of order 3 and an element of order 2 would have order 6.

CHAPTER 6

Think and you won't sink. **B. C. Forbes, *Epigrams***

1. Try $n \rightarrow 2n$.
3. $\phi(xy) = \sqrt{xy} = \sqrt{x}\,\sqrt{y} = \phi(x)\phi(y)$
5. Try $1 \rightarrow 1, 3 \rightarrow 5, 5 \rightarrow 7, 7 \rightarrow 11$.
7. D_{12} has elements of order 12 and S_4 does not.
9. The mapping $h \rightarrow xhx^{-1}$ is an isomorphism.
11. Let $G = Z_6$ and $H = Z_3$.
13. $axa^{-1} = aya^{-1}$ implies $x = y$, so ϕ_a is one-to-one. If $b \in G$, then $\phi_a(a^{-1}ba) = b$, so ϕ_a is onto; $\phi_a(xy) = axya^{-1} = axa^{-1}aya^{-1} = \phi_a(x)\phi_a(y)$ so ϕ_a is operation-preserving.
15. Let $\alpha \in \text{Aut}(G)$. We show α^{-1} is operation-preserving: $\alpha^{-1}(xy) = \alpha^{-1}(x)\alpha^{-1}(y)$ if and only if $\alpha(\alpha^{-1}(xy)) = \alpha(\alpha^{-1}(x)\alpha^{-1}(y))$, that is, if and only if $xy = \alpha(\alpha^{-1}(x))\alpha(\alpha^{-1}(y)) = xy$. So α^{-1} is operation-preserving. That $\text{Inn}(G)$ is a group follows from the equation $\phi_g\phi_h = \phi_{gh}$.
17. Use the fact that $x^4 = 1$ for all x in $U(16)$. In general, $x \rightarrow x^n$ is an automorphism of $U(16)$ when n is odd.
19. If $\alpha \in S_8$, write

$$\alpha = \begin{bmatrix} 1 & 2 & \cdots & 7 & 8 \\ \alpha(1) & \alpha(2) & \cdots & \alpha(7) & 8 \end{bmatrix}.$$

Then map α to $\begin{bmatrix} 1 & 2 & \cdots & 7 \\ \alpha(1) & \alpha(2) & \cdots & \alpha(7) \end{bmatrix}.$

21. Compare orders of elements.
23. Apply the appropriate definitions.
25. Show that Q is not cyclic.
27. Try $a + bi \rightarrow \begin{bmatrix} a & -b \\ b & a \end{bmatrix}.$
29. Yes, by Cayley's Theorem.
31. $\log_{10}(xy) = \log_{10} x + \log_{10} y$. The marks on a slide rule represent logarithmic lengths. Manipulating the slide rule corresponds to adding (or subtracting) these lengths. The given equation shows this corresponds to multiplying (or dividing) the marks.
33. $\phi_g = \phi_h$ implies $gxg^{-1} = hxh^{-1}$ for all x. This implies $h^{-1}gx(h^{-1}g)^{-1} = x$ and therefore $h^{-1}g \in Z(G)$.
35. The elements of D_n are permutations on the vertices of a regular n-gon.

CHAPTER 7

There is always a right and a wrong way, and the wrong way always seems the more reasonable. **George Moore, *The Bending of the Bough***

1. $|(0, 0)| = 1$
   $|(0, 2)| = |(1, 2)| = |(1, 0)| = 2$
   $|(0, 1)| = |(0, 3)| = |(1, 1)| = |(1, 3)| = 4$
3. Every nonidentity element in the group has order 2. Each of these generates a subgroup of order 2.
5. Suppose $\langle (a, b) \rangle = Z \oplus Z$. If $a = b$, then $(1, 2) \notin \langle (a, b) \rangle$. If $a \neq b$, then $(1, 1) \notin \langle (a,b) \rangle$.

7. Try $(g_1, g_2) \to (g_2, g_1)$.
9. Yes, by Theorem 7.2.
11. Each of Z_8, Z_4, $Z_{8000000}$, and $Z_{4000000}$ has a unique subgroup of order 4. If $|(a,b)| = 4$, then a and b both belong to the unique subgroup of order 4. So the number of choices for a and b (actually 12 in all) is the same in either group.
13. Try $(a, b) \to a + bi$.
17. 3.

19. Map $\begin{bmatrix} a & b \\ c & d \end{bmatrix}$ to (a, b, c, d). Let $\mathbf{R}^k$ denote $\mathbf{R} \oplus \mathbf{R} \oplus \cdots \oplus \mathbf{R}$ (k factors). Then the group of $n \times n$ matrices under addition is isomorphic to $\mathbf{R}^{n^2}$.

21. $(g, g)(h, h)^{-1} = (gh^{-1}, gh^{-1})$
    When $G = \mathbf{R}$, $G \oplus G$ is the plane and H is the line $y = x$.
23. $\langle (3, 0) \rangle$, $\langle (3, 1) \rangle$, $\langle (3, 2) \rangle$, $\langle (0, 1) \rangle$
25. 60
27. $\{0, 400\} \oplus \{0, 50, 100, 150\}$
29. Try $g \to (g, e)$.
31. Try $3^m 6^m \to (m, n)$.
33. D_{24} has elements of order 24 whereas $D_3 \oplus D_4$ does not.
35. Use the observation that $(h, k)^n = (h^n, k^n)$.
37. 12
39. 48; 6
41. $U(165) \approx U(11) \oplus U(15) \approx U(5) \oplus U(33) \approx U(3) \oplus U(55)$
43. Mimic the analysis for elements of order 12 in $U(720)$ at the end of this chapter.
45. 60
47. They are both isomorphic to $Z_{10} \oplus Z_4$.
49. That $U(n)^2$ is a subgroup follows from exercise 13 of supplementary exercises for Chapters 1–4. $1^2 = (n - 1)^2$ shows that it is a proper subgroup.
51. 275
53. $U(117) \approx U(9) \oplus U(13) \approx Z_6 \oplus Z_{12}$, which contains $\langle (2, 0) \rangle + \langle (0, 4) \rangle$.
55. Consider $U(49)$.
57. Consider $U(65)$.
59. NO

SUPPLEMENTARY EXERCISES FOR CHAPTERS 5–7

For those who keep trying failure is temporary. **Frank Tyger**

1. Consider the finite and infinite cases separately. In the finite case, note that $|H| = |\phi(H)|$. Now use Theorem 4.3. For the infinite case, use exercise 2 of Chapter 6.
3. Observe that $\phi(x^{-1}y^{-1}xy) = (\phi(x))^{-1}(\phi(y))^{-1}\phi(x)\phi(y)$ so ϕ carries commutators to commutators.
5. All nonidentity elements of G and H have order 3. $G \approx H$.
7. Certainly the set HK has $|H||K|$ symbols. However, not all symbols need represent distinct group elements. That is, we may have $hk = h'k'$ although $h \neq h'$ and $k \neq k'$. We must determine the extent to which this happens. For every t in $H \cap K$, $hk = (ht)(t^{-1}k)$ so each group element in HK is represented by at least $|H \cap K|$ products in HK. But $hk = h'k'$ implies $t = h^{-1}h' = k(k')^{-1} \in H \cap K$ so that $h' = ht$ and $k' = t^{-1}k$. Thus each element in HK is represented by exactly $|H \cap K|$ products. So, $|HK| = |H||K|/|H \cap K|$.
9. $U(n)$ where $n = 4, 8, 3, 6, 12, 24$.

11. Hint:

$$3 + 2i = \sqrt{13}\left(\frac{3}{\sqrt{13}} + \frac{2}{\sqrt{13}}i\right)$$

13. Suppose $\phi: Q \to R$ is an isomorphism. Let $\phi(1) = x_0$. Show that $\phi(a/b) = (a/b)x_0$ for all integers.

15. In Q, equation $2x = a$ has a solution for all a. The corresponding equation $x^2 = b$ in Q^+ does not have a solution for all b.

17. $Z_5 \oplus Z_5$

19. $\langle 3 \rangle \oplus \langle 4 \rangle$

21. See exercise 1 of Chapter 6.

23. m is a multiple of 6 but not a multiple of 30.

25. Count elements of order 2.

27. Count elements of order 2.

29. $x = \phi_a(x) = axa^{-1}$ so that $xa = ax$. Conversely, if G is Abelian, ϕ_a is the identity.

31. $U_{50}(450)$

33. $(4, 10)$

35. Count elements of order 2.

37. $20; (8, 7, (3251))$

CHAPTER 8

Failure is the path of least persistence. **Author unknown.**

1. $H = \{\alpha_1, \alpha_2, \alpha_3, \alpha_4\}$, $\alpha_5 H = \{\alpha_5, \alpha_8, \alpha_6, \alpha_7\}$, $\alpha_9 H = \{\alpha_9, \alpha_{11}, \alpha_{12}, \alpha_{10}\}$

3. $H, 1 + H, 2 + H$

5. a. yes   b. yes   c. no

7. $8/2 = 4$ so there are four cosets. Let $H = \{1, 11\}$. The cosets are $H, 7H, 13H, 19H$.

9. Since $|a^4| = 15$ there are two cosets: $\langle a^4 \rangle$ and $a\langle a^4 \rangle$.

11. The correspondence $ah \to bh$ is one-to-one and onto.

13. Say the points in H lie on the line $y = mx$. Then $(a, b) + H = \{(a + x, b + mx) \mid x \in R\}$. This set is the line $y - b = m(x - a)$.

15. The subgroup is the solution set of the system

$$3x + 2y - 3z = 0$$
$$5x + y + 4z = 0.$$

17. 1, 2, 3, 4, 5, 6, 10, 12, 15, 20, 30, 60

19. Use Lagrange's Theorem and one of its corollaries.

21. By exercise 20 we have $5^6 = 1 \bmod 7$. So $5^{15} = 5^6 \cdot 5^6 \cdot 5^2 \cdot 5 = 1 \cdot 1 \cdot 4 \cdot 5 = 6 \bmod 7$.

23. By Corollary 1 of Theorem 8.1 we know that any nonidentity element of a non-Abelian group G of order 10 has order 2 or 5. By exercise 38 of Chapter 2, there must be an element of order 5. Call it a. Then for any $b \notin \langle a \rangle$ we have $G = \langle a \rangle \cup b\langle a \rangle$. Cancellation shows that $b^2 \notin b\langle a \rangle$. Now show $b^2 = e$. A non-Abelian group of order $2p$ has p elements of order 2. An Abelian group of order $2p$ has one element of order 2.

25. Use the coset representatives $(0, 1), (1, 1), (2, 1), (3, 1)$.

27. Let H be the subgroup of order p and K be the subgroup of order q. Then $H \cup K$ has $p + q - 1 < pq$ elements. Let a be any element in G that is not in $H \cup K$. By Lagrange's Theorem, $|a| = p, q,$ or pq. But $|a| \neq p$ for if so, then $\langle a \rangle = H$. Similarly, $|a| \neq q$.

29.   1, 3, 11, 33. If $|x| = 33$, then $|x^{11}| = 3$. Elements of order 11 occur in multiples of 10.

31.   Use the facts that every real number is an nth power when n is odd and every positive real number is an nth power when n is even.

33.   Certainly, $a \in \text{orb}_G(a)$. Now suppose $c \in \text{orb}_G(a) \cap \text{orb}_G(b)$. Then $c = \alpha(a)$ and $c = \beta(b)$ for some α and β and therefore $(\beta^{-1}\alpha)(a) = b$. So, if $x \in \text{orb}_G(b)$, then $x = \gamma(b) = (\gamma\beta^{-1}\alpha)(a)$ for some γ. This proves $\text{orb}_G(b) \subseteq \text{orb}_G(a)$. By symmetry, $\text{orb}_G(a) \subseteq \text{orb}_G(b)$.

35.   a. $\text{stab}_G(1) = \{(1), (24)(56)\}$; $\text{orb}_G(1) = \{1,2,3,4\}$
      b. $\text{stab}_G(3) = \{(1), (24)(56)\}$; $\text{orb}_G(3) = \{3,4,1,2\}$
      c. $\text{stab}_G(5) = \{(1), (12)(34), (13)(24), (14)(23)\}$; $\text{orb}_G(5) = \{5,6\}$

37.   Let F_1, F_2, F_3, F_4, F_5 be the five reflections in D_5. The subgroup lattice of D_5 is

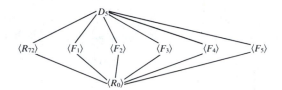

39.   2520

41.   Suppose $x^2 = y^2$ and let $|G| = 2k + 1$. Then $x = xe = xx^{2k+1} = x^{2k+2} = (x^2)^{k+1} = (y^2)^{k+1} = y^{2k+2} = yy^{2k+1} = ye = y$.

43.   It is the set of all permutations that carry face 2 to face 1.

45. $aH = bH$ if and only if $\det(a) = \pm\det(b)$.

47.   50

49.   The order of the symmetry group would have to be $6 \cdot 20 = 120$.

51.   24

CHAPTER 9

> There's a mighty big difference between good, sound reasons
> and reasons that sound good.   **Burton Hillis**

1.   no

3.   Every element of $\alpha A_n \alpha^{-1}$ is even for all α in S_n.

5.   Recall that if A and B are matrices, then $\det(ABA^{-1}) = (\det A)(\det B)(\det A)^{-1}$.

7.   Let $x \in G$. If $x \in H$, then $xH = H = Hx$. If $x \notin H$, then xH is the set of elements in G, not in H. But Hx is also the elements in G, not in H.

9. $G/H \approx Z_4$
     $G/K \approx Z_2 \oplus Z_2$

11.   6

13.   2

15. $H = \{0 + \langle 20\rangle, 4 + \langle 20\rangle, \ 8 + \langle 20\rangle, \ 12 + \langle 20\rangle, \ 16 + \langle 20\rangle\}$.
     $G/H = \{0 + \langle 20\rangle + H, 1 + \langle 20\rangle + H, 2 + \langle 20\rangle + H, 3 + \langle 20\rangle + H\}$.

17. $40/10 = 4$

19. Z_4

21. ∞; no, $(6, 3) + \langle(4, 2)\rangle$ has order 2.

23. Z_8

25.   yes; no

27.   Since $\det(AB^{-1}) = (\det A)(\det B)^{-1}$, H is a subgroup.
     Since $\det(CAC^{-1}) = (\det C)(\det A)(\det C)^{-1} = \det A$, H is normal.

29. Certainly, every nonzero real number is of the form $\pm r$ where r is a positive real. Real numbers commute and $\mathbf{R}^+ \cap \{1, -1\} = \{1\}$.

31. No. If $G = H \times K$, then $|g| = \operatorname{lcm}(|h|, |k|)$ provided that $|h|$ and $|k|$ are finite. If $|h|$ or $|k|$ is infinite, so is $|g|$.

33. For the first question note that $\langle 3 \rangle \cap \langle 6 \rangle = \{1\}$ and $\langle 10 \rangle \cap \langle 3 \rangle \langle 6 \rangle = \{1\}$. For the second question observe that $12 = 3^{-1}6^2$.

35. Take $G = Z_6$, $H = \{0, 3\}$, $a = 1$, $b = 4$.

37. $aHbH = abH = baH = bHaH$

39. Because xHx^{-1} and H are both subgroups of the same order.

41. Use the "G/Z Theorem."

43. If H is normal in G, then $xNhN(xN)^{-1} = xhx^{-1}N \in H/N$ so H/N is normal in G/N. Now assume H/N is normal in G/N. Then $xNhN(xN)^{-1} = xhx^{-1}N \in H/N$. Thus $xhx^{-1}N = h'N$ for some $h' \in H$. So, $xhx^{-1} = h'n$ for some $n \in N$.

45. Use exercise 7 and observe $VK \neq KV$.

47. $x(H \cap N)x^{-1} = xHx^{-1} \cap xNx^{-1} = H \cap N$

49. Use Theorem 9.4.

51. $\gcd(|x|, |G/H|) = 1$ implies $\gcd(|xH|, |G/H|) = 1$. But $|xH|$ divides $|G/H|$. Thus $|xH| = 1$ and therefore $xH = H$.

53. Note that G/H is a group and use Corollary 3 of Theorem 8.1.

55. Use Theorems 9.4 and 9.3.

57. Say $|gH| = n$. Then $|g| = nt$ (by exercise 40) and $|g^t| = n$. For the second part consider $Z/\langle k \rangle$.

59. It is not a group table. No, because $\mathcal{H}$ is not normal in D_4.

61. Use Theorem 9.3 and Example 4 of Chapter 8.

63. By exercise 62, A_5 would have an element of the form $(ab)(cd)$ that commutes with every element of A_5. Try (abc).

65. Observe that $xg^2x^{-1} = (xgx^{-1})^2$.

CHAPTER 10

Sixty minutes of thinking of any kind is bound to lead to
confusion and unhappiness. **James Thurber**

1. Observe that $(xy)' = x'y'$.

3. Observe that $\det(AB) = (\det A)(\det B)$.

5. Observe that $(f + g)' = f' + g'$.

7. $\phi((g, h), (g', h')) = \phi((gg', hh')) = gg' = \phi((g, h))\phi((g', h'))$

9. Consider $\phi: Z \oplus Z \to Z_a \oplus Z_b$ given by $\phi((x, y)) = (x \bmod a, y \bmod b)$ and use Theorem 10.2.

11. $(a, b) \to b$ is a homomorphism from $A \oplus B$ onto B with kernel $A \oplus \{e\}$.

13. 3, 13, 23

15. Use the First Isomorphism Theorem.

17. Use the First Isomorphism Theorem.

19. $\langle 5 \rangle$

21. 4 onto; 10 to

23. For each k with $0 \le k \le n - 1$ the mapping $1 \to k$ determines a homomorphism.

25. Use properties 10, 11, and 12 of Theorem 10.1.

27. $\phi^{-1}(7) = 7$ Ker $\phi = \{7, 17\}$

29. 11 Ker ϕ

31. $\phi((a, b) + (c, d)) = \phi((a + c, b + d)) = (a + c) - (b + d) = a - b + c - d = \phi((a, b)) + \phi((c, d))$. Ker $\phi = \{(a, a) \mid a \in Z\}$. $\phi^{-1}(3) = \{(a + 3, a) \mid a \in Z\}$.

33. $\phi(xy) = (xy)^6 = x^6y^6 = \phi(x)\phi(y)$. Ker $\phi = \langle \cos 60° + i \sin 60° \rangle$.

35. Show that the mapping from K to KN/N given by $k \to kN$ is an onto homomorphism with kernel $K \cap N$.
37. For each divisor d of k there is a unique subgroup of Z_k of order d and this subgroup is generated by $\phi(d)$ elements. A homomorphism from Z_n to a subgroup of Z_k must carry 1 to a generator of the subgroup. Furthermore, the order of the image of 1 must divide n so we need only consider those divisors d of k that also divide n.
39. $D_4, \{e\}, Z_2, Z_2 \oplus Z_2$
41. It is divisible by 10.
43. It is infinite.
45. Let ϕ be the natural homomorphism from G onto G/N. Let $\overline{H}$ be a subgroup of G/N and let $\phi^{-1}(\overline{H}) = H$. Then H is a subgroup of G and $H/N = \phi(H) = \phi(\phi^{-1}(\overline{H})) = \overline{H}$.
47. The mapping $g \to \phi_g$ is a homomorphism with kernel $Z(G)$.
49. $(f + g)(3) = f(3) + g(3)$. The kernel is the set of elements in $Z[x]$ whose graph passes through the point $(3, 0)$.
51. Use exercise 47 of Chapter 9 and exercise 39 to prove the first assertion. To verify $G/H \cap K$ is not cyclic, observe that it has two subgroups of order 2.

CHAPTER 11

Think before you think!   **Stanislaw J. Lec,** *Unkempt Thoughts*

1. $n = 4$
   $Z_4, Z_2 \oplus Z_2$
3. $n = 36$
   $Z_9 \oplus Z_4, Z_3 \oplus Z_3 \oplus Z_4, Z_9 \oplus Z_2 \oplus Z_2, Z_3 \oplus Z_3 \oplus Z_2 \oplus Z_2$
5. The only Abelian groups of order 45 are Z_{45} and $Z_3 \oplus Z_3 \oplus Z_5$. In the first group $|3| = 15$, in the second one $|(1, 1, 1)| = 15$. $Z_3 \oplus Z_3 \oplus Z_5$ does not have an element of order 9.
7. $Z_9 \oplus Z_3 \oplus Z_4; Z_9 \oplus Z_3 \oplus Z_2 \oplus Z_2$
9. $Z_4 \oplus Z_2 \oplus Z_3 \oplus Z_5$
11. $360 = 8 \cdot 9 \cdot 5$
    $$Z_8 \oplus Z_9 \oplus Z_5$$
    $$Z_4 \oplus Z_2 \oplus Z_9 \oplus Z_5$$
    $$Z_2 \oplus Z_2 \oplus Z_2 \oplus Z_9 \oplus Z_5$$
    $$Z_8 \oplus Z_3 \oplus Z_3 \oplus Z_5$$
    $$Z_4 \oplus Z_2 \oplus Z_3 \oplus Z_3 \oplus Z_5$$
    $$Z_2 \oplus Z_2 \oplus Z_2 \oplus Z_3 \oplus Z_3 \oplus Z_5$$
13. $Z_2 \oplus Z_2$
15. a. 1   b. 1   c. 1   d. 1   e. 1   f. There is a unique Abelian group of order n if and only if n is not divisible by the square of any prime.
17. $Z_2 \oplus Z_2$
19. $Z_3 \oplus Z_3$
21. n is square-free (no prime factor of n occurs more than once).
23. Among the first 11 elements in the table there are 9 elements of order 4. None of the other isomorphism classes has this many.
25. $Z_4 \oplus Z_2 \oplus Z_2$; One internal direct product is $\langle 7 \rangle \times \langle 101 \rangle \times \langle 199 \rangle$.
27. 3; 6; 12
29. $Z_4 \oplus Z_4$
31. Use Theorems 11.1, 7.1, and 4.3.
33. If $|C| = p^n$, use the Fundamental Theorem and Theorem 9.6. If every element has order a power of p use the corollary to the Fundamental Theorem.

35. By the Fundamental Theorem of Finite Abelian Groups it suffices to show that every group of the form $Z_{p_1^{n_1}} \oplus Z_{p_2^{n_2}} \cdots \oplus Z_{p_k^{n_k}}$ is a subgroup of a U-group. Consider first a group of the form $Z_{p_1^{n_1}} \oplus Z_{p_2^{n_2}}$ (p_1 and p_2 need not be distinct). By Dirichlet's Theorem, for some s and t there are distinct primes q and r such that $q = tp_1^{n_1} + 1$ and $r = sp_2^{n_2} + 1$. Then $U(qr) = U(q) \oplus U(r) \approx Z_{tp_1^{n_1}} \oplus Z_{sp_2^{n_2}}$ and this latter group contains a subgroup isomorphic to $Z_{p_1^{n_1}} \oplus Z_{p_2^{n_2}}$. The general case follows in the same way.

SUPPLEMENTARY EXERCISES FOR CHAPTERS 8–11

It took me so long to find out.   **John Lennon and Paul McCartney,** *Day Tripper,* **single**

1. Say $aH = Hb$. Then $a = hb$ for some h in H. Then $Ha = Hhb = Hb = aH$.
3. Observe that $x^{-1}N = yN$ so that $N = xyN$.
5. Say $aH = bK$. Then $H = a^{-1}bK$ and $K = b^{-1}aH$. Thus $a^{-1}b \in H$ and $(a^{-1}b)^{-1} = b^{-1}a \in H$. So, $K = b^{-1}aH = H$. For the example try A_4.
7. $aH = bH$ implies $a^{-1}b \in H$. So $(a^{-1}b)^{-1} = b^{-1}a \in H$. Thus, $Hb^{-1}a = H$ or $Hb^{-1} = Ha^{-1}$. These steps are reversible.
9. Suppose diag(G) is normal. Then $(e, a)(b, b)(e, a)^{-1} = (b, aba^{-1}) \in$ diag(G). Thus $b = aba^{-1}$. If G is Abelian, $(g, h)(b, b)(g, h)^{-1} = (gbg^{-1}, hbh^{-1}) = (b, b)$. When $G = \mathbf{R}$, diag(G) is the line $y = x$. The index of diag(G) is $|G|$.
11. Let $\alpha \in$ Aut(G) and $\phi_a \in$ Inn(G). Then $(\alpha\phi_a\alpha^{-1})(x) = (\alpha\phi_a)\alpha^{-1}(x) = \alpha(a\alpha^{-1}(x)a^{-1}) = \alpha(a)x(\alpha(a))^{-1} = \phi_{\alpha(a)}(x)$.
13. $\mathbf{R}^*$ (See Example 2 of Chapter 10.)
15. a.

$$Z(H) = \left\{ \begin{bmatrix} 1 & 0 & b \\ 0 & 1 & 0 \\ 0 & 0 & 1 \end{bmatrix} \,\middle|\, b \in Q \right\}$$

The mapping

$$\begin{bmatrix} 1 & 0 & b \\ 0 & 1 & 0 \\ 0 & 0 & 1 \end{bmatrix} \to b$$

is an isomorphism.
b. The mapping

$$\begin{bmatrix} 1 & a & b \\ 0 & 1 & c \\ 0 & 0 & 1 \end{bmatrix} \to (a, c)$$

is a homomorphism with $Z(H)$ as the kernel.
c. The proofs are valid with $\mathbf{R}$ and Z_p.
17. $b(a/b + Z) = a + Z = Z$
19. Use exercise 6 of the supplemental exercises for Chapters 1–4. For the example, try $G = D_4$.
21. $Z_2 \oplus Z_2$ has 3 subgroups of order 2.
    $Z_3 \oplus Z_3$ has 4 subgroups of order 3.

$Z_p \oplus Z_p$ has $p + 1$ subgroups of order p.
It suffices to count elements of order p and divide by $p - 1$. Thus there are $(p^2 - 1)/(p - 1) = p + 1$ subgroups of order p.

23. Observe that $hkh^{-1}k^{-1} = (hkh^{-1})k^{-1} \in K$ and $hkh^{-1}k^{-1} = h(kh^{-1}k^{-1}) \in H$.

25. The mapping $g \to g^n$ is a homomorphism from G onto G^n with kernel G_n.

27. Use Theorem 8.2 and exercise 7 of Chapter 9.

29. The number is m in all cases.

31. For each x with $|x| > 2$, both x and x^{-1} are factors in the product and therefore cancel out. This reduces the problem to groups of the form $Z_2 \oplus Z_2 \oplus \cdots \oplus Z_2$ of order at least 4. Now appeal to exercise 30. If the subgroup of order 2^n is cyclic, the product is the unique element of order 2.

33. First observe that $\phi((4, 0, 0)) = \phi(4(1, 0, 0)) = 4\phi(1, 0, 0) = (0, 0)$ so that Ker $\phi = \{(0, 0, 0), (4, 0, 0)\}$. But then $(Z_8 \oplus Z_2 \oplus Z_2)/$Ker ϕ has more than three elements of order 2 whereas $Z_4 \oplus Z_4$ has only three.

35. Use Example 4 of Chapter 8 together with the fact that S_4 has no element of order 6.

CHAPTER 12

> Think for yourself.   **Title of a song by George Harrison,**
> *Rubber Soul*

1. For any $n > 1$, the ring $M_2(Z_n)$ of 2×2 matrices with entries from Z_n is a finite non-commutative ring. The set $M_2(2Z)$ of 2×2 matrices with even integer entries is an infinite noncommutative ring that does not have a unity.

3. The only property that is not immediate is closure under multiplication:

$$(a + b\sqrt{2})(c + d\sqrt{2}) = (ac + 2bd) + (ad + bc)\sqrt{2}.$$

5. The proof given for a group applies to a ring as well.

7. In Z_p nonzero elements have multiplicative inverses. Use them.

9. Note that $(na)(ma) = (nm)a^2 = (mn)a^2 = (ma)(na)$.

11. Part 3: $0 = 0(-b) = (a + (-a))(-b) = a(-b) + (-a)(-b) = -(ab) + (-a)(-b)$. So, $ab = (-a)(-b)$.
    Part 4: $a(b - c) = a(b + (-c)) = ab + a(-c) = ab + (-(ac)) = ab - ac$.
    Part 5: Use Part 2.
    Part 6: Use Part 3.

13. Hint: Z is a cyclic group under addition, and every subgroup of a cyclic group is cyclic.

15. Use exercise 19 of Chapter 2.

17. If a and b belong to the intersection, then they belong to each member of the intersection. Thus $a - b$ and ab belong to each member of the intersection. So, $a - b$ and ab belong to the intersection.

19. Let a, b belong to the center. Then $(a - b)x = ax - bx = xa - xb = x(a - b)$. Also, $(ab)x = a(bx) = a(xb) = (ax)b = (xa)b = x(ab)$.

21. $(x_1, \ldots, x_n)(a_1, \ldots, a_n) = (x_1, \ldots, x_n)$ for all x_i in R_i if and only if $x_i a_i = x_i$ for all x_i in R_i and $i = 1, \ldots, n$.

23. $\{1, i\}$

25. $f(x) = 1$ and $g(x) = -1$.

27. If a is a unit, then $b = a(a^{-1}b)$.

29. Consider $a^{-1} - a^{-2}b$.

31. Try the ring $M_2(Z)$.

33. The equation has nontrivial solutions if and only if $p = 1 \bmod 4$.

35. Every subgroup of Z_n is closed under multiplication.

37. $ara - asa = a(r - s)a.$ $(ara)(asa) = ara^2sa = arsa.$ $a1a = a^2 = 1$ so $1 \in S.$
39. The subring test is satisfied.
41. Look at $(1, 0, 1)$ and $(0, 1, 1).$
43. $2Z \cup 3Z$ contains 2 and 3, but not $2 + 3.$
45. $\{m/2^n \mid m \in Z, n \in Z^+\}$
47. $(a + b)(a - b) = a^2 + ba - ab - b^2 = a^2 - b^2$ if and only if $ba - ab = 0.$
49. Say $n = 2m.$ Then $-a = (-a)^n = (-a)^{2m} = [(-a)^2]^m = (a^2)^m = a^n = a.$

CHAPTER 13

Work now or wince later. **B. C. Forbes, *Epigrams***

3. Let $ab = 0$ and $a \neq 0.$ Then $ab = a \cdot 0$ so $b = 0.$
5. Let $k \in Z_n.$ If $\gcd(k, n) = 1,$ then k is a unit. If $\gcd(k, n) = d > 1,$ write $k = sd.$ Then $k(n/d) = sd(n/d) = sn = 0 \bmod n.$
7. Let $s \in R, s \neq 0.$ Consider the set $S = \{sr \mid r \in R\}.$ If $S = R,$ then $sr = 1$ (the unity) for some $r.$ If $S \neq R,$ then there are distinct r_1 and r_2 such that $sr_1 = sr_2.$ In this case, $s(r_1 - r_2) = 0.$ To see what happens when the "finite" condition is dropped consider $Z.$
9. $(a_1 + b_1\sqrt{d}) - (a_2 + b_2\sqrt{d}) = (a_1 + a_2) - (b_1 + b_2)\sqrt{d};$
   $(a_1 + b_1\sqrt{d})(a_2 + b_2\sqrt{d}) = (a_1a_2 + b_1b_2d) + (a_1b_2 + a_2b_1)\sqrt{d}.$ Thus the set is a ring. Since $Z[\sqrt{d}]$ is a subring of the ring of real numbers, it has no zero-divisors.
11. The even integers.
15. Suppose $a \neq 0$ and $a^n = 0$ (where we take n to be as small as possible). Then $0 = a \cdot 0 = a^n = a \cdot a^{n-1},$ so by cancellation, $a^{n-1} = 0.$
17. If $a^2 = a$ and $b^2 = b,$ then $(ab)^2 = a^2b^2 = ab.$
19. $Z_4 \oplus Z_4$
21. $a^2 = a$ implies $a(a - 1) = 0.$ So if a is a unit, $a - 1 = 0$ and $a = 1.$
23. See Theorems 3.1 and 12.3.
25. Note that $ab = 1$ implies $aba = a.$ Thus $0 = aba - a = a(ba - 1).$ So, $ba - 1 = 0.$
27. A subdomain of an integral domain D is a subset of D that is an integral domain under the operations of $D.$ To show that P is a subdomain show that it is a subring and contains 1. Every subdomain contains 1 and is closed under addition and subtraction so every subdomain contains $P.$ $|P| = $ char $D.$
29. No, $(1, 0) \cdot (0, 1) = (0, 0).$
31. b. Use the fact that there exist integers s and t such that $1 = sn + tm,$ but remember that you cannot use negative exponents in a ring.
35. Z_8
37. Let $S = \{a_1, a_2, \ldots, a_n\}$ be the *nonzero* elements of the ring. First show that $S = \{a_1a_1, a_1a_2, \ldots, a_1a_n\}.$ Thus, $a_1 = a_1a_i$ for some $i.$ Then a_i is the unity, for if a_k is any element of $S,$ we have $a_1a_k = a_1a_ia_k$ so that $a_1(a_k - a_ia_k) = 0.$
39. Say $|x| = n$ and $|y| = m$ with $n < m.$ Consider $n(xy) = (nx)y = x(ny).$
41. Use exercise 39.
43. Try $Z_2.$
45. $n\begin{bmatrix} a & b \\ c & d \end{bmatrix} = \begin{bmatrix} 0 & 0 \\ 0 & 0 \end{bmatrix}$ for all members of $M_2(R)$ if and only if $na = 0$ for all a in $R.$
47. Use exercise 46.
49. a. 2 b. 2, 3 c. 2, 3, 6, 11 d. 2, 3, 9, 10
51. 2
53. See Example 9.
55. Use exercise 23 and part a of exercise 42.

57. Choose $a \neq 0$ and $a \neq 1$ and consider the image of $1 + a$.
59. $\phi(x) = \phi(x \cdot 1) = \phi(x) \cdot \phi(1)$ so $\phi(1) = 1$. Also, $1 = \phi(1) = \phi(xx^{-1}) = \phi(x)\phi(x^{-1})$.

CHAPTER 14

> Not one student in a thousand breaks down from overwork.
> **William Allan Neilson**

1. $\sqrt{2} \cdot 1$ is not rational.
3. Let $a + bi, c + di \in S$. Then $(a + bi) - (c + di) = a - c + (b - d)i$ and $b - d$ is even. Also, $(a + bi)(c + di) = ac - bd + (ad + cb)i$ and $ad + cb$ is even. Finally, $(1 + 2i)(1 + i) = -1 + 3i \notin S$.
5. $ar_1 - ar_2 = a(r_1 - r_2); (ar_1)r = a(r_1r)$
   $4R = \langle 8 \rangle$
7. Mimic exercise 17 of Chapter 12.
9. a. $a = 1$ b. $a = 3$ c. $a = \gcd(m, n)$
11. a. $a = 12$
    b. $a = 48$. To see this, note that every element of $\langle 6 \rangle \langle 8 \rangle$ has the form $6t_1 8k_1 + 6t_2 8k_2 + \cdots + 6t_n 8k_n = 48s \in \langle 48 \rangle$. So, $\langle 6 \rangle \langle 8 \rangle \subseteq \langle 48 \rangle$. Also, since $48 \in \langle 6 \rangle \langle 8 \rangle$, we have $\langle 48 \rangle \subseteq \langle 6 \rangle \langle 8 \rangle$.
    c. $a = mn$
13. By exercise 12, we have $AB \subseteq A \cap B$. So, let $x \in A \cap B$. To show that $x \in AB$, start by writing $1 = a + b$ where $a \in A, b \in B$.
15. Let $u \in I$ be a unit and let $r \in R$. Then $r = r(u^{-1}u) = (ru^{-1})u \in I$.
17. Use the observation that every member of R can be written in the form
$$\begin{bmatrix} 2q_1 + r_1 & 2q_2 + r_2 \\ 2q_3 + r_3 & 2q_4 + r_4 \end{bmatrix}.$$
19. $(br_1 + a_1) - (br_2 + a_2) = b(r_1 - r_2) + (a_1 - a_2) \in B; r'(br + a) = b(r'r) + r'a \in B$
21. $(b + A)(c + A) = bc + A = cb + A = (c + A)(b + A)$. If 1 is the unity of R, then $1 + A$ is the unity of R/A.
23. Use Theorems 14.4 and 14.3.
25. Suppose $f(x) + A \neq A$. Then $f(x) + A = f(0) + A$ and $f(0) \neq 0$. Thus,
$$(f(x) + A)^{-1} = \frac{1}{f(0)} + A.$$
27. Since $(3 + i)(3 - i) = 10$, $10 + \langle 3 + i \rangle = 0 + \langle 3 + i \rangle$. Also, $i + \langle 3 + i \rangle = -3 + \langle 3 + i \rangle = 7 + \langle 3 + i \rangle$. So, $Z[i]/\langle 3 + i \rangle = \{k + \langle 3 + i \rangle \mid k = 0, 1, \ldots, 9\}$.
29. Use Theorems 14.3 and 14.4.
31. Let $f, g \in I$ and $h \in Z[x]$. Then $(f - g)(0) = f(0) - g(0)$ and $(f \cdot h)(0) = f(0)h(0)$ so that I is an ideal. Also, if $h(0)k(0)$ is even then one of $h(0)$ or $k(0)$ is even. Yes, it is maximal.
33. $3x + 1 + I$
35. Let $a, b \in I_p$. Say $|a| = p^n$ and $|b| = p^m$. Then $p^{n+m}(a - b) = 0$ so $|a - b|$ divides p^{n+m}. Also, $p^n(ra) = r(p^na) = 0$ so $|ra|$ divides p^n.
37. Say $b, c \in \text{Ann}(A)$. Then $(b - c)a = ba - ca = 0 - 0 = 0$. Also, $(rb)a = (ba)r = 0 \cdot r = 0$.
39. a. $\langle 3 \rangle$ b. $\langle 3 \rangle$ c. $\langle 3 \rangle$
41. Suppose $(x + \sqrt{\langle 0 \rangle})^n = 0 + \sqrt{\langle 0 \rangle}$. We must show that $x \in \sqrt{\langle 0 \rangle}$. We know that $x^n + \sqrt{\langle 0 \rangle} = 0 + \sqrt{\langle 0 \rangle}$ so that $x^n \in \sqrt{\langle 0 \rangle}$. Then, for some m, $(x^n)^m = 0$ and $x \in \sqrt{\langle 0 \rangle}$.

43. The set $Z_2[x]/\langle x^2 + x + 1 \rangle$ has only four elements and each of the nonzero ones has a multiplicative inverse. For example,

$$(x + \langle x^2 + x + 1 \rangle)(x + 1 + \langle x^2 + x + 1 \rangle) = 1 + \langle x^2 + x + 1 \rangle.$$

45. $x + 2 + \langle x^2 + x + 1 \rangle$ is not zero but its square is.
47. If f and $g \in A$, then $(f - g)(0) = f(0) - g(0)$ is even and $(f \cdot g)(0) = f(0) \cdot g(0)$ is even. $f(x) = \sqrt{2} \in R$ and $g(x) = 2 \in A$ but $f(x)g(x) \notin A$.
49. Hint: Any ideal of R/I has the form A/I where A is an ideal of R.
51. Use the fact that R/I is an integral domain to show that $R/I = \{I, 1 + I\}$.

SUPPLEMENTARY EXERCISES FOR CHAPTERS 12–14

> If at first you don't succeed, try, try, again. Then quit. There's no use being a damn fool about it.　**W. C. Fields**

1. In Z_{10} they are 0, 1, 5, 6.
3. We must show that $a^n = 0$ implies $a = 0$. First show for the case when n is a power of 2. If n is not a power of 2, say 13 for example, note that $a^{13} = 0$ implies $a^{16} = 0$.
5. Suppose $A \not\subseteq C$ and $B \not\subseteq C$. Pick $a \in A$ and $b \in B$ so that $a, b \notin C$. But $ab \in C$ and C is prime.
7. Suppose I is an ideal and $a \in I$ with $a \neq 0$. Then $1 = a^{-1}a \in I$ so $I = F$.
9. Observe that $(Z \oplus Z)/A = \{(b, 0) \mid b = 0, 1, \ldots, p - 1\} \approx Z_p$, a field.
11. Suppose $a_1, a_2 \in A$ but $a_1 \notin B$ and $a_2 \notin C$. Use $a_1 + a_2$ to derive a contradiction.
13. Clearly $\langle a \rangle$ contains the right-hand side. Now show that the right-hand side contains a and is an ideal.
15. Since A is an ideal, $ab \in A$. Since B is an ideal, $ab \in B$. So $ab \in A \cap B = \{0\}$.
17. 6
19. Use exercise 4.
21. Consider $x^2 + 1 + \langle x^4 + x^2 \rangle$.
23. Consider Z_8.
25. Say char $R = p$ (remember p must be prime). Then char $R/A =$ the additive order of $1 + A$. But $|1 + A|$ divides $|1| = p$.
27. Use Theorems 13.2, 14.3, and 14.4.
29. $\begin{bmatrix} 1 & 0 \\ 0 & 0 \end{bmatrix}\begin{bmatrix} 1 & 1 \\ 1 & 1 \end{bmatrix} = \begin{bmatrix} 1 & 1 \\ 0 & 0 \end{bmatrix}$, which is not in the set.
31. $Z[i]/A$ has two elements. (From this it follows that A is maximal. See Theorem 14.4.)
33. A finite subset of a field is a subfield if it contains a nonzero element and is closed under addition and multiplication.
35. Observe that an element a in Z_n has a multiplicative inverse if and only if $\gcd(a, n) = 1$. But then the additive group generated by a is Z_n.
37. 5

CHAPTER 15

> A sign in a Pentagon office reads: If at first you don't succeed, forget it.

1. $\phi(2 + 4) = \phi(1) = 5$ whereas $\phi(2) + \phi(4) = 0 + 0 = 0$.
3. a. No. Suppose $2 \rightarrow a$. Now consider $2 + 2$ and $2 \cdot 2$.
   b. no.
5. Apply the definition.
7. Multiplication is not preserved.
9. yes

11. The set of all polynomials passing through the point $(1, 0)$.
13. The group A/B is cyclic of order 4. The ring A/B has no unity.
15. The zero map and the identity map.
17. If $2 + 8k = x^3$ for some k, then $2 = x^3 \bmod 8$ has a solution. But direct substitution of $0, 1, 2, 3, 4, 5, 6, 7$ shows that there is no solution.
19. Observe that an idempotent must map to an idempotent. It follows that $(a, b) \rightarrow a$, $(a, b) \rightarrow b$, and $(a, b) \rightarrow 0$ are the only ring homomorphisms.
21. Say $m = a_k a_{k-1} \ldots a_1 a_0$ and $n = b_k b_{k-1} \ldots b_1 b_0$. Then $m - n = (a_k - b_k)10^k + (a_{k-1} - b_{k-1})10^{k-1} + \cdots + (a_1 - b_1)10 + (a_0 - b_0)$. Now use the test for divisibility by 9.
23. Use the appropriate divisibility tests.
25. Mimic Example 8.
27. Observe that $2 \cdot 10^{75} + 2 = 1 \bmod 3$ and $10^{100} + 1 = 2 = -1 \bmod 3$.
29. This follows directly from Theorem 13.3 and Theorem 10.1, part 7.
31. No. The kernel must be an ideal.
33. a. Suppose $ab \in \phi^{-1}(A)$. Then $\phi(a)\phi(b) \in A$ so that $a \in \phi^{-1}(A)$ or $b \in \phi^{-1}(A)$.
     b. Consider the natural homomorphism from R to S/A. Then use Theorems 15.3 and 14.4.
35. c. Use $(r, s) \rightarrow (s, r)$.
37. Observe $x^4 = 1$ has two solutions in **R** but four in C.
39. Use exercises 32 and 38.
41. To check that multiplication is operation preserving, observe that $xy \rightarrow a(xy) = a^2xy = axay$.
43. First note that any field containing Z and i must contain $Q[i]$. Then prove $(a + bi)/(c + di) \in Q[i]$.
45. The subfield of E is $\{ab^{-1} \mid a, b \in D, b \neq 0\}$.
47. The set of even integers is a subring of the rationals.
49. Try $ab^{-1} \rightarrow a/b$.
51. Say 1_R is the unity of R and 1_S is the unity of S. Pick $a \in R$ such that $\phi(a) \neq 0$. Then $1_S\phi(a) = \phi(1_R a) = \phi(1_R)\phi(a)$. Now cancel. For the example, consider the mapping from Z to $2Z$ that sends n to $2n$.
53. Certainly, the unity 1 is contained in every subfield. So, if a field has characteristic p, the subfield $\{0, 1, \ldots, p - 1\}$ is contained in every subfield. If a field has characteristic 0, then $\{(n1)(m1)^{-1} \mid n, m \in Z, m \neq 0\}$ is a subfield contained in every subfield. This subfield is isomorphic to Q (map $(n1)(m1)^{-1}$ to n/m).

CHAPTER 16

> You know my methods, apply them! **Sherlock Holmes,** *The Hound of the Baskervilles*

1. $f + g = 3x^4 + 2x^3 + 2x + 2$
    $f \cdot g = 2x^7 + 3x^6 + x^5 + 2x^4 + 3x^2 + 2x + 2$
3. Let $f(x) = x^4 + x$ and $g(x) = x^2 + x$. Then $f(0) = 0 = g(0); f(1) = 2 = g(1); f(2) = 0 = g(2)$.
5. Use Corollary 1 of Theorem 16.2.
7. Since R is isomorphic to the subring of constant polynomials, char $R \leq$ char $R[x]$. On the other hand, char $R = c$ implies

$$c(a_n x^n + \cdots + a_0) = (ca_n)x^n + \cdots + (ca_0) = 0$$

so that char $R[x] \leq$ char R.

9. Use exercise 8 and observe that $\phi(a_n)x^n + \cdots + \phi(a_0) = \phi(b_n)x^n + \cdots + \phi(b_0)$ if and only if $\phi(a_i) = \phi(b_i)$ for all i.
11. quotient $2x^2 + 2x + 1$; remainder 2
13. It is its own inverse.
15. No (see exercise 16).
17. Observe that $Z[x]/\langle x \rangle$ is isomorphic to Z. Now use Theorems 14.3 and 14.4.
19. Use Corollary 3 of Theorem 16.2.
21. If $f(x) \neq g(x)$, then $\deg[f(x) - g(x)] < \deg p(x)$. But the minimum degree of any member of $\langle p(x) \rangle$ is $\deg p(x)$.
23. Start with $(x - 1/2)(x + 1/3)$ and clear fractions.
25. "Long divide" $x - a$ into $f(x)$ and induct on $\deg f(x)$.
27. By the Corollary to Theorem 16.3, $I = \langle x - 1 \rangle$.
29. Use the Factor Theorem.
31. For any a in $U(p)$, $a^{p-1} = 1$ so every member of $U(p)$ is a zero of $x^{p-1} - 1$. Now use the Factor Theorem and a degree argument.
33. Use exercise 31.
35. Observe that, modulo 101, $(50!)^2 = (50!)(-1)(-2) \cdots (-50) = (50!)(100)(99) \cdots (51) = 100!$ and use exercise 32.
37. Take $R = Z$ and $I = \langle 2 \rangle$.
39. Mimic Example 3.
41. Write $f(x) = (x - a)g(x)$. Use the product rule to compute $f'(x)$.

CHAPTER 17

> Experience enables you to recognize a mistake when you make it again. **Franklin P. Jones**

1. $f(x)$ factors over D as $ah(x)$ where a is not a unit in D.
3. Apply Eisenstein's Criterion.
5. a. If $f(x) = g(x)h(x)$, then $af(x) = ag(x)h(x)$.
   b. If $f(x) = g(x)h(x)$, then $f(ax) = g(ax)h(ax)$.
   c. If $f(x) = g(x)h(x)$, then $f(x + a) = g(x + a)h(x + a)$.
   d. Try $a = 1$.
7. Find an irreducible polynomial $p(x)$ of degree 2 over Z_5. Then $Z_5[x]/\langle p(x) \rangle$ is a field of order 25.
9. Note that -1 is a root. No, since 4 is not a prime.
11. Use direct calculations to show that it has no zeros.
13. $(x + 3)(x + 5)(x + 6)$
15. a. Consider the number of distinct expressions of the form $(x - c)(x - d)$.
    b. Reduce the problem to (a).
17. Use exercise 16.
19. $x^n + p$ where p is prime is irreducible over Q.
21. $x^2 + 1, x^2 + x + 2, x^2 + 2x + 2$
23. 1 has multiplicity 1, 3 has multiplicity 2.
25. We know $a_n(r/s)^n + a_{n-1}(r/s)^{n-1} + \cdots + a_0 = 0$. So $a_n r^n + a_{n-1} s r^{n-1} + \cdots + s^n a_0 = 0$. This shows that $s \mid a_n r^n$ and $r \mid s^n a_0$. Now use Euclid's Lemma and the fact that r and s are relatively prime.
27. Use exercise 5a and clear fractions.
29. If there is an a in Z_p such that $a^2 = -1$, then $x^4 + 1 = (x^2 + a)(x^2 - a)$.
    If there is an a in Z_p such that $a^2 = 2$, then $x^4 + 1 = (x^2 + ax + 1)(x^2 - ax + 1)$.
    If there is an a in Z_p such that $a^2 = -2$, then $x^4 + 1 = (x^2 + ax - 1)(x^2 - ax - 1)$.
    Now show that one of these three cases must occur.

31. Since $(f + g)(a) = f(a) + g(a)$ and $(f \cdot g)(a) = f(a)g(a)$ the mapping is a homomorphism. Clearly, $p(x)$ belongs to the kernel. By Theorem 17.5, $\langle p(x) \rangle$ is a maximal ideal so $\langle p(x) \rangle$ = kernel.

33. Use the Corollary to Theorem 17.4 and exercise 5b with $a = -1$.

35. The analysis is identical except $0 \le q, r, t, u \le n$. Now just as when $n = 2$, we have $q = r = t = 1$. But this time $0 \le u \le n$. However, when $u > 2$, $P(x) = x(x + 1) \cdot (x^2 + x + 1)(x^2 - x + 1)^u$ has $(-u + 2)x^{2u+3}$ as one of its terms. Since the coefficient of x^{2u+3} represents the number of dice with the label $2u + 3$, the coefficient cannot be negative. Thus, $u \le 2$, as before.

37. Although the probability of rolling any particular sum is the same with either pair of dice, the probability of rolling doubles is different (1/6 with ordinary dice, 1/9 with Sicherman dice). Thus the probability of going to jail is different. Other probabilities are also affected. For example, if in jail one cannot land on Virginia by rolling a pair of 2's with Sicherman dice but one is twice as likely to land on St. James with a pair of 3's with the Sicherman dice as with ordinary dice.

CHAPTER 18

> He thinks things through very carefully before going off half-cocked.   **General Carl Spaatz, in *Presidents Who Have Known Me*, George E. Allen**

1. Say r is irreducible and u is a unit. If $ru = ab$ where a and b are not units, then $r = a(bu^{-1})$ where a and bu^{-1} are not units.

3. Clearly, $\langle ab \rangle \subseteq \langle b \rangle$. If $\langle ab \rangle = \langle b \rangle$, then $b = rab$ so that $1 = ra$ and a is a unit.

5. Say $x = a + bi$ and $y = c + di$. Then

$$xy = (ac - bd) + (bc + ad)i.$$

So

$$d(xy) = (ac - bd)^2 + (bc + ad)^2 = (ac)^2 + (bd)^2 + (bc)^2 + (ad)^2.$$

On the other hand,

$$d(x)d(y) = (a^2 + b^2)(c^2 + d^2) = a^2c^2 + b^2d^2 + b^2c^2 + a^2d^2.$$

7. Suppose $a = bu$ where u is a unit. Then $d(b) \le d(bu) = d(a)$. Also, $d(a) \le d(au^{-1}) = d(b)$.

9. $3 \cdot 7$ and $(1 + 2\sqrt{-5})(1 - 2\sqrt{-5})$. Mimic Example 7 to show that these are irreducible.

11. Mimic Example 7 and observe $10 = 2 \cdot 5$ and $10 = (2 - \sqrt{-6})(2 + \sqrt{-6})$. A PID is a UFD.

13. Suppose $3 = \alpha\beta$ where $\alpha, \beta \in Z[i]$ and neither is a unit. Then $9 = d(3) = d(\alpha)d(\beta)$ so that $d(\alpha) = 3$. But there are no integers such that $a^2 + b^2 = 3$. Observe that $2 = -i(1 + i)^2$ and $5 = (1 + 2i)(1 - 2i)$.

15. Say $a = bu$ where u is a unit. Then $ra = rbu = (ru)b \in \langle b \rangle$ so that $\langle a \rangle \subseteq \langle b \rangle$. By symmetry, $\langle b \rangle \subseteq \langle a \rangle$. If $\langle a \rangle = \langle b \rangle$, then $a = bu$ and $b = av$. Thus $a = avu$ and $vu = 1$.

17. Use exercise 14 with $d = -1$. 5 and $1 + 2i$; 13 and $3 + 2i$; 17 and $4 + i$.

19. Mimic Example 1.

21. Use the fact that x is a unit if and only if $N(x) = 1$.

23. See Example 2.
25. $\pm 1, \pm i$
27. $(-1 + \sqrt{5})(1 + \sqrt{5}) = 4 = 2 \cdot 2$. Now use exercise 20.
29. Use exercise 28 and Theorem 14.4.
31. Suppose R satisfies the ascending chain condition and there is an ideal I of R that is not finitely generated. Then pick $a_1 \in I$. Since I is not finitely generated, $\langle a_1 \rangle$ is a proper subset of I so we may choose $a_2 \in I$ but $a_2 \notin \langle a_1 \rangle$. As before $\langle a_1, a_2 \rangle$ is proper so we may choose $a_3 \in I$ but $a_3 \notin \langle a_1, a_2 \rangle$. Continuing in this fashion, we obtain a chain of infinite length $\langle a_1 \rangle \subset \langle a_1, a_2 \rangle \subset \langle a_1, a_2, a_3 \rangle \subset \cdots$.

    Now suppose every ideal of R is finitely generated and there is a chain $I_1 \subseteq I_2 \subseteq I_3 \subset \cdots$. Let $I = \cup I_i$. Then $I = \langle a_1, a_2, \ldots, a_n \rangle$. Since $I = \cup I_i$ each a_i belongs to some member of the union, say $I_{i'}$. Letting $k = \max\{i' \mid i = 1, \ldots, n\}$, we see that all $a_i \in I_k$. Thus $I \subseteq I_k$ and the chain has length at most k.
33. Say $I = \langle a + bi \rangle$. Then $a^2 + b^2 + I = (a + bi)(a - bi) + I = I$ and $a^2 + b^2 \in I$. For any $c, d \in Z$, let $c = q_1(a^2 + b^2) + r_1$ and $d = q_2(a^2 + b^2) + r_2$ where $0 \leq r_1$, $r_2 < a^2 + b^2$. Then $c + di + I = r_1 + r_2 i + I$.

SUPPLEMENTARY EXERCISES FOR CHAPTERS 15–18

> Yes I get by with a little help from my friends.   **John Lennon and Paul McCartney, *With a Little Help from My Friends, Sgt. Pepper's Lonely Hearts Club Band***

1. Use Theorem 15.3, supplementary exercise 8 for Chapters 12–14, Theorem 14.4, and Example 10 of Chapter 14.
3. To show the isomorphism, use the First Isomorphism Theorem.
5. Use the First Isomorphism Theorem.
7. Consider the obvious homomorphism from $Z[x]$ onto $Z_2[x]$. Then use the First Isomorphism Theorem and Theorem 14.3.
9. As in exercise 13 of Chapter 6, the mapping is onto, 1-1, and preserves multiplication. Also $a(x + y)a^{-1} = axa^{-1} + aya^{-1}$ so that it preserves addition as well.
11. $Z[i]/\langle 2 + i \rangle = \{0 + \langle 2 + i \rangle, 1 + \langle 2 + i \rangle, 2 + \langle 2 + i \rangle, 3 + \langle 2 + i \rangle, 4 + \langle 2 + i \rangle\}$
    Note that

    $$5 + \langle 2 + i \rangle = (2 + i)(2 - i) + \langle 2 + i \rangle = 0 + \langle 2 + i \rangle.$$

13. Observe that $(3 + 2\sqrt{2})(3 - 2\sqrt{2}) = 1$.
15. We are given $(k + 1)^2 = k + 1 \bmod n$. So, $k^2 + 2k + 1 = k + 1 \bmod n$ or $k^2 = -k = n - k \bmod n$. Also, $(n - k)^2 = n^2 - 2nk + k^2 = k^2 \bmod n$ so $(n - k)^2 = n - k \bmod n$.
17. Suppose $0 \leq a < p^k$ and $a^2 = a \bmod p^k$. Then $p^k \mid a(a - 1)$ and since a and $a - 1$ are relatively prime $p^k \mid a$ or $p^k \mid a - 1$. Thus, $a = 0$ or $a - 1 = 0$.
19. Use the Mod 2 Irreducibility Test.
21. Use Theorem 14.4.
23. Use Theorem 14.4.
25. Say $a/b, c/d \in R$. Then $(ad - bc)/bd$ and $ac/bd \in R$ by Euclid's Lemma. The field of quotients is Q.
27. $Z[i]/\langle 3 \rangle$ is a field and $Z_3 \oplus Z_3$ is not.

CHAPTER 19

> When I was young I observed that nine out of every ten things I
> did were failures, so I did ten times more work.   **George Bernard Shaw**

1. $\mathbf{R}^n$ has basis $(1, 0, \ldots, 0), (0, 1, 0, \ldots, 0), \ldots, (0, 0, \ldots, 1)$.
    $M_2(Q)$ has basis $\begin{bmatrix} 1 & 0 \\ 0 & 0 \end{bmatrix}, \begin{bmatrix} 0 & 1 \\ 0 & 0 \end{bmatrix}, \begin{bmatrix} 0 & 0 \\ 1 & 0 \end{bmatrix}, \begin{bmatrix} 0 & 0 \\ 0 & 1 \end{bmatrix}$.
    $Z_p[x]$ has basis $1, x, x^2, \ldots$. **C** has basis $1, i$.

3. $(a_2x^2 + a_1x + a_0) - (a_2'x^2 + a_1'x + a_0') = (a_2 - a_2')x^2 + (a_1 - a_1')x + (a_0 - a_0')$ and
    $a(a_2x^2 + a_1x + a_0) = aa_2x^2 + aa_1x + aa_0$. A basis is $\{1, x, x^2\}$. Yes.

5. Linearly dependent since $-3(2, -1, 0) - (1, 2, 5) + (7, -1, 5) = (0, 0, 0)$.

7. Suppose $au + b(u + v) + c(u + v + w) = 0$. Then $(a + b + c)u + (b + c)v + cw = 0$. Since $\{u, v, w\}$ are linearly independent, we obtain $c = 0$, $b + c = 0$, and $a + b + c = 0$. So, $a = b = c = 0$.

9. If the set is linearly independent, it is a basis. If not, then delete one of the vectors that is a linear combination of the others (see exercise 8). This new set still spans V. Repeat this process until we obtain a linearly independent subset. This subset will still span V since we only deleted vectors that are linear combinations of the remaining ones.

11. Let u_1, u_2, u_3 be a basis for U and w_1, w_2, w_3 be a basis for W. Use the fact that $u_1, u_2, u_3, w_1, w_2, w_3$ are linearly dependent over F. In general, if dim U + dim $W >$ dim V, then $U \cap W \neq \{0\}$.

13. no

15. yes; 2

17. $\begin{bmatrix} a & a + b \\ a + b & b \end{bmatrix} - \begin{bmatrix} a' & a' + b' \\ a' + b' & b' \end{bmatrix} = \begin{bmatrix} a - a' & a + b - a' - b' \\ a + b - a' - b' & b - b' \end{bmatrix}$
    and $c\begin{bmatrix} a & a + b \\ a + b & b \end{bmatrix} = \begin{bmatrix} ac & ac + bc \\ ac + bc & bc \end{bmatrix}$

19. If v and $v' \in U \cap W$ and a is a scalar, then $v + v' \in U$, $v + v' \in W$, $av \in U$, and $av \in W$. So $U \cap W$ is a subspace. (See exercise 11.) If $u_1 + w_1$ and $u_2 + w_2 \in U + W$, then $(u_1 + w_1) - (u_2 + w_2) = (u_1 - u_2) + (w_1 - w_2) \in U + W$ and $a(u_1 + w_1) = au_1 + aw_1 \in U + W$.

21. p^n

23. 2

25. Yes, because Z_7 is a field and, therefore, $1/2, -2/3,$ and $-1/6$ exist in Z_7. Specifically, $1/2 = 4, -2/3 = 4, -1/6 = 1$.

27. If V and W are vector spaces over F, then the mapping must preserve addition and scalar multiplication. That is, $T: V \to W$ must satisfy $T(u + v) = T(u) + T(v)$ for all vectors u and v in V, and $T(au) = aT(u)$ for all vectors u in V and scalars a in F. A vector space isomorphism from V to W is a one-to-one linear transformation from V onto W.

29. Suppose v and u belong to the kernel and a is a scalar. Then $T(v + u) = T(v) + T(u) = 0 + 0 = 0$ and $T(av) = aT(u) = a \cdot 0 = 0$.

31. Let $\{v_1, v_2, \ldots, v_n\}$ be a basis for V. Map $a_1v_1 + a_2v_2 + \cdots + a_nv_n$ to $(a_1, a_2, \ldots, a_n)$.

CHAPTER 20

> Well here's another clue for you all.   **John Lennon and Paul McCartney, *Glass Onion, The Beatles* (The White Album)**

1. Compare with exercise 24 of the supplementary exercises for Chapters 12–14.

3. $Q(\sqrt{-3})$

5. $Q(\sqrt{-3})$
7. Note that $x = \sqrt{1 + \sqrt{5}}$ implies $x^4 - 2x^2 - 4 = 0$.
9. $a^5 = a^2 + a + 1; a^{-2} = a^2 + a + 1; a^{100} = a^2$
11. The set of all expressions of the form

$$(a_n\pi^n + a_{n-1}\pi^{n-1} + \cdots + a_0)/(b_m\pi^m + b_{m-1}\pi^{m-1} + \cdots + b_0)$$

where $b_m \neq 0$.

13. Hint: Note that any isomorphism must act as the identity on the rationals.
15. Hint: Use exercise 42 of Chapter 13.
17. $a = 4/3, b = 2/3, c = 5/6$
19. Use the fact that $1 + i = -(4 - i) + 5$ and $4 - i = 5 - (1 + i)$.
21. If the zeros of $f(x)$ are $a_1, a_2, \ldots, a_n$, then the zeros of $f(x + a)$ are $a_1 - a$, $a_2 - a, \ldots, a_n - a$. Now use exercise 20.
23. Q and $Q(\sqrt{2})$
25. Let $F = Z_3[x]/\langle x^3 + 2x + 1 \rangle$ and denote the cosets $x + \langle x^3 + 2x + 1 \rangle$ by β and $2 + \langle x^3 + 2x + 1 \rangle$ by 2. Then $x^3 + 2x + 1 = (x - \beta)(x - \beta - 1)(x + 2\beta + 1)$.
27. Suppose that $\phi \colon Q(\sqrt{-3}) \rightarrow Q(\sqrt{3})$ is an isomorphism. Since $\phi(1) = 1$, we have $\phi(-3) = -3$. Then $-3 = \phi(-3) = \phi(\sqrt{-3}\sqrt{-3}) = (\phi(\sqrt{-3}))^2$. This is impossible since $\phi(\sqrt{-3})$ is a real number.

CHAPTER 21

> Work is the greatest thing in the world, so we should always
> save some of it for tomorrow.   **Don Herald**

1. It follows from Theorem 21.1 that if $p(x)$ and $q(x)$ are both monic irreducible polynomials in $F[x]$ with $p(a) = q(a) = 0$, then $\deg p(x) = \deg q(x)$. If $p(x) \neq q(x)$, then $(p - q)(a) = p(a) - q(a) = 0$ and $\deg(p(x) - q(x)) < \deg p(x)$, contradicting Theorem 21.1. To prove Theorem 21.3, use the Division Algorithm (Theorem 16.2).
3. If $f(x) \in F[x]$ does not split in E, then it has a nonlinear factor $q(x)$ which is irreducible over E. But then $E[x]/\langle q(x) \rangle$ is a proper algebraic extension of E.
5. Use exercise 4.
7. Suppose $Q(\sqrt{a}) = Q(\sqrt{b})$. If $\sqrt{b} \in Q$, then $\sqrt{a} \in Q$ and we may take $c = \sqrt{a}/\sqrt{b}$. If $\sqrt{b} \notin Q$, then $\sqrt{a} \notin Q$. Write $\sqrt{a} = r + s\sqrt{b}$. It follows that $r = 0$ and $a = bs^2$. The other half follows from exercise 20 of Chapter 20.
9. Observe that $[F(a):F]$ must divide $[E:F]$.
11. Note that $[F(a, b):F]$ is divisible by both $m = [F(a):F]$ and $n = [F(b):F]$.
13. Note that a is a zero of $x^3 - a^3$ over $F(a^3)[x]$. For the second part take $F = Q, a = 1$; $F = Q, a = (-1 + i\sqrt{3})/2; F = Q, a = \sqrt[3]{2}$.
15. Suppose $E_1 \cap E_2 \neq F$. Then $[E_1 : E_1 \cap E_2][E_1 \cap E_2 : F] = [E_1 : F]$ implies $[E_1 : E_1 \cap E_2] = 1$ so that $E_1 = E_1 \cap E_2$. Similarly, $E_2 = E_1 \cap E_2$.
17. E must be an algebraic extension of R so that $E \subseteq C$. But then $[C:E][E:R] = [C:R] = 2$.
19. Let a be a zero of $p(x)$ in some extension of F. First note $[E(a):E] \leq [F(a):F] = \deg p(x)$. Then observe that $[E(a):F(a)][F(a):F] = [E(a):E][E:F]$. This implies $\deg p(x)$ divides $[E(a):E]$ so that $\deg p(x) = [E(a):E]$.
21. Hint: if $\alpha + \beta$ and $\alpha\beta$ are algebraic, then so is $\sqrt{(\alpha + \beta)^2 - 4\alpha\beta}$.
23. $\sqrt{b^2 - 4ac}$
25. Use the Factor Theorem.
27. Say a is a generator of F^*. Then $F = Z_p(a)$ and it suffices to show a is algebraic over Z_p. If $a \in Z_p$, we are done. Otherwise, $1 + a = a^k$ for some $k \neq 0$. If $k > 0$, we are done. If $k < 0$, then $a^{-k} + a^{1-k} = 1$ and we are done.

29. Let $f(x)$ be the minimal polynomial for a over ϕ, and let $r = m/n$ where m and n are integers. Consider $g(x) = f(x^n)$.
31. Let $f(x)$ be the minimum polynomial for b over F. Use the facts: deg $f(x) \leq [F(a, b):F]$; $F(a, b) = F(a)(b)$; $f(x) \in F(a)[x]$.

CHAPTER 22

> Tell me tell me tell me come on tell me the answer.
> **John Lennon and Paul McCartney, *Helter Skelter, The Beatles***
> **(The White Album)**

1. $[GF(729):GF(9)] = 3$
   $[GF(64):GF(8)] = 2$
3. Use Theorem 22.2.
5. The only possibilities for $f(x)$ are $x^3 + x + 1$ and $x^3 + x^2 + 1$. See exercise 10 of Chapter 20 for the first case. For the second case, let a be a zero of $x^3 + x^2 + 1$ in some extension of Z_2. Then use the fact that $a^3 = a^2 + 1$ to test the elements of $Z_2(a)^\# = \{1, a, a^2, \ldots, a^6\}$ for zeros of $f(x)$.
7. Use the fact that $|GF(8):GF(2)| = 3$ and GF(8) is the splitting field of $x^8 - x$.
9. Direct calculations show that given $x^3 + 2x + 1 = 0$, we have $x^2 \neq 1$ and $x^{13} \neq 1$.
11. Note that the group has prime order.
13. Find a cubic irreducible polynomial $p(x)$ over Z_3, then $Z_3[x]/\langle p(x)\rangle$ is a field of order 27.
15. Use long division.
17. Theorem 22.3 reduces the problem to constructing the subgroup lattices for Z_{18} and Z_{30}.
19. identical
21. Consider $g(x) = x^2 - a$. Note that $|GF(p)[x]/\langle g(x)\rangle| = p^2$ so that $g(x)$ has a zero in $GF(p^2)$. Now use Theorem 22.3.
23. Use exercise 22.
25. Since $F^\#$ is a cyclic group of order 124, it has a unique subgroup of order 2.
27. Use exercise 42 of Chapter 13.

CHAPTER 23

> Why, sometimes I've believed as many as six impossible things before breakfast. **Lewis Carroll**

1.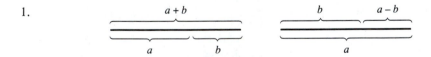

5. Use $\sin^2 \theta + \cos^2 \theta = 1$.
7. Use $\cos 2\theta = 2\cos^2 \theta - 1$.
9. Observe that $\sin 15° \in Q(\sqrt{2}, \sqrt{3})$.
11. Solving two linear equations with coefficients from F involves only the operations of F.
13. Use Theorem 17.1 and exercise 25 of Chapter 17.
15. Try $\theta = 90°$.
17. If so, then an angle of $40°$ is constructible. Now use exercise 10.

19. Use exercises 5, 6, and 7.
21. No, since $[Q(\sqrt[3]{3}):Q] = 3$.
23. No, since $[Q(\sqrt[3]{\pi}):Q]$ is infinite.

SUPPLEMENTARY EXERCISES FOR CHAPTERS 19–23

> The things taught in colleges and schools are not an education,
> but the means of education.   **Ralph Waldo Emerson, *Journals***

1. Use Theorem 20.5.
3. Suppose b is one solution of $x^n = a$. Since $F^{\#}$ is a cyclic group of order $q - 1$, it has a cyclic subgroup of order n, say $\langle c \rangle$. Then each member of $\langle c \rangle$ is a solution to the equation $x^n = 1$. It follows that $b\langle c \rangle$ is the solution set of $x^n = a$.
5. $(5a^2 + 2)/a = 5a + 2a^{-1}$. Now observe that since $a^2 + a + 1 = 0$ we know $a(-a - 1) = 1$ so that $a^{-1} = -a - 1$. Thus $(5a^2 + 2)/a = -2 + 3a$.
7. 5
9. Since $F(a) = F(a^{-1})$, we have degree of $a = [F(a):F] = [F(a^{-1}):F] =$ degree of a^{-1}.
11. If ab is a zero of $c_n x^n + \cdots + c_1 x + c_0 \in F[x]$, then a is a zero of $c_n b^n x^n + \cdots + c_1 bx + c_0 \in F(b)[x]$.
13. Every element of $F(a)$ can be written in the form $f(a)/g(a)$ where $f(x), g(x) \in F[x]$. If $f(a)/g(a)$ is algebraic and not in F, then there is some $h(x) \in F[x]$ such that $h(f(a)/g(a)) = 0$. By clearing fractions and collecting like powers of a, we obtain a polynomial in a with coefficients from F equal to 0. But then a would be algebraic over F.
15. Use Corollary 2 to Theorem 22.2.

CHAPTER 24

> All wish to possess knowledge, but few, comparatively
> speaking, are willing to pay the price.   **Juvenal**

1. $a = eae^{-1}; cac^{-1} = b$ implies $a = c^{-1}bc = c^{-1}b(c^{-1})^{-1}; a = xbx^{-1}$ and $b = ycy^{-1}$ imply $a = xycy^{-1}x^{-1} = xyc(xy)^{-1}$.
3. $\text{cl}(a) = \{a\}$ if and only if for all x in G, $xax^{-1} = a$. This is equivalent to $a \in Z(G)$.
5. See Example 3 of Chapter 5.
7. Use exercise 7 of Supplementary Exercises for Chapters 5–7.
9. Use exercise 46 of Chapter 9 and exercise 7 of the Supplementary Exercises for Chapters 5–7.
11. 8
13. 15
15. By exercise 14, G has seven subgroups of order 3.
17. 10; $\langle(123)\rangle, \langle(234)\rangle, \langle(134)\rangle, \langle(345)\rangle, \langle(245)\rangle$
19. If $p \nmid q - 1$ and $q \nmid p^2 - 1$, then a group of order $p^2 q$ is Abelian.
21. 21
23. By Sylow, $n_7 = 1$ or 15, and $n_5 = 1$ or 21. Counting elements reveals that at least one of these must be 1. Then the product of the Sylow 7-subgroup and the Sylow 5-subgroup is a subgroup of order 35.
25. By Sylow, $n_{17} = 1$ or 35. Assume $n_{17} = 35$. Then the union of the Sylow 17-subgroups has 561 elements. By Sylow, $n_5 = 1$. Thus we may form a cyclic subgroup of order 85 (exercise 46 of Chapter 9 and Theorem 24.7). But then there are 64 elements of order 85. This gives too many elements.

27. Use the "G/Z Theorem" (9.3).
29. Let H be the Sylow 3-subgroup and suppose the Sylow 5-subgroups are not normal. By Sylow, there must be six Sylow 5-subgroups, call them $K_1, \ldots, K_6$. These subgroups have 24 elements of order 5. Also, the cyclic subgroups $HK_1, \ldots, HK_6$ each have eight generators. Thus there are 48 elements of order 15.
31. Mimic the proof of Sylow's First Theorem.
33. Let $x \in G$ have maximum order, say $|x| = p^t$. Now let $y \in G$. Then $|y| = p^s \leq p^t$. Since $\langle x \rangle$ has a subgroup of order p^s, we have $\langle y \rangle \subseteq \langle x \rangle$.
35. Automorphisms preserve order.
37. Use the fact that HK is a cyclic group of order 15.
39. Normality of H implies $cl(h) \subseteq H$ for h in H. Now observe that $h \in cl(h)$. This is true only when H is normal.
41. The mapping from H to xHx^{-1} given by $h \to xhx^{-1}$ is an isomorphism.
43. $|C(a)|/|G|$
45. $\Pr(D_4) = 5/8$, $\Pr(S_3) = 1/2$, $\Pr(A_4) = 1/3$

CHAPTER 25

> Learn to reason forward and backward on both sides of a question. **Thomas Blandi**

1. Use the $2 \cdot$ odd Test.
3. Use the Index Theorem.
5. Suppose G is a simple group of order 525. Let L_7 be a Sylow 7-subgroup of G. It follows from Sylow's theorems that $|N(L_7)| = 35$. Let L be a subgroup of $N(L_7)$ of order 5. Since $N(L_7)$ is cyclic (Theorem 24.7), $N(L) \geq N(L_7)$ so that 35 divides $|N(L)|$. But L is contained in a Sylow 5-subgroup (Theorem 24.4) which is Abelian (see the corollary of Theorem 24.2). Thus 25 divides $|N(L)|$ as well. It follows that 175 divides $|N(L)|$. The Index Theorem now yields a contradiction.
7. $n_{11} = 12$. Use the N/C Theorem (Example 14 of Chapter 10) to show that there is an element of order 22, but A_{12} has no element of order 22.
9. A_{12} has no element of order 33.
11. If we can find a pair of distinct Sylow 2-subgroups A and B so that $|A \cap B| = 8$, then $N(A \cap B) \geq AB$ so that $N(A \cap B) = G$. Now let H and K be any distinct pair of Sylow 2-subgroups. Then $16 \cdot 16/|H \cap K| = |HK| \leq 112$ (supplementary exercise 7 for Chapters 5–7) so that $|H \cap K|$ is at least 4. If $|H \cap K| = 4$, then $N(H \cap K)$ picks up at least eight elements from H and at least eight from K (see exercise 32 of Chapter 24). Thus $|N(H \cap K)| \geq 16$ and is divisible by 8. So $|N(H \cap K)| = 16, 56,$ or 112. The last two cases give us a normal subgroup. If $|N(H \cap K)| = 16$, then $N(H \cap K)$ and H have at least eight elements in common and so do $N(H \cap K)$ and K. This guarantees the A and B.
15. Use the Index Theorem.
17. The Sylow Test for Nonsimplicity forces $n_5 = 6$ and $n_3 = 4$ or 10. The Index Theorem rules out $n_3 = 4$. Now appeal to Sylow's Third Theorem (24.5).
19. Let α be as in the proof of the Generalized Cayley Theorem. Then $\text{Ker } \alpha \leq H$ and $|G/\text{Ker } \alpha|$ divides $|G{:}H|!$ Now show $|\text{Ker } \alpha| = |H|$.
21. $n_5 = 6$ and $n_3 = 10$ or 40. If there are two Sylow 2-subgroups L_2 and L_2' whose intersection has order 4, show that $N(L_2 \cap L_2')$ has index at most 5. Now use the Embedding Theorem. If $n_3 = 40$, the union of all the Sylow subgroups has more than 120 elements. If $n_3 = 10$, use the N/C Theorem to show that there is an element of order 6. But A_6 has no element of order 6.
23. Mimic the proof that A_5 is simple.

25. By direct computation, show that PSL(2, 7) has more than four Sylow 3-subgroups, more than one Sylow 7-subgroup, and more than one Sylow 2-subgroup. Hint: Observe that $\begin{bmatrix} 2 & 1 \\ 2 & 3 \end{bmatrix}$ has order 3. Now use conjugation to find four other subgroups of order 3; observe that $\left| \begin{bmatrix} 5 & 5 \\ 1 & 4 \end{bmatrix} \right| = 7$ and use conjugation to find another subgroup of order 7; observe $\left| \begin{bmatrix} 5 & 1 \\ 3 & 5 \end{bmatrix} \right| = 8$ and use conjugation to find another subgroup of order 8. Now, argue as we did to show that A_5 is simple. In the cases that the supposed normal subgroup N has order 2 or 4, show that in G/N, the Sylow 7-subgroup is normal. But then, G has a normal subgroup of order 14 or 28, which were already ruled out.

27. Mimic exercise 26.

29. Let $p = [G{:}H]$ and $q = [G{:}K]$. It suffices to show that $p = q$. But if $p < q$, say, then $q \nmid p!$ so $|G| \nmid p!$ This contradicts the Index Theorem.

CHAPTER 26

> The dictionary is the only place where success comes before work. **Arthur Brisbane**

1. Let a be any reflection in D_n and let $b = aR_{360/n}$. Then $aZ(D_n)$ and $bZ(D_n)$ have order 2 and generate $D_n/Z(D_n)$. Now use Theorem 26.5 and the fact that $|D_n/Z(D_n)| = n = |D_{n/2}|$.

3. Since $b = b^{-1}$, we have $bab = a^2$. Then $a = a^6 = (bab)^3 = ba^3b$ so that $ba = a^3b$. Thus $a^3b = a^2b$ and $a = e$.

5. Let F be the free group on $\{a_1, a_2, \ldots, a_n\}$. Let N be the smallest normal group containing $\{w_1, w_2, \ldots, w_t\}$ and M be the smallest normal subgroup containing $\{w_1, w_2, \ldots, w_t, w_{t+1}, \ldots, w_{t+k}\}$. Then $F/N \approx G$ and $F/M \approx \overline{G}$. The mapping from F/N to F/M given by $aN \rightarrow aM$ induces a homomorphism from G onto $\overline{G}$.

   To prove the corollary observe that the theorem shows that K is a homomorphic image of G so that $|K| \leq |G|$.

7. Clearly, a and ab belong to $\langle a, b \rangle$, so $\langle a, ab \rangle \subseteq \langle a, b \rangle$. Now show that a and b belong to $\langle a, ab \rangle$.

9. Use Theorem 26.5.

11. Since $x^2 = y^2 = e$, we have $(xy)^{-1} = y^{-1}x^{-1} = yx$. Also $xy = z^{-1}yz$ so that $(xy)^{-1} = (z^{-1}yz)^{-1} = z^{-1}y^{-1}z = z^{-1}yz = xy$.

13. a. $b^6a^3 = a$ b. b^7a

15. Center is $\langle x^2 \rangle$. $|xy| = 8$.

17. Use the fact that the mapping from G onto G/N by $x \rightarrow xN$ is a homomorphism.

19. This is equivalent to showing that every left coset of $\langle b \rangle$ has the form $a^i\langle b \rangle$. But $\langle b \rangle$ absorbs all powers of b and commutes with all powers of a. So, if w is a word in a and b, then $w\langle b \rangle = a^i\langle b \rangle$ where i is the sum of all the exponents of the a terms in w.

21. 6; The given relations imply that $a^2 = e$. G is isomorphic to Z_6.

23. 1, 2, and ∞

25. $ab = c \Rightarrow abc^{-1} = e$
    $cd = a \Rightarrow (abc^{-1})cd = ae \Rightarrow bd = e \Rightarrow d = b^{-1}$
    $da = b \Rightarrow bda = b^2 \Rightarrow ea = b^2 \Rightarrow a = b^2$
    $ab = c \Rightarrow b^3 = c$
    So $G = \langle b \rangle$.
    $bc = d \Rightarrow bb^3 = b^{-1} \Rightarrow b^5 = e$. So $|G| = 1$ or 5.
    But Z_5 satisfies the generators and relations with $b = 3$, $c = 4$, and $d = 2$.

CHAPTER 27

> If at first you don't succeed—that makes you about average.
> Bradenton, *Florida Herald*

1. If T is a distance-preserving function and the distance between points a and b is positive, then the distance between $T(a)$ and $T(b)$ is positive.
3. See Figure 1.4.
5. 12
7. $4n$
9. a. Z_2
   b. $Z_2 \oplus Z_2$
   c. $G \oplus Z_2$ where G is isomorphic to the plane symmetry group of a circle.
11. 6
13. An inversion in $\mathbf{R}^3$ leaves only a single point fixed while a rotation leaves a line fixed.
15. In $\mathbf{R}^4$, a plane is fixed. In $\mathbf{R}^n$, a subspace of dimension $n-2$ is fixed.
17. Let T be an isometry, p, q, and r the three noncolinear points, and s any other point in the plane. Then the quadrilateral determined by $T(p)$, $T(q)$, $T(r)$, and $T(s)$ is congruent to the one formed by p, q, r, and s. Thus $T(s)$ is uniquely determined by $T(p)$, $T(q)$, and $T(r)$.
19. a rotation

CHAPTER 28

> The thing that counts is not what we know but the ability to use what we know. **Leo L. Spears**

1. Try $x^n y^m \rightarrow (n, m)$.
3. xy
5. Use Figure 28.9.
7. $x^2 yzxz = x^2 yx^{-1} = x^2 x^{-1} y = xy$
   $x^{-3} zxzy = x^{-3} x^{-1} y = x^{-4} y$
9. A subgroup of index 2 is normal.
11. a. V b. I c. II d. VI e. VII f. III
13. *cmm*
15. a. *p4m* b. *p3* c. *p31m* d. *p6m*
17. No. This print has S_4 as a subgroup but "Flying Fish" does not. The elements of $Z \oplus Z \oplus Z$ do not commute with the elements of S_4.
19. *p4g* (ignoring shading)
21. a. VI b. V c. I d. III e. IV f. VII g. IV

CHAPTER 29

> With every mistake we must surely be learning.
> George Harrison, *While My Guitar Gently Weeps, The Beatles*
> **(The White Album)**

1. 6
3. 30
5. 13
7. 45
9. For the first part, see exercise 35 of Chapter 2. For the second part, try D_4.

CHAPTER 30

Oh when you were young, did you question all the answers.
Graham Nash, *Wasted on the Way,* **single**

1. $4 * (b, a)$
3. $(m/2) * \{3 * [(a, 0), (b, 0)], (a, 0), (e, 1), 3 * (a, 0), (b, 0), 3 * (a, 0), (e, 1)\}$
5. $a^3 b$
7. Both yield paths from e to $a^3 b$.
11. Say we start at x. Then we know the vertices $x, xs_1, xs_1 s_2, \ldots, xs_1 s_2 \ldots s_{n-1}$ are distinct and $x = xs_1 s_2 \ldots s_n$. So if we apply the same sequence beginning at y, then cancellation shows that $y, ys_1, ys_1 s_2, \ldots, ys_1 s_2 \ldots s_{n-1}$ are distinct and $y = ys_1 s_2 \ldots s_n$.
13. If there were a Hamiltonian path from $(0, 0)$ to $(2, 0)$, there would be a Hamiltonian circuit in the digraph since $(2, 0) + (1, 0) = (0, 0)$.
15. a. If $s_1, s_2, \ldots, s_{n-1}$ traces a Hamiltonian path and $s_i s_{i+1} \ldots s_j = e$, then the vertex $s_1 s_2 \ldots s_{i-1}$ appears twice. Conversely, if $s_i s_{i+1} \ldots s_j \neq e$, then the sequence $e, s_1, s_1 s_2, \ldots, s_1 s_2 \ldots s_{n-1}$ yields the n vertices (otherwise, cancellation gives a contradiction).
    b. This is immediate from (1).
17. The sequence traces the digraph in a clockwise fashion.
19. Abbreviate $(a, 0)$, $(b, 0)$, and $(e, 1)$ by a, b, and 1, respectively. A circuit is $4 * (4 * 1, a), 3 * a, b, 7 * a, 1, b, 3 * a, b, 6 * a, 1, a, b, 3 * a, b, 5 * a, 1, a, a, b, 3 * a, b, 4 * a, 1, 3 * a, b, 3 * a, b, 3 * a, b$.
21. Abbreviate $(R_{90}, 0)$, $(H, 0)$, and $(R_0, 1)$ by R, H, and 1, respectively. A circuit is $3 * (R, 1, 1), H, 2 * (1, R, R), R, 1, R, R, 1, H, 1, 1$.
23. Abbreviate $(a, 0), (b, 0)$, and $(e, 1)$ by a, b, and 1, respectively. A circuit is $2 * (1, 1, a), a, b, 3 * a, 1, b, b, a, b, b, 1, 3, * a, b, a, a$.
25. Abbreviate $(r, 0)$, $(f, 0)$, and $(e, 1)$ by r, f, and 1, respectively. Then the sequence is $r, r, f, r, r, 1, f, r, r, f, r, 1 r, f, r, r, f, 1, r, r, f, r, r, 1, f, r, r, f, r, 1, r, f, r, r, f, 1$.
27. $m * ((n - 1) * (0, 1), (1, 1))$
29. Abbreviate $(r, 0)$, $(f, 0)$, and $(e, 1)$ by r, f, and 1, respectively. A circuit is $1, r, 1, 1, f, r, 1, r, 1, r, f, 1$.
31. In the proof of Theorem 30.3, we used the hypothesis that G is Abelian in two places: We needed H to satisfy the induction hypothesis, and we needed to form the factor group G/H. Now, if we assume only that G is Hamiltonian, then H also is Hamiltonian and G/H exists.

CHAPTER 31

We must view with profound respect the infinite capacity of the human mind to resist the introduction of useful knowledge.
Thomas R. Lounsbury

1. $wt(0001011) = 3$; $wt(0010111) = 4$; $wt(0100101) = 3$
3. 1000110; 1110100
5. 000000, 100011, 010101, 001110, 110110, 101101, 011011, 111000
7. Not all single errors can be detected.
9. Observe that a vector has even weight if and only if it can be written as a sum of an even number of vectors of weight 1.
11. No, by Theorem 31.3.
13. 0000000, 1000111, 0100101, 0010110, 0001011, 1100010, 1010001, 1001100, 0110011, 0101110, 0011101, 1110100, 1101001, 1011010, 0111000, 1111111

$$H = \begin{bmatrix} 1 & 1 & 1 \\ 1 & 0 & 1 \\ 1 & 1 & 0 \\ 0 & 1 & 1 \\ 1 & 0 & 0 \\ 0 & 1 & 0 \\ 0 & 0 & 1 \end{bmatrix}$$

yes

15. Suppose u is decoded as v and x is the coset leader of the row containing u. Coset decoding means v is at the head of the column containing u. So, $x + v = u$ and $x = u - v$. Now suppose $u - v$ is a coset leader and u is decoded as y. Then y is at the head of the column containing u. Since v is a code word, $u = u - v + v$ is in the row containing $u - v$. Thus $u - v + y = u$ and $y = v$.

17. $000000, 100110, 010011, 001101, 110101, 101011, 011110, 111000$

$$H = \begin{bmatrix} 1 & 1 & 0 \\ 0 & 1 & 1 \\ 1 & 0 & 1 \\ 1 & 0 & 0 \\ 0 & 1 & 0 \\ 0 & 0 & 1 \end{bmatrix}$$

001001 is decoded as 001101 by all four methods.
011000 is decoded as 111000 by all four methods.
000110 is decoded as 100110 by all four methods.
Since there are no code words whose distance from 100001 is 1 and three whose distance is 2, the nearest neighbor method will not decode or will arbitrarily choose a code word; parity-check matrix decoding does not decode 100001; the standard-array and syndrome methods decode 100001 as $000000, 110101,$ or 101011, depending on which of these vectors is used as coset leaders.

19. For any received word w, there are only eight possibilities for wH. But each of these eight possibilities satisfies condition 2 or the first portion of condition $3'$ of the decoding procedure, so decoding assumes no error was made or one error was made.

21. There are 3^4 code words and 3^6 possible received words.

23. No; row 3 is twice row 1.

25. No. For if so, nonzero code words would be all words with weight at least 5. But this set is not closed under addition.

27. Use exercise 24 together with the fact that the set of codewords is closed under addition.

29. Abbreviating the coset $a + \langle x^2 + x + 1 \rangle$ with a, the following generating matrix will produce the desired code:

$$\begin{bmatrix} 1 & 0 & 1 & 1 & x \\ 0 & 1 & x & x+1 & x+1 \end{bmatrix}.$$

31. Use exercise 14.

33. Let $c, c' \in C$. Then, $c + (v + c') = v + c + c' \in v + C$ and $(v + c) + (v + c') = c + c' \in C$, so the set $C \cup (v + C)$ is closed under addition.

35. If the ith component of both u and v is 0, then so is the ith component of $u - v$ and au where a is a scalar.

CHAPTER 32

> Wisdom rises upon the ruins of folly.   **Thomas Fuller,**
> *Gnomologia*

1. Note that $\phi(1) = 1$. Thus $\phi(n) = n$. Also, $1 = \phi(1) = \phi(nn^{-1}) = \phi(n)\phi(n^{-1}) = n\phi(n^{-1})$ so that $1/n = \phi(n^{-1})$.
3. If a and b are fixed by elements of H, so are $a + b$, $a - b$, $a \cdot b$, and a/b.
5. It suffices to show that each member of $G(K/F)$ defines a permutation on the a_i's. Let $\alpha \in G(K/F)$ and write

$$f(x) = c_n x^n + c_{n-1}x^{n-1} + \cdots + c_0 = c_n(x - a_1)(x - a_2)\cdots(x - a_n).$$

Then $f(x) = \alpha(f(x)) = c_n(x - \alpha(a_1))(x - \alpha(a_2))\cdots(x - \alpha(a_n))$. Thus $f(a_i) = 0$ implies $a_i = \alpha(a_j)$ for some j so that α permutes the a_i's.

7. By exercise 6 of Chapter 17, the splitting field is $Q(\sqrt{2}, i)$. Since $[Q(\sqrt{2}, i):Q] = 4$, $|G(E/Q)| = 4$. It follows that $G(E/Q) = \{\epsilon, \alpha, \beta, \alpha\beta\}$ where $\alpha(\sqrt{2}) = -\sqrt{2}$ and $\alpha(i) = i$, $\beta(\sqrt{2}) = \sqrt{2}$, and $\beta(i) = -i$ and the proper subfields of E are Q, $Q(\sqrt{2})$, $Q(\sqrt{-2})$, $Q(i)$. β has fixed field $Q(\sqrt{2})$, α has fixed field $Q(i)$, and $\alpha\beta$ has fixed field $Q(\sqrt{-2})$. No automorphism of E has fixed field Q.

9. $|G(E/Q)| = |E:Q| = 4$
    $|G(Q(\sqrt{10})/Q)| = [Q(\sqrt{10}):Q] = 2$

11. Recall A_4 has no subgroup of order 6. (See Example 13 of Chapter 9.)
13. Use Sylow's Theorem.
15. Let $\omega = \cos(360°/n) + i\sin(360°/n)$. Then $Q(\omega)$ is the splitting field. Every α in $G(K/F)$ has the form $\alpha(\omega) = \omega^i$. Since such elements commute, the group is Abelian.
17. Use the lattice of Z_{10}.
19. Z_6 (Be sure you know why the group is cyclic.)
21. See exercise 24 of Chapter 25.
23. Use exercise 31 of Chapter 24.
25. Let $\{e\} = H_0 \subset H_1 \subset \cdots \subset H_n = G$ be the series that shows that G is solvable. Then $H_0 \cap H \subset H_1 \cap H \subset \cdots \subset H_n \cap H$ shows that H is solvable.

CHAPTER 33

> Won't you please help me.   **John Lennon and Paul McCartney,**
> *Help,* **single**

1. a. $a \vee (a' \wedge b) = (a \vee a') \wedge (a \vee b) = 1 \wedge (a \vee b) = a \vee b$

    b. $(a \wedge b) \vee (a \wedge b') \vee (a' \wedge b') = [a \wedge (b \vee b')] \vee (a' \wedge b')$
        $= (a \wedge 1) \vee (a' \wedge b')$
        $= a \vee (a' \wedge b')$
        $= (a \vee a') \wedge (a \vee b')$
        $= 1 \wedge (a \vee b') = a \vee b'$

    c. This is the dual to (b).
3. Use Theorem 33.2.
5. $f_2 + f_8 = f_{10}; f_2 \cdot f_8 = f_0$
    $f_7 + f_{14} = f_{15}; f_7 \cdot f_{14} = f_6$
    $f_7' = f_8, f_{14}' = f_1$
7. $a'b + b'c + c'd$
9. $ab + (a + b')c(a + d) + a + b' + d$

11.

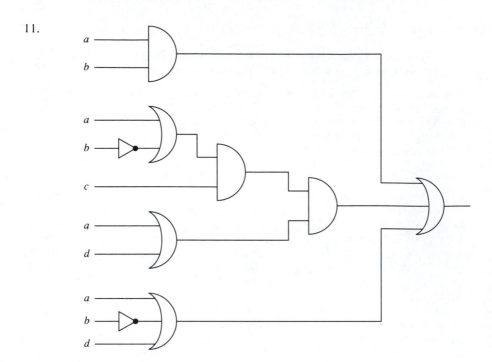

13. This follows directly from the fact that x and $21/x$ are relatively prime.
15. Let $a = \{1\}$, $b = \{2\}$, $c = \{1, 2\}$, and $\vee$ be set union, then $a \vee c = b \vee c$ but $a \neq b$.
17. Suppose $(B_1, +, \cdot, ')$ and $(B_2, \vee, \wedge, -)$ are Boolean algebras. An isomorphism ϕ from B_1 to B_2 is a one-to-one, onto mapping from B_1 to B_2 such that $\phi(a + b) = \phi(a) \vee \phi(b)$; $\phi(a \cdot b) = \phi(a) \wedge \phi(b)$; $\phi(a') = \overline{\phi(a)}$.

19.

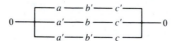

21.

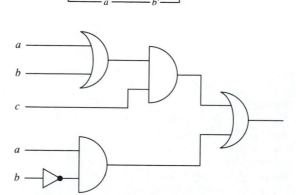

SUPPLEMENTARY EXERCISES FOR CHAPTERS 24–33

Give me your answer.   **John Lennon and Paul McCartney,**
When I'm Sixty-four, Sgt. Pepper's Lonely Hearts Club Band

1. Z_6
3. Let $|G| = 315$ and let H be a Sylow 3-subgroup and K a Sylow 5-subgroup. If $H \lhd G$, then $HK = 45$. If H is not normal, then by Sylow's Third Theorem, $|G/N(H)| = 7$, so that $|N(H)| = 45$.
5. Observe that $K \subseteq N(H)$ implies HK is a group of order 245. Now, use the previous exercise.
7. Note that $gKg^{-1} \subseteq gHg^{-1} = H$. Now use the corollary to Sylow's Third Theorem.
9. Use the same proof as for exercise 46 of Chapter 9.
11. Since $n_7 = 8$, we know by the Embedding Theorem (Chapter 25) that $G \leq A_8$. But A_8 does not have an element of order 21.
13. Let G be a non-Abelian group of order 105. By Theorem 9.3, $G/Z(G)$ is not cyclic. So $|Z(G)| \neq 3, 7, 15, 21,$ or 35. This leaves only 1 or 5 for $|Z(G)|$. Let $H, K,$ and L be Sylow 3-, Sylow 5- and Sylow 7-subgroups of G, respectively. Now, counting shows that $K \lhd G$ or $L \lhd G$. Thus, $|KL| = 35$ and KL is a cyclic subgroup of G. But, KL has 24 elements of order 35 (since $|U(Z_{35})| = 24$). Thus, a counting argument shows that $K \lhd G$ and $L \lhd G$. Now, $|HK| = 15$ and HK is a cyclic subgroup of G. Thus, $HK \subseteq C(K)$ and $KL \subseteq C(K)$. This means that 105 divides $|C(K)|$. So $K \subseteq Z(G)$.

15.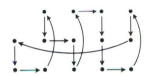

17. It suffices to show that x travels by a implies xab^{-1} travels by a (for we may successively replace x by xab^{-1}). If xab^{-1} traveled by b then the vertex xa would appear twice in the circuit.
19. a. $\{00, 11\}$
    b. $\{000, 111\}$
    c. $\{0000, 1100, 1010, 1001, 0101, 0110, 0011, 1111\}$
    d. $\{0000, 1100, 0011, 1111\}$
21. The mapping $T_v: F^n \to \{0, 1\}$ given by $T_v(u) = u \cdot v$ is an onto homomorphism. So $|F^n/\text{Ker } T_v| = 2$.
23. It follows from exercise 18 that if C is an (n, k) linear code, then $C^\perp$ is an $(n, n - k)$ linear code. Thus, in this problem, $k = n - k$. To prove the second claim, use exercise 21, the definition of $C^\perp$, and the hypothesis that $C^\perp = C$.
25. Let the two normal series with Abelian factors be

$$\{e\} = N_0 \lhd N_1 \lhd \cdots \lhd N_t \lhd H_1 \lhd \cdots \lhd H_s = G.$$

and

$$N/N = H_0/N \lhd H_1/N \lhd \cdots \lhd H_s/N = G/N.$$

Then

$$\{e\} = N_0 \lhd N_1 \lhd \cdots \lhd N_t \lhd H_1 \lhd \cdots \lhd H_s = G.$$

27. $a \vee b \vee c = a \wedge b \wedge c$ implies that $a' \wedge [a \vee (b \vee c)] = a' \wedge a \wedge b \wedge c = 0$. Thus, $(a' \wedge a) \vee [a' \wedge (b \vee c)] = 0$ so that $a' \wedge (b \vee c) = 0$. Then $a \vee [a' \wedge (b \vee c)] = a \vee 0 = a$. This implies that $(a \vee a') \wedge (a \vee b \vee c) = a$. It follows that $a \vee b \vee c = a$. By symmetry, $a \vee b \vee c = b$ and $a \vee b \vee c = c$.

Notations

(The number after the item indicates where the notation is defined.)

SET THEORY

$\bigcap_{i \in I} S_i$	intersection of sets S_i, $i \in I$
$\bigcup_{i \in I} S_i$	union of sets S_i, $i \in I$
$[a]$	$\{x \in S \mid x \sim a\}$, equivalence class of S containing a, 13

SPECIAL SETS

Z	integers, additive group of integers, ring of integers
Q	rational numbers, field of rational numbers
Q^+	multiplicative group of positive rational numbers
$F^\#$	set of nonzero elements of F
$\mathbf{R}$	real numbers, field of real numbers
$\mathbf{R}^+$	multiplicative group of positive real numbers
$\mathbf{C}$	complex numbers

FUNCTIONS AND ARITHMETIC

f^{-1}	the inverse of the function f
$t \mid s$	t divides s, 3
$t \nmid s$	t does not divide s, 3
$\gcd(m, n)$	greatest common divisor of integers m and n, 5
$\text{lcm}(s, t)$	least common multiple of the integers s and t, 7
gf, $\alpha\beta$	composite function, 15
$\phi(a)$	image of a under ϕ, 15
$\phi: A \to B$	mapping of A to B, 15

ALGEBRAIC SYSTEMS

D_4	group of symmetries of a square, dihedral group of order 8, 25
D_n	dihedral group of order $2n$, 26
e	identity element, 35
Z_n	group $\{0, 1, \ldots, n - 1\}$ under addition modulo n, 36

det A	the determinant of A, 37
$GL(2, F)$	2×2 matrices of nonzero determinant with coefficients from the field F (the general linear group), 37
$U(n)$	group of units modulo n (that is, the set of integers less than n and relatively prime to n under multiplication modulo n), 38
$\mathbf{R}^n$	$\{(a_1, a_2, \ldots, a_n) \mid a_1, a_2, \ldots, a_n \in \mathbf{R}\}$, 39
$SL(2, F)$	group of 2×2 matrices over F with determinant 1, 39
a^{-1}	multiplicative inverse of a, 43
$-a$	additive inverse of a, 43
$\|G\|$	order of the group G, 55
$\|g\|$	order of the element g, 55
$H \leq G$	subgroup inclusion, 56
$H < G$	subgroup $H \neq G$, 56
$\langle a \rangle$	$\{a^n \mid n \in Z\}$, cyclic group generated by a, 58
$Z(G)$	$\{x \in G \mid xy = yx$ for all $y \in G\}$, the center of G, 60
$C(a)$	$\{g \in G \mid ga = ag\}$, the centralizer of a in G, 62
$C(H)$	$\{x \in G \mid xh = hx$ for all $h \in H\}$, the centralizer of H, 64
$\phi(n)$	Euler phi function of n, 74
$N(H)$	$\{x \in G \mid x^{-1}Hx = H\} = \{x \in G \mid Hx = xH\}$, the normalizer of H in G, 82
cl(a)	conjugacy class of a, 82, 349
G^n	$\{g^n \mid g \in G\}$, 83
S_n	group of one-to-one functions from $\{1, 2, \cdots, n\}$ to itself, 87
A_n	alternating group of degree n, 93
$G \approx H$	G and H are isomorphic, 103
ϕ_a	mapping given by $\phi_a(x) = axa^{-1}$ for all x, 105, 109
Aut(G)	group of automorphisms of the group G, 109
Inn(G)	group of inner automorphisms of G, 109
$G_1 \oplus G_2 \oplus \cdots \oplus G_n$	external direct product of groups $G_1, G_2, \ldots, G_n$, 117
$U_k(n)$	$\{x \in U(n) \mid x = 1 \bmod k\}$, 121
G'	commutator subgroup, 131
$\oplus_{i=1}^n G_i$	$G_1 \oplus G_2 \oplus \cdots \oplus G_n$, 132
xH	$\{xh \mid h \in H\}$, 134
$\|G{:}H\|$	the index of H in G, 137
stab$_G(a)$	$\{\alpha \in G \mid \alpha(a) = a\}$, the stabilizer of a under the permutation group G, 139
orb$_G(i)$	$\{\alpha(i) \mid \alpha \in G\}$, the orbit of i under the permutation group G, 139
$H \triangleleft G$	H is a normal subgroup of G, 150
G/N; R/A	factor group; factor ring, 152, 223
HK	$\{hk \mid h \in H, k \in K\}$, 159
$H \times K$	internal direct product of H and K, 159
$H_1 \times H_2 \times \cdots \times H_n$	internal direct product of $H_1, \ldots, H_n$, 161
Ker ϕ	kernel of the homomorphism ϕ, 171
$\phi^{-1}(g')$	inverse image of g' under ϕ, 172
$\phi^{-1}(K)$	inverse image of K under ϕ, 172

$Z[x]$	ring of polynomials with integer coefficients, 202
$M_2(Z)$	ring of all 2×2 matrices with integer entries, 202
$R_1 \oplus R_2 \oplus \cdots \oplus R_n$	direct sum of rings, 203
nZ	ring of multiples of n, 205
$Z[i]$	ring of Gaussian integers, 205
$U(R)$	group of units of the ring R, 207
$Z_n[i]$	ring of Gaussian integers modulo n, 209
$\langle a \rangle$	principal ideal generated by a, 223
$A + B$	sum of ideals A and B, 228
AB	product of ideals A and B, 228
$\mathrm{Ann}(A)$	annihilator of A, 229
$\sqrt{A}$	nil radical of A, 229
$F(x)$	field of quotients of $F[x]$, 243
$R[x]$	ring of polynomials over R, 248
$\deg f(x)$	degree of the polynomial $f(x)$, 250
$g(x) \mid f(x)$	$g(x)$ divides $f(x)$, 252
$\Phi_p(x)$	pth cyclotomic polynomial, 261
$M_2(Q)$	ring of 2×2 matrices over Q, 296
$\langle v_1, v_2, \ldots, v_n \rangle$	subspace spanned by $v_1, v_2, \ldots v_n$, 297
$F(a_1, a_2, \ldots, a_n)$	extension of F by $a_1, a_2, \ldots, a_n$, 305
$f'(x)$	the derivative of $f(x)$, 309
$[E{:}F]$	index of E over F, 317
$\mathrm{GF}(p^n)$	Galois field of order p^n, 328
$\mathrm{GF}(p^n)^{\#}$	nonzero elements of $\mathrm{GF}(p^n)$, 329
$\mathrm{cl}(a)$	$\{x^{-1}ax \mid x \in G\}$, the conjugacy class of a, 349
$\mathrm{Pr}(G)$	probability that two elements from G commute, 351
n_p	the number of Sylow p-subgroups of a group, 355
$PSL(n, q)$	$SL(n, q)/Z(SL(n, q))$, the projective special linear group, 374
$W(S)$	set of all words from S, 386
$\langle x, y, \ldots, z \mid w_1 = w_2 = \cdots = w_t \rangle$	group with generators $x, y, \ldots, z$ and relations $w_1 = w_2 = \cdots = w_t$, 389
Q_4	quaternions, 392
Q_6	dicyclic group of order 12, 392
D_∞	infinite dihedral group, 393
$\mathrm{fix}(\phi)$	$\{i \in S \mid \phi(i) = i\}$, elements fixed by ϕ, 437
$\mathrm{Cay}(S{:}G)$	Cayley digraph of the group G with generating set S, 447
$k * (a, b, \ldots, c)$	concatenation of k copies of $(a, b, \ldots, c)$, 455
(n, k)	linear code, k dimensional subspace of F^n, 476
F^n	$F \times F \times \cdots \times F$, direct product of n copies of the field F, 476
$d(u, v)$	Hamming distance between vectors u and v, 477
$\mathrm{wt}(u)$	the number of nonzero components of the vector u (the Hamming weight of u), 477
$G(E/F)$	the automorphism group of E fixing F, 497
E_H	fixed field of H, 497
$a \wedge b$	a and b, 513
$a \vee b$	a or b, 513

Index of Mathematicians

(Biographies appear on pages in boldface.)

Index of Terms